p53 in the Clinics

Pierre Hainaut • Magali Olivier • Klas G. Wiman
Editors

p53 in the Clinics

Springer

Editors
Pierre Hainaut
International Prevention Research Institute
International Agency for Research
on Cancer
Lyon, France

Magali Olivier
Group of Molecular Mechanisms
and Biomarkers
International Agency
for Research on Cancer (IARC/CIRC)
Lyon Cedex 08, France

Klas G. Wiman
Department of Oncology-Pathology
Cancer Center Karolinska
Karolinska Institutet
Stockholm, Sweden

ISBN 978-1-4614-3675-1 ISBN 978-1-4614-3676-8 (eBook)
DOI 10.1007/978-1-4614-3676-8
Springer New York Heidelberg Dordrecht London

Library of Congress Control Number: 2012942477

Printed on acid-free paper

Springer is part of Springer Science+Business Media (www.springer.com)

Foreword

Initially ignored by many researchers as an outsider molecule of minor importance in carcinogenesis, p53 took over 10 years to rise to the status of "guardian of the genome (1992)" and "molecule of the year" (1993). Yet, despite all the hype of the 1990s, the path from molecular discoveries to clinical application has been frustratingly long and winding for p53. It is only in recent years that breakthroughs have started to accumulate, opening doors to applications of the huge amount of knowledge accumulated on p53 into clinical practice. If p53 is not yet part of the everyday cancer management, with the rapid diversification of translational research on p53 detection and treatment, the road towards major clinical applications is now rapidly clearing up.

Discovered in 1979, the p53 protein has now reached the full maturity age of 33—which does not necessarily mean that it is ready to deliver healing miracles in public health care. In 2005, the first p53 book entitled "25 Years of p53 Research" highlighted the development of the p53 field and its profound impact on concepts in cancer research. The book emphasized the role of p53 as a regulator of cell cycle checkpoints and cell death in response to multiple cellular stresses. It also reviewed experimental models for studying p53, as well as knowledge on the prognostic and predictive value of tumor-associated p53 mutations and emerging therapeutic strategies for restoring p53 function in tumors. Since the publication of the first p53 book, the p53 field has developed at an ever-increasing pace and p53 researchers have uncovered novel and entirely unexpected functions and aspects of p53. There has been substantial progress on the significance of the genetic diversity of p53 and on mutations as biomarkers in molecular pathology. Research on p53-based therapy has intensified dramatically with the development of several drugs that target the p53 pathway. Thus, there have been considerable advances on the potential applications of p53 in the clinic.

The goal of this new p53 book is to capture these developments as the field moves into a next phase with strong emphasis on translational research. The present book covers the most striking advances in p53 research and clinical applications that emerged over the last few years. These advances include:

1. Establishing a role for p53 in biological processes such as energy metabolism and fertility: Although p53 was previously thought to act mainly as a cellular trigger for cell cycle arrest or cell death, accumulating evidence has shown that p53 also regulates cell oxidative metabolism and the cellular response to nutrient deprivation. In addition, p53 has a role in fertility. These and other findings have led to the notion that p53's main task is to ensure the fidelity of a wide range of biological and physiological processes.
2. Uncovering the fundamental role of p53 in stem cell biology and in generation of induced pluripotent stem cells (iPS): Stem cells and regenerative medicine is a dynamic research area with great promises for clinical application. Studies have demonstrated that p53 controls the division and renewal of stem cells and that p53 influences the efficiency of induction of pluripotent stem cells from differentiated somatic cells. Therefore, p53 is a key protein to be considered in new strategies for implementing stem cell-based therapy in the clinic.
3. Understanding the interconnections between p53 family members, p63 and p73, and particularly their roles in cancer: Although not frequently mutated in cancer, both p63 and p73 are involved in cancer development through altered regulation and expression of isoforms that lack the N-terminal transactivation domain. Novel insights into p63 and p73 may lead to improved diagnosis and better prediction of prognosis and therapy response in cancer.
4. Pioneering translational research on the impact of p53 status on prognosis and clinical outcome in human cancer: It is now well established that p53 mutation is associated with poor prognosis in most cancers, particularly in breast cancer. Current challenges are (1) to better assess the function of p53 in tumors through new powerful techniques such as genome-wide analysis of the p53 pathway; (2) to determine in which context p53 status is clinically predictive of response to therapy and disease outcome; and (3) to develop diagnostic tests adapted to routine clinical practice.
5. Identifying germline *TP53* mutations not just as the basis of a rare form of familial cancer, but as one of the main forms of inherited predisposition to cancer: Germline *TP53* mutations are more common than previously considered (occurring in about 1 to 2,500 to 5,000 births) and are also extremely diverse in their impact on cancer risk. Furthermore, the discovery of a common founder mutation predisposing to cancer in Brazil provides a paradigm for searching for other population clusters in which germline *TP53* mutations might occur at a high frequency.
6. Developing p53-based methods for cancer therapy and bringing them to clinical trials: These approaches include long-awaited developments in gene therapy using viral vectors and their evaluation into several clinical contexts. Moreover, p53 researchers have identified small molecules that either activate wild-type p53 in wild-type p53-carrying tumors or restore wild-type function to mutant p53 in mutant p53-carrying tumors. In most cases these molecules are still at a preclinical

stage. However, at least two compounds, one that targets the p53-Mdm2 complex and induces wild-type p53 expression, and one that targets mutant p53 and restores its conformation and pro-apoptotic activity, have recently been tested in phase I clinical trials. Further studies should clarify whether these molecules have potent anticancer efficacy in larger cohorts of patients. Also, a number of novel molecules that reactivate wild-type or mutant p53 are under way.

This short survey makes it clear that p53 is ripe for clinical application. In fact, p53 is already in the clinic, although not yet included in routine analysis and still far from being used for targeted therapy. Exactly how could p53 be exploited for improved cancer diagnosis and treatment in oncology clinics around the world? This book shall address this and its related questions and hopefully provide information that will be useful and inspiring to both basic cancer researchers and clinicians. We strongly believe that the coming 10 years will be the decade of "p53 in the clinic," with major benefits for early detection, prognosis, allocation of personalized treatments, and, ultimately, cancer patient survival.

8 Nov 2011

Lyon, France — Pierre Hainaut
Lyon, France — Magali Olivier
Stockholm, Sweden — Klas G. Wiman

Contents

Chapter 1
TP53: Coordinator of the Processes That Underlie the Hallmarks of Cancer

Pierre Hainaut

1 Introduction

Over 10 years ago, Hanahan and Weinberg developed the concept of "Hallmarks of Cancer" as an organizing principle for rationalizing the molecular and mechanistic complexity of human cancer (Hanahan and Weinberg 2000). Initially, these Hallmarks included six major biological characteristics acquired during the process of multistep cancer development: sustaining proliferative signaling, evading growth suppressors, resisting cell death, enabling replicative immortality, inducing angiogenesis, and activating invasion and metastasis. During the last decade, technical advances in the comprehensive analysis of the patterns of genetic and epigenetic changes in cancers have led to a deeper understanding of these Hallmarks as well as to an extension of the concept to include new, emerging Hallmark processes such as reprogramming of energy metabolism and evading destruction by the immune system. Moreover, the biological changes that contribute to each Hallmark are promoted and accelerated by genome instability and inflammation, two conditions that can be seen as the cornerstones of cancer development (Hanahan and Weinberg 2011). Further, the development of Hallmark capabilities takes place within a complex network of interactions between neoplastic and normal cells, defining the concept of "tumor microenvironment." Not the least merit of the Hallmark concept is to provide a rationale for developing novel, molecular approaches to cancer therapy: in principle, each Hallmark has as molecular "signature" a number of molecular targets that can be specifically addressed using selective pharmacological agents, thus allowing for a radical departure from the classical approaches of cancer eradication by conventional cytotoxic therapies.

P. Hainaut, Ph.D. (✉)
International Prevention Research Institute,
International Agency for Research on Cancer,
95 Cours Lafayette, 69006 Lyon, France
e-mail: pierre.hainaut@i-pri.org

P. Hainaut et al. (eds.), *p53 in the Clinics*, DOI 10.1007/978-1-4614-3676-8_1,

The *TP53* tumor suppressor gene, encoding the p53 phosphoprotein, occupies a special place in the gallery of molecular factors that contribute to the Hallmarks of Cancer. This multifunction gene encodes a jewel of molecular integration, with much functional complexity packed up in a small amino acid sequence (393 residues) and simple molecular architecture (Hollstein and Hainaut 2010). One can only marvel at the simple beauty of this extremely efficient, almost minimalistic, use of molecular structures to provide multiple and adaptative functional diversity. These characteristics endow p53 with the capability to play critical roles in each of the ten Hallmarks of Cancer. As a result, the p53 signaling pathway can be seen as a molecular device integrating the Hallmark processes into a coherent biological program. A direct consequence of this role is that p53 function is at least partially disabled in most, if not all, cancers. To date, mutation in *TP53* remains the most frequent genetic change identified in human cancers, irrespective of organ site, histology, or natural history (Petitjean et al. 2007). In cancers that retain normal *TP53* alleles, the expression, stability, or activity of the p53 protein is often altered by multiple mechanisms. Such is the central role of p53 in carcinogenesis that it may be considered that alteration of p53 function is required, although not sufficient, for any form of cancer to develop. In the global world of molecular carcinogenesis, p53 occupies the main market place, marking the closest spot we currently know to the point where all cancer trade routes begin or end. This unique place among cancer genes makes it an attractive focal point for directing interventions aimed at controlling cancer—in particular in the clinics.

This introduction chapter briefly describes how p53 may operate as growth suppressor within the general framework of the ten Hallmarks of Cancer as recently redefined by Hanahan and Weinberg (2011). The aim of this outline is not to scoop through the daunting amount of literature accumulated on p53 since its discovery in 1979. Rather, my approach will be to use the beautiful conceptual framework of the Hallmarks as a magnifying lens to scrutinize aspects of p53 biology that might be amenable to clinical intervention.

2 The Long Road to p53 Function(s)

The p53 protein came under the limelight in a modest way, almost as a supporting role in the cast of the big show of molecular cancer research. It first emerged in 1979 as a protein of cellular origin which coprecipitated with the large T antigen of the Simian Virus SV40. At about the same time, it was identified as an antigen recognized by antisera raised against chemically transformed mouse cells (which have accumulated the p53 protein as the result of a chemically induced mutation in the *TP53* gene). In the late 1980s, the emergence of the concept of tumor-suppressor gene, largely based on studies of the retinoblastoma gene, paved the way for a paradigm shift that totally changed our understanding of p53 and of the molecular mechanisms of carcinogenesis (Hainaut and Wiman 2009; Lane and Levine 2010). First, it was shown that wild-type p53 cDNA clones could suppress transformation of rodent cells in culture, whereas point mutant versions of p53 were transforming (Levine et al. 2004; Eliyahu et al. 1989). Second, from 1989, there has been a rapid accumulation of reports showing that missense mutations in *TP53* and/or loss of *TP53* alleles were common in colorectal, lung, breast, liver, and many other cancers. Third, inherited *TP53* mutations were found to be the underlying genetic defect of Li–Fraumeni syndrome (LFS), a familial syndrome of predisposition to multiple early cancers (Malkin et al. 1990).

Today, the vision of p53 biochemical roles and functions has considerably diversified. A selection of landmark p53 publications that trace the main advances in the field is given in Hainaut and Wiman (2009). During the early 1990s, convergent studies established that p53 was activated in response to DNA damage, justifying its 1992 nickname of "guardian of the genome," an expression that captures its capacity to arrest the proliferation of cells with genomes damaged by carcinogens by forcing them to undergo cell-cycle arrest or apoptosis (Lane 1992). At the turn of 2012, there are over 60,000 articles containing "p53" as keyword indexed in PubMed, demonstrating the extraordinary success and diversification of p53 research. In addition to its role in response to many forms of DNA damage, p53 has now gained recognition as a critical oncogene-induced barrier against progression of cancer beyond its early stages and as an important inducer of replicative senescence. In 2007, a study by Scott Lowe and collaborators provided one of the most compelling examples to date of p53 function as a critical tumor suppressor (Xue et al. 2007). To determine the consequences of reactivating the p53 pathway in tumors, these authors introduced into liver progenitor cells an inducible RNA interference system that conditionally regulates endogenous p53 expression. Their results show that switching off p53 function could activate the growth of large liver tumors and that even brief (24 h) reactivation of endogenous p53 could produce complete tumor regression. Furthermore, they found that the primary response to p53 reactivation was not apoptosis, but the induction of a cellular senescence program associated with differentiation, upregulation of inflammatory cytokines, and activation of innate immune response contributing to the clearance of tumor cells (Xue et al. 2007). Although there is no doubt that cancer involves complex combinations of genetic and epigenetic alterations in multiple genes, the extent to which loss of p53 function is required to sustain cancer growth suggests it has a huge potential as target for clinical cancer control (Martins et al. 2006).

3 An Evolutionary Perspective on p53

For over about 20 years, *TP53* has appeared to stand alone among oncogenes and tumor suppressor genes by not being a member of a multigene family. Until the late 1990s, it was even unclear whether a true equivalent to *TP53* was present in organisms other than vertebrates. This view has considerably evolved over the past 15 years with the discovery of two homologues of *TP53*, *TP63*, and *TP73* (Aylon and Oren 2011; Allocati et al. 2012). Although p53, p63, and p73 share structural and sequence similarities, p63 and p73 are rarely mutated in tumors, in contrast with p53. Instead, the *TP63* locus is amplified in 20–30 % of squamous cell carcinomas of the head and neck, esophagus, or lung, and p73 is overexpressed in diverse types of cancers. While mice deficient for p53 function develop early multiple cancers, p63- or p73-deficient mice die tumor free from complex and diverse developmental and physiological defects (Melino et al. 2003). Although experimental, ectopic expression of p63 or p73 in cultured cells can activate roughly equivalent biological effects as p53, these two proteins play different roles than p53 in physiological conditions. Due to the presence of internal promoters and/or alternative splicing mechanisms, *TP63* and *TP73* encode two different classes of protein isoforms: those containing an N-terminal transactivation (TA) domain, similar to the one of p53, and those that lack this transactivation domain, commonly termed Delta-N isoforms. High expression of Delta-Np63, for example, is essential for the morphogenesis and maintenance of proliferative and renewal capabilities of many tissues, including the epidermis, squamous, respiratory or urothelial epithelia, as well as basal/myoepithelial cell compartments of salivary, lachrymal, mammary, and prostate mucosa. Moreover, Delta-Np63 is critical for the asymmetric division and proliferative potential of epidermal stem cells (Nekulova et al. 2011).

Sequences homologous to p53/p63/p73 have now been identified in organisms other than vertebrates, including, for example, *Drosophila*. Strikingly, the pattern of expression of all family members into two classes of isoforms, TA and Delta-N, appears to be highly conserved. The existence of Delta-N isoforms of p53, some of them extremely similar to Delta-Np63 or Delta-Np73 in their mechanisms of production, their architecture, and their biochemical activities, suggests that the ancestral function of the family might be related to stem cell maintenance, renewal, and proliferation (Marcel et al. 2011). According to this view, p53 might have evolved from a blueprint of genes essential for stem cell maintenance to specialize in functions that are not essential for development per se, such as sensing environmental changes in the cell ecosystem. Taking into account the role of p53 in response to oxidative stress and its capacity to control many aspects of oxidative metabolism (Hafsi and Hainaut 2011), it is interesting to speculate that *TP53* has evolved to adapt cells to hosting mitochondria, a supreme source of damaging radicals, as their main energy production system (Wang et al. 2012).

4 Implications of p53 in the "Hallmarks of Cancer"

Studies on genome-wide mutation patterns in cancer cells have confirmed that *TP53* is one of the most, if not the most, consistently altered gene in human cancer (Pfeifer and Hainaut 2011). Taking into account the multiplicity of mechanisms other than mutation that may inactivate p53 functions (such as enhanced protein degradation through increased expression of cellular or viral proteins that target p53 for destruction by the proteasome), it may be considered that p53 function is inactivated, or somehow made deficient, in the vast majority of cancers. The fact that this deficiency

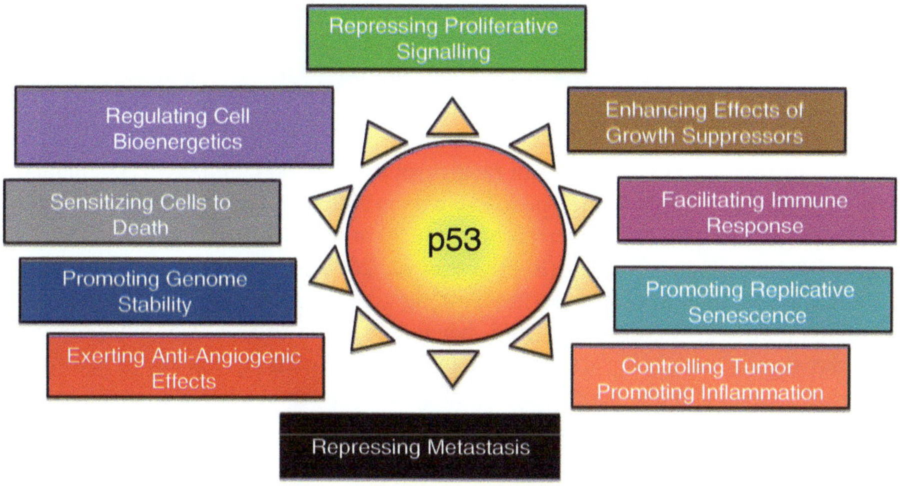

Fig. 1.1 A Copernican view of p53 in the "Hallmarks of Cancer." This model depicts p53 as the central point of a solar system in which each Hallmark process represents a zodiac sign. The Hallmark boxes are colored according to the original scheme given in "Hallmarks of Cancer: The next Generation" (Hanahan and Weinberg 2011)

may occur—or may show its effects—at any step in the carcinogenic process from precancer condition to highly aggressive, metastatic lesions, makes it very complex to address the prognostic or predictive significance of measurable p53 alterations. Furthermore, such a sweeping role for a single molecule in such a mechanistically complex, multifactorial disease raises the suspicion that p53 does not act as a front-line driver but as a kind of facilitator of carcinogenesis. In the following section, I briefly discuss some of the mechanisms by which p53 interferes in each of the ten Hallmarks of Cancer processes, supporting the view that loss of p53 function not only make cells permissive to the acquisition of Hallmark capabilities but also removes a critical brake that prevents the expression of Hallmark capabilities. In this model, p53 is not a specific component of any particular Hallmark process but rather a molecular device that organizes the Hallmark processes into a coherent biological program (Fig. 1.1).

4.1 Repressing Proliferative Signaling

Cancer cells have systematically acquired the capability to sustain proliferative signaling, and they can do so in dazzling number of ways. Although p53 is not on its own a component of proliferation signal transduction cascades, it influences and controls them by entertaining dialogues with regulators of growth and survival. Loss of p53 function thus enhances proliferation by at least three main mechanisms. First, p53 regulates the transcription of *PTEN* (phosphatase and tensin homolog), a well-established tumor suppressor gene that is mutated, deleted, or epigenetically silenced in a variety of cancers (Yin and Shen 2008). Induction of *PTEN* induces apoptosis and controls cell growth by inhibiting the PI3K/AKT pathway. Second, downstream of the same pathway, p53 is inhibiting mTOR (mammalian target of rapamycin) (Huang et al. 2001). In metazoans, TOR is a central integrator that receives signals arising from growth factors, nutrients, and cellular energy metabolism. The mechanisms by which p53 regulates mTOR involves AMP kinase activation and requires TSC1/2 (tuberous sclerosis 1 and 2 complex), two factors that respond to energy deprivation (Matthew et al. 2009). Thus, cross-talks between mTOR and p53 interconnect growth factor signaling and sensing nutrient bioavailability with growth suppression. Third, when faced with excessive or untimely proliferative signals such as the activation of an oncogene, cells are capable of activating antiproliferative signals acting as a safeguard against excessive proliferation (Lomazzi et al. 2002). Excess proliferative signaling, as for example in response to oncogene activation, increases the expression of $p14^{ARF}$, the alternative reading frame product of the *INK4a/CDKN2a* locus, encoding p16, another canonical tumor suppressor gene and located on chromosome 9p22. In turn, $p14^{ARF}$ sequester Mdm2 and prevents it from binding and degrading p53, thus causing an increase in nuclear levels and suppressive activity of p53. $p14^{ARF}$ also exerts p53-independent function as tumor suppressor. In this context, loss of p53 helps cells to evade senescence and to maintain a high proliferative capacity. Deletion of *INK4a/CDKN2a* may represent a double blow, with ablation of both p16, a negative regulator of G1 phase of the cell cycle, and of $p14^{ARF}$, a tumor suppressor that links cell-cycle progression with p53-mediated growth suppression (Dominguez-Brauer et al. 2010).

4.2 Enhancing the Effects of Growth Suppressors

The main mechanisms by which loss of p53 functions helps cells to evade growth suppressor are by impairing the expression of *CDKN1/p21WAF1*, encoding the $p21^{WAF1}$ cyclin kinase inhibitor with multiple roles in controlling G1/S, G2, and M phases. This gene was the first one to be identified as a transcriptional target of p53 and is a central effector of the growth suppressive effect of p53 in response to intracellular stress signals such as DNA damage, levels of nucleotide pools, or availability of oxygen (el-Deiry et al. 1993; Sherr and Roberts 1999). However, *p21WAF1* has complex activities, both pro- and antioncogenic, and its genetic or epigenetic inactivation is not common in human cancer.

In cultured cells, one of the mechanisms that suppresses proliferation is cell-to-cell contacts (contact inhibition). This mechanism involves the reorganization of cell surface molecules and of their coupling with cell surface receptors, the extracellular matrix and the cytoskeleton. One of the main factors involved in these processes is Merlin, the product of the *NF2* (neurofibromatosis type 2) tumor suppressor gene. Merlin orchestrates contact inhibition by coupling cell surface adhesion molecules to receptor tyrosine kinases and by strengthening the adhesivity of cadherin-mediated cell-to-cell attachments (Curto and McClatchey 2008). Merlin increases p53 stability and activity by inducing the degradation of Mdm2, thus leading to p53-dependent apoptosis in response to serum starvation or to DNA damage (Kim et al. 2004).

Yet another important role of p53 in regulating responses to growth suppressors is underlined by its connection with the tumor suppressor *LKB1*, which is genetically inactivated in patients with Peutz–Jeghers syndrome and somatically mutated in a variety of cancers. *LKB1* is responsible for the maintenance of epithelial cell polarity and can, for example, overrule the mitogenic effects of the *MYC* oncogene. In contrast, ablation of *LKB1* leads to disruption of epithelial structures and enhanced cell proliferation mediated at least in part by a decline of p53 and p16 growth suppressor pathways (Liang et al. 2010).

Perhaps one of the most elaborate antiproliferative pathways is the one controlled by TGF-beta. Far more than just an inhibitory pathway, TGF beta can induce profound re-programming of cell fate, as for example during the Epithelial to Mesenchymal Transition (EMT) (Wendt et al. 2009; Xu et al. 2009). The TGF-beta signaling pathway has long been considered as largely independent upon p53 but recent results have highlighted roles for p53, not only in the modulation of distal biological responses such as EMT but also in more proximal aspects of TGF-beta-mediated growth inhibition. For example, the *SnoN* proto-oncogene product negatively regulates TGF-beta signaling through its interactions with Smad complexes. The same factor also interacts with the promyelocytic leukemia (PML) protein and the accumulation of SnoN in PML nuclear bodies induces the stabilization and activation of p53, leading to premature senescence (Lamouille and Derynck 2009).

4.3 Suppressing Replicative Immortality Through Senescence

When maintained in culture, normal cells undergo a phase of expansion through division until they enter senescence, reaching a crisis during which cells die unless they acquire an unlimited replicative potential that confer them immortality in culture. The rheostat that determines how many division cycles a cell can achieve before reaching replicative senescence is a structure located at the end of chromosomes and composed of tandem repeats of six nucleotides, the telomeres. Each normal division cycle is accompanied by telomere erosion. In human cancer, telomere erosion is compensated by activation of telomerase, a specialized DNA polymerase that adds telomere repeats. This enzymatic activity is almost absent in normal, postmitotic cells but is expressed at high levels in 90 % of immortalized and transformed human cells (Shay and Wright 2001). The p53 protein exerts at least three complementary roles in the regulation of telomeres.

First, over-expression of wild-type p53 down-regulates the enzymatic activity of telomerase through transcriptional repression of the gene encoding its catalytic subunit, human telomerase reverse transcriptase (*hTERT*) (Kanaya et al. 2000). This repression appears to require p53 DNA binding activity and to involve $p21^{WAF1}$. Furthermore, it is attenuated by inactivation of the *RB* family or by dominant-negative mutation of *E2F1*. Mutation of *TP53* eliminates this repression and may therefore contribute to high telomerase activity in cancer cells.

Second, telomere erosion activates a complex growth suppression program which is controlled by p53. Mice deficient for telomerase acquire dysfunctional telomeres and develop a dwarf phenotype associated with p53 activation of senescence or apoptosis in several tissues including skin. This dwarf phenotype is rescued by ablation of p53, which also restitutes skin renewal and wound healing responses (Flores and Blasco 2009). There is evidence that telomere erosion induces a form of DNA damage detected by p53, thus inducing a senescence response in stem/progenitor cells that controls their capacity to contribute to tissue growth and regeneration, thus affecting tissue and organismal fitness (Sahin et al. 2011). Furthermore, p53 deficiency enhances the efficiency of somatic cell reprogramming to a pluripotent state. Cells with mutant *TP53* have to undergo reprogramming much more efficiently than cells lacking *TP53*. Moreover, reprogrammed cells expressing mutant p53 tend to give rise to malignant tumors much more than cells lacking *TP53*, suggesting that mutant p53 exerts a gain-of-function effect on reprogramming (Sarig et al. 2010).

Third, p53 regulates cell-cycle progression in response to telomere uncapping. In normal cells, telomeres are protected by protein complexes that specifically bind to telomeric DNA. Uncapping occurs transiently in each G2 phase of the cell cycle, following DNA replication. Telomere caps are reassembled at the G2/M transition under the control of the p53/p21^{WAF1} DNA damage response pathway. Human or mouse cells lacking p53 or p21^{WAF1} progress into mitosis prematurely with uncapped telomeres which make chromosomes prone to religate into deleterious end-to-end fusions (Thanasoula et al. 2010).

4.4 Sensitizing Cells to Death

Induction of apoptosis is one of the best characterized biological effects of p53 (Oren 2003). It has long been recognized that wild-type p53 directly controls the transcription of several classes of genes involved at different levels in apoptosis signaling. The apoptotic circuitry is commonly described as composed of two complementary programs: the extrinsic and the intrinsic apoptotic programs (Moffitt et al. 2010). The extrinsic program involves extracellular signaling through death-mediating receptors. The p53 protein directly activates the transcription of genes encoding death receptors such as *APO1/FAS/CD95* and *KILLER/DR5* (Wu et al. 2000). It also controls the expression of genes regulating the bioavailability of survival factors, such as *IGF-BP3*, which encodes a protein that binds and neutralizes IGF1 and IGF2 (Buckbinder et al. 1995; Butt et al. 1999). The intrinsic program integrates multiples signals of intracellular origin, with the mitochondria as centerpiece. The role of p53 in this program involves both transcription-dependent and independent aspects. Among the products of genes transcriptionally regulated by p53, several interact with the antiapoptotic factor Bcl2 and/or its family members. These p53 target genes include *PUMA*, *NOXA*, and *BAX*, encoding proteins with BH3 motifs that interact with Bcl2 and disrupt the integrity of the outer mitochondrial membrane, leading to the release of apoptosis signaling factors among which the most important is cytochrome C (Shamas-Din et al. 2011). Nontranscriptional functions may involve indirect effects such as those responsible for down-regulating aspects of base-excision repair (BER), therefore sensitizing cells to death in response to DNA damage (Gatz and Wiesmuller 2006).

Separately from its functions as a transcription factor, p53 is also capable to be translocated to the mitochondria (Vaseva et al. 2009), presumably by interacting with Tid1, a mitochondrial DnaJ-like chaperone. In breast cancer cells, depletion of Tid1 by short hairpin RNA (shRNA) leads to absence of p53 accumulation at mitochondria and resistance to apoptosis under hypoxic or genotoxic stress (Ahn et al. 2010; Trinh et al. 2010).

Distinct from its role in sensitizing cells to apoptosis, p53 is also active in the control of autophagy, an intracellular catabolic process during which cells degrade their own components through the lysosomal machinery. Autophagy is a double-edged sword. It is a stress-induced cell death mechanism, in particular in response to nutrient deprivation. It is also a metabolic switch through which cells reallocate energetic resources to different biological processes (Eng and Abraham 2011). Deletion, depletion, or inhibition of p53 induces autophagy in human, mouse, and nematode cells. In the latter organism, deletion of the p53 homologue cep-1 increases life span through an increase in baseline autophagy. The effects of p53 on induction of autophagy may be, at least in part, a consequence of the cross-talks between p53 and the PTEN/PI3K/AKT/mTOR pathway, which plays a critical role in nutrient sensing (Galluzzi et al. 2010; Feng 2010). Other mechanisms by which p53 induces autophagy have emerged with the discovery that p53 directly transactivates *SESN2*, encoding Sestrin2, a controverted cysteine sulfinic acid reductase thought to be involved in responses to stress, to accumulate with age, and to inhibit mTOR activation (Maiuri et al. 2009). On the other hand, p53 also transactivates *DRAM* (damage-regulated autophagy modulator), which encodes a protein with six putative transmembrane domains that co-localizes with cathepsin D in the lysosome. Silencing *DRAM* inhibits both p53-mediated autophagy and apoptosis, suggesting these two outcomes are mechanistically linked (Crighton et al. 2006, 2007).

4.5 Regulating Cell Bioenergetics

One of the most spectacular developments in our understanding of p53 functions is the discovery that it plays a fundamental role in cell bioenergetics. Loss of p53 activity may be pivotal for the Warburg effect, or aerobic glycolysis, the long-known capacity of many cancer cells to generate energy mainly through glycolysis even in the presence of nonlimiting levels of oxygen (Gottlieb and Vousden 2010; Hafsi and Hainaut 2011). Overall, p53 appears to down-regulate glucose usage and to favor oxidative metabolism. These effects include down-regulation of glucose transporters at the plasma membrane through direct inhibition of GLUT-1 and GLUT-4

(Schwartzenberg-Bar-Yoseph et al. 2004) and indirect inhibition of GLUT-3 by a mechanism involving NF-KappaB (Kawauchi et al. 2008). Further downstream, p53 regulates the synthesis of two rate limiting enzymes in the glycolytic pathway. It induces *TIGAR* (*TP53*-induced glycolysis and apoptosis regulator), an enzyme with fructose bi-phosphatase activities that counteracts the activity of 6-phosphofructo-1 kinase (Bensaad et al. 2006). In contrast, it down-regulates *PGM* (phosphoglycerate mutase), which converts 3-phosphoglycerate into 2-phosphoglycerate during the late ATP-generating steps of glycolysis (Kondoh et al. 2005). In parallel with its capacity to limit glycolysis, p53 inhibits the pentose phosphate pathway (PPP). The p53 protein binds to glucose-6-phosphate dehydrogenase (G6PD), the first and rate-limiting enzyme of the PPP, and prevents the formation of the active dimer. Tumor-associated mutant p53 mutants lack this G6PD-inhibitory activity (Jiang et al. 2011). p53 also promotes oxidative phosphorylation through at least two mechanisms. First, it activates *AIF*, encoding apoptosis inducing factor, a bifunctional protein with oxido-reductase function contributing to the assembly and function of complex I of the respiratory chain (Vahsen et al. 2004). Second, p53 transactivates *SCO2* (encoding synthesis of cytochrome c oxidase 2), a copper-dependent chaperone protein that is required for the assembly of complex IV of the respiratory chain (Wanka et al. 2011).

Another critical effect of p53 on energy metabolism and antioxidant defenses is mediated by its transactivation of *GLS2*, encoding a mitochondrial glutaminase catalyzing the hydrolysis of glutamine to glutamate (Hu et al. 2010). GLS2 regulates energy metabolism by increasing the production of glutamate and alpha-ketoglutarate, which in turn results in citric acid cycle, mitochondrial respiration, and ATP generation. GLS2 also contributes to antioxidant defenses by increasing reduced glutathione (GSH) which in turn protects cells from oxidative stress. Overall, loss of p53 through mutation in cancer cells may thus promote a profound switch in metabolism, facilitating the capacity of cancer cells to generate energy and to maintain high pools of reduced nucleotides for DNA synthesis under oxygen-poor conditions.

4.6 Promoting Genetic and Genomic Stability

Following the recognition that it regulates cell-cycle progression and apoptosis, p53 has been rapidly identified as a main contender in maintaining genetic stability, a property that earned it its nickname of "guardian of the genome." Indeed, by arresting cell cycle and controlling DNA replication in cells with genetic damage, p53 protects cells from acquiring further DNA defects that may initiate cancer or drive its progression.

Nucleotide excision repair (NER), the mechanism which removes the vast majority of UV-induced DNA damage, is regulated by the ATR/p53 checkpoint via modulation of XPA (*Xeroderma pigmentosum* group A protein) nuclear import in a cell cycle-dependent manner in G1 and S phases. XPA, one of eight factors implicated in XP disorders, is an indispensable factor for both transcription-coupled NER (TC-NER) and global genome NER (GG-NER) which play roles in verifying DNA damage, stabilizing repair intermediates, and recruiting other NER factors to the damage site. The DNA damage-induced response of XPA nuclear import is significantly slower in p53-deficient cells than in p53-proficient cells, consistent with the notion that loss of p53 function may decrease the elimination of cyclopentane pyrimidine dimers, a typical UV photoproduct, thus enhancing UV-induced mutagenesis (Li et al. 2011a, b).

In contrast to NER which recognizes bulky distortions of the DNA helix, BER is a mechanism that removes smaller base lesions that could otherwise cause mispairing or strand breaks during DNA replication. BER is initiated by DNA glycosylases that remove specific damaged bases. The resulting apurinic/apyriminidic (AP) sites are cleaved by the AP endonuclease APE1/Ref1 and the resulting single strands are processed by either short-patch (where a single nucleotide is replaced) or long-patch BER. The p53 protein contributes to BER at several levels, both in a positive and in a negative fashion (Gatz and Wiesmuller 2006; Offer et al. 2001). First, p53 regulates the transcription of *hOGG1*, encoding the glycosylase responsible for the excision of 7,8-dihydro-8-oxoguanine (8-oxoG; the main product of guanine oxidation by reactive oxygen species) (Chatterjee et al. 2006). Second, it regulates the transcription of *MGMT* (O^6-methyl-guanine-DNA-methyl-transferase), the primary enzyme that repairs alkyl adducts at the O^6 position of guanine. The effect of p53 on *MGMT* expression is somehow biphasic: while wild-type p53 is required for enhanced *MGMT* expression, transfection and over-expression of p53 represses *MGMT* (Grombacher et al. 1998). Third, p53 induces *PPM1D* encoding protein phosphatase 1D, a serine/threonine phosphatase that interacts with Ung2 (uracil DNA glycosylase 2) and may suppress BER probably via dephosphorylation of Ung2 (Lu et al. 2004a, b). Fourth, p53 activates the expression of genes such as *GADD45* (growth arrest and DNA damage inducible gene) which product interferes with BER by controlling nuclear localization and activity of APE1/Ref1. Finally, p53 interacts with APE1/Ref1 and may modulate its stability and activity (Seemann and Hainaut 2005). The general message emerging from these observations is that p53 may enhance BER as part of its genome-stabilizing, tumor-suppressive activities. However, after activation of p53 in response to stress, high levels of p53 may repress BER to return DNA repair to the deactivated state and therefore participate to maintenance of basal DNA damage capacity. In conditions where p53 becomes activated to very high levels, such as for example in response to severe DNA damage, p53 may stop BER to accelerate apoptosis of cells that have accumulated DNA damage beyond repair capacity.

The involvement of p53 in mismatch repair (MMR) is demonstrated by the identification of *MSH2* (MutS homologue 2), MLH1 (MutL homologue 1), and *PMS2* (postmeiotic segregation 2) as transcriptional p53 target genes [reviewed in Gatz and Wiesmuller (2006)]. MMR is the main mechanism that corrects DNA following DNA polymerase errors, removing mismatches in heteroduplex DNA during recombination and preventing homologous recombination (HR). *TP53* mutation increases HR by several orders of magnitude and experiments with *TP53* hotspot mutants in cultured cells have revealed severe HR inhibitory defects.

Despite its wide implication in DNA repair, there is only limited evidence that impaired p53 function leads to an increased mutation load (mutator phenotype). Spontaneous tumors in a mouse model of LFS show only modest increase in mutation loads (Hill et al. 2006). In humans, there is a highly significant increase in copy number variations among carriers of germline *TP53* mutations with a familial cancer history (Shlien et al. 2008). However, *TP53* mutation carriers do not appear to be hyper-sensitive to mutagens although radiosensitivity has been demonstrated in cultured cells and observed in some LFS patients who received radiation-based treatments [discussed in Palmero et al. (2010)].

4.7 Controlling Tumor Promoting Inflammation

Inflammation is a complex biological response involving vascular structures, immune system and tissue microenvironment aimed at removing injurious stimuli and at initiating healing processes. This is achieved by mobilizing a wide set of bioactive molecules, including reactive oxygen and nitrogen species acting both as chemical cleansers and as signaling molecules, cytokines, growth factors, and survival factors that promote cell regeneration, and angiogenic and extracellular matrix remodeling factors that restructure the microenvironment towards wound healing. Excess or persistent inflammation, however, turns these protective responses into a cancer-promoting process. Many aspects of chronic inflammation involve p53 in a direct or indirect way (Kamp et al. 2011). First, p53 is activated in response to DNA damage inflicted by reactive oxygen and nitrogen species generated during inflammation, thus mediating growth suppressive responses that characterize several chronic inflammatory lesions [reviewed in Hafsi and Hainaut (2011)]. Second, p53 regulates the expression of genes encoding enzymes involved in the production or detoxification of reactive species. These genes include up-regulation of *COX2* (cyclooxygenase 2) (de Moraes et al. 2007), down-regulation of *SOD2* (mitochondrial superoxide dismutase) (Forrester et al. 1996), and activation of *GPX1* (glutathione peroxidase) (Tan et al. 1999) and of *ALDH4A1* (aldehyde dehydrogenase 4 A1) (Yoon et al. 2004). Third, p53 represses the expression of *NOS2* (inducible nitric oxide synthase) (Ambs et al. 1998; Forrester et al. 1996), and *TP53* mutation is correlated with increased expression of *NOS2* and enhanced damage by nitrogen species in cancers developing in a context of chronic inflammation (Ambs et al. 1999; Vaninetti et al. 2008). Fourth, p53 entertains a complex network of functional cross-talks with NF-KappaB, the main transcription factor involved in the regulation of inflammatory responses. For example, p53 and NF-kappaB cooperate in the regulation of *COX2*, which provides a survival mechanism in chronically inflamed tissues (de Moraes et al. 2007; Ryan et al. 2000). Recently, p53 has been found to attenuate lipopolysaccharide (LPS)-induced NF-kappaB activation and acute lung injury in mice (Liu et al. 2009). Moreover, loss of p53 function impairs the repression of NF-kappaB target gene transcription by glucocorticoids and severely impairs glucocorticoid rescue of death in a mouse model of LPS shock (Liu et al. 2009).

The functional cross-talk between p53 and NF-kappaB may play an important role in regulating organismal senescence and life span: it seems that the efficiency of p53 signaling declines during aging whereas that of NF-kappaB is clearly enhanced.

Overall, the involvement of p53 in inflammatory responses may be such that p53 operates as a natural brake against excess DNA damage and abnormal cell proliferation in the context of chronic inflammation. Loss of p53 function may therefore upset the balance of inflammatory signals towards tumor-promoting effects, with excessive cell proliferation, escape from senescence and loss of control over genetic and genomic stability in the face of high levels of damaging reactive oxygen species.

4.8 Repressing Metastasis

The invasion–metastasis cascade is taking place according to a sequence of discrete steps encompassing local invasion, intravasation by cancer cells and transit into blood and lymphatic vessels, extravasation into the parenchyma of distant tissues, formation of micrometastases and their growth into macroscopic tumors, this last step being termed "colonization." Many aspects of this multistep process are directly or indirectly influenced by p53 and mutation of *TP53* is commonly observed in metastatic cancers. In recent years, several studies have uncovered an important role of p53 in regulating epithelial to mesenchymal transition (EMT), a developmental morphogenesis and wound healing program that underlies the capacity of epithelial cancer cells to acquire metastatic properties. EMT is activated by multiple pathways (beta-Catenin, Notch, EGFR, Ras/MAPK) and is controlled by transcription factors such as Snail/Slug and ZEB family members (Brabletz 2012; Jing et al. 2011; Xu et al. 2009). EMT confers a combined stemness and motility phenotype to cancer cells, defining a subpopulation of migrating cancer stem cells as potential source of metastasis. The miR-200 family of microRNAs consists of two gene clusters (miR-200a, b and 429, Chr1; miR-141 and 200c, Chr12) acting as orchestrators of EMT through their capacity to inhibit EMT activators such as ZEB factors, thereby inducing a reverse process to EMT, the mesenchymal to epithelial transition (MET). MiR-200 family members not only counteract EMT but also suppress stem cells factors, such as Bmi1. The p53 protein regulates both EMT and EMT-associated stem cell properties through transcriptional activation of the miR-200c promoter (Chang et al. 2011). Loss of p53 correlates with a decrease in the level of miR-200c, associated with an increase in EMT and stemness markers, and development of a high tumor grade in a cohort of breast tumors.

Another important connection between p53, EMT, and metastasis involves the Twist1 protein, a regulator of embryogenesis. Twist1 has been shown to induce EMT and is over-expressed in a large fraction of human cancers (Ansieau et al. 2008). A common cancer-derived mutant p53 protein, p.R175H, up-regulates Twist1 expression in several cancer cell lines, suggesting that increased Twist1 might occur in cancer cells as the consequence of gain-of-function mutations in *TP53* (Kogan-Sakin et al. 2011).

4.9 Exerting Antiangiogenic Effects

Tumor progression almost systematically involves the activation of an "angiogenic switch" causing the normally quiescent vasculature to resume sprouting new vessels that supply cancer cells in nutrients, growth factors, and oxygen. Angiogenesis is induced by hypoxic conditions and regulated by the hypoxia-inducible factor 1 (HIF-1), an hetero-dimer transcription factor which regulates the expression of vascular endothelial cell growth factor (VEGF) degradation and activity. p53 interacts with the alpha subunit of HIF1 (HIF-1alpha) and regulates its degradation. Mutant p53, in contrast, promotes the stabilization of HIF-1aplha and the expression of VEGF, thus enhancing angiogenesis (Khromova et al. 2009; Choi et al. 2003).

Antiangiogenic drugs targeting the VEGF pathway have shown clinical effects in delaying or slowing down metastatic disease in some patients. Antiangiogenic therapy appears to be sensitive to p53 status in tumors, implicating a role for p53 in the regulation of angiogenesis. One of the mechanisms of this regulation may involve the transcriptional activation by p53 of *alpha(II)PH*, the gene encoding alpha(II) collagen prolyl-4-hydroxylase. This activation leads to the extracellular release of antiangiogenic fragments of collagen type 4 and 18. As a result, conditioned medium of cells expressing high p53 levels inhibits the growth of human endothelial cells (Teodoro et al. 2006). Another collagen-derived antiangiogenic factor, Arresten, is processed from alpha 1 collagen 4 (encoded by *COL4A1*). Recent studies show that p53 induces the expression of *COL4A1* and the release of fragments of extracellular matrix containing Arresten. The p53 protein directly activates the transcription of *COL4A1* and also increases the metalloproteinase-mediated release of Arresten, thus controlling the production of an important antiangiogenic factor (Assadian et al. 2012).

4.10 Facilitating Innate and Adaptative Immune Response

Recently, the panel of genes regulated by p53 has extended beyond the already large set of cell cycle, DNA repair, and apoptosis regulators to include the control of human Toll-like receptor (*TLR*) gene expression (Menendez et al. 2011). The *TLR* gene family mediates innate immunity to a wide variety of pathogenic agents through recognition of conserved pathogen-associated molecular motifs. The role of p53 in regulating *TLR* expression appears to be extremely diverse and complex. The promoters of most *TLR* family members contain both canonical and noncanonical p53 response elements, thus defining a potentially rich repertoire of dose- and context-dependent stress responses to p53 activation (Menendez et al. 2011). Furthermore, several lines of evidence suggest that p53 regulates antiviral immunity. Mice lacking p53 have impaired and delayed antiviral response to influenza A virus (IAV), caused by disruption of both innate and adaptive immunity (Munoz-Fontela et al. 2011). In cells infected by hepatitis C virus (HCV), the HCV core protein induces the p53-dependent expression of *TAP1* (encoding transporter associated with antigen processing 1) and consecutive major histocompatibility complex (MHC) class I up-regulation, leading to a significant down-regulation of the cytotoxic activity of natural killer (NK) cells against HCV-infected cells and facilitating the establishment of a chronic infection (Herzer et al. 2003).

Some other p53 activities might also directly contribute to help cancer cells to escape immune destruction. Many early transformed cells express ligands for the natural killer cell immunoreceptor NKG2D, which sensitizes them to recognition and elimination by cytotoxic lymphocytes (Textor et al. 2011). The expression of the NKG2D ligand ULBP2 is controlled by the tumor-suppressive microRNAs (miRNA) miR-34a and miR-34c, which are themselves regulated by p53. Therefore, p53 represses ULBP2 through miR-34a and miR-34c and loss of this function may contribute to eliminate an innate barrier against tumor development (Heinemann et al. 2012).

5 p53 and the Coordination of Hallmark Processes

This rapid and selective survey of the multiple roles of p53 in processes underlying the Hallmarks of Cancer show that these processes are deeply interconnected, representing nodes in a web-like network rather than independent functional units. Its implication in each of these nodes supports that p53 represents a unifying factor between Hallmark processes. While acquisition of one or several Hallmark capabilities drives cells towards cancer, loss of p53 disintegrates the coherence between the Hallmark processes and removes molecular obstacles to malignancy. Thus, in cancer cells having lost p53 function, DNA damage can accumulate without inducing cell growth arrest, accompanied by a metabolic switch that confers energetic autonomy, bypassing replicative senescence, activating embryonic developmental programs such as EMT, promoting tumor inflammation and avoiding immune destruction. In cells with normal p53 competence, transient activation of any of these capabilities will be compensated by p53-dependent down-regulation of other capabilities, thus maintaining homeostasis.

5.1 A Pivotal Connection: From Telomeres to Stem Cell Maintenance and Aging

This pivotal role of p53 in coordinating Hallmark processes is illustrated in the sequence of molecular events that links telomere dysfunction, DNA damage, p53 activation, metabolic adaptation, progressive tissue functional decline and atrophy, and aging (Fig. 1.2) (Sahin et al. 2011; Sahin and Depinho 2010). Mice with constitutively dysfunctional telomeres accumulate relatively low levels of persistent DNA damage in a variety of organs including hematopoietic stem cells, heart, and liver.

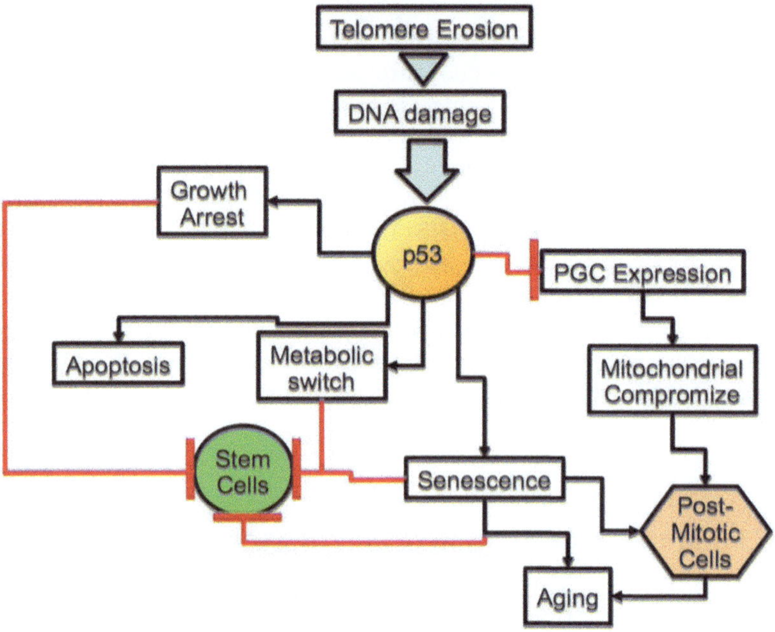

Fig. 1.2 Roles of p53 in the molecular and cellular circuitry linking telomere erosion, stem cell maintenance, senescence, energy metabolism, and aging. This model shows how DNA damage resulting from telomere attrition or dysfunction induces p53, which itself activates coordinated growth arrest and metabolic responses that reduce stem and progenitor cell capabilities and compromises the metabolic fitness of postmitotic cells, leading to organismal aging. Based on Sahin et al. (2011), adapted and augmented

This DNA damage results in p53 activation. In such conditions, p53 induces the repression of *PGC-1alpha* and *PGC-1beta*, encoding peroxisome proliferator-activated receptor gamma coactivators 1 alpha and 1 beta (Sahin et al. 2011). These two factors are master regulators of mitochondrial physiology and metabolism. As a result, p53 interconnects telomere dysfunction with impaired mitochondrial biogenesis and function, decreased gluconeogenesis, cardiomyopathy, and increased production of reactive oxygen species. These conditions favor replicative senescence and the depletion of pools of stem/progenitor cells responsible for tissue renewal and, ultimately, organismal aging. Ablation of *TP53* substantially restores *PGC* expression and the functionality of the subsequent redox and energy metabolic network, restituting mitochondrial respiration, cardiac function, and gluconeogenesis in liver cells. These observations demonstrate that p53 couples telomere attrition and subsequent DNA damage with metabolic effects, senescence, and aging. However, even in the face of normal telomere function, lack of p53 function has, on its own, major effects on mitochondrial determinants of fitness and exercise capacity. p53 interacts with TFAM (mitochondrial transcription factor A), a nuclear-encoded gene important for mitochondrial DNA (mtDNA) transcription and maintenance

(Park et al. 2009). p53-deficient mice show decreased mtDNA content compared to their p53-competent counterparts, with reduction in aerobic versus glycolytic skeletal muscle capacity. Such molecular mechanisms may provide a basis for the inverse correlation between cancer incidence and cardiorespiratory fitness observed in population studies (Wang et al. 2012).

5.2 p53 and Orchestration of the Microenvironment

Hallmark processes form a highly coherent biological program that operates within cells and also between cells, providing an orchestration score for choral interactions between multiple cell types and the extracellular matrix. Acquisition of Hallmark capabilities distorts these interactions and modifies the microenvironment in a way conducive for cancer development. Maintaining the integrity of the microenvironment is therefore a potentially critical tumor suppressive mechanism. Indeed, it is still debated whether cancer cells become metastatic because of their acquired endogenous Hallmark capabilities or because of permissive conditions occurring in the microenvironment. Understanding the tumor suppressive role of the microenvironment may therefore hold the key to tumor reversion and control of metastasis, the holy grail of cancer therapeutic research.

How p53 participates in the control and orchestration of microenvironment signaling is starting to emerge. An interesting lead in this respect is the identification of *TSAP-6* (tumor suppression associated pathway 6) as a target gene of p53 (Amzallag et al. 2004). The product of *TSAP-6* is a 5–6 transmembrane domain protein which interacts with and enhances the secretion of TCTP (translationally controlled tumor protein, also called histamine-releasing factor). This secreted protein participates in inflammatory responses by promoting the release of histamine. TSAP-6 promotes the incorporation of secreted factors into exosomes (Lespagnol et al. 2008). Exosomes are a family of secreted microvesicles which have recently been shown to play a role in remodeling the tumor microenvironment and in priming the metastatic niche, perhaps carrying addressing molecules that target specific signals to particular cells or structures in the microenvironment (Peinado et al. 2011). These observations identify *TSAP-6* as a "multipass" membrane protein with a general role in the regulation of vesicular trafficking, secretion and, beyond, signaling within the microenvironment.

A recent study by Carol Prives and collaborators has provided a very impressive demonstration of the role of p53 as orchestrator of cell–cell and cell–matrix interactions (Freed-Pastor et al. 2012). These authors have used three-dimensional culture models of epithelial breast cells in which nonmalignant cells form spheroids reminiscent of normal breast acinar structures, whereas cancer cells with mutant *TP53* form disrupted structures. In this model, depletion of mutant p53 protein phenotypically reverts breast cancer cells to a more acinar-like morphology. Furthermore, through genome-wide expression analysis, they identified the mevalonate pathway as significantly up-regulated by mutant p53. This pathway (also known as the isoprenoid pathway) serves as the basis for the biosynthesis of molecules used in isoprenoid synthesis, protein anchoring into membranes, and also initiates the steroid biosynthesis pathways. Mutant p53 associates with sterol gene promoters at least partly via SREBP transcription factors and mutation in *TP53* correlates with highly expressed sterol biosynthesis genes in human breast tumors. These observations provide a mechanism linking p53, essential aspects of breast cell metabolism, cell membrane dynamics, and cell-to-cell as well as cell–matrix communications in forming organized, microenvironmental structures.

6 Conclusions: Controlling Hallmark Capabilities Through p53-Based Therapy

One of the main attractions of the Hallmarks of Cancer is that this paradigm provides a rationale and a framework for developing and combining molecular targeted therapeutic interventions. Each Hallmark capability may be approached as a potential drug target. Furthermore, combining drugs that target different Hallmark capabilities is expected to result into increased therapeutic efficacy. So far, most of the promising small-drug based therapies have been designed to target particular factor within defined signaling pathways. Typical examples are tyrosine kinase inhibitors specific for mutant, activated EGFR or monoclonal antibodies to V-EGFR that inhibit angiogenesis signaling. Redundancy between pathways and the existence of multiple bypass mechanisms across the web of Hallmarks are major obstacles for the long-term efficacy of such forms of treatment.

The identification of p53 as a coordinator of Hallmark processes points to a new paradigm for developing combined therapeutic approaches. When associated with drugs addressed at neutralizing or correcting specific Hallmark capabilities, drugs targeting the p53 pathway may restitute critical interconnections between biological processes which, together, cooperate towards effective tumor suppression. The molecular methods to target the p53 pathways are many and diverse: they range from gene-based therapies to small drugs that bind to mutant p53 and restore its activities, or to specific drugs that enhance wild-type p53 function through neutralization of its degradation. The most promising options, however, may stem from recent studies in which p53 has lead to re-discover, under a new light, long-known aspects of biochemistry and bioenergetics. Given the wide possibilities for identifying simple molecules that interferes with metabolic intermediates in these p53-dependent biochemical pathways, the potential for new and effective drugs has never been brighter. Developing and clinically testing such new drugs will require mastering the measurement of a wide range of p53 targets as biomarkers of effects.

Acknowledgment I acknowledge that this review is based on a personal and selective review of literature that omits to cite many important papers, which are nevertheless part of the background of this chapter. I apologize to these authors.

References

Ahn BY, Trinh DL, Zajchowski LD, Lee B, Elwi AN, Kim SW (2010) Tid1 is a new regulator of p53 mitochondrial translocation and apoptosis in cancer. Oncogene 29:1155–1166

Allocati N, Di Ilio C, De Laurenzi V (2012) p63/p73 in the control of cell cycle and cell death. Exp Cell Res 318:1285–1290. PMID: 22326462

Ambs S, Ogunfusika MO, Merriam WG, Bennett WP, Billiar TR, Harris CC (1998) Up-regulation of inducible nitric oxide synthase expression in cancer-prone p53 knockout mice. Proc Natl Acad Sci USA 95:8823–8828

Ambs S, Bennett WP, Merriam WG, Ogunfusika MO, Oser SM, Harrington AM, Shields PG, Felley-Bosco E, Hussain SP, Harris CC (1999) Relationship between p53 mutations and inducible nitric oxide synthase expression in human colorectal cancer. J Natl Cancer Inst 91:86–88

Amzallag N, Passer BJ, Allanic D, Segura E, Thery C, Goud B, Amson R, Telerman A (2004) TSAP6 facilitates the secretion of translationally controlled tumor protein/histamine-releasing factor via a nonclassical pathway. J Biol Chem 279:46104–46112

Ansieau S, Bastid J, Doreau A, Morel AP, Bouchet BP, Thomas C, Fauvet F, Puisieux I, Doglioni C, Piccinin S, Maestro R, Voeltzel T, Selmi A, Valsesia-Wittmann S, Caron de Fromentel C, Puisieux A (2008) Induction of EMT by twist proteins as a collateral effect of tumor-promoting inactivation of premature senescence. Cancer Cell 14:79–89

Assadian S, El-Assaad W, Wang XQ, Gannon PO, Barres V, Latour M, Mes-Masson AM, Saad F, Sado Y, Dostie J, Teodoro JG (2012) p53 inhibits angiogenesis by inducing the production of Arresten. Cancer Res 72(5):1270–1279

Aylon Y, Oren M (2011) New plays in the p53 theater. Curr Opin Genet Dev 21:86–92

Bensaad K, Tsuruta A, Selak MA, Vidal MN, Nakano K, Bartrons R, Gottlieb E, Vousden KH (2006) TIGAR, a p53-inducible regulator of glycolysis and apoptosis. Cell 126:107–120

Brabletz T (2012) MiR-34 and SNAIL: another double-negative feedback loop controlling cellular plasticity/EMT governed by p53. Cell Cycle 11:215

Buckbinder L, Talbott R, Velasco-Miguel S, Takenaka I, Faha B, Seizinger BR, Kley N (1995) Induction of the growth inhibitor IGF-binding protein 3 by p53. Nature 377:646–649

Butt AJ, Firth SM, Baxter RC (1999) The IGF axis and programmed cell death. Immunol Cell Biol 77:256–262

Chang CJ, Chao CH, Xia W, Yang JY, Xiong Y, Li CW, Yu WH, Rehman SK, Hsu JL, Lee HH, Liu M, Chen CT, Yu D, Hung MC (2011) p53 regulates epithelial-mesenchymal transition and stem cell properties through modulating miRNAs. Nat Cell Biol 13:317–323

Chatterjee A, Mambo E, Osada M, Upadhyay S, Sidransky D (2006) The effect of p53-RNAi and p53 knockout on human 8-oxoguanine DNA glycosylase (hOgg1) activity. FASEB J 20:112–114

Choi KS, Bae MK, Jeong JW, Moon HE, Kim KW (2003) Hypoxia-induced angiogenesis during carcinogenesis. J Biochem Mol Biol 36:120–127

Crighton D, Wilkinson S, O'Prey J, Syed N, Smith P, Harrison PR, Gasco M, Garrone O, Crook T, Ryan KM (2006) DRAM, a p53-induced modulator of autophagy, is critical for apoptosis. Cell 126:121–134

Crighton D, Wilkinson S, Ryan KM (2007) DRAM links autophagy to p53 and programmed cell death. Autophagy 3:72–74

Curto M, McClatchey AI (2008) Nf2/Merlin: a coordinator of receptor signalling and intercellular contact. Br J Cancer 98:256–262

de Moraes E, Dar NA, de Moura Gallo CV, Hainaut P (2007) Cross-talks between cyclooxygenase-2 and tumor suppressor protein p53: balancing life and death during inflammatory stress and carcinogenesis. Int J Cancer 121:929–937

Dominguez-Brauer C, Brauer PM, Chen YJ, Pimkina J, Raychaudhuri P (2010) Tumor suppression by ARF: gatekeeper and caretaker. Cell Cycle 9:86–89

el-Deiry WS, Tokino T, Velculescu VE, Levy DB, Parsons R, Trent JM, Lin D, Mercer WE, Kinzler KW, Vogelstein B (1993) WAF1, a potential mediator of p53 tumor suppression. Cell 75:817–825

Eliyahu D, Michalovitz D, Eliyahu S, Pinhasi-Kimhi O, Oren M (1989) Wild-type p53 can inhibit oncogene-mediated focus formation. Proc Natl Acad Sci USA 86:8763–8767
Eng CH, Abraham RT (2011) The autophagy conundrum in cancer: influence of tumorigenic metabolic reprogramming. Oncogene 30:4687–4696
Feng Z (2010) p53 regulation of the IGF-1/AKT/mTOR pathways and the endosomal compartment. Cold Spring Harb Perspect Biol 2:a001057
Flores I, Blasco MA (2009) A p53-dependent response limits epidermal stem cell functionality and organismal size in mice with short telomeres. PLoS One 4:e4934
Forrester K, Ambs S, Lupold SE, Kapust RB, Spillare EA, Weinberg WC, Felley-Bosco E, Wang XW, Geller DA, Tzeng E, Billiar TR, Harris CC (1996) Nitric oxide-induced p53 accumulation and regulation of inducible nitric oxide synthase expression by wild-type p53. Proc Natl Acad Sci USA 93:2442–2447
Freed-Pastor WA, Mizuno H, Zhao X, Langerod A, Moon SH, Rodriguez-Barrueco R, Barsotti A, Chicas A, Li W, Polotskaia A, Bissell MJ, Osborne TF, Tian B, Lowe SW, Silva JM, Borresen-Dale AL, Levine AJ, Bargonetti J, Prives C (2012) Mutant p53 disrupts mammary tissue architecture via the mevalonate pathway. Cell 148:244–258
Galluzzi L, Morselli E, Kepp O, Maiuri MC, Kroemer G (2010) Defective autophagy control by the p53 rheostat in cancer. Cell Cycle 9:250–255
Gatz SA, Wiesmuller L (2006) p53 in recombination and repair. Cell Death Differ 13:1003–1016
Gottlieb E, Vousden KH (2010) p53 regulation of metabolic pathways. Cold Spring Harb Perspect Biol 2:a001040
Grombacher T, Eichhorn U, Kaina B (1998) p53 is involved in regulation of the DNA repair gene O6-methylguanine-DNA methyltransferase (MGMT) by DNA damaging agents. Oncogene 17:845–851
Hafsi H, Hainaut P (2011) Redox control and interplay between p53 isoforms: roles in the regulation of basal p53 levels, cell fate, and senescence. Antioxid Redox Signal 15:1655–1667
Hainaut P, Wiman KG (2009) 30 years and a long way into p53 research. Lancet Oncol 10:913–919
Hanahan D, Weinberg RA (2000) The hallmarks of cancer. Cell 100:57–70
Hanahan D, Weinberg RA (2011) Hallmarks of cancer: the next generation. Cell 144:646–674
Heinemann A, Zhao F, Pechlivanis S, Eberle J, Steinle A, Diederichs S, Schadendorf D, Paschen A (2012) Tumor suppressive microRNAs miR-34a/c control cancer cell expression of ULBP2, a stress-induced ligand of the natural killer cell receptor NKG2D. Cancer Res 72:460–471
Herzer K, Falk CS, Encke J, Eichhorst ST, Ulsenheimer A, Seliger B, Krammer PH (2003) Upregulation of major histocompatibility complex class I on liver cells by hepatitis C virus core protein via p53 and TAP1 impairs natural killer cell cytotoxicity. J Virol 77:8299–8309
Hill KA, Buettner VL, Heidt A, Chen LL, Li W, Gonzalez KD, Wang JC, Scaringe WA, Sommer SS (2006) Most spontaneous tumors in a mouse model of Li-Fraumeni syndrome do not have a mutator phenotype. Carcinogenesis 27:1860–1866
Hollstein M, Hainaut P (2010) Massively regulated genes: the example of TP53. J Pathol 220:164–173
Hu W, Zhang C, Wu R, Sun Y, Levine A, Feng Z (2010) Glutaminase 2, a novel p53 target gene regulating energy metabolism and antioxidant function. Proc Natl Acad Sci USA 107:7455–7460
Huang S, Liu LN, Hosoi H, Dilling MB, Shikata T, Houghton PJ (2001) p53/p21(CIP1) cooperate in enforcing rapamycin-induced G(1) arrest and determine the cellular response to rapamycin. Cancer Res 61:3373–3381
Jiang P, Du W, Wang X, Mancuso A, Gao X, Wu M, Yang X (2011) p53 regulates biosynthesis through direct inactivation of glucose-6-phosphate dehydrogenase. Nat Cell Biol 13:310–316
Jing Y, Han Z, Zhang S, Liu Y, Wei L (2011) Epithelial-Mesenchymal Transition in tumor microenvironment. Cell Biosci 1:29
Kamp DW, Shacter E, Weitzman SA (2011) Chronic inflammation and cancer: the role of the mitochondria. Oncology (Williston Park) 25(5):400–410, 413
Kanaya T, Kyo S, Hamada K, Takakura M, Kitagawa Y, Harada H, Inoue M (2000) Adenoviral expression of p53 represses telomerase activity through down-regulation of human telomerase reverse transcriptase transcription. Clin Cancer Res 6:1239–1247

Kawauchi K, Araki K, Tobiume K, Tanaka N (2008) p53 regulates glucose metabolism through an IKK-NF-kappaB pathway and inhibits cell transformation. Nat Cell Biol 10:611–618
Khromova NV, Kopnin PB, Stepanova EV, Agapova LS, Kopnin BP (2009) p53 hot-spot mutants increase tumor vascularization via ROS-mediated activation of the HIF1/VEGF-A pathway. Cancer Lett 276:143–151
Kim H, Kwak NJ, Lee JY, Choi BH, Lim Y, Ko YJ, Kim YH, Huh PW, Lee KH, Rha HK, Wang YP (2004) Merlin neutralizes the inhibitory effect of Mdm2 on p53. J Biol Chem 279:7812–7818
Kogan-Sakin I, Tabach Y, Buganim Y, Molchadsky A, Solomon H, Madar S, Kamer I, Stambolsky P, Shelly A, Goldfinger N, Valsesia-Wittmann S, Puisieux A, Zundelevich A, Gal-Yam EN, Avivi C, Barshack I, Brait M, Sidransky D, Domany E, Rotter V (2011) Mutant p53(R175H) upregulates Twist1 expression and promotes epithelial-mesenchymal transition in immortalized prostate cells. Cell Death Differ 18:271–281
Kondoh H, Lleonart ME, Gil J, Wang J, Degan P, Peters G, Martinez D, Carnero A, Beach D (2005) Glycolytic enzymes can modulate cellular life span. Cancer Res 65:177–185
Lamouille S, Derynck R (2009) Oncogene and tumour suppressor: the two faces of SnoN. EMBO J 28:3459–3460
Lane DP (1992) Cancer. p53, guardian of the genome. Nature 358:15–16
Lane D, Levine A (2010) p53 Research: the past thirty years and the next thirty years. Cold Spring Harb Perspect Biol 2:a000893
Lespagnol A, Duflaut D, Beekman C, Blanc L, Fiucci G, Marine JC, Vidal M, Amson R, Telerman A (2008) Exosome secretion, including the DNA damage-induced p53-dependent secretory pathway, is severely compromised in TSAP6/Steap3-null mice. Cell Death Differ 15:1723–1733
Levine AJ, Finlay CA, Hinds PW (2004) P53 is a tumor suppressor gene. Cell 116:S67–S69, 1 p following S69
Li Z, Musich PR, Zou Y (2011a) Differential DNA damage responses in p53 proficient and deficient cells: cisplatin-induced nuclear import of XPA is independent of ATR checkpoint in p53-deficient lung cancer cells. Int J Biochem Mol Biol 2:138–145
Li Z, Musich PR, Serrano MA, Dong Z, Zou Y (2011b) XPA-mediated regulation of global nucleotide excision repair by ATR Is p53-dependent and occurs primarily in S-phase. PLoS One 6:e28326
Liang X, Wang P, Gao Q, Xiang T, Tao X (2010) Endogenous LKB1 knockdown accelerates G(1)/S transition through p53 and p16 pathways. Cancer Biol Ther 9:156–160
Liu G, Park YJ, Tsuruta Y, Lorne E, Abraham E (2009) p53 Attenuates lipopolysaccharide-induced NF-kappaB activation and acute lung injury. J Immunol 182:5063–5071
Lomazzi M, Moroni MC, Jensen MR, Frittoli E, Helin K (2002) Suppression of the p53- or pRB-mediated G1 checkpoint is required for E2F-induced S-phase entry. Nat Genet 31: 190–194
Lu X, Bocangel D, Nannenga B, Yamaguchi H, Appella E, Donehower LA (2004a) The p53-induced oncogenic phosphatase PPM1D interacts with uracil DNA glycosylase and suppresses base excision repair. Mol Cell 15:621–634
Lu X, Nguyen TA, Appella E, Donehower LA (2004b) Homeostatic regulation of base excision repair by a p53-induced phosphatase: linking stress response pathways with DNA repair proteins. Cell Cycle 3:1363–1366
Maiuri MC, Malik SA, Morselli E, Kepp O, Criollo A, Mouchel PL, Carnuccio R, Kroemer G (2009) Stimulation of autophagy by the p53 target gene Sestrin2. Cell Cycle 8:1571–1576
Malkin D, Li FP, Strong LC, Fraumeni JFJ, Nelson CE, Kim DH, Kassel J, Gryka MA, Bischoff FZ, Tainsky MA et al (1990) Germ line p53 mutations in a familial syndrome of breast cancer, sarcomas, and other neoplasms. Science 250:1233–1238
Marcel V, Dichtel-Danjoy ML, Sagne C, Hafsi H, Ma D, Ortiz-Cuaran S, Olivier M, Hall J, Mollereau B, Hainaut P, Bourdon JC (2011) Biological functions of p53 isoforms through evolution: lessons from animal and cellular models. Cell Death Differ 18:1815–1824
Martins CP, Brown-Swigart L, Evan GI (2006) Modeling the therapeutic efficacy of p53 restoration in tumors. Cell 127:1323–1334

Matthew EM, Hart LS, Astrinidis A, Navaraj A, Dolloff NG, Dicker DT, Henske EP, El-Deiry WS (2009) The p53 target Plk2 interacts with TSC proteins impacting mTOR signaling, tumor growth and chemosensitivity under hypoxic conditions. Cell Cycle 8:4168–4175

Melino G, Lu X, Gasco M, Crook T, Knight RA (2003) Functional regulation of p73 and p63: development and cancer. Trends Biochem Sci 28:663–670

Menendez D, Shatz M, Azzam K, Garantziotis S, Fessler MB, Resnick MA (2011) The Toll-like receptor gene family is integrated into human DNA damage and p53 networks. PLoS Genet 7:e1001360

Moffitt KL, Martin SL, Walker B (2010) From sentencing to execution – the processes of apoptosis. J Pharm Pharmacol 62:547–562

Munoz-Fontela C, Pazos M, Delgado I, Murk W, Mungamuri SK, Lee SW, Garcia-Sastre A, Moran TM, Aaronson SA (2011) p53 serves as a host antiviral factor that enhances innate and adaptive immune responses to influenza A virus. J Immunol 187:6428–6436

Nekulova M, Holcakova J, Coates P, Vojtesek B (2011) The role of p63 in cancer, stem cells and cancer stem cells. Cell Mol Biol Lett 16:296–327

Offer H, Milyavsky M, Erez N, Matas D, Zurer I, Harris CC, Rotter V (2001) Structural and functional involvement of p53 in BER in vitro and in vivo. Oncogene 20:581–589

Oren M (2003) Decision making by p53: life, death and cancer. Cell Death Differ 10:431–442

Palmero EI, Achatz MI, Ashton-Prolla P, Olivier M, Hainaut P (2010) Tumor protein 53 mutations and inherited cancer: beyond Li-Fraumeni syndrome. Curr Opin Oncol 22:64–69

Park JY, Wang PY, Matsumoto T, Sung HJ, Ma W, Choi JW, Anderson SA, Leary SC, Balaban RS, Kang JG, Hwang PM (2009) p53 improves aerobic exercise capacity and augments skeletal muscle mitochondrial DNA content. Circ Res 105:705–712, 11 p following 712

Peinado H, Lavotshkin S, Lyden D (2011) The secreted factors responsible for pre-metastatic niche formation: old sayings and new thoughts. Semin Cancer Biol 21:139–146

Petitjean A, Achatz MI, Borresen-Dale AL, Hainaut P, Olivier M (2007) TP53 mutations in human cancers: functional selection and impact on cancer prognosis and outcomes. Oncogene 26:2157–2165

Pfeifer GP, Hainaut P (2011) Next-generation sequencing: emerging lessons on the origins of human cancer. Curr Opin Oncol 23:62–68

Ryan KM, Ernst MK, Rice NR, Vousden KH (2000) Role of NF-kappaB in p53-mediated programmed cell death. Nature 404:892–897

Sahin E, Depinho RA (2010) Linking functional decline of telomeres, mitochondria and stem cells during ageing. Nature 464:520–528

Sahin E, Colla S, Liesa M, Moslehi J, Muller FL, Guo M, Cooper M, Kotton D, Fabian AJ, Walkey C, Maser RS, Tonon G, Foerster F, Xiong R, Wang YA, Shukla SA, Jaskelioff M, Martin ES, Heffernan TP, Protopopov A, Ivanova E, Mahoney JE, Kost-Alimova M, Perry SR, Bronson R, Liao R, Mulligan R, Shirihai OS, Chin L, DePinho RA (2011) Telomere dysfunction induces metabolic and mitochondrial compromise. Nature 470:359–365

Sarig R, Rivlin N, Brosh R, Bornstein C, Kamer I, Ezra O, Molchadsky A, Goldfinger N, Brenner O, Rotter V (2010) Mutant p53 facilitates somatic cell reprogramming and augments the malignant potential of reprogrammed cells. J Exp Med 207:2127–2140

Schwartzenberg-Bar-Yoseph F, Armoni M, Karnieli E (2004) The tumor suppressor p53 down-regulates glucose transporters GLUT1 and GLUT4 gene expression. Cancer Res 64:2627–2633

Seemann S, Hainaut P (2005) Roles of thioredoxin reductase 1 and APE/Ref-1 in the control of basal p53 stability and activity. Oncogene 24:3853–3863

Shamas-Din A, Brahmbhatt H, Leber B, Andrews DW (2011) BH3-only proteins: orchestrators of apoptosis. Biochim Biophys Acta 1813:508–520

Shay JW, Wright WE (2001) Telomeres and telomerase: implications for cancer and aging. Radiat Res 155:188–193

Sherr CJ, Roberts JM (1999) CDK inhibitors: positive and negative regulators of G1-phase progression. Genes Dev 13:1501–1512

Shlien A, Tabori U, Marshall CR, Pienkowska M, Feuk L, Novokmet A, Nanda S, Druker H, Scherer SW, Malkin D (2008) Excessive genomic DNA copy number variation in the Li-Fraumeni cancer predisposition syndrome. Proc Natl Acad Sci USA 105:11264–11269

Tan M, Li S, Swaroop M, Guan K, Oberley LW, Sun Y (1999) Transcriptional activation of the human glutathione peroxidase promoter by p53. J Biol Chem 274:12061–12066

Teodoro JG, Parker AE, Zhu X, Green MR (2006) p53-mediated inhibition of angiogenesis through up-regulation of a collagen prolyl hydroxylase. Science 313:968–971

Textor S, Fiegler N, Arnold A, Porgador A, Hofmann TG, Cerwenka A (2011) Human NK cells are alerted to induction of p53 in cancer cells by upregulation of the NKG2D ligands ULBP1 and ULBP2. Cancer Res 71:5998–6009

Thanasoula M, Escandell JM, Martinez P, Badie S, Munoz P, Blasco MA, Tarsounas M (2010) p53 prevents entry into mitosis with uncapped telomeres. Curr Biol 20:521–526

Trinh DL, Elwi AN, Kim SW (2010) Direct interaction between p53 and Tid1 proteins affects p53 mitochondrial localization and apoptosis. Oncotarget 1:396–404

Vahsen N, Cande C, Briere JJ, Benit P, Joza N, Larochette N, Mastroberardino PG, Pequignot MO, Casares N, Lazar V, Feraud O, Debili N, Wissing S, Engelhardt S, Madeo F, Piacentini M, Penninger JM, Schagger H, Rustin P, Kroemer G (2004) AIF deficiency compromises oxidative phosphorylation. EMBO J 23:4679–4689

Vaninetti NM, Geldenhuys L, Porter GA, Risch H, Hainaut P, Guernsey DL, Casson AG (2008) Inducible nitric oxide synthase, nitrotyrosine and p53 mutations in the molecular pathogenesis of Barrett's esophagus and esophageal adenocarcinoma. Mol Carcinog 47:275–285

Vaseva AV, Marchenko ND, Moll UM (2009) The transcription-independent mitochondrial p53 program is a major contributor to nutlin-induced apoptosis in tumor cells. Cell Cycle 8:1711–1719

Wang PY, Zhuang J, Hwang PM (2012) p53: exercise capacity and metabolism. Curr Opin Oncol 24:76–82

Wanka C, Brucker DP, Bahr O, Ronellenfitsch M, Weller M, Steinbach JP, Rieger J (2011) Synthesis of cytochrome c oxidase 2: a p53-dependent metabolic regulator that promotes respiratory function and protects glioma and colon cancer cells from hypoxia-induced cell death. Oncogene (2011 Nov 28. doi: 10.1038/onc.2011.530. [Epub ahead of print])

Wendt MK, Allington TM, Schiemann WP (2009) Mechanisms of the epithelial-mesenchymal transition by TGF-beta. Future Oncol 5:1145–1168

Wu GS, Kim K, el-Deiry WS (2000) KILLER/DR5, a novel DNA-damage inducible death receptor gene, links the p53-tumor suppressor to caspase activation and apoptotic death. Adv Exp Med Biol 465:143–151

Xu J, Lamouille S, Derynck R (2009) TGF-beta-induced epithelial to mesenchymal transition. Cell Res 19:156–172

Xue W, Zender L, Miething C, Dickins RA, Hernando E, Krizhanovsky V, Cordon-Cardo C, Lowe SW (2007) Senescence and tumour clearance is triggered by p53 restoration in murine liver carcinomas. Nature 445:656–660

Yin Y, Shen WH (2008) PTEN: a new guardian of the genome. Oncogene 27:5443–5453

Yoon KA, Nakamura Y, Arakawa H (2004) Identification of ALDH4 as a p53-inducible gene and its protective role in cellular stresses. J Hum Genet 49:134–140

Chapter 2
The Inheritance of p53

Lukasz F. Grochola, Jorge Zeron-Medina, Emmanouela Repapi, Alexander E. Finlayson, Ying Cai, Gurinder Singh Atwal, and Gareth L. Bond

1 Introduction

1.1 p53 in the Cell

Since its first description in 1979, the p53 protein and its cellular network of genes has constituted one of the most extensively studied areas in cancer biology (Lane and Levine 2010). In more than three decades of intense research, compelling evidence has established the role of the p53 pathway as a major cellular tumor suppressor network (Lane and Levine 2010). The p53 protein has been demonstrated to suppress cancer primarily through the induction of cell cycle arrest, apoptosis, or senescence in response to a wide range of different cellular stresses in a stimulus- and cell-type dependent manner (Lane and Levine 2010; Vousden and Prives 2009). The stress signals include genotoxic damage, hypoxia, loss of normal cell contact, nutrient deprivation, mitotic spindle damage, heat or cold shock, telomere shortening, unfolded proteins, and oncogene activation (Vousden and Prives 2009; Lane and Levine 2010). Upon cellular stress, the p53 protein undergoes a plethora of post-translational modifications, which can affect the stability and activity of the protein (Vousden and Prives 2009; Lane and Levine 2010). To elicit the cellular stress responses p53 primarily acts as a sequence-specific transcription factor, but it can also directly regulate proteins from the Bcl-2 family (Moll et al. 2005), microRNA processing (Suzuki et al. 2009), and lincRNA expression (Huarte et al. 2010).

The complex regulation of p53 enables it to integrate a variety of stress signals into specific cellular responses in a stimulus- and cell-type dependent manner

L.F. Grochola • J. Zeron-Medina • E. Repapi • A.E. Finlayson • G.L. Bond (✉)
The Ludwig Institute for Cancer Research, ORCRB, University of Oxford,
Old Road Campus, Off Roosevelt Drive, Oxford OX3 7DQ, UK
e-mail: gareth.bond@ndm.ox.ac.uk

Y. Cai • G.S. Atwal
Cold Spring Harbor Laboratory, Cold Spring Harbor, New York, NY, USA

P. Hainaut et al. (eds.), *p53 in the Clinics*, DOI 10.1007/978-1-4614-3676-8_2,

Lane and Levine 2010; Lu 2010. An interesting component of this regulation is a multitude of feedback loops, which positively or negatively affect the activity of p53 (Lu 2010). Hereby, the MDM2-p53 feedback loop is thought to be the key regulatory component of the p53 stress response (Lu 2010). It controls the level of expression and the activity of p53 by different mechanisms (Lu 2010; Manfredi 2010). Murine double minute 2 (Mdm2) is an E3 ubiquitin ligase, which can bind the p53 protein and target it for proteosome-mediated degradation through ubiquitination (Manfredi 2010; Lu 2010). Furthermore, Mdm2 directly binds to the N-terminus of p53, thereby masking its transactivation domain (Manfredi 2010). In addition, Mdm2 has been demonstrated to shuttle p53 out of the nucleus, where it can no longer function as a transcriptional activator (Wade et al. 2010). Interestingly, *MDM2* itself is a transcriptional target of p53, thus resulting in a negative feedback loop between p53 and Mdm2 (Manfredi 2010; Lu 2010).

1.2 p53 in the Organism: Cancer

The impairment of this tightly regulated network of genes centered on p53 is a hallmark of tumorigenesis (Lane and Levine 2010; Lu 2010). In fact, it has been estimated that 50 % of human cancers carry somatic mutations of the *TP53* gene, which inactivate the normal function of p53 (Lane and Levine 2010; Olivier et al. 2010). In addition, human tumors that retain a *TP53* wild-type gene frequently show a partial abrogation of downstream effectors (e.g., *BAX*, *BAK*, and *APAF-1*) or upstream (e.g., *ATM*, *CHK2*, *MDM2*, and *p19ARF*) regulators of this network (Lane and Levine 2010; Wade et al. 2010).

Genes that constitute the p53 network do not only affect the susceptibility to cancer, but are also crucial in the mediation of the cellular response to treatment with DNA-damaging therapeutic agents. Those agents, such as radio- and chemotherapy, lead to the activation of the upstream effectors of p53 (e.g., Atm, Atr, Chk1, Chk2) (Zhou and Elledge 2000; Bolderson et al. 2009) that activate p53 and can ultimately lead to cancer cell clearance via activation of apoptotic or senescence pathways (Johnstone et al. 2002; Lowe and Lin 2000). Indeed, mutations in the *TP53* gene can produce multidrug resistance in cells and in mice, and reintroduction of wild-type p53 can reconstitute chemosensitivity (Johnstone et al. 2002; Wallace-Brodeur and Lowe 1999). By the same token, mutations or altered expression of other p53 network genes, such as *APAF-1* or *BCL-2*, can significantly alter drug sensitivity in experimental models and are associated with multidrug resistance in human cancers (Johnstone et al. 2002; Reed 1999; Schmitt et al. 2000a; Wei et al. 2001; Zhang et al. 2000).

1.3 p53 in the Organism: Reproduction

Besides its crucial role in maintaining genomic stability and preventing tumor formation, a growing body of evidence suggests that p53 also regulates reproductive processes and fecundity (Hu et al. 2008). *TP53* is conserved from invertebrates to mammals (Belyi et al. 2010) and homologs of *TP53* have been described in many different organisms, such as sea anemone, clams, *Caenorhabditis elegans*, Drosophila, frogs, and zebra fish (Hu 2009; Hu et al. 2008). Interestingly, one of the primary functions of the *TP53* ancestral gene in worms and flies, which are short-lived, cancer-free organisms, seems to be to ensure germline genomic integrity and the fidelity of the developmental process by regulating germ cell replication in order to eliminate defective offspring from the population (Hu et al. 2008; Hu 2009). These primordial p53 activities in the surveillance of germ cell replication and fecundity are also retained in higher organisms (Hu et al. 2008; Hu 2009). For example, in mice and rats, p53 levels are very high during spermatogenesis and mice with reduced levels of p53 show germ cell degeneration during the meiotic prophase (Hu et al. 2008; Hu 2009). Furthermore, p53 has also been shown to regulate reproduction in mice, whereby loss of p53 causes a significant decrease in fertility particularly in female animals (Hu et al. 2007a). Recently, Hu et al. have shown that one crucial mechanism by which p53 regulates maternal reproduction in mice is through its target gene leukemia inhibitory factor (*LIF*), a multifunctional cytokine, which plays a crucial role in blastocyst implantation in a p53-dependent manner (Hu et al. 2007a). Taken together, the observations made in invertebrates and mammals suggest that, besides its important role in maintaining genomic stability in somatic cells, p53 is also responsible to ensure faithful development, germline genomic integrity and fertility, and thus the production of normal offspring that will survive and reproduce.

2 Low-Frequency, Highly Penetrant Inherited Mutations in Humans

2.1 *Introduction*

As mentioned earlier, evidence from over 30 years of intense research efforts has established that the p53 network of genes is crucial in suppressing cancer in vertebrates. In fact, it has been well documented that low-frequency, highly penetrant inherited mutations in genes that are either directly in the p53 signaling pathway, or in closely interacting pathways, underlie many known cancer predisposition syndromes. These include Ataxia telangiectasia (caused by a homozygous mutation in the *ATM* gene) (McKinnon 2004), Cowden disease, Lhermitte–Duclos disease, and Bannayan–Zonana syndrome (all caused by mutations in the *PTEN* gene) (Liaw et al. 1997; Marsh et al. 1997; Di Cristofano et al. 1998), Xeroderma pigmentosum (caused by mutations in the *XP* genes) (Kraemer et al. 1994), Tuberous sclerosis complex 1 and 2 (caused by disruptions in the *TSC1* and *TSC2* genes) (Miyoshi et al. 2002; Orlova and Crino 2010), or the Peutz–Jeghers Syndrome (caused by disruptions in the *LKB1* gene) (Miyoshi et al. 2002).

Many of the same somatic mutations of the *TP53* gene that are found at such high frequencies in cancer cells can also be found as inherited mutations in individuals of the Li–Fraumeni cancer syndrome (LFS) (Varley 2003; Varley et al. 1997). This autosomal-dominant inherited cancer predisposition disorder with an estimated prevalence between 1:5,000 and 1:20,000 (0.02–0.005 %) in the United Kingdom and the U.S.A (Lalloo et al. 2003; Gonzalez et al. 2009) is characterized by a familial clustering of early onset tumors including osteosarcomas, soft-tissue sarcomas, breast cancers, leukemia, and brain tumors (Varley 2003; Varley et al. 1997). The causative role of mutant p53 in this syndrome has been supported by many mouse models. Mice that harbor germline mutations or deletions of the *TP53* gene display cancer phenotypes that closely resemble the human syndrome (Donehower and Lozano 2009; Lozano 2010). Specifically, animals with a germline deletion of one *TP53* allele ($p53^{+/-}$) or carrying hot-spot mutations in the DNA-binding domain of p53 (corresponding to the human R175H and R273H loci) develop a broad spectrum of tumors closely resembling LFS, particularly lymphomas, osteosarcomas, soft-tissue sarcomas, and breast tumors significantly earlier than their wild-type counterparts (Donehower and Lozano 2009; Lozano 2010). Importantly, the median tumor incidence in the p53-targeting murine models is 18 months, and thus, for a mammal with a lifespan of 3 years, is similar to the human tumor onset in p53-mutation carriers, whereby the average cancer incidence is of 50 % by the age of 30 years (Donehower and Lozano 2009).

2.2 *Genetic Testing*

Genetic screening for inherited *TP53* mutations and close surveillance of carriers in order to improve patient prognosis has been a subject of great debate and is riddled with challenges (American College of Medical Genetics Board of Directors 1995; Li 1995; Committee on Bioethics 2001; Evans et al. 2010; Villani et al. 2011). Some of the challenges have been the wide range of tumors, the variability of the penetrance of the mutations, limited preventive options for most of the cancers, and ethical issues concerning newborn screening (American College of Medical Genetics Board of Directors 1995; Li 1995; Committee on Bioethics 2001; Evans et al. 2010; Villani et al. 2011; Achatz et al. 2009). However, females with a germline *TP53* mutation can be given the option of prophylactic mastectomy to reduce the risk of breast cancer (Hartmann et al. 1999; Thull and Vogel 2004; Schneider and Garber 1993). Furthermore, it has been recommended that radiation treatment in germline *TP53* mutation carriers should be assessed with scrutiny, as ionizing radiation has been shown to significantly increase the risk of second malignancies in this patient group (Varley 2003; Hisada et al. 1998; Limacher et al. 2001). However, a recent prospective observational study provides strong support for the utilization of genetic screening for inherited *TP53* mutations and subsequent presymptomatic surveillance of carriers in order to identify low-grade and premalignant cancers before they progress to a more malignant state and, therefore, improve outcome (Villani et al. 2011). In this study, asymptomatic *TP53* mutation carriers were surveyed using a protocol of noninvasive biochemical and imaging techniques. Of the 33 *TP53* mutation carriers identified, 18 underwent surveillance. Remarkably, 3-year survival was 100 % in the surveillance group and only 21 % in the nonsurveillance group.

Indeed, mass newborn screening for a relatively prevalent germline p53 mutation has been adopted in the state of Parana, Brazil (Achatz et al. 2007, 2009). This p53 mutation, R337H, is at an unusually high frequency (1:300 individuals, 0.3 %) in the Brazilian population (Achatz et al. 2007, 2009). It associates with a 25 % cancer risk at the age of 30 years compared to the estimated 50 % by 30 years for the above-described p53 mutations. Although the genetic screening remains controversial due to the above-mentioned criticisms associated with genetic testing (Achatz et al. 2009), it is the first mass-implemented approach to utilize the inherited genetics of the p53 pathway to identify population groups at higher risk for cancer with the goal of early detection and improving the prognosis of cancer.

3 High Frequency, Lesser Penetrant Human Polymorphisms

3.1 Lessons from Mice

Interestingly, results from multiple mouse models suggest that less penetrant alleles of p53 network genes could also significantly alter p53 signaling and affect cancer onset and progression (Mendrysa et al. 2003, 2006; Alimonti et al. 2010). For example, haploinsufficiency for the murine double minute proteins Mdm2 or Mdm4 leads to increased p53 activity exhibited as increased sensitivity to DNA damage, decreased transformation potential, and tumor development in mice (Terzian et al. 2007). Moreover, mouse models designed to express a hypomorphic Mdm2 allele, which resulted in an approximately 30 % reduction of overall Mdm2 expression, had significantly increased p53 transcriptional activation and apoptotic activities (Mendrysa et al. 2003). Furthermore, the mice also had a significantly reduced formation of cancer compared to wild-type Mdm2 animals in an intestinal tumor model (Mendrysa et al. 2006). Similar observations have been recently made for the tumor suppressor PTEN, a key component of the PTEN/PI3K/Akt pathway implicated in the regulation of multiple biological processes such as apoptosis, metabolism, cell proliferation, and cell growth (Blanco-Aparicio et al. 2007). The PTEN/PI3K/Akt pathway is tightly integrated within the p53 network and has been shown to affect the activity of p53 via multiple mechanisms (Mayo and Donner 2001; Ogawara et al. 2002; Zhou et al. 2003; Freeman et al. 2003; Li et al. 2006; Zhou et al. 2001). In a recent paper, Alimonti et al. engineered a hypomorphic allele for murine PTEN that reduced PTEN expression by only 20 % (Alimonti et al. 2010). Interestingly, the authors demonstrate that animals expressing only 80 % of normal PTEN levels show an increased susceptibility to developing various types of cancer (Alimonti et al. 2010). Together, these mouse models suggest that less penetrant alleles of p53 network genes could also significantly alter p53 signaling and affect cancer onset. Indeed, extensive study of human high frequency genetic variants in the p53 network of genes supports this hypothesis.

3.2 High Frequency Genetic Variation

Human genetic variation, defined as the differences in DNA sequence within the genome of individuals, can be broadly classified into two different nucleotide composition classes, the single nucleotide variants (SNVs), which include single nucleotide polymorphisms (SNPs), point mutations, and single nucleotide insertions or deletions (indels), and the structural variants (Feuk et al. 2006; Frazer et al. 2009). One single nucleotide variant, the SNP, constitutes the most frequently studied form of high frequency human genetic variation (Feuk et al. 2006; Frazer et al. 2009). They can be distinguished from point mutations by their higher frequency in a population, whereby the minor allele frequency (MAF) of a SNP has to exceed the arbitrary cutoff of at least 1 % in any population (Feuk et al. 2006; Frazer et al. 2009). SNPs are usually bi-morphic, with two out of four possible nucleotide bases (i.e., Guanine, Cytosine, Adenine, Thymine) present at a particular allelic locus in a population (http://www.ncbi.nlm.nih.gov/snp). They arise from single historical mutational events in the human DNA, most frequently through physiological errors in DNA replication during cell division or alternatively induced by exogenous DNA-damaging agents (Ku et al. 2010; International HapMap Consortium 2003). The new mutation, or "allele," is termed the derived allele and, the former, the ancestral allele (International HapMap Consortium 2003).

Each allele is initially associated with a set of alleles of other SNPs that were present on the particular chromosomal background on which the SNP arose (International HapMap Consortium 2003; Kelley and Swanson 2008). The specific set of alleles observed on a single chromosome is called a haplotype (International HapMap Consortium 2003; Kelley and Swanson 2008). The coinheritance of SNP alleles in haplotypes results in associations between the respective alleles in a given population that is called linkage disequilibrium (LD) (International HapMap Consortium 2003; Neale 2010). Over time, chromosomal recombination during meiosis results in the reduction of length of each haplotype, and thus the LD between SNPs, whereby the probability of recombination between two SNPs increases with the distance between them (Slatkin 2008). The level of linkage disequilibrium is influenced by a number of factors, including the local rate of recombination, non-random mating, population subdivision and population bottlenecks, mutation, genetic drift, and natural selection (Slatkin 2008).

3.3 p53 Pathway High Frequency Genetic Variants in Mice and Humans

Two well-characterized SNPs in the p53 pathway are found in the *TP53* and *MDM2* genes (p53 codon72, rs1042522, C/G; MDM2 SNP309, rs2279744, T/G) and have been the subject of recent reviews (Vazquez et al. 2008; Grochola et al. 2010; Whibley et al. 2009). Both SNPs can be found at very high frequencies in certain populations, but also vary significantly in their frequencies in different racial and ethnic groups. The p53 codon72 SNP can be as low as 33 % (in a Sub-Saharan African population, HapMap project) and as high as 77 % (in a Northern European population, dbSNP database). The MDM2 SNP309 can be as low as 10 % (in African Americans) and as high as 50 % (in Ashkenazi Jewish individuals) (Vazquez et al. 2008). Allele-specific natural selection has been proposed to explain theses differences and will be discussed below in more detail.

3.4 p53 Codon72 SNP

The p53 codon72 SNP encodes either a proline (p53-codon72-Pro) or arginine (p53-codon72-Arg) residue in a polyproline region of p53 that is located between the transactivation and the DNA-binding domains (Buchman et al. 1988) (Table 2.1). This polymorphism has been the subject of intense study and it is clear that the two different p53 variants differ in their activities significantly. For example, over 12 years ago, the first study to provide evidence that the two different p53 isoforms encoded by the p53 codon72 SNP are not functionally equivalent was published (Thomas et al. 1999). One of the differences the authors observed in cell culture experiments was that the codon72-Pro variant possesses increased ability to induce cell cycle arrest and that p53-codon72-Arg was more efficient at inducing apoptosis (Thomas et al. 1999). However, the precise molecular mechanisms by which these differences occur are still the subject of great debate in the literature (Vazquez et al. 2008; Grochola et al. 2010; Whibley et al. 2009).

Recent studies utilizing genetically engineered mice and mouse cells have shed light on the possible physiological effects of the p53 codon72 alleles (Frank et al. 2011; Zhu et al. 2010; Reinbold et al. 2008; Phang and Sabapathy 2007). The first major challenge these studies have had to overcome is the fact that the region surrounding codon 72 is not highly conserved in mice and encodes an alanine residue. Zhu et al. utilize mice with a humanized exon 4, which includes the sequence that encodes codon 72 (Arg/Pro). Frank et al. utilize mice with humanized exons 4 to 9, which they refer to as *hu*manized *p*53 *k*nock-*i*n gene (Hupki). Zhu et al. observe that their mice carrying the arginine residue associate with an increased rate of apoptosis in MEFs and in the small intestines after radiation compared to mice carrying the proline residue (Zhu et al. 2010). Frank et al. also observe that their mice carrying the arginine residue associate with an increased rate of apoptosis in those tissues after radiation compared to their mice carrying the proline residue (Frank et al. 2011). These data from both models are consistent with observations made in human tumor-derived cell culture systems (Grochola et al. 2010; Vazquez et al. 2008).

Frank et al. go on to show that the allelic differences in apoptotic potential for p53 codon72 are tissue specific and that in another tissue it is the proline isoform that associates with more apoptosis. Specifically, in thymocytes, p53 codon72-Pro mice have increased apoptosis following ionizing radiation along with an increased transactivation of a subset of p53 target genes (Frank et al. 2011). The authors

Table 2.1 Single nucleotide polymorphisms in the p53 pathway

Gene			Single nucleotide polymorphism			
Hugo name	Function	Association with the p53 pathway	Description	db SNP rs#ID	Clinical association	Molecular description
TP53	Tumor suppressor; functions as transcription factor, responds to diverse cellular stresses. Regulates target genes that induce cell cycle arrest, apoptosis, senescence, DNA repair, or changes in metabolism	Central node of the p53 pathway	p53 codon72	rs1042522	Many examples, such as allelic differences in response to chemotherapeutic treatment of head and neck carcinomas	Proline to arginine residue change in a region important in mediating the apoptotic response affects the level of apoptosis
MDM2	Key negative regulator of the p53 protein	Transcriptional target of p53; inhibits p53 activity by binding the transactivation domain of the p53 protein and promoting its ubiquitination as well as regulating its cellular location	MDM2 SNP309	rs2279744	Increased risk for and earlier age of onset of various cancer types, particularly in younger females	T to G change in the promoter region of intron 1 results in an enhanced binding affinity of the transcription factor SP1 and increased MDM2 gene transcription
			MDM2 SNP285	N.A.[a]	Modifier of MDM2 SNP309 phenotype	SNP285G>C diminishes Sp1 transcription factor binding to the MDM2 promoter

CASP8	An upstream protease of the activation cascade of caspases responsible for the death receptor-induced apoptosis	*TP53* activates caspase 8 gene expression after induction by death effector domains or exposure to cytotoxic drugs. Positive feedback loop with p53	D302H	rs1045485	Minor allele associates with reduced incidence of breast cancer	Aspartic acid to histidine substitution in codon 302
ATM	Serine/threonine protein kinase, which activates checkpoint signaling upon genotoxic stresses	Regulates the phosphorylation of p53 on various residues and activates MDM2 as well as the checkpoint kinase CHK2	ATM Ser49Cys	rs1800054	Minor allele suggested to associate with increased cancer risk, particularly breast carcinoma	Serine to cystein substitution in codon 49
CDKN1A	Binds and inhibits the activity of cyclin-dependent kinases, and thus functions as a regulator of cell cycle progression at G1	Tightly controlled by p53, through which this protein mediates the p53-dependent cell cycle G1 phase arrest in response to a variety of stress stimuli	p21 codon31	rs1801270	Allelic differences reported for various cancer types	Serine to arginine substitution in a highly conserved region of the gene; alleles suggested to differ in transcriptional efficiency
CDKN1B	Cyclin-dependent kinase inhibitor; binds to and prevents the activation of cyclin E-CDK2 or cyclin D-CDK4 complexes, and thus controls the cell cycle at G1	Interacts with p53 to modulate the cell cycle and suppress tumorigenesis	−79C/T	rs34330	The T-allele associates with increased risk of breast cancer	C to T transition 5′ UTR at the site of nucleotide −79
TP73	Member of p53 family of transcription factors, involved in cellular responses to stress and development	TP73 transactivates p53-responsive genes causing cell cycle arrest and apoptosis; some isoforms can directly inhibit p53 function	G4C14-to-A4T14	rs2273953 rs1801173	Allelic differences for various cancer types	Two linked intronic SNPs, just upstream of the initiating AUG of exon 2 in position 4 and 14 of the gene

(continued)

Table 2.1 (continued)

Gene			Single nucleotide polymorphism			
Hugo name	Function	Association with the p53 pathway	Description	db SNP rs#ID	Clinical association	Molecular description
MDM4	Key negative regulator of the p53 protein	Transcriptional target of p53; inhibits p53 activity by binding the transactivation domain of the p53 protein. Heterodimerizes with the Mdm2 protein	Selected MDM4 haplotype	rs2369244; rs1563828	Selected haplotype associates with decreased risk and/or earlier age of breast and ovarian cancers	N.A.[a]
PPP2R5E	Regulatory subunit B, beta of protein phosphatase 2A	Major cellular phosphatase, regulates the phosphorylation of p53 and MDM2	epsilon-SNP2	rs11158491	Altered cancer risk and survival of soft-tissue sarcomas	N.A.[a]
PPP2R2B	Regulatory subunit B, beta of protein phosphatase 2A	Major cellular phosphatase, regulates the phosphorylation of p53 and MDM2	rs319217	rs319217	Altered breast cancer risk and recurrence, cellular chemosensitivities	N.A.[a]
TSC1	Encodes a growth inhibitory protein thought to play a role in the stabilization of tuberin	p53 regulation of mTOR activity depends on TSC1/2 complex formation	rs7874234	rs7874234	Allelic differences in age of ER-positive breast cancer tumor onset	T-allele of TSC1 SNP rs7874234 may create an estrogen receptor element (ERE) site

[a]*N.A.* not assigned

provide a possible explanation for the tissue specificity that is based on their observations that a group of genes induced by irradiation in the thymus are different than in other tissues, such as MEFs and the small intestines (Frank et al. 2011). This group is made up of genes that have known roles in immunity and inflammation (Frank et al. 2011). Interestingly, the authors go on to explore the possible differences in the p53 isoforms directly in the inflammatory response. They were able to demonstrate that upon inflammatory stimulation of the mice through endotoxin lipopolysaccharide, those mice harboring the proline residue are significantly more likely to succumb to septic shock than animals carrying the arginine residue (Frank et al. 2011).

Intriguingly, both Frank et al. and Zhu et al. were unable to find any differences between the p53 isoforms in cancer risk or outcome. Frank et al. crossed the proline and arginine mice with a mouse model that develops lymphomas (the Eμ-myc mouse), and with $p53^{+/-}$ mice that develop different cancers such as lymphomas and sarcomas (Donehower et al. 1992; Schmitt et al. 2000b; Frank et al. 2011). They observed no significant difference in the survival or tumor spectrum between the mice harboring the different alleles (Frank et al. 2011). Zhu et al. exposed their mice to chronic UVB treatment to induce skin cancer (Zhu et al. 2010). However, they also observed no statistically significant differences in susceptibility to skin cancer development in the animals (Zhu et al. 2010). Together, these data clearly demonstrate tissue-specific differences between the p53 proline and arginine isoforms in their abilities to induce apoptosis after DNA damage in vivo. Both models suggest that the p53 codon72 SNP alone might not have a direct impact on cancer development. However, a trend is emerging in association studies in human cancer patients that this SNP might work in conjunction with another functional p53 pathway SNP in the *MDM2* oncogene (MDM2 SNP309) to affect cancer risk (Wan et al. 2011).

3.5 MDM2 SNP309

MDM2 SNP309 is found at position 309 in the first intron of the *MDM2* oncogene, which serves as a transcriptional enhancer region (Bond et al. 2004) (Table 2.1). The SNP results in either a thymine (T) or a guanine (G) base. The G-allele increases the affinity of the transcription factor Sp1, which leads to the increased transcription and expression of Mdm2 and inhibition of the p53 stress response (Bond et al. 2004). Importantly, evidence from patient populations has also lent support to this model (Vazquez et al. 2008). For example, *TP53* mutation carriers (Li–Fraumeni individuals) with the G-allele of MDM2 SNP309 were shown to be diagnosed with tumors on average 7 years earlier than those that were T/T in genotype (Bond et al. 2004). This observation has been reproduced in three independent studies, in which *TP53* mutation carriers with the G-allele of MDM2 SNP309 were diagnosed with cancer on average 10, 16, and 12.5 years earlier than those who were homozygous for the T-allele (Bougeard et al. 2006; Ruijs et al. 2007; Marcel et al. 2009). Earlier ages of onset associated with individuals with the G-allele, but no known *TP53* mutations, were also demonstrated in soft-tissue sarcomas, lymphoma, leukemia, head, neck, and oral squamous cell carcinomas, cancers of the colon, breast, bladder, ovary, brain, melanoma, and liver (Phillips et al. 2010; Whibley et al. 2009; Grochola et al. 2010).

To date, a plethora of case-control studies have explored the association of MDM2 SNP309 with overall cancer risk (Wan et al. 2011; Hu et al. 2007b). While many have shown a significant association of increased risk with the G-allele of SNP309 and various cancer types, such as lung, breast, and endometrial cancer, others have failed to provide support for these observations (Hu et al. 2007b; Wan et al. 2011; Ueda et al. 2009; Joshi et al. 2011; Fang et al. 2011). However, a recent meta-analysis of 66 papers that in total included over 25,000 cases and 30,000 controls comprising 30 tumor types provided evidence for an association of the G-allele of MDM2 SNP309 with increased overall cancer risk (Wan et al. 2011). The authors have shown that SNP309 G-allele carriers were associated with a significantly increased risk for various tumor types, such as breast, colorectal, and lung cancer, with a combined odds ratio of up to 1.25 (95 % confidence interval (CI) = 1.13–1.37), providing supportive evidence to the model that MDM2 SNP309 serves as a low-penetrance tumor susceptibility marker (Wan et al. 2011).

Similar to p53 codon72, a recent study evaluated the effects of MDM2 SNP309 on the p53 pathway and cancer risk by creating a mouse model (Post et al. 2010). The authors were motivated by being able to explore its impact on cancer risk in a system that excluded the genetic and environmental heterogeneity of the cancer phenotype inherent to human association studies. To do this, Post et al. generated mice that carry the humanized intron 1 of the *MDM2* gene, containing either the G- or the T-allele of MDM2 SNP309 (Post et al. 2010). Interestingly, the authors find that cells from animals with a G/G genotype of MDM2 SNP309 have elevated Mdm2 levels, reduced p53 levels, and decreased apoptosis. Moreover, and in contrast to the p53 codon72 mice, MDM2 SNP309 G/G mice have a shorter tumor latency and decreased survival, both in animals with two copies of wild-type p53 and in animals with one copy mutated ($p53^{515A/+}$) (Post et al. 2010). These data provide strong evidence that the G-allele of MDM2 SNP309 has a direct impact on cancer risk.

Interestingly, further studies in humans and human-derived material have demonstrated that the effects of the G-allele of MDM2 SNP309 on cancer can be modified by additional variables, such as gender, estrogen, and other p53 pathway SNPs (Bond et al. 2006a, b; Bond and Levine 2007; Lind et al. 2006; Alhopuro et al. 2005). Specifically, MDM2 SNP309 T/G has been repeatedly shown to associate with allele- and gender-specific differences in tumor diagnosis in various malignancies (Bartel et al. 2008; Bond et al. 2006a, b). Interestingly, female carriers of the G-allele have been shown to be diagnosed earlier in life with various cancers, such as colorectal cancer, diffuse large B-cell lymphoma, lung cancer, and for highly estrogen receptor positive (>50 % of tumor cells), but not for estrogen receptor negative, invasive ductal carcinoma of the breast (Bond et al. 2006a, b; Lind et al. 2006). This was shown to result in the enrichment of individuals with the G-allele in premenopausal women with these cancers, when compared to either

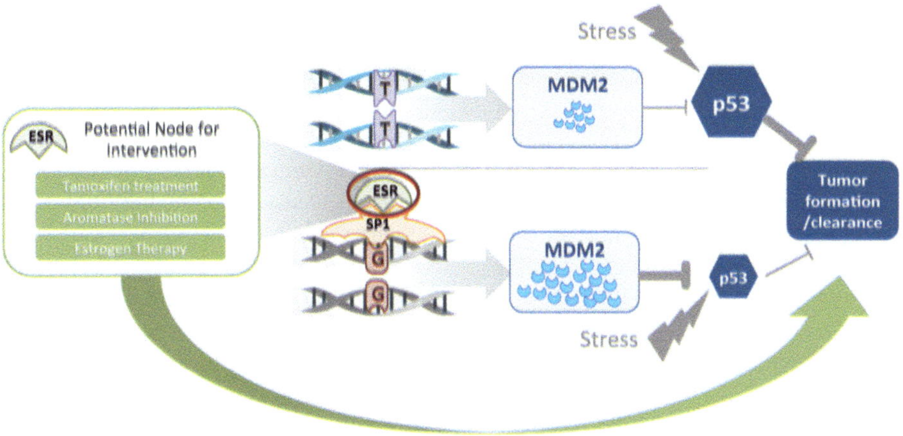

Fig. 2.1 The MDM2 SNP309 genotype could aid in the identification of individuals whose p53 pathway is attenuated in response to estrogen. MDM2 SNP309 is located in an enhancer region of *MDM2*. Multiple lines of evidence support a model that the G-allele has a higher affinity towards ER-alpha-Sp1-dependent activation of Mdm2 expression that leads to an attenuation of p53 activity. This model predicts that patients who carry a G-allele of MDM2 SNP309 would be more sensitive to estrogen signaling manipulation, such as Tamoxifen treatment or estrogen substitution, which is already being successfully used in both cancer prevention and treatment strategies

postmenopausal women or men with the same cancers. Subsequently, similar trends have been observed in melanoma and osteosarcoma patients (Firoz et al. 2009; Toffoli et al. 2009), while other studies suggest these trends will not be seen in every cancer and could be restricted to specific racial and ethnic backgrounds (Park et al. 2006; Bittenbring et al. 2008; Grochola et al. 2009). Molecular and functional studies suggest that the enhancer region of *MDM2* with the G-allele has a higher affinity towards ER-alpha-Sp1-dependent activation of Mdm2 expression (Grochola et al. 2010). This model requires further testing, however, as estrogen signaling manipulation is being successfully used in both cancer prevention and treatment strategies (Jordan 2006), the identification of a genetically defined population of women whose p53 pathway is attenuated in response to estrogen activation could positively affect clinical decisions (Fig. 2.1).

A recent report has provided evidence that a lower frequency SNP in the p53 network that resides in close proximity to SNP309 in intron 1 of the *MDM2* gene can affect the G-allele of MDM2 SNP309 in its ability to interact with Sp1 and affect cancer risk (Knappskog et al. 2011). Specifically, Knappskog et al. report that the C-allele of MDM2 SNP285 (G/C) is found only in Caucasians and with a minor allele frequency of 3.6–4.1 % compared to the 33.7–43.7 % frequency of MDM2 SNP309 (Table 2.1). They showed that the C-allele of MDM2 SNP285 strongly reduces the binding of Sp1 to the G-allele of MDM2 SNP309 in vitro (Knappskog et al. 2011). Moreover, the authors of the study show in a case-control study comprising 1,993 ovarian, 1,973 breast cancer patients, and 3,646 healthy controls that the C-allele of MDM2 SNP285 reduces the risk of both ovarian (OR 0.74;

CI 0.58–0.94) and breast cancers (OR 0.79; CI 0.62–1.00) among SNP309 G-allele carriers. Finally, in line with those results, they identified a population of women (MDM2 SNP309 G-allele and MDM2 SNP285 G-allele carriers) that had an increased risk for breast cancer (OR 1.27; CI 1.03–1.55) (Knappskog et al. 2011).

Evidence for a possible interaction between a higher frequency SNP in the p53 network is provided by another recent publication (Wan et al. 2011). Specifically, the above-mentioned meta-analysis of MDM2 SNP309 case-control studies by Wan et al., comprising 66 papers with over 25,000 cases and 30,000 controls suggests a significant interaction between MDM2 SNP309 and the p53 codon72 polymorphism (Wan et al. 2011). Hereby, the authors show that the combination of a SNP309G/G-codon72Pro/Pro genotype associated with an increased overall risk for cancer compared to the SNP309T/T-codon72Arg/Arg genotype (OR = 3.38, 95 % CI = 1.77–6.47), suggesting that there is an association between MDM2 SNP309 and p53 Arg72Pro regarding tumor susceptibility. Taken together, these reports suggest that in the future knowing the gender, estrogen exposures, and the genotypes of interacting SNP loci, such as MDM2 SNP309 and SNP285, or SNP309 and p53 codon72, could help to more precisely predict an individual's cancer risk and could influence future cancer therapies.

3.6 Haplotype Structures of p53 Pathway Genes and Natural Selection

Most SNPs are thought to arise due to mutations that do not influence the fitness of the organism and the relative proportion (frequency) of the alleles in the population changes over time by random genetic drift. Therefore, they may achieve significant frequencies in the human population simply by chance (Kelley and Swanson 2008; MacCallum and Hill 2006). However, some mutational events will give rise to true functional alleles. These changes in function could confer a fitness advantage to individuals causing the beneficial alleles to increase, on average, in frequency in the population (positive selection). However, should the changes in function be detrimental to the organism, the frequency of the mutation would decrease (negative selection) (Kelley and Swanson 2008; MacCallum and Hill 2006).

One consequence of the positive selection of an allele of a SNP and the subsequent selective sweep in a population is the increase in frequency of neutral alleles in other SNPs closely linked to the selected allele, an effect called the hitchhiking effect (Kelley and Swanson 2008; MacCallum and Hill 2006). This effect gives rise to hallmarks of allele-specific positive selection, namely, a reduction of nucleotide variation in the region of the genome surrounding the beneficial allele and a region characterized by a mix of long haplotypes harboring the selected allele and short haplotypes with the ancestral allele (Kelley and Swanson 2008; MacCallum and Hill 2006). Therefore, long high-frequency haplotypes indicate the action of positive selection, whereas nonselected haplotypes occur at lower frequencies and vary in length (Kelley and Swanson 2008; MacCallum and Hill 2006). In light of the effect of natural selection on haplotype distributions, a haplotype-based statistical test was devised that detected evidence of natural selection on individual SNPs, and it was shown that SNP309 deviates significantly from the standard assumptions of models of selective neutrality (Atwal et al. 2007). The haplotype structure of *MDM2* was analyzed by combining genotype data generated from the International Hapmap Consortium, Celera Diagnostics, and Sheba Cancer Research Center in Israel (Atwal et al. 2007). A low level of haplotypic variability was observed in Northern European populations and Ashkenazi Jewish populations, reflecting strong linkage disequilibrium across the whole gene. The level of linkage disequilibrium was perceived to be stronger than that due to a population bottleneck effect, arising from a founder population out of Africa. The effect of a founder population bottleneck effect would reduce the genetic variability of the entire genome and not just a single gene. By comparisons with extensive computer simulations of molecular evolution it was suggested that the G-allele of MDM2 SNP309 had experienced a positive selection sweep in human populations. The authors reasoned that an evolutionary selection pressure could act on the human p53 pathway and a growing body of evidence now suggests that p53 regulates other human conditions that are under selective pressures, such as germline maintenance, fertility, and reproduction. Indeed, p53 codon72 polymorphism and MDM2 SNP309 have been shown to associate with recurrent implantation failure, missed abortion, and allelic differences in the outcome of in vitro fertilization treatment (Kang et al. 2009; Kay et al. 2006; Fang et al. 2009; Firouzabadi et al. 2009; Hu et al. 2007a; Atwal et al. 2007).

The finding that the G-allele of MDM2 SNP309 had experienced a positive selection sweep gave rise to the hypothesis that other functional SNPs in the p53 pathway could be determined by haplotype studies that looked for evidence of natural selection. Recent work has shown that using signatures of natural selection to identify putatively functional sites and variants has been a fruitful exercise in the study of human cancer. One of the first studies to demonstrate this was an investigation of the haplotype distribution of *MDM4* (Atwal et al. 2009), a structural homologue of *MDM2*, that binds to the amino terminus of p53, functioning as a major inhibitor of p53 activity (Wade et al. 2010). The haplotype structure was analyzed in several human populations. Genotype data for SNPs were collected from three sources: (1) a collection of 84 lymphoblastoid cell lines of African American and Caucasian ethnicity; (2) the Hapmap Project repository including the Caucasian (CEU) and African (YRI) populations; and (3) a cohort of 299 Ashkenazi Jewish controls from a breast cancer association study at Memorial Sloan Kettering Cancer Center. The low diversity of haplotypes from each dataset, resulting in atypical patterns of linkage disequilibrium, indicated the presence of candidate SNPs that may also modify the efficacy of the p53 pathway. In particular, one derived haplotype across the gene had increased to a high frequency (68 % in the Caucasian lymphoblastoid dataset, compared to 30 % in the African American population) in a relatively short amount of time in human history. Subsequent association studies in five different patient populations revealed that these SNPs in *MDM4* conferred an increased risk for, or early onset of, human breast and ovarian cancers in Ashkenazi Jewish and European cohorts, respectively (Atwal et al. 2009). This association was subsequently validated in two patient cohorts, whereby the *MDM4* SNPs have been shown to associate with an accelerated age of onset of estrogen receptor negative breast cancer (Kulkarni et al. 2009) (Table 2.1). These observations not only suggest that, like *MDM2*, *MDM4* harbors SNPs that affect cancer in humans, but also that other SNPs in the p53 network that affect cancer could also be under selection pressures. Indeed multiple studies have lent support to this hypothesis and have identified candidate functional polymorphic loci in *PPP2R5E* (Grochola et al. 2009), *PPP2R2B* (Vazquez et al. 2011), *TSC1* (Mehta et al. 2011), *TSC2* (Mehta et al. 2011), and *HAUSP* (Kang et al. 2009) genes (Table 2.1).

With the onslaught of whole-genome sequences now being generated it will soon be possible to investigate the fine-detailed haplotype structure of the p53 pathway for large populations. The pilot phase of so-called The 1000 Genome Project has recently been completed with the aim of providing a deep characterization of the vast majority (over 95 %) of the common variations that are in genomic regions accessible to current high-throughput sequencing technologies and that have allele frequencies of 1 % or higher in a number of major population groups (The 1000 Genome Project Consortium). Genetic variations with an allele frequency as low as 0.1 % were also cataloged in coding regions since these regions tend to have reduced allele frequencies. This dataset furnishes us with the most fine-scale map to date of genotypes throughout the human genome across multiple different ethnicities. Preliminary haplotype studies on the dataset from the 1000 Genomes Project have illustrated that the *TP53* gene network is highly polymorphic, strengthening the hypothesis that there may be numerous subtle inherited genetic changes that impact the normal function of these different proteins (Cai and Atwal, unpublished). Despite the greater amount of newly discovered genetic variants in the *TP53* gene network, the genes appear to have slightly reduced levels of haplotypic diversity, indicating evidence of natural selection, corroborating the aforementioned earlier studies.

The above-described studies suggest that the p53 stress response pathway could harbor more functional inherited genetic variants. Indeed, other less studied polymorphic variants have been reported for the p53 network genes *CDKN1A*, *CDKN1B*, *TP73*, *ATM*, and *CASP8* (Grochola et al. 2010; Whibley et al. 2009) (Table 2.1). The identification of other functional SNPs that mediate the p53 stress response will prove challenging, as there are over 50,000 SNPs in the NCBI SNP repository (dbSNP) in genes that have been implicated in mediating and regulating the p53 response (Vazquez et al. 2008). However, recently, approaches have been described that could help identify potential functional p53 pathway SNPs (Smirnov et al. 2009; Tomso et al. 2005; Noureddine et al. 2009; Bandele et al. 2011; Grochola et al. 2010). Specifically, these strategies identify SNPs, or haplotypes, that demonstrate allelic differences in characteristics similar, but not limited to, functional p53 pathway SNPs, such as differences in cellular chemosensitivities and the ability to change p53-dependent transcriptional activation of genes. The more in-depth study of these and other genetic variants in the p53 network should help further define populations in their abilities to respond to stress, suppress tumor formation, and respond to DNA damaging therapies.

4 Conclusion

Taken together, mounting evidence suggests that the inherited genetics of p53 pathway have a great potential to further define populations in their abilities to react to stress, suppress tumor formation, and respond to therapies. Various cancer predisposition syndromes are caused by low-frequency, highly penetrant inherited mutations in the p53 network, the knowledge of which can be used to individually tailor cancer prevention and treatment strategies. In addition, it has been demonstrated that the p53 tumor suppressor pathway also harbors functional high-frequency, low-penetrance genetic variants that affect p53 signaling in cells, resulting in allelic differences in cancer risk, prognosis, and responses to chemotherapy. Observations such as these suggest that the insight gained from a thorough analysis of the functional human genetics of this important network of genes could offer novel cancer treatment and prevention strategies as well as personalize those that already exist.

References

Achatz MI, Hainaut P, Ashton-Prolla P (2009) Highly prevalent TP53 mutation predisposing to many cancers in the Brazilian population: a case for newborn screening? Lancet Oncol 10:920–925. doi:S1470-2045(09)70089-0 [pii] 10.1016/S1470-2045(09)70089-0

Achatz MI, Olivier M, Le Calvez F et al (2007) The TP53 mutation, R337H, is associated with Li-Fraumeni and Li-Fraumeni-like syndromes in Brazilian families. Cancer Lett 245:96–102. doi:S0304-3835(06)00005-X [pii] 10.1016/j.canlet.2005.12.039

Alhopuro P, Ylisaukko-Oja SK, Koskinen WJ et al (2005) The MDM2 promoter polymorphism SNP309T->G and the risk of uterine leiomyosarcoma, colorectal cancer, and squamous cell carcinoma of the head and neck. J Med Genet 42:694–698

Alimonti A, Carracedo A, Clohessy JG et al (2010) Subtle variations in Pten dose determine cancer susceptibility. Nat Genet 42:454–458. doi:ng.556 [pii] 10.1038/ng.556

American College of Medical Genetics Board of Directors (1995) Points to consider: ethical, legal, and psychosocial implications of genetic testing in children and adolescents. American Society of Human Genetics Board of Directors, American College of Medical Genetics Board of Directors. Am J Hum Genet 57:1233–1241

Atwal GS, Bond GL, Metsuyanim S et al (2007) Haplotype structure and selection of the MDM2 oncogene in humans. Proc Natl Acad Sci USA 104:4524–4529

Atwal GS, Kirchhoff T, Bond EE et al (2009) Altered tumor formation and evolutionary selection of genetic variants in the human MDM4 oncogene. Proc Natl Acad Sci USA 106:10236–10241

Bandele OJ, Wang X, Campbell MR et al (2011) Human single-nucleotide polymorphisms alter p53 sequence-specific binding at gene regulatory elements. Nucleic Acids Res 39:178–189

Bartel F, Jung J, Bohnke A et al (2008) Both germ line and somatic genetics of the p53 pathway affect ovarian cancer incidence and survival. Clin Cancer Res 14:89–96

Belyi VA, Ak P, Markert E et al (2010) The origins and evolution of the p53 family of genes. Cold Spring Harb Perspect Biol 2:a001198. doi:cshperspect.a001198 [pii] 10.1101/cshperspect.a001198

Bittenbring J, Parisot F, Wabo A et al (2008) MDM2 gene SNP309 T/G and p53 gene SNP72 G/C do not influence diffuse large B-cell non-Hodgkin lymphoma onset or survival in central European Caucasians. BMC Cancer 8:116

Blanco-Aparicio C, Renner O, Leal JF et al (2007) PTEN, more than the AKT pathway. Carcinogenesis 28:1379–1386. doi:bgm052 [pii] 10.1093/carcin/bgm052

Bolderson E, Richard DJ, Zhou BB et al (2009) Recent advances in cancer therapy targeting proteins involved in DNA double-strand break repair. Clin Cancer Res 15:6314–6320. doi:1078-0432.CCR-09-0096 [pii] 10.1158/1078-0432.CCR-09-0096

Bond GL, Hirshfield KM, Kirchhoff T et al (2006a) MDM2 SNP309 accelerates tumor formation in a gender-specific and hormone-dependent manner. Cancer Res 66:5104–5110

Bond GL, Hu W, Bond EE et al (2004) A single nucleotide polymorphism in the MDM2 promoter attenuates the p53 tumor suppressor pathway and accelerates tumor formation in humans. Cell 119:591–602

Bond GL, Levine AJ (2007) A single nucleotide polymorphism in the p53 pathway interacts with gender, environmental stresses and tumor genetics to influence cancer in humans. Oncogene 26:1317–1323

Bond GL, Menin C, Bertorelle R et al (2006b) MDM2 SNP309 accelerates colorectal tumour formation in women. J Med Genet 43:950–952

Bougeard G, Baert-Desurmont S, Tournier I et al (2006) Impact of the MDM2 SNP309 and p53 Arg72Pro polymorphism on age of tumour onset in Li-Fraumeni syndrome. J Med Genet 43:531–533

Buchman VL, Chumakov PM, Ninkina NN et al (1988) A variation in the structure of the protein-coding region of the human p53 gene. Gene 70:245–252

Committee on Bioethics (2001) Ethical issues with genetic testing in pediatrics. Pediatrics 107:1451–1455

Di Cristofano A, Pesce B, Cordon-Cardo C et al (1998) Pten is essential for embryonic development and tumour suppression. Nat Genet 19:348–355

Donehower LA, Harvey M, Slagle BL et al (1992) Mice deficient for p53 are developmentally normal but susceptible to spontaneous tumours. Nature 356:215–221

Donehower LA, Lozano G (2009) 20 years studying p53 functions in genetically engineered mice. Nat Rev Cancer 9:831–841

Evans DG, Lunt P, Clancy T et al (2010) Childhood predictive genetic testing for Li-Fraumeni syndrome. Fam Cancer 9:65–69. doi:10.1007/s10689-009-9245-9

Fang F, Yu XJ, Yu L et al (2011) MDM2 309 T/G polymorphism is associated with colorectal cancer risk especially in Asians: a meta-analysis. Med Oncol 28(4):981–985

Fang Y, Kong B, Yang Q et al (2009) MDM2 309 polymorphism is associated with missed abortion. Hum Reprod 24:1346–1349

Feuk L, Carson AR, Scherer SW (2006) Structural variation in the human genome. Nat Rev Genet 7:85–97. doi:nrg1767 [pii] 10.1038/nrg1767

Firouzabadi RD, Ghasemi N, Rozbahani MA et al (2009) Association of p53 polymorphism with ICSI/IVF failure and recurrent pregnancy loss. Aust NZ J Obstet Gynaecol 49:216–219

Firoz EF, Warycha M, Zakrzewski J et al (2009) Association of MDM2 SNP309, age of onset, and gender in cutaneous melanoma. Clin Cancer Res 15:2573–2580

Frank AK, Leu JI, Zhou Y et al (2011) The codon 72 polymorphism of p53 regulates interaction with NF-{kappa}B and transactivation of genes involved in immunity and inflammation. Mol Cell Biol 31:1201–1213

Frazer KA, Murray SS, Schork NJ et al (2009) Human genetic variation and its contribution to complex traits. Nat Rev Genet 10:241–251. doi:nrg2554 [pii] 10.1038/nrg2554

Freeman DJ, Li AG, Wei G et al (2003) PTEN tumor suppressor regulates p53 protein levels and activity through phosphatase-dependent and -independent mechanisms. Cancer Cell 3:117–130

Gonzalez KD, Noltner KA, Buzin CH et al (2009) Beyond Li Fraumeni syndrome: clinical characteristics of families with p53 germline mutations. J Clin Oncol 27:1250–1256. doi:JCO.2008.16.6959 [pii] 10.1200/JCO.2008.16.6959

Grochola LF, Vazquez A, Bond EE et al (2009) Recent natural selection identifies a genetic variant in a regulatory subunit of protein phosphatase 2a that associates with altered cancer risk and survival. Clin Cancer Res 15(19):6301–6308

Grochola LF, Muller TH, Bond GL et al (2010) MDM2 SNP309 associates with accelerated pancreatic adenocarcinoma formation. Pancreas 39(1):76–80

Grochola LF, Zeron-Medina J, Meriaux S et al (2010) Single-nucleotide polymorphisms in the p53 signaling pathway. Cold Spring Harb Perspect Biol 2:a001032. doi:cshperspect.a001032 [pii] 10.1101/cshperspect.a001032

Hartmann LC, Schaid DJ, Woods JE et al (1999) Efficacy of bilateral prophylactic mastectomy in women with a family history of breast cancer. N Engl J Med 340:77–84. doi:10.1056/NEJM199901143400201

Hisada M, Garber JE, Fung CY et al (1998) Multiple primary cancers in families with Li-Fraumeni syndrome. J Natl Cancer Inst 90:606–611

Hu W (2009) The role of p53 gene family in reproduction. Cold Spring Harb Perspect Biol 1:a001073. doi:cshperspect.a001073 [pii] 10.1101/cshperspect.a001073

Hu W, Feng Z, Atwal GS et al (2008) p53: a new player in reproduction. Cell Cycle 7:848–852

Hu W, Feng Z, Teresky AK et al (2007a) p53 regulates maternal reproduction through LIF. Nature 450:721–724

Hu Z, Jin G, Wang L et al (2007b) MDM2 promoter polymorphism SNP309 contributes to tumor susceptibility: evidence from 21 case-control studies. Cancer Epidemiol Biomarkers Prev 16:2717–2723

Huarte M, Guttman M, Feldser D et al (2010) A large intergenic noncoding RNA induced by p53 mediates global gene repression in the p53 response. Cell 142:409–419. doi:S0092-8674(10)00730-0 [pii] 10.1016/j.cell.2010.06.040

International HapMap Consortium (2003) The International HapMap Project. Nature 426:789–796. doi:10.1038/nature02168 nature02168 [pii]

Johnstone RW, Ruefli AA, Lowe SW (2002) Apoptosis: a link between cancer genetics and chemotherapy. Cell 108:153–164

Jordan VC (2006) Tamoxifen (ICI46,474) as a targeted therapy to treat and prevent breast cancer. Br J Pharmacol 147(Suppl 1):S269–276. doi:0706399 [pii] 10.1038/sj.bjp. 0706399

Joshi AM, Budhathoki S, Ohnaka K et al (2011) TP53 R72P and MDM2 SNP309 polymorphisms and colorectal cancer risk: the Fukuoka colorectal cancer study. Jpn J Clin Oncol 41:232–238

Kang HJ, Feng Z, Sun Y et al (2009) Single-nucleotide polymorphisms in the p53 pathway regulate fertility in humans. Proc Natl Acad Sci USA 106:9761–9766

Kay C, Jeyendran RS, Coulam CB (2006) p53 tumour suppressor gene polymorphism is associated with recurrent implantation failure. Reprod Biomed Online 13:492–496

Kelley JL, Swanson WJ (2008) Positive selection in the human genome: from genome scans to biological significance. Annu Rev Genomics Hum Genet 9:143–160. doi:10.1146/annurev.genom.9.081307.164411

Knappskog S, Bjornslett M, Myklebust LM et al (2011) The MDM2 promoter SNP285C/309G haplotype diminishes Sp1 transcription factor binding and reduces risk for breast and ovarian cancer in Caucasians. Cancer Cell 19:273–282

Kraemer KH, Lee MM, Andrews AD et al (1994) The role of sunlight and DNA repair in melanoma and nonmelanoma skin cancer. The xeroderma pigmentosum paradigm. Arch Dermatol 130:1018–1021

Ku CS, Loy EY, Salim A et al (2010) The discovery of human genetic variations and their use as disease markers: past, present and future. J Hum Genet 55:403–415. doi:jhg201055 [pii] 10.1038/jhg.2010.55

Kulkarni DA, Vazquez A, Haffty BG et al (2009) A polymorphic variant in human MDM4 associates with accelerated age of onset of estrogen receptor negative breast cancer. Carcinogenesis 30:1910–1915

Lalloo F, Varley J, Ellis D et al (2003) Prediction of pathogenic mutations in patients with early-onset breast cancer by family history. Lancet 361:1101–1102. doi:S0140-6736(03)12856-5 [pii] 10.1016/S0140-6736(03)12856-5

Lane D, Levine A (2010) p53 Research: the past thirty years and the next thirty years. Cold Spring Harb Perspect Biol 2:a000893. doi:cshperspect.a000893 [pii] 10.1101/cshperspect.a000893

Li AG, Piluso LG, Cai X et al (2006) Mechanistic insights into maintenance of high p53 acetylation by PTEN. Mol Cell 23:575–587. doi:S1097-2765(06)00456-4 [pii] 10.1016/j.molcel.2006.06.028

Li FP (1995) Identification and management of inherited cancer susceptibility. Environ Health Perspect 103(Suppl 8):297–300

Liaw D, Marsh DJ, Li J et al (1997) Germline mutations of the PTEN gene in Cowden disease, an inherited breast and thyroid cancer syndrome. Nat Genet 16:64–67

Limacher JM, Frebourg T, Natarajan-Ame S et al (2001) Two metachronous tumors in the radiotherapy fields of a patient with Li-Fraumeni syndrome. Int J Cancer 96:238–242. doi:10.1002/ijc.1021 [pii]

Lind H, Zienolddiny S, Ekstrom PO et al (2006) Association of a functional polymorphism in the promoter of the MDM2 gene with risk of nonsmall cell lung cancer. Int J Cancer 119:718–721

Lowe SW, Lin AW (2000) Apoptosis in cancer. Carcinogenesis 21(3):485–495

Lozano G (2010) Mouse models of p53 functions. Cold Spring Harb Perspect Biol 2:a001115. doi:cshperspect.a001115 [pii] 10.1101/cshperspect.a001115

Lu X (2010) Tied up in loops: positive and negative autoregulation of p53. Cold Spring Harb Perspect Biol 2:a000984. doi:cshperspect.a000984 [pii] 10.1101/cshperspect.a000984

MacCallum C, Hill E (2006) Being positive about selection. PLoS Biol 4:e87. doi:06-PLBI-ED-0101 [pii] 10.1371/journal.pbio.0040087

Manfredi JJ (2010) The Mdm2-p53 relationship evolves: Mdm2 swings both ways as an oncogene and a tumor suppressor. Genes Dev 24:1580–1589. doi:24/15/1580 [pii] 10.1101/gad.1941710

Marcel V, Palmero EI, Falagan-Lotsch P et al (2009) TP53PIN3 and MDM2 SNP309 polymorphisms as genetic modifiers in the Li-Fraumeni syndrome: impact on age at first diagnosis. J Med Genet 46:766–772

Marsh DJ, Dahia PL, Zheng Z et al (1997) Germline mutations in PTEN are present in Bannayan-Zonana syndrome. Nat Genet 16:333–334

Mayo LD, Donner DB (2001) A phosphatidylinositol 3-kinase/Akt pathway promotes translocation of Mdm2 from the cytoplasm to the nucleus. Proc Natl Acad Sci USA 98:11598–11603

McKinnon PJ (2004) ATM and ataxia telangiectasia. EMBO Rep 5:772–776. doi:10.1038/sj.embor.7400210 7400210 [pii]

Mehta MS, Vazquez A, Kulkarni DA et al (2011) Polymorphic variants in TSC1 and TSC2 and their association with breast cancer phenotypes. Breast Cancer Res Treat 125:861–868. doi:10.1007/s10549-010-1062-1

Mendrysa SM, McElwee MK, Michalowski J et al (2003) mdm2 Is critical for inhibition of p53 during lymphopoiesis and the response to ionizing irradiation. Mol Cell Biol 23:462–472

Mendrysa SM, O'Leary KA, McElwee MK et al (2006) Tumor suppression and normal aging in mice with constitutively high p53 activity. Genes Dev 20:16–21

Miyoshi H, Nakau M, Ishikawa TO et al (2002) Gastrointestinal hamartomatous polyposis in Lkb1 heterozygous knockout mice. Cancer Res 62:2261–2266

Moll UM, Wolff S, Speidel D et al (2005) Transcription-independent pro-apoptotic functions of p53. Curr Opin Cell Biol 17:631–636. doi:S0955-0674(05)00146-8 [pii] 10.1016/j.ceb.2005.09.007

Neale BM (2010) Introduction to linkage disequilibrium, the HapMap, and imputation. Cold Spring Harb Protoc 2010: pdb top74. doi:2010/3/pdb.top74[pii]10.1101/pdb.top74

Noureddine MA, Menendez D, Campbell MR et al (2009) Probing the functional impact of sequence variation on p53-DNA interactions using a novel microsphere assay for protein-DNA binding with human cell extracts. PLoS Genet 5:e1000462

Ogawara Y, Kishishita S, Obata T et al (2002) Akt enhances Mdm2-mediated ubiquitination and degradation of p53. J Biol Chem 277:21843–21850. doi:10.1074/jbc.M109745200 M109745200 [pii]

Olivier M, Hollstein M, Hainaut P (2010) TP53 mutations in human cancers: origins, consequences, and clinical use. Cold Spring Harb Perspect Biol 2:a001008. doi:10.1101/cshperspect.a001008

Orlova KA, Crino PB (2010) The tuberous sclerosis complex. Ann NY Acad Sci 1184:87–105. doi:NYAS5117 [pii]10.1111/j.1749-6632.2009.05117.x

Park SH, Choi JE, Kim EJ et al (2006) MDM2 309T>G polymorphism and risk of lung cancer in a Korean population. Lung Cancer 54:19–24

Phang BH, Sabapathy K (2007) The codon 72 polymorphism-specific effects of human p53 are absent in mouse cells: implications on generation of mouse models. Oncogene 26:2964–2974

Phillips CL, Gerbing R, Alonzo T et al (2010) MDM2 polymorphism increases susceptibility to childhood acute myeloid leukemia: a report from the Children's Oncology Group. Pediatr Blood Cancer 55:248–253. doi:10.1002/pbc.22519

Post SM, Quintas-Cardama A, Pant V et al (2010) A high-frequency regulatory polymorphism in the p53 pathway accelerates tumor development. Cancer Cell 18:220–230

Reed JC (1999) Dysregulation of apoptosis in cancer. J Clin Oncol 17(9):2941–2953

Reinbold M, Luo JL, Nedelko T et al (2008) Common tumour p53 mutations in immortalized cells from Hupki mice heterozygous at codon 72. Oncogene 27:2788–2794

Ruijs MW, Schmidt MK, Nevanlinna H et al (2007) The single-nucleotide polymorphism 309 in the MDM2 gene contributes to the Li-Fraumeni syndrome and related phenotypes. Eur J Hum Genet 15:110–114

Schmitt CA, Rosenthal CT, Lowe SW (2000a) Genetic analysis of chemoresistance in primary murine lymphomas. Nat Med 6:1029–1035. doi:10.1038/79542

Schmitt CA, Wallace-Brodeur RR, Rosenthal CT et al (2000b) DNA damage responses and chemosensitivity in the E mu-myc mouse lymphoma model. Cold Spring Harb Symp Quant Biol 65:499–510

Schneider K, Garber J (1993) Li-Fraumeni Syndrome. doi:NBK1311 [bookaccession]

Slatkin M (2008) Linkage disequilibrium—understanding the evolutionary past and mapping the medical future. Nat Rev Genet 9:477–485. doi:nrg2361 [pii] 10.1038/nrg2361

Smirnov DA, Morley M, Shin E et al (2009) Genetic analysis of radiation-induced changes in human gene expression. Nature 459:587–591

Suzuki HI, Yamagata K, Sugimoto K et al (2009) Modulation of microRNA processing by p53. Nature 460:529–533. doi:nature08199 [pii] 10.1038/nature08199

Terzian T, Wang Y, Van Pelt CS et al (2007) Haploinsufficiency of Mdm2 and Mdm4 in tumorigenesis and development. Mol Cell Biol 27:5479–5485

Thomas M, Kalita A, Labrecque S et al (1999) Two polymorphic variants of wild-type p53 differ biochemically and biologically. Mol Cell Biol 19:1092–1100

The 1000 Genomes Project Consortium, Nature (2010) 467, 1061–1073

Thull DL, Vogel VG (2004) Recognition and management of hereditary breast cancer syndromes. Oncologist 9:13–24

Toffoli G, Biason P, Russo A et al (2009) Effect of TP53 Arg72Pro and MDM2 SNP309 polymorphisms on the risk of high-grade osteosarcoma development and survival. Clin Cancer Res 15:3550–3556

Tomso DJ, Inga A, Menendez D et al (2005) Functionally distinct polymorphic sequences in the human genome that are targets for p53 transactivation. Proc Natl Acad Sci USA 102:6431–6436

Ueda M, Yamamoto M, Nunobiki O et al (2009) Murine double-minute 2 homolog single nucleotide polymorphism 309 and the risk of gynecologic cancer. Hum Cell 22:49–54

Varley JM (2003) Germline TP53 mutations and Li-Fraumeni syndrome. Hum Mutat 21: 313–320

Varley JM, Evans DG, Birch JM (1997) Li-Fraumeni syndrome—a molecular and clinical review. Br J Cancer 76:1–14

Vazquez A, Bond EE, Levine AJ et al (2008) The genetics of the p53 pathway, apoptosis and cancer therapy. Nat Rev Drug Discov 7:979–987

Vazquez A, Kulkarni D, Grochola LF et al (2011) A genetic variant in a PP2A regulatory subunit encoded by the PPP2R2B gene associates with altered breast cancer risk and recurrence. Int J Cancer 128:2335–2343. doi:10.1002/ijc.25582

Villani A, Tabori U, Schiffman J et al (2011) Biochemical and imaging surveillance in germline TP53 mutation carriers with Li-Fraumeni syndrome: a prospective observational study. Lancet Oncol 12:559–567. doi:S1470-2045(11)70119-X [pii] 10.1016/S1470-2045(11)70119-X

Vousden KH, Prives C (2009) Blinded by the light: the growing complexity of p53. Cell 137:413–431. doi:S0092-8674(09)00459-0 [pii] 10.1016/j.cell.2009.04.037
Wade M, Wang YV, Wahl GM (2010) The p53 orchestra: Mdm2 and Mdmx set the tone. Trends Cell Biol 20:299–309. doi:S0962-8924(10)00033-4 [pii] 10.1016/j.tcb.2010.01.009
Wallace-Brodeur RR, Lowe SW (1999) Clinical implications of p53 mutations. Cell Mol Life Sci 55:64–75
Wan Y, Wu W, Yin Z et al (2011) MDM2 SNP309, gene-gene interaction, and tumor susceptibility: an updated meta-analysis. BMC Cancer 11:208
Wei MC, Zong WX, Cheng EH et al (2001) Proapoptotic BAX and BAK: a requisite gateway to mitochondrial dysfunction and death. Science 292:727–730. doi:10.1126/science.1059108 292/5517/727 [pii]
Whibley C, Pharoah PD, Hollstein M (2009) p53 polymorphisms: cancer implications. Nat Rev Cancer 9:95–107
Zhang L, Yu J, Park BH et al (2000) Role of BAX in the apoptotic response to anticancer agents. Science 290:989–992. doi:8940 [pii]
Zhou BB, Elledge SJ (2000) The DNA damage response: putting checkpoints in perspective. Nature 408:433–439. doi:10.1038/35044005
Zhou BP, Liao Y, Xia W et al (2001) HER-2/neu induces p53 ubiquitination via Akt-mediated MDM2 phosphorylation. Nat Cell Biol 3:973–982. doi:10.1038/ncb1101-973 ncb1101-973 [pii]
Zhou M, Gu L, Findley HW et al (2003) PTEN reverses MDM2-mediated chemotherapy resistance by interacting with p53 in acute lymphoblastic leukemia cells. Cancer Res 63:6357–6362
Zhu F, Dolle ME, Berton TR et al (2010) Mouse models for the p53 R72P polymorphism mimic human phenotypes. Cancer Res 70:5851–5859

Chapter 3
p53: Guardian of the Metabolome

Masha V. Poyurovsky and Carol Prives

1 p53 Regulates Aerobic Respiration

In the context of normal growth and development, energy demands of any cell depend chiefly on its physiological environment and proliferative status, varying greatly depending on the organ types and immediate energy requirements. Conversion of glucose into ATP is the major mechanism of cellular energy generation (Holley and Kiernan 1974). Glycolysis refers to oxygen-independent breakdown of glucose to pyruvate in the cytoplasm. Pyruvate is subsequently transported to the mitochondria where it is oxidized to produce over 30 molecules of ATP. At first glance this seems to be a remarkably efficient process, but it is rather slow and is made even more so by the fact that tricarboxylic acid cycle (TCA cycle) intermediates are continually withdrawn for use in anabolic processes within the cell (DeBerardinis et al. 2008a, b). Thus, in situations of rapid proliferation or elevated energy demand associated with exercise using glycolysis for ATP production may become preferable (Yeung et al. 2008). In the 1920s, Otto Warburg observed that tumor cells break down glucose via anaerobic glycolysis, which only produces two ATPs per molecule of glucose (instead of the 26 ATPs that are generated by oxidative phosphorylation) and generates lactic acid as the carbon end-product, a phenomenon now known as the "Warburg effect" (Warburg et al. 1927). Aerobic glycolysis not only allows for rapid generation of high levels of ATP, by secreting lactate cancer cells acidify their extracellular environment, providing survival advantage to tumors as lactic acid is toxic to nontransformed cells (Jones and Thompson 2009). Molecular mechanisms governing the propensity of cancer cells to take up glucose and convert it almost exclusively to lactate are currently being reexamined from the perspective of regulation of these processes by tumor suppressors and oncoproteins (Gillies and Gatenby 2007).

M.V. Poyurovsky • C. Prives (✉)
Department of Biological Sciences, Columbia University, New York 10027 NY, USA
e-mail: clp3@columbia.edu

P. Hainaut et al. (eds.), *p53 in the Clinics*, DOI 10.1007/978-1-4614-3676-8_3,

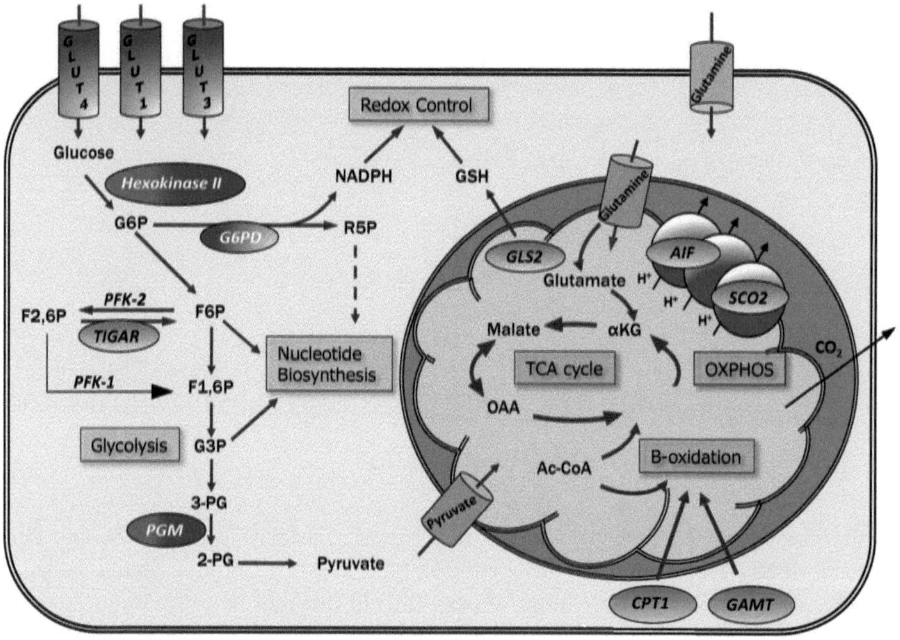

Fig. 3.1 p53 signaling in cellular respiration. p53-regulated gene names are represented by *shaded ovals*. p53 is able to suppress glycolysis in favor of ox-PPP by inducing the transcription of *TIGAR* under conditions of oxidative stress, while also suppressing the PPP by direct binding to G6PD in the cytoplasm during homeostasis. p53 represses transcription of *GLUT1/4* transporters as well as GLUT3 via regulation of NF-κB. p53 regulates HKII and PGM in context-dependent manners, leading to stimulation or repression of glycolysis. p53 enhances mitochondrial respiration via positive regulation of AIF and SCO2 which support the function of the electron transport chain. p53 stimulates the TCA cycle by transcriptional activation of GLS2 and fatty acid oxidation by positively regulating CPT1 and GAMT

Tumor suppressor p53 is implicated in the regulation of multiple steps of ATP generation and glucose metabolism and has been shown to do so both under stress and during cellular homeostasis (Ma et al. 2007). The surprisingly large number of different processes directly affected by p53 underscores its importance in the maintenance of cellular respiration (Bensaad and Vousden 2007).

p53 controls several stages of glucose catabolism (Fig. 3.1). The earliest stage involves the transport of glucose into the cell via a family of glucose transporters (GLUTs). p53 regulates cellular uptake of glucose by direct repression of *GLUT1* and *GLUT4* gene transcription. Transcriptionally impaired tumor-derived mutants of p53 lose their ability to repress these genes. GLUT4 is a tissue-specific and insulin responsive transporter, while GLUT1 performs a more "housekeeping" function; coincidently the inhibitory effect of p53 is relatively more extensive on the *GLUT4* gene (Schwartzenberg-Bar-Yoseph et al. 2004). Additionally, p53 indirectly reduces cellular levels of GLUT3 through its regulation of the IKK–NF-κB pathway (Kawauchi et al. 2008a, b). Cellular uptake of glucose is responsive to insulin signals and the repression of the insulin receptor promoter by p53 further negatively regulates sugar uptake (Webster et al. 1996).

Once glucose is transported into the cell it is either converted into glycogen for storage or into Glucose-6-Phosphate (G6P) for use in glycolysis and the pentose phosphate pathway (PPP) that converts glucose into the precursors used in nucleotide biosynthesis. The increased flux of glucose into cancer cells is attributed to the increase in glucose transport and increase in the activity of type II hexokinase (HKII) (Rempel et al. 1996) that catalyzes conversion of glucose into G6P. The promoter of the *HKII* gene contains a p53 response element. The transcription of the *HKII* gene can be induced by p53 in a hepatoma cell line, although these cells were later discovered to have mutant p53 (Mathupala et al. 1997). The *presence* of p53 response element suggests that the regulation of HKII by p53 could be a part of its pro-survival function and could aid normal cells experiencing mild stress or hypoxia, which activates p53 and increases HKII expression, to recover from metabolic perturbation (Mathupala et al. 2001). Because HKs are often upregulated in cancers, positive regulation by p53 seems contradictory to normal p53 function; however, it is also possible that by increasing the levels of HKII p53 promotes the PPP, which is critical for cellular redox control (Smith 2000; Maddocks and Vousden 2011).

G6P is either converted into fructose-6-phosphate (F6P) or used in the PPP for the generation of nucleotide precursors and NADPH (Tennant et al. 2009). There are multiple starting points of entry into the PPP and subsequent products of glycolysis (including F6P and G3P) can also be directed into the PPP to generate ribose-5-phosphate (R5P) (Vander Heiden et al. 2009). That said, only one of the PPP pathways leads to the synthesis of NADPH as a major cellular antioxidant; this branch of the PPP is thus referred to as ox-PPP while the other pathways constitute the non-ox-PPP. Intriguingly, it appears that p53 can either induce the PPP by activation of the transcription of *TP53*-induced glycolysis and apoptosis regulator (TIGAR) or inhibit the PPP by direct binding to and inactivation of the glucose-6-phosphate dehydrogenase (G6PDH) protein (Gottlieb 2011). Regarding the latter, Jiang et al. reported that under homeostatic conditions, cytoplasmic p53 inhibits the rate-limiting step in the ox-PPP, thus facilitating the generation of trioses via glycolysis and the generation of nucleotides via the nonoxidative PPP (Jiang et al. 2011). They showed that a large fraction of cytosolic p53 is bound to G6PDH (which itself is in great excess to p53) and inhibits the activity of the enzyme by a catalytic mechanism. Interestingly, the activity of Myc oncoprotein, which stimulates glycolysis, also elevates the levels of glucose diverted into the PPP. Myc stimulates expression of the pyruvate kinase 2 (PMK2), which inhibits the conversion of phosphoenolpyruvate (PEP) into pyruvate, thus blocking the rate-limiting step of glycolysis and diverting more glucose into the PPP (Mazurek et al. 2005; David et al. 2010). It would be of interest to investigate the activity of PMK2 in $p53^{-/-}$ MEFs perhaps lack of p53, similarly to overexpression of Myc, could block pyruvate production also contributing to the observed phenotype. Negative regulation of G6PDH by p53 should favor completion of glycolysis in order to generate NADPH and pyruvate for the TCA cycle (discussed in detail later in this chapter) as well as the use of non-ox-PPP for nucleotide biosynthesis. Conversely, elevated levels of p53 following DNA damage do not further inhibit the ox-PPP; in fact p53 upregulates ox-PPP under these conditions via transactivation of *TIGAR* and repression of phosphoglycerate mutase (PGM) (Kondoh et al. 2005; Bensaad et al. 2006).

TIGAR is not directly involved in glycolysis, yet its activation by p53 following DNA damage inhibits glycolysis and favors increased flux of glucose into the ox-PPP. TIGAR expression results in decreased levels of Fructose-2,6-Phosphate (F-2,6-P) which is a very potent allosteric activator of 6-Phosphofructo-1-kinase (PFK-1). Thus TIGAR increases the levels of F6P, inhibiting glycolysis and diverting glucose metabolism toward the ox-PPP. This in turn slows energy generation and cell cycle progression while allowing for the generation of nucleotides needed for DNA repair and increased levels of NADPH for antioxidant defense (Bensaad and Vousden 2007). p53 activation of the PPP and inhibition of glycolysis is believed to be a potent mechanism allowing for cells to recover from minor stresses and damage (Levine and Puzio-Kuter 2010).

p53 is also implicated in regulation of one of the later glycolytic reactions. Specifically, PGM, which converts 3-phosphoglycerate (3PG) to 2-phosphoglycerate (2PG) is downregulated by p53 through a posttranscriptional mechanism in fibroblasts (Kondoh et al. 2005). This activity of 53 is consistent with the repression of glycolysis in favor of nucleotide biosynthesis and lowering ROS levels. Yet the PGM gene has a p53 response element and in cardiomyocytes the transcription of *PGM* is activated in p53-dependent manner (Ruiz-Lozano et al. 1999). Upregulation of PGM increases the rate of glycolysis, and while under conditions of genomic instability this might be undesirable, it is possible that in muscle tissues p53 could positively regulate pathways responsible for rapid ATP generation. Taken together, the multiple p53-regulated steps in glycolysis outlined earlier suggest a model whereby p53 is able to survey and help cells survive and repair minor damage associated with normal growth conditions (Gottlieb and Vousden 2010).

Glycolysis is not the only energy generating process regulated by p53. Indeed, p53 is also able to directly impact mitochondrial metabolic processes via its regulation of glutamine metabolism and oxidative phosphorylation, a hallmark of normally functioning cells growing in aerobic conditions (Matoba et al. 2006) (Fig. 3.1). Focusing first on oxidative phosphorylation itself, there are multiple modes by which p53 regulates this important process. First, the synthesis of Cytochrome c Oxidase 2 (*SCO2*,) gene, a copper chaperon protein required for assembly of the mitochondrial cytochrome c oxidase complex IV, is directly induced by p53. Decreased levels of aerobic respiration in cells lacking p53 are rescued by overexpression of SCO2 (Jaksch et al. 2001; Matoba et al. 2006). Consistently, HCT116 human colon cancer cells that lack p53 compensate for the reduction in the levels of SCO2 and overall decrease in ATP production via oxidative phosphorylation by increasing glycolysis (Ma et al. 2007). Second, p53 positively regulates transcription of the mitochondrial apoptosis-inducing factor (*AIF*) gene that is essential for the function of complex I in the electron transport chain (Vahsen et al. 2004; Stambolsky et al. 2006). Third, p53 controls mitochondrial DNA copy number and mitochondrial biogenesis via its transactivation of the ribonucleotide reductase (*p53R2*) gene (Bourdon et al. 2007; Kulawiec et al. 2009). Finally, consistent with other data demonstrating positive effects of p53 on mitochondrial oxidative phosphorylation, depletion of carbohydrate responsive element-binding protein (ChREBP), which promotes aerobic glycolysis by inducing glycolytic and lipogenic genes in a glucose depended matter, results in a p53-mediated increase of oxygen consumption (Yamashita et al. 2001; Tong et al. 2009).

p53 also regulates the expression of glutaminase (*GLS2*), a key mitochondrial enzyme (Hu et al. 2010; Suzuki et al. 2010). GSL2 converts glutamine to glutamate, which can then be converted to α-ketoglutarate (α-KG) or used in the synthesis of reduced glutathione (GSH) to function in the regulation of cellular ROS levels. The ability of p53 to regulate GSL2 implies that p53 is able to regulate glutamine metabolism. While many tumors rely on aerobic glycolysis for their ATP needs, rapidly growing cells require additional intermediates for protein synthesis. Glutamine is able to serve as an important source of carbon and nitrogen for the proliferating cell (Wise and Thompson 2010). While it was previously assumed that the tumor-associated increase in glycolysis reflects mitochondrial dysfunction, it is becoming more evident that rapidly dividing cells not only maintain mitochondrial respiratory capacity, but in some tumors it is even increased (Moreno-Sanchez et al. 2007). In this regard, glutamine can be used as the key carbon source for the TCA cycle. In fact, some tumors are dependent on glutamine and display a significantly higher rate of glutamine uptake compared to normal cells; moreover, glutamine carbons can be found in the carbon dioxide released by tumor cells (Eagle 1955; Kovacevic 1971).

We have discussed how p53 negatively regulates glycolysis via multiple mechanisms in favor of PPP and production of ribonucleotides. However, p53 also stimulates mitochondrial respiration by maintaining the levels of oxidative phosphorylation through SCO2 and, via GLS2, increases the flow of glutamine into the TCA cycle. GSL2 promotes oxidative phosphorylation by increasing the influx of α-Keto Glutarate (α-KG) into the TCA cycle (Fig. 3.1). Influx of α-KG promotes the use of pyruvate (and thus efficient glucose utilization), because pyruvate is needed to supplement the TCA cycle with Acetyl-coA (Hu et al. 2010). Since glutamine can be used as either carbon or nitrogen sources in proliferating cells, it would seem paradoxically that by inducing GSL2 p53 may be functioning to promote rather than limit proliferation (DeBerardinis et al. 2007). To add further complexity, the Myc oncoprotein also stimulates glutamine catabolism in part via repression of miR-23a and miR-23b and consequent upregulation of GSL1 (Gao et al. 2009). How could Myc and p53 have similar effects on the mitochondrial function?

The apparent inconsistency of both Myc and p53 inducing expression of glutaminases may be at least partly resolved when considering that there are subtle differences in the activities of GSL1 and 2 and that there are different functions of glutamine. Although GSL1 and GLS2 exhibit a high degree of sequence and structure similarity, overexpression of GSL1 is associated with increased proliferation and positively correlates with malignancy (Lora et al. 2004; Perez-Gomez et al. 2005) while, as discussed later, GLS2 levels are reduced in a significant number of hepatic and glioblastomas. GSL1 is strongly inhibited by the end-product glutamate, and so even with elevated levels of expression it is limited in its ability to supply the TCA cycle (Campos-Sandoval et al. 2007). Myc also induces glutamine import via its upregulation of two high affinity glutamine transporters SLC38A5 and SLC1A5, thereby increasing the intracellular glutamine concentration but with only limited increase in the ability to divert it into the TCA cycle (Wise et al. 2008). SLC1A5 is a transporter required for mammalian target of Rapamycin (mTOR) activation. mTOR, central regulator of cell growth, activates protein translation in response to elevated levels of amino acids and growth factor signaling. Glutamine (a nonessential amino acid) after importation into cells via SLC1A5 is rapidly exported via a bidirectional amino acid transporter (Lat1) in order to facilitate the uptake of essential amino acids (EAAs) such as leucine leading to mTOR activation (Nicklin et al. 2009). Thus, while GSL2 activation by p53 favors mitochondrial respiration and increased rates of oxidative phosphorylation and oxygen consumption, Myc induction of glutamine import not only allows for the influx of glutamate into the TCA cycle but it also activates mTOR. Additionally, Myc functions to attenuate conversion of pyruvate to acetyl-CoA as well as to increase the expression of multiple glycolic enzymes and thereby elevate the rate of glycolysis (Kim and Dang 2006; Dang 2010). So while both p53 and Myc induce glutaminases, when examined in the context of the overall activity of each protein it is not surprising that their respective induction of these functionally related enzymes lead to very different outcomes.

p53 can stimulate the influx of carbon sources into energy generating pathways by increasing the levels of β-oxidation (Buzzai et al. 2005). β-oxidation is the process by which fatty acids are broken down by the mitochondria or the peroxisomes to generate acetyl-CoA, feeding the TCA cycle and generating energy during periods of nutrient deprivation (Jensen 2003). Fatty acid β-oxidation is diminished in livers of p53 null mice, when compared to their wild-type littermates, and in tumor cells lacking p53 (Buzzai et al. 2007). These effects are attributed to two proteins whose levels are upregulated in a p53-dependent manner: guanidinoacetate methyltransferase (GAMT) and carnitine palmitoyltransferase (CPT1) (Deberardinis et al. 2006; Ide et al. 2009). CPT1 is localized to the outer membrane of the mitochondria and functions in the import of fatty acids by binding them to carnitine. Akt pathways are reported to reduce the expression of CPT1; p53, by blocking Akt signaling, thereby stimulates fatty acids import (Deberardinis et al. 2006).

GAMT, a p53 inducible gene, converts guanidineacetate to creatine, which can then be directly utilized in maintaining ATP homeostasis or in stimulation of β-oxidation. Increased levels of creatine are associated with phosphorylation of AMPK and acetyl-CoA carboxylase (ACC), an indicator of fatty acid catabolism (Zhu and Prives 2009). Regulation of creatine metabolism by p53 is somewhat complicated by the fact that the creatine kinase gene is also transcriptionally upregulated by p53 (Zhao et al. 1996). GAMT and creatine kinase may act in concert to further elevate the levels of fatty acid oxidation under conditions of nutrient deprivation or during heightened energy demands of the cell, or possibly to increase the apoptotic potential of p53 when such conditions become too extreme (Ide et al. 2010). It is also possible that the two enzymes are functionally antagonistic and so additional studies are needed to fully establish a role for p53-mediated regulation of creatine kinase in metabolism.

2 p53 and the Regulation of Reactive Oxygen Species

In addition to regulating respiratory energy generating cellular processes, p53 is also involved in the regulation of biosynthetic pathways and modulation of oxidative stress, an unavoidable consequence of proliferation. Aerobically growing organisms are unavoidably subjected to the production of reactive oxygen species (ROS), radicals produced by a plethora of reactions taking place during the normal course of growth of any cell. These oxygen derivatives are highly reactive and capable of modifying a large number of biological substrates, as well as modulating their activities. ROS levels are commonly thought to be a major by-product of mitochondrial respiration and fatty acid oxidation. Furthermore, new data suggest that ROS can be generated in response to various signals and to act as second messengers (Circu and Aw 2010). Such redox signals differ in their site of generation, spatial distribution, intensity, and duration, thus allowing for specificity of the message (Forman 2007; Menon and Goswami 2007).

ROS levels are very closely linked to the aging process, cancer and cardiac disease. They fuel other deleterious outcomes for a growing organism due to the damage caused by the oxidation of lipids, proteins, and nucleic acids (Ames et al. 1993; Essick and Sam 2010). DNA damage is one of the most well-characterized activators of p53 response and the ability of p53 to respond to perturbation in genome integrity is central to its tumor suppressive activity (Vousden and Prives 2009). Thus, p53 activation following elevation of ROS levels and the subsequent upregulation of the apoptotic cascade seems to fall in line with the function of p53 as the gatekeeper of the genome. The situation is made far more complex, however, by the fact that p53 is not only regulated by ROS levels, conversely p53 is able to positively and negatively regulate ROS depending on cellular context.

High ROS levels form a feed-forward loop with p53, causing eventual cellular demise (Fig. 3.2). Treatment of cells with H_2O_2 causes massive oxidative damage and upregulates transcription of a number of genes, a large proportion of which are p53 targets (Desaint et al. 2004). A search for p53 inducible genes (*PIG*s) identified a number of redox active proteins and ROS generating enzymes (Polyak et al. 1997). These include quinone oxidoreductase (*PIG3, NQO1*), proline oxidase (*PIG6, POX*), and ferredoxin reductase (*FDXR*). All these target genes were shown to elevate intracellular ROS levels and increase oxidative stress and subsequent apoptosis (Polyak et al. 1997; Liu and Chen 2002; Rivera and Maxwell 2005). While two other p53 target genes, BCL-2 associates X protein (*BAX*) and p53 upregulated modulator of apoptosis (*PUMA*) do not directly elevate ROS levels by their enzymatic activity, these proteins facilitate the uncoupling of mitochondria (Liu et al. 2005; Sablina et al. 2005). Less efficient electron transport causes more ROS to leak out of the mitochondria and also inhibits the function of many mitochondrial enzymes by oxidizing iron-sulfur centers and causing degradation of mitochondrial components and in so doing induce apoptosis (Raha and Robinson 2001). Another p53 target, p66^{Shc} oxidizes cytochrome c, which leads to increased mitochondrial permeability and production of H_2O_2, also triggering apoptosis (Giorgio et al. 2005).

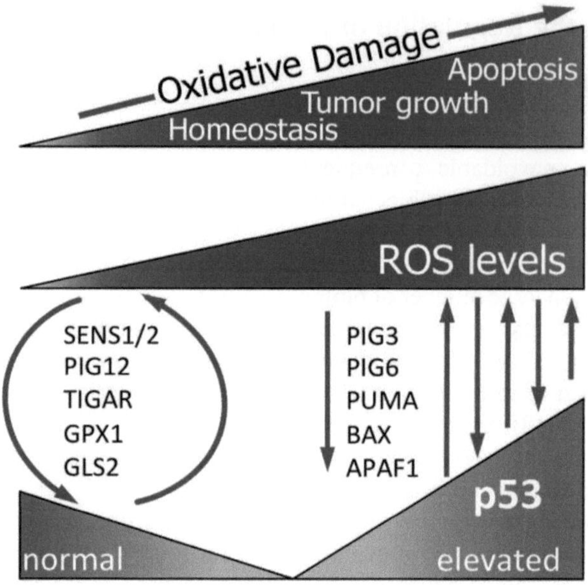

Fig. 3.2 Reciprocal control of p53 activity and intracellular ROS levels. The activity of p53 under conditions of homeostasis and mild oxidative stress allows for transcription of antioxidant genes and defense against oxidative damage. Activities of antioxidant genes in turn lower ROS levels. Loss of p53 activity through deletion or mutation is associated with elevated levels of ROS, increased rate of mutagenesis, and promotion of tumorigenicity. Elevated levels of ROS lead to detrimental oxidative stress and activate p53 transcription of pro-oxidant and pro-apoptotic genes, setting up a feed-forward look with p53 leading to cell death

In the above cases, following elevation in ROS levels, p53 initiates a transcriptional program that results in further increase in ROS, with the consequent oxidative modification of cytochome c and mitochondrial outer membrane permeabilization (Liu et al. 2008). Release of cytochome c from mitochondria and activation of the apoptosome commit the cells to the execution of apoptosis (Circu and Aw 2010).

The pro-oxidant functions of p53 and the associated enhancement of apoptosis seem to be in direct functional opposition to a growing number of reports that describe the antioxidant activity of p53 and a role for p53 in the reduction of ROS. A growing cell will carefully maintain the ROS concentration by the activity of non-enzymatic antioxidants and antioxidant enzymes. Among the different ROS eliminating pathways are three distinct thiol/disulfide pairs: (1) glutathione/glutathione disulfide (GSH/GSSH), (2) thioredoxin (Trx1) (–SH_2/–SS–), and (3) cysteine/cystine, which mediate cellular responses to redox stimuli (Jones 2006a, b). In addition to these circuits, NADPH plays a critical part in modulating ROS levels by providing the reducing power for both GSH and Trx1 systems as they scavenge cellular ROS and facilitate the repair of oxidative damage (Martindale and Holbrook 2002).

At physiological levels p53 is required for optimal expression of multiple antioxidant genes and their expression is compromised in cells lacking p53 (Tomko et al. 2006). p53 regulates the expression of SESN1 and SESN2, two members of the sestrin family of proteins (Velasco-Miguel et al. 1999; Budanov et al. 2002). Sestrins function in the cellular ROS defense by reducing peroxiredoxin, which overoxidizes in the presence of elevated H_2O_2 (Budanov et al. 2004). Suppression of p53 activity reduces the expression of SESN1/2 without affecting the expression of p53 pro-oxidant targets (Sablina et al. 2005). p53-mediated induction of aldehyde dehydrogenase (*ALDH4*), a mitochondrial NAD^+ dependent enzyme involved in proline degradation, also lowers ROS levels (Yoon et al. 2004). Further, p53 regulates glutathione peroxidase-1 (*GPX1*) (Tan et al. 1999) and, as mentioned earlier, *GLS2* genes. GPX1 functions in the reduction of ROS by scavenging hydrogen peroxide or organic hydroperoxides, while, by converting glutamine to glutamate, GLS2 provides critical maintenance of the GSH/GSSH balance in cells.

GSH is involved in many enzymatic and nonenzymatic antioxidative reactions as an electron donor, this oxidation of GSH leads to the formation of the glutathione dimer (GSSG) (Nakamura et al. 1997). In order to insure proper redox control and maximal protection from oxidative stress, a cell needs to efficiently reduce GSSG back to GSH (Circu and Aw 2008). Reduction of GSSG is catalyzed by glutathione reductase and requires NADPH, which is primarily generated during the first step of the ox-PPP (Scott et al. 1991). By inducing expression of TIGAR, p53 facilitates the channeling of glucose into the PPP, this not only allows for the synthesis of nucleotide intermediates required for DNA damage repair, but it also modulates the redox state of the cell (Bensaad et al. 2006). It is of interest that under homeostatic conditions p53 actually represses the PPP (perhaps to control the rate of proliferation) but under conditions of mild oxidative stress p53 promotes the PPP in order to maintain the redox state of the cell and to protect DNA from oxidative damage (Jiang et al. 2011). Both control of proliferation and protection of the DNA are antitumorigenic activities consistent with p53s known function as a tumor suppressor.

Glutamate generated by the activity of GLS2 can be used both to fuel the TCA cycle and for generation of GSH (DeBerardinis et al. 2008a, b). Thus, induction of expression of GLS2 by p53 favors efficient antioxidant activity. Conversely, loss of p53 or depletion of GLS2 by siRNA increases the intercellular ROS levels, augments the GSH/GSSH balance, and further sensitizes the cells to ROS-associated apoptosis (Hu et al. 2010; Suzuki et al. 2010). Further, loss of GLS2 expression was documented in hepatocellular carcinomas, presumably due to the oncogenic effects of elevated ROS levels (Hu et al. 2010; Suzuki et al. 2010). GLS2 expression is also lost in brain tumors and the restoration of GLS2 in glioblastomas inhibits tumor cell migration and proliferation (Szeliga et al. 2009). In line with other reports describing the antioxidant activity of p53, GLS2 is also regulated by p53 in unstressed cells, providing additional evidence for the regulation of intercellular ROS by p53 in normal conditions.

Just as DNA damage signals to p53 through the actions of multiple upstream kinases, mechanisms by which cellular levels of ROS modulate p53 activity have also been proposed. Crosstalk between the cellular levels of ROS and p53 is needed to keep the fluctuations of ROS and p53 within physiological ranges (Liu et al. 2008). It is also worth mentioning that p53 is positively regulated by redox factor-1 (ref-1), which stimulates p53 DNA binding via a redox-dependent mechanism (Abate et al. 1990; Jayaraman et al. 1997; Gaiddon et al. 1999). On the other hand, p53 DNA binding is inhibited after its oxidation by thioredoxin reductase (TRR), thus indicating a complex mechanism between cellular levels of ROS and p53 (Seemann and Hainaut 2005).

3 p53 and the Control of Biosynthesis Rates

It is well accepted that the primary function of p53 is to initiate a transcriptional program aimed at halting cellular proliferation as a response to stress (Kastan et al. 1992; el-Deiry et al. 1993). In addition to restricting cell cycle progression, p53 works by downregulating anabolic processes and protein synthesis. P53 plays a central role in repression of protein translation, induction of β-oxidation of fatty acids and autophagy, key mechanisms of cellular response to metabolic and hypoxic stess. As with regulation of cellular respiration, at first glance these functions of p53 are inconsistent with its role as a tumor suppressor as they may be considered as pro-survival activities. However, we may also consider the function of p53 in maintaining damage free survival. In this capacity, the ability of p53 to downregulate cellular anabolic processes becomes an additional mechanism to make sure that cells can weather environmentally damaging conditions, unperturbed.

Abundant nutrient supplies and growth factor signaling converge on and stimulate the activity of the mammalian target of rapamaycin (mTOR) kinase complex via Akt (Protein Kinase B) (Edinger and Thompson 2002). The importance of these pathways is underscored by the frequent activation of cell growth signaling cascades in tumors, allowing for cellular proliferation in less than optimal conditions (DeBerardinis et al. 2008a, b). Transcriptional targets of p53 regulate a number of steps in the cellular system that controls cellular biosynthetic rates. During periods of limited energy availability, elevated levels of AMP activate the serine-threonine liver kinase B1 (LKB1)-AMPK signaling axis, which in turn reduces metabolic rates, and increases aerobic respiration and catabolic processes (Shackelford and Shaw 2009). Thus, AMPK and AKT function as nutrient sensors and opposing kinases delivering the signals for the execution of the appropriate metabolic program. Interestingly, the activity of p53 is affected by both of these signaling networks while p53 itself exerts additional levels of control on the biosynthetic rates in the cell.

AMPK is a heterotrimeric protein kinase made up of one catalytic subunit and two regulatory subunits. The function of p53 and the AMPK network is closely linked. Stress activates p53-mediated transcription of the gene encoding the β-regulatory subunit of AMPK (*AMPK β1*), which stabilizes and controls the localization of the heterotrimer (Warden et al. 2001; Feng et al. 2007). p53 in turn is positively regulated by AMPK as well as LKB1, which is a master upstream kinase, responsible for the activation of AMPK and 12 other related proteins, all involved in cell growth and metabolism. Activated AMPK very rapidly (within 15 min of removing glucose) phosphorylates p53 at Ser15, leading to increased p53 stability and activity (Feng et al. 2005; Jones et al. 2005). There is also evidence that AMPK positively regulates

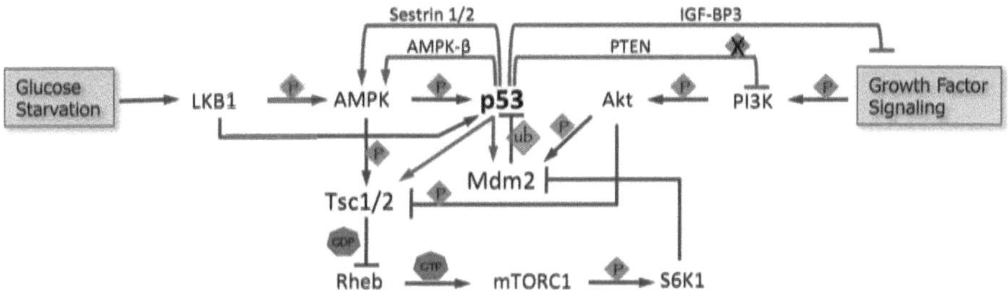

Fig. 3.3 p53 integrates upstream metabolic signaling and responds to cellular energy and nutrient levels. Low levels of AMP caused by glucose deprivation activate LKB1 and AMPK, which in turn phosporylates and activates p53 and downregulates mTORC1. p53 induces transcription of genes encoding AMPK subunit AMPK-β and PTEN phosphatase that downregulate pro-proliferative signaling activated by nutrients and growth factors. In opposition, IGF signaling downregulates the activity of p53 via AKT-dependent phosphorylation, stabilization, and activation of Mdm2, forming a negative feedback loop to negatively regulate p53. *Arrows* represent upregulation.

the transcription of the p53 gene (Okoshi et al. 2008). In fact LKB1 reportedly is found together with p53 at p53 responsive promoters where it is able to stimulate transcriptional activation by phosphorylation of the components of chromatin and transcriptional machinery as well as to induce p53 phosphorylation at Ser15 and Ser392 (Karuman et al. 2001; Zeng and Berger 2006). p53 further upregulates AMPK via transcriptional induction of Sesn1 and Sesn2, which are also involved in the p53 antioxidant response (Budanov and Karin 2008). Sestrins promote AMP- and LKB1-independent phosphorylation and activation of AMPK by binding to the catalytic α-subunit of the trimer. Supporting a functional connection between p53 and LKB1, mice heterozygous for LKB1 in the p53 null background have accelerated tumor development (Wei et al. 2005; Takeda et al. 2006). Additionally, knock-in mice expressing p53 mutated at Ser15 exhibit a defect in glucose metabolism and increased metabolic stress (Armata et al. 2010).

The PI3K/Akt/mTOR pathway is a conserved signaling network used in cells to respond to growth factors (Franke et al. 2003) and p53 is functionally relevant to this cellular pathway as well. p53 is both activated by and reinforces the cellular systems associated with starvation and low energy levels and also negatively regulates pro-survival signaling initiated by growth factors and hormones and transmitted through the Akt phosphorylation cascade (Fig. 3.3). Growth factors through their respective receptor tyrosine kinases activate PI3K, resulting in phosphorylation of phosphatidyl-inositol lipids (PIP_3) at the plasma membrane. In tumor cells, aberrant signaling may constitutively activate PI3K and its downstream effector, Akt, a protein kinase that activates mTOR via inhibition of TSC-1 protein (Bader et al. 2005). The induction of Akt by PI3K is reversed by the action of phosphatase and tensin homolog (PTEN), a lipid phosphatase, which dephosphorylates PIP_3 (Carracedo and Pandolfi 2008). *PTEN* is a p53 target gene, thus providing an additional feedback loop between growth signaling and p53 (Stambolic et al. 2001). Akt also phosphorylates Mdm2 leading to enhanced p53 ubiquitination and degradation

(Zhou et al. 2001; Shaw and Cantley 2006). By attenuating the activity of Akt, p53 further stimulates mitochondrial respiration as Akt activity leads to increased association of HKII with mitochondria as well as higher concentrations of the glucose transporters GLUT1 and GLUT4 at the plasma membrane (Hill et al. 1999; Robey and Hay 2006; Young and Anderson 2008).

p53 can further negatively regulate the PI3K/Akt/mTor pathway through upregulation of the IGF-BP3 gene (Buckbinder et al. 1995). IGF-BP3 is a potent inhibitor of IGF-1 and thereby controls the excretion of IGF-1 into the extravascular compartment. Thus p53-mediated upregulation of IGF-BP3 not only controls growth factor signaling in the same cell, but has extracellular effects as well (Ferry et al. 1999). Further, p53 negatively regulates mTORC1 by activating the tuberous sclerosis 1/2 (TSC1/2) complex, which acts as a GTPase activating protein (GAP) for Rheb (Ras homologue enriched in brain) through two different mechanisms. First, the p53-inducible sestrins bind to TSC2 and induce its interaction with AMKP leading to targeted inactivation of the mTOR pathways during the condition of metabolic stress (Budanov and Karin 2008). Second, *TSC2* is a p53-inducible gene whose expression is increased by conditions that stimulate the activity of p53 (Feng et al. 2005, 2007). During hypoxic conditions, p53 induces transcription of *REDD1*, which inhibits the activity of mTOR through activation of TSC1/2 complex (Ellisen et al. 2002; Brugarolas et al. 2004). However, the effects of p53 on the mTOR activity seem to be tissue specific and p53 does not always function to downregulate protein synthesis in cells, suggesting that under some circumstances metabolic changes may require protein synthesis (Jones et al. 2005; Maddocks and Vousden 2011).

It is interesting to observe that the pathways that regulate p53 and cellular commitment to growth and division are modulated by a number of feedback loops (Fig. 3.3). First, Akt phosphorylates Mdm2 at Ser166/186, activating Mdm2 and decreasing the levels and activity of p53. p53 in turn activates transcription of *PTEN*, which inhibits activation of Akt, thus forming a p53-PTEN-Akt-Mdm2 signaling loop (Zhou et al. 2001). Additionally, p53 phosphorylation by the AMPK cascade on Ser15 can be reversed by the enzymatic activity of Protein Phosphatase 2A (PP2A). mTORC1 phosphorylates and activates the α-4 subunit of PP2A, thus establishing another loop whereby mTOR and AMPK integrate environmental signaling and initiate the appropriate p53 response (Imamura et al. 2001; Kong et al. 2004). It appears that at least following DNA damage signaling, mTOR is able to positively regulate the p53 apoptotic response. One of the mTOR primary downstream effectors, S6K1, is able to interact with and inhibit Mdm2, causing accumulation of p53 following p38 MAPK activation (Lai et al. 2010).

4 Summary

The existence of these multiple regulatory pathways between p53 and proteins regulating cellular metabolism and growth suggests the capacity of these stress pathways and survival pathways to cooperate in order to assure the fidelity of progression

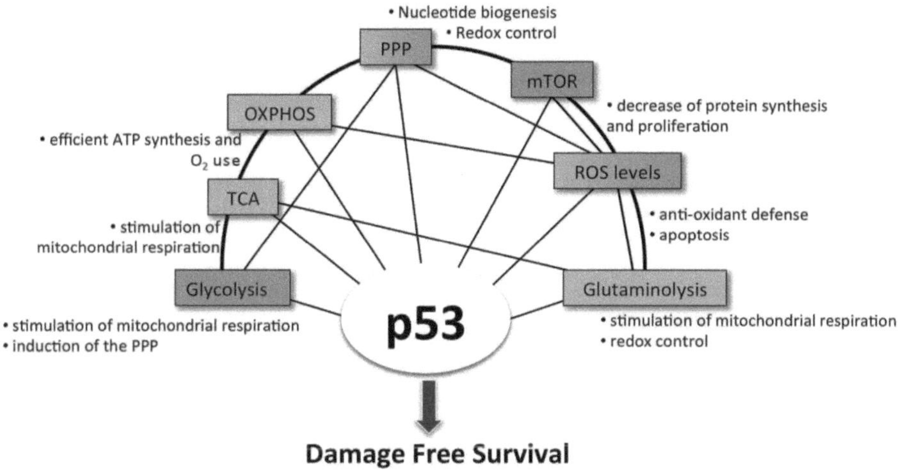

Fig. 3.4 p53 and the metabolic network. p53 is able to influence the central metabolic processes in the cell through direct regulation of one or more proteins involved. Due to the fact that these metabolic pathways are themselves interconnected, regulation of any one of the pathways by p53 also affects the activity of all connected respiratory nodules. Collectively, negative regulation of proliferation and control of respiration and ROS levels establish the critical role of p53 as a protector of genetic integrity

through cell cycle, growth, and DNA replication (Fig. 3.4). It is not surprising then, that p53 would have a pivotal role in the surveillance of these processes as well as protecting the genome integrity during the conditions of genotoxic stress.

References

Abate C, Patel L, Rauscher FJ 3rd et al (1990) Redox regulation of fos and jun DNA-binding activity in vitro. Science 249(4973):1157–1161

Ames BN, Shigenaga MK, Hagen TM (1993) Oxidants, antioxidants, and the degenerative diseases of aging. Proc Natl Acad Sci USA 90(17):7915–7922

Armata HL, Golebiowski D, Jung DY et al (2010) Requirement of the ATM/p53 tumor suppressor pathway for glucose homeostasis. Mol Cell Biol 30(24):5787–5794

Bader AG, Kang S, Zhao L et al (2005) Oncogenic PI3K deregulates transcription and translation. Nat Rev Cancer 5(12):921–929

Bensaad K, Vousden KH (2007) p53: new roles in metabolism. Trends Cell Biol 17(6):286–291

Bensaad K, Tsuruta A, Selak MA et al (2006) TIGAR, a p53-inducible regulator of glycolysis and apoptosis. Cell 126(1):107–120

Bourdon A, Minai L, Serre V et al (2007) Mutation of RRM2B, encoding p53-controlled ribonucleotide reductase (p53R2), causes severe mitochondrial DNA depletion. Nat Genet 39(6):776–780

Brugarolas J, Lei K, Hurley RL et al (2004) Regulation of mTOR function in response to hypoxia by REDD1 and the TSC1/TSC2 tumor suppressor complex. Genes Dev 18(23):2893–2904

Buckbinder L, Talbott R, Velasco-Miguel S et al (1995) Induction of the growth inhibitor IGF-binding protein 3 by p53. Nature 377(6550):646–649

Budanov AV, Karin M (2008) p53 target genes sestrin1 and sestrin2 connect genotoxic stress and mTOR signaling. Cell 134(3):451–460
Budanov AV, Shoshani T, Faerman A et al (2002) Identification of a novel stress-responsive gene Hi95 involved in regulation of cell viability. Oncogene 21(39):6017–6031
Budanov AV, Sablina AA, Feinstein E et al (2004) Regeneration of peroxiredoxins by p53-regulated sestrins, homologs of bacterial AhpD. Science 304(5670):596–600
Buzzai M, Bauer DE, Jones RG et al (2005) The glucose dependence of Akt-transformed cells can be reversed by pharmacologic activation of fatty acid beta-oxidation. Oncogene 24(26): 4165–4173
Buzzai M, Jones RG, Amaravadi RK et al (2007) Systemic treatment with the antidiabetic drug metformin selectively impairs p53-deficient tumor cell growth. Cancer Res 67(14): 6745–6752
Campos-Sandoval JA, Lopez de la Oliva AR, Lobo C et al (2007) Expression of functional human glutaminase in baculovirus system: affinity purification, kinetic and molecular characterization. Int J Biochem Cell Biol 39(4):765–773
Carracedo A, Pandolfi PP (2008) The PTEN-PI3K pathway: of feedbacks and cross-talks. Oncogene 27(41):5527–5541
Circu ML, Aw TY (2008) Glutathione and apoptosis. Free Radic Res 42(8):689–706
Circu ML, Aw TY (2010) Reactive oxygen species, cellular redox systems, and apoptosis. Free Radic Biol Med 48(6):749–762
Dang CV (2010) Rethinking the Warburg effect with Myc micromanaging glutamine metabolism. Cancer Res 70(3):859–862
David CJ, Chen M, Assanah M et al (2010) HnRNP proteins controlled by c-Myc deregulate pyruvate kinase mRNA splicing in cancer. Nature 463(7279):364–368
Deberardinis RJ, Lum JJ, Thompson CB (2006) Phosphatidylinositol 3-kinase-dependent modulation of carnitine palmitoyltransferase 1A expression regulates lipid metabolism during hematopoietic cell growth. J Biol Chem 281(49):37372–37380
DeBerardinis RJ, Mancuso A, Daikhin E et al (2007) Beyond aerobic glycolysis: transformed cells can engage in glutamine metabolism that exceeds the requirement for protein and nucleotide synthesis. Proc Natl Acad Sci USA 104(49):19345–19350
DeBerardinis RJ, Lum JJ, Hatzivassiliou G et al (2008a) The biology of cancer: metabolic reprogramming fuels cell growth and proliferation. Cell Metab 7(1):11–20
Deberardinis RJ, Sayed N, Ditsworth D et al (2008b) Brick by brick: metabolism and tumor cell growth. Curr Opin Genet Dev 18(1):54–61
Desaint S, Luriau S, Aude JC et al (2004) Mammalian antioxidant defenses are not inducible by H2O2. J Biol Chem 279(30):31157–31163
Eagle H (1955) Nutrition needs of mammalian cells in tissue culture. Science 122(3168):501–514
Edinger AL, Thompson CB (2002) Akt maintains cell size and survival by increasing mTOR-dependent nutrient uptake. Mol Biol Cell 13(7):2276–2288
el-Deiry WS, Tokino T, Velculescu VE et al (1993) WAF1, a potential mediator of p53 tumor suppression. Cell 75:817–825
Ellisen LW, Ramsayer KD, Johannessen CM et al (2002) REDD1, a developmentally regulated transcriptional target of p63 and p53, links p63 to regulation of reactive oxygen species. Mol Cell 10(5):995–1005
Essick EE, Sam F (2010) Oxidative stress and autophagy in cardiac disease, neurological disorders, aging and cancer. Oxid Med Cell Longev 3(3):168–177
Feng Z, Zhang H, Levine AJ et al (2005) The coordinate regulation of the p53 and mTOR pathways in cells. Proc Natl Acad Sci USA 102(23):8204–8209
Feng Z, Hu W, de Stanchina E et al (2007) The regulation of AMPK beta1, TSC2, and PTEN expression by p53: stress, cell and tissue specificity, and the role of these gene products in modulating the IGF-1-AKT-mTOR pathways. Cancer Res 67(7):3043–3053
Ferry RJ Jr, Cerri RW, Cohen P (1999) Insulin-like growth factor binding proteins: new proteins, new functions. Horm Res 51(2):53–67

Forman HJ (2007) Use and abuse of exogenous H_2O_2 in studies of signal transduction. Free Radic Biol Med 42(7):926–932
Franke TF, Hornik CP, Segev L et al (2003) PI3K/Akt and apoptosis: size matters. Oncogene 22(56):8983–8998
Gaiddon C, Moorthy NC, Prives C (1999) Ref-1 regulates the transactivation and pro-apoptotic functions of p53 in vivo. EMBO J 18(20):5609–5621
Gao P, Tchernyshyov I, Chang TC et al (2009) c-Myc suppression of miR-23a/b enhances mitochondrial glutaminase expression and glutamine metabolism. Nature 458(7239):762–765
Gillies RJ, Gatenby RA (2007) Adaptive landscapes and emergent phenotypes: why do cancers have high glycolysis? J Bioenerg Biomembr 39(3):251–257
Giorgio M, Migliaccio E, Orsini F et al (2005) Electron transfer between cytochrome c and p66Shc generates reactive oxygen species that trigger mitochondrial apoptosis. Cell 122(2): 221–233
Gottlieb E (2011) p53 guards the metabolic pathway less travelled. Nat Cell Biol 13:292–302
Gottlieb E, Vousden KH (2010) p53 regulation of metabolic pathways. Cold Spring Harb Perspect Biol 2(4):a001040
Hill MM, Clark SF, Tucker DF et al (1999) A role for protein kinase Bbeta/Akt2 in insulin-stimulated GLUT4 translocation in adipocytes. Mol Cell Biol 19(11):7771–7781
Holley RW, Kiernan JA (1974) Control of the initiation of DNA synthesis in 3T3 cells: low-molecular weight nutrients. Proc Natl Acad Sci USA 71(8):2942–2945
Hu W, Zhang C, Wu R et al (2010) Glutaminase 2, a novel p53 target gene regulating energy metabolism and antioxidant function. Proc Natl Acad Sci USA 107(16):7455–7460
Ide T, Brown-Endres L, Chu K et al (2009) GAMT, a p53-inducible modulator of apoptosis, is critical for the adaptive response to nutrient stress. Mol Cell 36(3):379–392
Ide T, Chu K, Aaronson SA et al (2010) GAMT joins the p53 network: branching into metabolism. Cell Cycle 9(9):1706–1710
Imamura K, Ogura T, Kishimoto A et al (2001) Cell cycle regulation via p53 phosphorylation by a 5'-AMP activated protein kinase activator, 5-aminoimidazole- 4-carboxamide-1-beta-D-ribofuranoside, in a human hepatocellular carcinoma cell line. Biochem Biophys Res Commun 287(2):562–567
Jaksch M, Paret C, Stucka R et al (2001) Cytochrome c oxidase deficiency due to mutations in SCO2, encoding a mitochondrial copper-binding protein, is rescued by copper in human myoblasts. Hum Mol Genet 10(26):3025–3035
Jayaraman L, Murthy KG, Zhu C et al (1997) Identification of redox/repair protein Ref-1 as a potent activator of p53. Genes Dev 11(5):558–570
Jensen MD (2003) Fate of fatty acids at rest and during exercise: regulatory mechanisms. Acta Physiol Scand 178(4):385–390
Jiang P, Du W, Wang X et al (2011) p53 regulates biosynthesis through direct inactivation of glucose-6-phosphate dehydrogenase. Nat Cell Biol 13(3):310–316
Jones DP (2006a) Disruption of mitochondrial redox circuitry in oxidative stress. Chem Biol Interact 163(1–2):38–53
Jones DP (2006b) Redefining oxidative stress. Antioxid Redox Signal 8(9–10):1865–1879
Jones RG, Thompson CB (2009) Tumor suppressors and cell metabolism: a recipe for cancer growth. Genes Dev 23(5):537–548
Jones RG, Plas DR, Kubek S et al (2005) AMP-activated protein kinase induces a p53-dependent metabolic checkpoint. Mol Cell 18(3):283–293
Karuman P, Gozani O, Odze RD et al (2001) The Peutz-Jegher gene product LKB1 is a mediator of p53-dependent cell death. Mol Cell 7(6):1307–1319
Kastan MB, Zhan Q, el-Deiry WS et al (1992) A mammalian cell cycle checkpoint pathway utilizing p53 and GADD45 is defective in ataxia-telangiectasia. Cell 71(4):587–597
Kawauchi K, Araki K, Tobiume K et al (2008a) Activated p53 induces NF-kappaB DNA binding but suppresses its transcriptional activation. Biochem Biophys Res Commun 372(1):137–141
Kawauchi K, Araki K, Tobiume K et al (2008b) p53 regulates glucose metabolism through an IKK-NF-kappaB pathway and inhibits cell transformation. Nat Cell Biol 10(5):611–618

Kim JW, Dang CV (2006) Cancer's molecular sweet tooth and the Warburg effect. Cancer Res 66(18):8927–8930

Kondoh H, Lleonart ME, Gil J et al (2005) Glycolytic enzymes can modulate cellular life span. Cancer Res 65(1):177–185

Kong M, Fox CJ, Mu J et al (2004) The PP2A-associated protein alpha4 is an essential inhibitor of apoptosis. Science 306(5696):695–698

Kovacevic Z (1971) The pathway of glutamine and glutamate oxidation in isolated mitochondria from mammalian cells. Biochem J 125(3):757–763

Kulawiec M, Ayyasamy V, Singh KK (2009) p53 regulates mtDNA copy number and mitocheckpoint pathway. J Carcinog 8:8

Lai KP, Leong WF, Chau JF et al (2010) S6K1 is a multifaceted regulator of Mdm2 that connects nutrient status and DNA damage response. EMBO J 29(17):2994–3006

Levine AJ, Puzio-Kuter AM (2010) The control of the metabolic switch in cancers by oncogenes and tumor suppressor genes. Science 330(6009):1340–1344

Liu G, Chen X (2002) The ferredoxin reductase gene is regulated by the p53 family and sensitizes cells to oxidative stress-induced apoptosis. Oncogene 21(47):7195–7204

Liu Z, Lu H, Shi H et al (2005) PUMA overexpression induces reactive oxygen species generation and proteasome-mediated stathmin degradation in colorectal cancer cells. Cancer Res 65(5): 1647–1654

Liu B, Chen Y, St Clair DK (2008) ROS and p53: a versatile partnership. Free Radic Biol Med 44(8):1529–1535

Lora J, Alonso FJ, Segura JA et al (2004) Antisense glutaminase inhibition decreases glutathione antioxidant capacity and increases apoptosis in Ehrlich ascitic tumour cells. Eur J Biochem 271(21):4298–4306

Ma W, Sung HJ, Park JY et al (2007) A pivotal role for p53: balancing aerobic respiration and glycolysis. J Bioenerg Biomembr 39(3):243–246

Maddocks OD, Vousden KH (2011) Metabolic regulation by p53. J Mol Med 89(3):237–245

Martindale JL, Holbrook NJ (2002) Cellular response to oxidative stress: signaling for suicide and survival. J Cell Physiol 192(1):1–15

Mathupala SP, Heese C, Pedersen PL (1997) Glucose catabolism in cancer cells. The type II hexokinase promoter contains functionally active response elements for the tumor suppressor p53. J Biol Chem 272(36):22776–22780

Mathupala SP, Rempel A, Pedersen PL (2001) Glucose catabolism in cancer cells: identification and characterization of a marked activation response of the type II hexokinase gene to hypoxic conditions. J Biol Chem 276(46):43407–43412

Matoba S, Kang JG, Patino WD et al (2006) p53 regulates mitochondrial respiration. Science 312(5780):1650–1653

Mazurek S, Boschek CB, Hugo F et al (2005) Pyruvate kinase type M2 and its role in tumor growth and spreading. Semin Cancer Biol 15(4):300–308

Menon SG, Goswami PC (2007) A redox cycle within the cell cycle: ring in the old with the new. Oncogene 26(8):1101–1109

Moreno-Sanchez R, Rodriguez-Enriquez S, Marin-Hernandez A et al (2007) Energy metabolism in tumor cells. FEBS J 274(6):1393–1418

Nakamura H, Nakamura K, Yodoi J (1997) Redox regulation of cellular activation. Annu Rev Immunol 15:351–369

Nicklin P, Bergman P, Zhang B et al (2009) Bidirectional transport of amino acids regulates mTOR and autophagy. Cell 136(3):521–534

Okoshi R, Ozaki T, Yamamoto H et al (2008) Activation of AMP-activated protein kinase induces p53-dependent apoptotic cell death in response to energetic stress. J Biol Chem 283(7):3979–3987

Olovnikov IA, Kravchenko JE, Chumakov PM (2009) Homeostatic functions of the p53 tumor suppressor: regulation of energy metabolism and antioxidant defense. Semin Cancer Biol 19(1):32–41

Perez-Gomez C, Campos-Sandoval JA, Alonso FJ et al (2005) Co-expression of glutaminase K and L isoenzymes in human tumour cells. Biochem J 386(Pt 3):535–542
Polyak K, Xia Y, Zweier JL et al (1997) A model for p53-induced apoptosis. Nature 389(6648):300–305
Raha S, Robinson BH (2001) Mitochondria, oxygen free radicals, and apoptosis. Am J Med Genet 106(1):62–70
Rempel A, Mathupala SP, Griffin CA et al (1996) Glucose catabolism in cancer cells: amplification of the gene encoding type II hexokinase. Cancer Res 56(11):2468–2471
Rivera A, Maxwell SA (2005) The p53-induced gene-6 (proline oxidase) mediates apoptosis through a calcineurin-dependent pathway. J Biol Chem 280(32):29346–29354
Robey RB, Hay N (2006) Mitochondrial hexokinases, novel mediators of the antiapoptotic effects of growth factors and Akt. Oncogene 25(34):4683–4696
Ruiz-Lozano P, Hixon ML, Wagner MW et al (1999) p53 is a transcriptional activator of the muscle-specific phosphoglycerate mutase gene and contributes in vivo to the control of its cardiac expression. Cell Growth Differ 10(5):295–306
Sablina AA, Budanov AV, Ilyinskaya GV et al (2005) The antioxidant function of the p53 tumor suppressor. Nat Med 11(12):1306–1313
Schwartzenberg-Bar-Yoseph F, Armoni M, Karnieli E (2004) The tumor suppressor p53 down-regulates glucose transporters GLUT1 and GLUT4 gene expression. Cancer Res 64(7):2627–2633
Scott MD, Zuo L, Lubin BH et al (1991) NADPH, not glutathione, status modulates oxidant sensitivity in normal and glucose-6-phosphate dehydrogenase-deficient erythrocytes. Blood 77(9):2059–2064
Seemann S, Hainaut P (2005) Roles of thioredoxin reductase 1 and APE/Ref-1 in the control of basal p53 stability and activity. Oncogene 24(24):3853–3863
Shackelford DB, Shaw RJ (2009) The LKB1-AMPK pathway: metabolism and growth control in tumour suppression. Nat Rev Cancer 9(8):563–575
Shaw RJ, Cantley LC (2006) Ras, PI(3)K and mTOR signalling controls tumour cell growth. Nature 441(7092):424–430
Smith TA (2000) Mammalian hexokinases and their abnormal expression in cancer. Br J Biomed Sci 57(2):170–178
Stambolic V, MacPherson D, Sas D et al (2001) Regulation of PTEN transcription by p53. Mol Cell 8(2):317–325
Stambolsky P, Weisz L, Shats I et al (2006) Regulation of AIF expression by p53. Cell Death Differ 13(12):2140–2149
Suzuki S, Tanaka T, Poyurovsky MV et al (2010) Phosphate-activated glutaminase (GLS2), a p53-inducible regulator of glutamine metabolism and reactive oxygen species. Proc Natl Acad Sci USA 107(16):7461–7466
Szeliga M, Obara-Michlewska M, Matyja E et al (2009) Transfection with liver-type glutaminase cDNA alters gene expression and reduces survival, migration and proliferation of T98G glioma cells. Glia 57(9):1014–1023
Takeda H, Miyoshi H, Kojima Y et al (2006) Accelerated onsets of gastric hamartomas and hepatic adenomas/carcinomas in Lkb1+/–p53–/– compound mutant mice. Oncogene 25(12):1816–1820
Tan M, Li S, Swaroop M et al (1999) Transcriptional activation of the human glutathione peroxidase promoter by p53. J Biol Chem 274(17):12061–12066
Tennant DA, Duran RV, Boulahbel H et al (2009) Metabolic transformation in cancer. Carcinogenesis 30(8):1269–1280
Tomko RJ Jr, Bansal P, Lazo JS (2006) Airing out an antioxidant role for the tumor suppressor p53. Mol Interv 6(1), 23–25, 22
Tong X, Zhao F, Mancuso A et al (2009) The glucose-responsive transcription factor ChREBP contributes to glucose-dependent anabolic synthesis and cell proliferation. Proc Natl Acad Sci USA 106(51):21660–21665

Vahsen N, Cande C, Briere JJ et al (2004) AIF deficiency compromises oxidative phosphorylation. EMBO J 23(23):4679–4689

Vander Heiden MG, Cantley LC, Thompson CB (2009) Understanding the Warburg effect: the metabolic requirements of cell proliferation. Science 324(5930):1029–1033

Velasco-Miguel S, Buckbinder L, Jean P et al (1999) PA26, a novel target of the p53 tumor suppressor and member of the GADD family of DNA damage and growth arrest inducible genes. Oncogene 18(1):127–137

Vousden KH, Prives C (2009) Blinded by the light: the growing complexity of p53. Cell 137(3):413–431

Warburg O, Wind F, Negelein E (1927) The metabolism of tumors in the body. J Gen Physiol 8(6):519–530

Warden SM, Richardson C, O'Donnell J Jr et al (2001) Post-translational modifications of the beta-1 subunit of AMP-activated protein kinase affect enzyme activity and cellular localization. Biochem J 354(Pt 2):275–283

Webster NJ, Resnik JL, Reichart DB et al (1996) Repression of the insulin receptor promoter by the tumor suppressor gene product p53: a possible mechanism for receptor overexpression in breast cancer. Cancer Res 56(12):2781–2788

Wei C, Amos CI, Stephens LC et al (2005) Mutation of Lkb1 and p53 genes exert a cooperative effect on tumorigenesis. Cancer Res 65(24):11297–11303

Wise DR, Thompson CB (2010) Glutamine addiction: a new therapeutic target in cancer. Trends Biochem Sci 35(8):427–433

Wise DR, DeBerardinis RJ, Mancuso A et al (2008) Myc regulates a transcriptional program that stimulates mitochondrial glutaminolysis and leads to glutamine addiction. Proc Natl Acad Sci USA 105(48):18782–18787

Yamashita H, Takenoshita M, Sakurai M et al (2001) A glucose-responsive transcription factor that regulates carbohydrate metabolism in the liver. Proc Natl Acad Sci USA 98(16):9116–9121

Yeung SJ, Pan J, Lee MH (2008) Roles of p53, MYC and HIF-1 in regulating glycolysis—the seventh hallmark of cancer. Cell Mol Life Sci 65(24):3981–3999

Yoon KA, Nakamura Y, Arakawa H (2004) Identification of ALDH4 as a p53-inducible gene and its protective role in cellular stresses. J Hum Genet 49(3):134–140

Young CD, Anderson SM (2008) Sugar and fat—that's where it's at: metabolic changes in tumors. Breast Cancer Res 10(1):202

Zeng PY, Berger SL (2006) LKB1 is recruited to the p21/WAF1 promoter by p53 to mediate transcriptional activation. Cancer Res 66(22):10701–10708

Zhao J, Schmieg FI, Logsdon N et al (1996) p53 binds to a novel recognition sequence in the proximal promoter of the rat muscle creatine kinase gene and activates its transcription. Oncogene 13(2):293–302

Zhou BP, Liao Y, Xia W et al (2001) HER-2/neu induces p53 ubiquitination via Akt-mediated MDM2 phosphorylation. Nat Cell Biol 3(11):973–982

Zhu Y, Prives C (2009) p53 and Metabolism: The GAMT Connection. Mol Cell 36(3):351–352

Chapter 4
The p53 Family and Stem Cell Biology

Massimiliano Agostini, Alessandro Rufini, Edward T.W. Bampton, Francesca Bernassola, Gerry Melino, and Richard A. Knight

1 Introduction

The p53-family is composed of *TP63*, *TP73*, and *TP53* itself (Dötsch et al. 2010). These genes are characterized by their similar protein structure. The proteins contain three main domains, which are highly conserved between the family members: an amino-terminal transactivation domain (TA), a DNA-binding domain, and a carboxy-terminal oligomerization domain. Moreover, at the p63 and p73 carboxy-terminal are two additional domains: the SAM domain, involved in protein–protein interaction, and a transcription inhibitory domain.

These three proteins are transcription factors that, by binding to the DNA, modulate the expression of multiple genes that play pivotal functions in the regulation of proliferation, differentiation, cell death, stem cell renewal, and cell fate commitment (Melino et al. 2002; Vousden and Lane 2007; Yang et al. 2002).

All those cells within the human body characterized by the ability to renew themselves through mitotic cell division (self-renewal) and the ability to differentiate into a diverse range of specialized cell types are called stem cells. Generally, stem cells are classified into two broad types of mammalian cells: embryonic stem cells (ESCs), which are isolated from the inner cell mass of the

M. Agostini (✉) • A. Rufini (✉) • E.T.W. Bampton • R.A. Knight
Medical Research Council, Toxicology Unit, Leicester University, Leicester LE1 9HN, UK
e-mail: ma285@le.ac.uk; ar230@le.ac.uk

F. Bernassola
Biochemistry IDI-IRCCS Laboratory, Department of Experimental Medicine,
University of Rome Tor Vergata, 00133 Rome, Italy

G. Melino
Medical Research Council, Toxicology Unit, Leicester University, Leicester LE1 9HN, UK

Biochemistry IDI-IRCCS Laboratory, Department of Experimental Medicine,
University of Rome Tor Vergata, 00133 Rome, Italy

P. Hainaut et al. (eds.), *p53 in the Clinics*, DOI 10.1007/978-1-4614-3676-8_4,

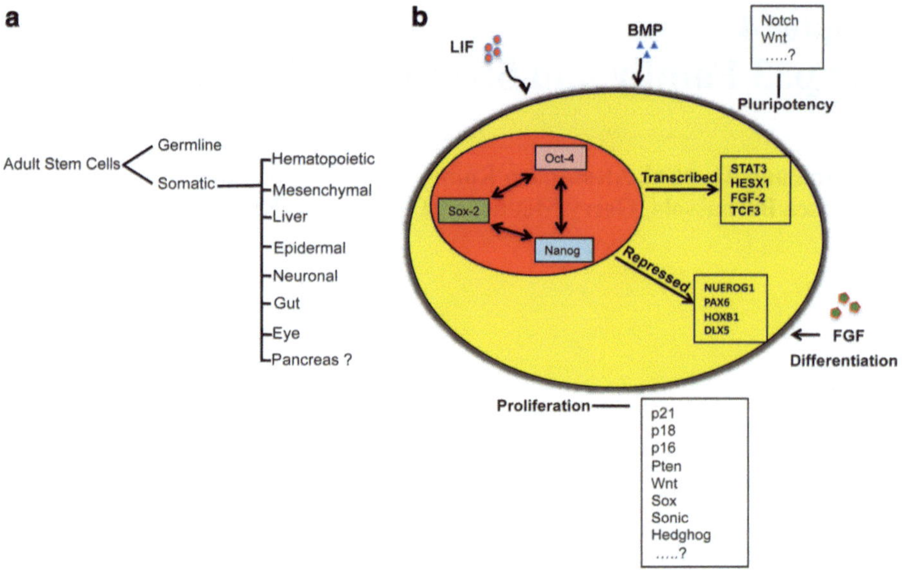

Fig. 4.1 (**a**) Adult stem cells classification. (**b**) Extrinsic and intrinsic factors involved in ESCs biology. Three transcription factors, Nanog, Sox-2, and Oct-4, are the master regulators of self-renewal and pluripotency in ESCs. They promote the expression of genes that preserves self-renewal and pluripotency while inhibiting genes that regulate differentiation. The Sox-2–Oct-4–Nanog network is fine-tuned regulated by leukemia inhibitor factor (LIF) and bone morphogenetic factor (BMP); both contribute to self-renewal inhibiting differentiation. Autocrine releases of fibroblast growth factor induce differentiation through activation of MAPK signaling

blastocyst, and adult stem cells. Adult stem cells are present in many tissues and they can be classified into different subtypes (Fig. 4.1a). The main difference between these two groups of stem cells is that while ESCs have unlimited self-renewal potential and the capability to generate all the cell types in the body, adult stem cells have limited self-renewal and pluripotency. In adult organisms, stem cells play a role in tissue repair and homeostasis. Because of their capacity to regenerate tissues, stem cells are under consideration in clinics for cell-based therapy (Daley 2010).

Adult stem cells and ESCs share many genes and signaling pathways for the control of self-renewal and cell fate decisions (Ivanova et al. 2002). This molecular signature is called “stemness.” It is well accepted that stem cells’ behavior (self-renewal + pluripotency) is regulated by the combination of intrinsic elements (i.e., transcription factors) and extrinsic signals [i.e., leukemia inhibitor factor (LIF), bone morphogenetic (BMP) factor, and FGF] from their microenviroment (Molofsky et al. 2004; Shenghui et al. 2009; Haojian and Wang 2008). This indicates that self-renewal and pluripotency in stem cells is a complicated process that

Table 4.1 Pathways involved in Stem Cell functions

Pathway	Role	References
Intrinsic factors		
Sox-2	Required for maintenance of pluripotency in ESCs *in vivo* and *in vitro*	Avilion et al. (2003) Graham et al. (2003)
Oct-4	Essential for maintenance of pluripotency in ESCs in the inner cell mass and *in vitro*	Nichols et al. (1998) Niwa et al. (2000)
Nanog	Required for maintenance of pluripotency in ESCs in the inner cell mass	Mitsui et al. (2003) Chambers et al. (2003)
Notch	Critical for maintenance of self-renewal through inhibition of differentiation	Hansen et al. (2004) Hitoshi et al. (2002)
Polycomb	Modulation of several developmental genes through histone modification	Bernstein et al. (2006)
Wnt	Promotes the self-renewal in a contextdipendent manner	Batlle et al. (2002) Reya et al. (2003)
Sonic Hedgehog	Maintenance of stem cells promoting proliferation and survival	Machold et al. (2003)
Extrinsic factors		
LIF (Leukemia Inhibitor factor)	Blocks the differentiation of mouse ESCs through activation of Jak/STAT3 signaling	Niwa et al. (1998) Dahéron et al. (2004)
BMP (Bone morphogenetic protein)	Involved trough SMAD protein in the maintenance of ESCs pluripotency	Ying et al. (2003a)
FGF (fibroblast growth factor)	Promotes differentiation of ESCs trough activation MAPK signaling	Ying et al. (2003b)

requires the coordination of multiple pathways involved in proliferation and the maintenance of an undifferentiated state. In a simplistic way, we can consider that three transcription factors form the core of a regulatory circuit, which maintains the stem cells' self-renewal and pluripotency. These transcription factors are the divergent homeodomain protein Nanog, the POU domain transcription factor Oct4, and the SRY-related HMG-box transcription factor Sox-2. A more detailed list of pathways involved in stem cell function is depicted in Fig. 4.1b and listed in Table 4.1. The Oct4, Sox2, and Nanog transcription factors control the expression of genes, including further transcription factors (*STAT3*, *HESX1*, *FGF-2*, and *TCF*) and signaling components necessary to maintain the stem cell state. Moreover, they also repress the expression of genes that, if expressed, would promote a differentiated cell state (*NUEROG1*, *PAX6*, *HOXB1*, *DLX5*) (Boyer et al. 2005; Loh et al. 2006). This triad also forms an autoregulatory circuit. In fact, by binding to their own promoters, as well as to the promoters of the genes encoding the two other factors, Oct4, Sox2, and Nanog collaborate to maintain their own expression.

2 p53 and Stem Cells

The first indication of a direct p53 involvement in the regulation of stem cell biology come from the observation that p53 is able to regulate *NANOG* expression (Lin et al. 2005). In particular, Lin et al. show that after induction of DNA damage in mouse embryonic stem cells (mESCs), p53 is phosphorylated at Ser315 and binds to the promoter of *NANOG*, suppressing its expression. The phosphorylation step also assists in the recruitment of the co-repressor mSin3a to the *NANOG* promoter and promotes the mSin3a–p53 interaction. In ESCs p53-dependent cell-cycle arrest and apoptosis are not efficient after DNA double strand breaks (Aladjem et al. 1998). Consequently, the result of p53 activation in this system is to bring mESCs into a more differentiated state where the cells can go efficiently to p53-dependent cell cycle arrest or apoptosis, promoting the preservation of genomic stability (Fig. 4.2a). This role of p53 is supported by the fact that the loss of p53 in human ESCs leads to an increase of genomic instability (Song et al. 2010).

It has also been demonstrated that p53 participates in the control of neural stem cell (NSC) behavior. Frisen's group (Meletis et al. 2006) found that p53 was expressed in the neurogenic niche of the lateral ventricle wall in the adult mouse brain. The expression is evident in NSCs expressing GFAP and musashi-1 markers. The expression is lost in the majority of differentiating cells, such as Doublecortin positive cells. *TP53* null mice display an elevated proliferation rate in the neurogenic niche of the adult lateral ventricle wall, as well as an increase in the self-renewal of in vitro propagated p53–/– NSCs. Indeed, neurospheres derived from

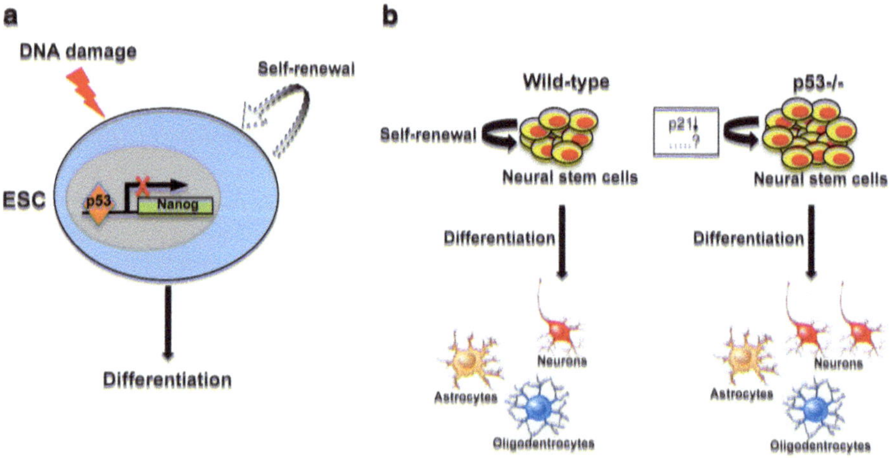

Fig. 4.2 (**a**) p53 regulates Nanog expression in ESCs. DNA damage in ESCs activates p53 expression. p53 binds to the promoter of Nanog and repress Nanog expression, leading to the differentiation of ESCs. (**b**) p53 negatively regulates self-renewal and differentiation of neuronal stem cells (NSCs). The lack of p53 leads to an increase of self-renewal that is mediated by an increase in proliferation and a decrease in cell death. NSCs differentiation is also affected by the loss of p53. Note the higher number of neurons derived from p53–/– NSCs

TP53–/– mice grow faster and are larger in size compared to wild-type neurospheres. Transcriptional analysis of *TP53*–/– neurospheres revealed that several genes are dysregulated, with *P*21 being most affected by the loss of p53. The role of p53 in NSC differentiation is still an open question due to the controversial data. Thus, while Meletis et al. found that the lack of p53 does not affect the potential of NSC to generate neurons, astrocytes, and oligodendrocytes, another group (Armesilla-Diaz et al. 2009) showed an impaired differentiation of *TP53*–/– NSCs. In particular, using NSCs derived from olfactory bulb, they show that the p53 deficiency promotes the differentiation of NSCs into neurons with a concomitant reduction of astrocytes (Fig. 4.2b).

Recently, an important step forward in stem cell biology has been achieved. Yamanaka and colleagues were able to reprogram somatic cells to an ES-like condition (Takahashi and Yamanaka 2006). In particular, the overexpression in mouse embryonic fibroblasts of four transcription factors (Oct-4, Sox-2, Klf4 and c-Myc), called Yamanaka factors, is sufficient to reprogram somatic cells into induced pluripotent stem cells (iPSCs). While Oct-4 and Sox-2 are required for reprogramming, a combination of other factors, such as Oct-4, Sox-2, Nanog, and Lin28, is also able to reprogram somatic cells into iPSCs. One limitation of this system is that the frequency of reprogramming is very low, suggesting that inside the cell are mechanisms that counteract the reprogramming process. Several observations indicate that p53 is responsible for the low efficiency in the reprogramming of somatic cells. Indeed, the overexpression of the oncogene c-myc induces the ARF/p53 pathway driving the cells to apoptosis or senescence (Pelengaris et al. 2002). In support of this, the efficiency of reprogramming is higher in a p53 null context (Zhao et al. 2008; Kawamura et al. 2009; Marión et al. 2009; Utikal et al. 2009; Hong et al. 2009). Because p53 inactivation seems to be critical for the reprogramming efficiency, this raises the concern about the tumorigenicity of iPSCs since chromosomal aberrations have been found in iPSCs (Chin et al. 2007). Moreover, injection of iPSCs into blastocysts led to an incidence (about 20 %) of tumors in the resultant chimeric mice, attributable to reactivation of the c-*MYC* transgene (Okita et al. 2007). Therefore, the optimism that iPSCs could be used in therapy as a source of autologous cells for transplantation into human patients should be reconsidered on the basis of their tumorigenic potential.

3 p63 and Stem Cells

Among the p53-family members, p63 has the strongest link with stem cells. This possibility has been evident since the production of two independent knockout mouse models (Yang et al. 1999; Mills et al. 1999), which revealed a crucial role for p63 in the development of epithelia and epithelial appendages. Indeed, *TP63* null mice are born with severe developmental defects, including absence of skin, skin appendages, and limbs and die shortly after birth, due to dehydration. Epidermis is a stratified epithelium composed of a basal proliferating layer and a suprabasal layer of transiently amplifying cells committed to differentiate to an external cornified envelope. The stem cell compartment is localized within the basal layer and in the hair follicle at the level of the hair bulge. Intriguingly, these regions robustly express the p63 N-terminal truncated isoform ΔNp63. It is noteworthy that a detailed analysis revealed the presence of "patches" of fully differentiated cells on the back of p63 null embryos (Mills et al. 1999; Candi et al. 2006). This compelling evidence proved that p63-depleted epithelial cells are still fully committed to stratification, but lack replenishing capability; i.e., they are defective in stemness. This possibility was also proved by in vitro cultures. Primary epithelial cells isolated from skin are able to clonally expand and produce three different kinds of clones: holoclones, which are the least represented, have a slow proliferative rate, but the highest reproductive capacity, as few colonies formed by the cells of a holoclone arrest their growth to terminally differentiate. In contrast, paraclones have a short replicative life span and rapidly undergo terminal differentiation. Finally, the meroclones are essentially intermediate stages between holoclones and paraclones, containing a mixture of cells of different growth and differentiation potential (Barrandon and Green 1985). In other words, holoclones originate from stem cells, while transient amplifying cells form paraclones and meroclones are mixtures of both cell types. Importantly, ΔNp63 is robustly expressed in stem cells originating holoclones, absent in paraclones and barely detectable in meroclones (Pellegrini et al. 2001).

Downregulation of the ΔNp63 protein level is a prerequisite to differentiation: suprabasal epithelial cells of the skin lose p63 expression and, similarly, induction of differentiation in in vitro cultured keratinocytes triggers rapid decline of p63. The mechanism underlying this event is still poorly defined, but lately microRNAs (miRs) have been brought to light as decisive players in regulating p63 levels in the skin. MicroRNAs are small noncoding RNAs that suppress gene expression by targeting the 3′-UTR of mRNA, which results in translational repression and/or mRNA degradation. MicroRNAs have a spectrum of functions spanning from development, differentiation, and organogenesis to stem cell proliferation, growth control, and apoptosis (Ambros 2004; Bartel 2004). MicroRNA-203 (miR-203) is expressed in the suprabasal layer of the skin or in in vitro cultured keratinocytes upon induction of differentiation and it negatively regulates proliferation of the epidermal stem cell compartment (Lena et al. 2008; Yi et al. 2008). Transgenic animals overexpressing miR-203 die shortly after birth, but embryos show thinning of the epidermis and reduced proliferation in the basal layers. In addition, primary keratinocytes isolated from these animals elicit mostly paraclones and relatively few holoclones when compared to wild type controls. Finally, overexpression of miR-203 restricts the clonogenic capacity of keratinocytes. Importantly, p63 is a target of miR-203 and is repressed in miR-203 overexpressing cells, consistent with a role for p63 in maintenance of stem cells in the epidermis (Fig. 4.3a) (Lena et al. 2008; Yi et al. 2008).

Thymus is another important organ where ΔNp63 plays a pivotal role in stem cell biology. *TP63*-null embryos display unusual hypomorphic thymuses, but substantially normal T-cell maturation (Senoo et al. 2007; Candi et al. 2007). Importantly, p63 is expressed in the thymic epithelial compartment. Thymic epithelial cells from p63 null embryos do not show any maturation defect, but have impaired colony

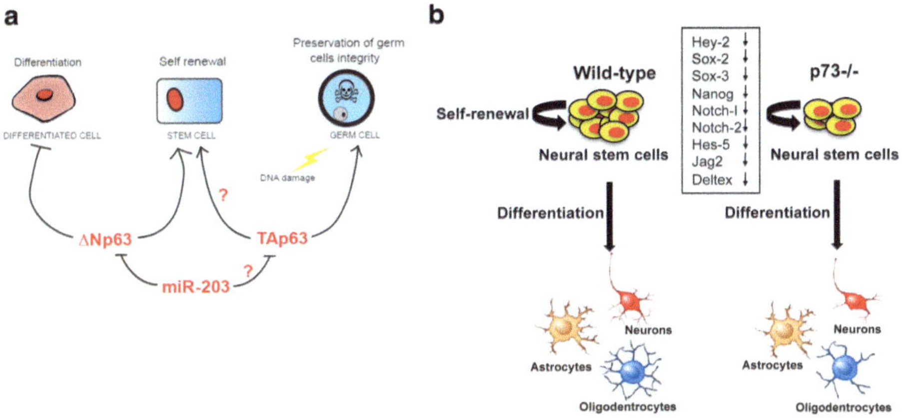

Fig. 4.3 (**a**) Expression of ΔNp63 acts to prevent differentiation and supports maintenance of "stemness." If TAp63 influences stem cells maintenance needs to be addressed. On the other side, DNA-damage induced apoptosis in germ cells is mediated by TAp63 and is essential for the preservation of genomic integrity in these cells. Finally, recent findings are unveiling upstream regulators of p63: miR-203 negatively affects ΔNp63 expression in the skin, resulting in antiproliferative and prodifferentiation outcome. (**b**) p73 positively regulates self-renewal of neuronal stem cells (NSCs). Neurospheres derived from p73–/– mice are smaller. This phenotype is associated with a reduced number of cells in S-phase and at molecular level by a transcriptional dysregulation of genes involved in the regulation of self-renewal pathway. Note that neurons and oligodentrocytes derived from p73–/– NSCs are not fully differentiated

formation ability in vitro, suggesting defective proliferation and renewal. In addition, higher levels of apoptosis and clearance of epithelial cells were observed in vivo, substantiating the idea that p63 is important for stem cells' self-renewal.

Although ΔNp63 is thought to play the most relevant role, two different selective TAp63 knockout mice have been reported which showed alterations in stem cells, albeit in the context of quite different and partially contradictory phenotypes (Suh et al. 2006; Su et al. 2009). Mice generated in Frank McKeon's lab are normal and do not show any of the abnormal developmental features triggered by loss of ΔNp63. Nonetheless, this mouse model unveiled a major role for TAp63 in depletion of germ cells bearing damaged DNA. Indeed, TAp63, but not ΔNp63, is expressed in oocytes arrested in prophase of meiosis-I and its expression persists in the primordial and primary follicle, to be lost when the mature follicle undergoes ovulation. When mice are challenged with DNA damaging agents germ cells die, thereby preventing propagation of irreversibly damaged cells. This event is dependent on TAp63, but independent of p53. In fact, TAp63 is phosphorylated immediately before induction of cell death and DNA damage-dependent death of germ cells is abrogated in *TAP63* null mice, but substantially normal in p53 knockout animals (Su et al. 2009).

On the other hand, *TAP63* null mice developed by Elsa Flores' lab show a severe phenotype affecting stem cell maintenance. These mice developed severe skin ulceration and loss of hair follicles and have a reduced median life span (333 days vs. 712 days in control mice), associated with signs of premature aging. In addition, a severe impairment of stem cells was reported. Indeed, TAp63 depleted animals show increased senescence and reduced proliferation of dermal and epidermal precursor cells, which results in in vivo depletion of the stem cell reservoir. The underlying molecular mechanism is still to be fully clarified, but the absence of TAp63 triggers a strong genetic instability, which could explain the associated phenotype (Su et al. 2009).

4 p73 and Stem Cells

The human *TP73* gene maps within chromosome 1p36, a region frequently deleted in tumors. Moreover, monosomy 1p36 is also associated with developmental brain abnormalities (Kaghad et al. 1997; Bagchi and Mills 2008; Campeau et al. 2008). When the first knockout mouse for *TP73* was generated, it was immediately evident that p73 could play a role in neural development (Yang et al. 2000). In fact, total p73 knockout mice show congenital hydrocephalus and hippocampal dysgenesis. In particular, the lower blade of the dentate gyrus is truncated and the CA1–CA3 pyramidal cell layer structure is impaired. These data suggested that p73 could have a role in hippocampal neurogenesis. Indeed, several experimental findings (Talos et al. 2010; Fujitani et al. 2010; Agostini et al. 2010; González-Cano et al. 2010) demonstrate that p73 can also play a role in stem cells. In particular, using the well-established in vitro model of embryonic NSCs, they show p73 to be a positive regulator of embryonic and adult NSCs. Indeed, neurospheres derived from *TP73* null mice grow slower and are smaller. This phenotype is associated with a reduced number of cells in S-phase and an increase in the senescent population (Talos et al. 2010). Apparently, apoptosis does not play a part in the size reduction of neurospheres since Talos et al. found no differences in apoptosis levels between p73–/– and control NSCs. The same phenotype has been observed in TAp73–/– NSCs, indicating that TA isoform is responsible for the NSCs' maintenance (Fujitani et al. 2010). To support this, TAp73 is the main isoform expressed in embryonic NSCs, and endogenous expression of TAp73 increases during differentiation of NSCs (Agostini et al. 2010).

The potential downstream candidates responsible for this phenotype are genes involved in the regulation of proliferation and/or self-renewal pathways (Molofsky et al. 2004). In fact, the loss of p73 leads to a transcriptional dysregulation of *SOX-2*, *SOX-3*, *NANOG*, *NOTCH-1*, *NOTCH-2*, *HES-5*, *JAG2*, *HEY-2*, and *DELTEX*. Although additional studies are required to address how p73 physiologically regulates these factors, so far, Hey-2 is the only factor shown to be a transcriptional target of TAp73 (Fujitani et al. 2010).

In addition, there is evidence that p73 also regulates the differentiation of NSCs, since it has been shown that neurons derived from p73–/– NSCs do not fully differentiate and they exhibit defects in arborization of the dendritic tree and in physical connectivity.

p73 has also been implicated in oligodendrocyte development (Billon et al. 2004) and oligodendrocytes derived from p73–/– NSCs are lower in number and display a poorer quality than those derived from wild-type NSCs. However, the loss of p73 does not affect the multipotency of NSCs because dissociated p73–/– NSCs preserve the ability to differentiate into all the neural lineages (neurons, astrocytes, and oligodendrocytes).

In conclusion, all this evidence indicates that p73 is a positive regulator of embryonic and adult NSCs (Fig. 4.3b).

These in vitro data were also confirmed in vivo since *TP73* null mice show a reduction in neurogenesis in the subgranular zone of the dentate gyrus and in the subventricolar zone.

5 Conclusions and Future Perspective

In the last decade the field of stem cells has attracted the attention of both scientists and clinicians. One particular focus has been the possibility of stem cell-based therapies for diseases that are refractory to conventional treatments, such as neurodegenerative diseases and immunodeficiency. In theory, two main therapeutic approaches are possible: stem cells as a target of small molecules that modulate their biology (proliferation, differentiation, and self-renewal) or as a source of cells for transplantation therapy. However, both approaches require more basic research in order to understand the molecular mechanisms that regulate proliferation, survival, and differentiation of stem cells after transplantation and in the pathological environment. However, in terms of the first approach, the p53 family is clearly a potential target for modulation of stem cell biology. As to the second approach, cell transplantation, one key aspect that we would like to point out is the possibility of using iPSCs in cell replacement therapy. If the inactivation of p53 is a prerequisite to reprogram the iPSCs successfully, as a range of experimental evidence shows, then their potential tumorigenicity will need much deeper exploration. Thus in the chapter reprising the role of the p53 family in the regulation of stem cell biology, we have highlighted not only some of the complexities of the system, but also where there may be therapeutic potential for modulation of intrinsic stem cells and where particular caution may be needed in terms of cell transplantation therapies.

Acknowledgments This work was supported by the Medical Research Council, and by grants from AIRC, Telethon, and Min Sanita' to G.M.

References

Agostini M, Tucci P, Chen H, Knight RA, Bano D, McKeon F, Nicotera P, Melino G (2010) p73 regulates maintenance of neural stem cell. Biochem Biophys Res Commun 403:13–17

Aladjem MI, Spike BT, Rodewald LW, Hope TJ, Klemm M, Jaenisch R, Wahl GM (1998) ES cells do not activate p53-dependent stress responses and undergo p53-independent apoptosis in response to DNA damage. Curr Biol 8:145–155

Ambros V (2004) The functions of animal microRNAs. Nature 431:350–355

Armesilla-Diaz A, Bragado P, Del Valle I, Cuevas E, Lazaro I, Martin C, Cigudosa JC, Silva A (2009) p53 regulates the self-renewal and differentiation of neural precursors. Neuroscience 158:1378–1389

Avilion AA, Nicolis SK, Pevny LH, Perez L, Vivian N, Lovell-Badge R (2003) Multipotent cell lineages in early mouse development depend on SOX2 function. Genes Dev 17(1):126–140

Bagchi A, Mills AA (2008) The quest for the 1p36 tumor suppressor. Cancer Res 68:2551–2556

Barrandon Y, Green H (1985) Cell size as a determinant of the clone-forming ability of human keratinocytes. Proc Natl Acad Sci USA 82:5390–5394

Bartel DP (2004) MicroRNAs: genomics, biogenesis, mechanism, and function. Cell 16:281–297

Batlle E, Henderson JT, Beghtel H, van den Born MM, Sancho E, Huls G, Meeldijk J, Robertson J, van de Wetering M, Pawson T, Clevers H (2002) Beta-catenin and TCF mediate cell positioning in the intestinal epithelium by controlling the expression of EphB/ephrinB. Cell 111(2): 251–263

Bernstein BE, Mikkelsen TS, Xie X, Kamal M, Huebert DJ, Cuff J, Fry B, Meissner A, Wernig M, Plath K, Jaenisch R, Wagschal A, Feil R, Schreiber SL, Lander ES (2006) A bivalent chromatin structure marks key developmental genes in embryonic stem cells. Cell 125(2):315–326

Billon N, Terrinoni A, Jolicoeur C, McCarthy A, Richardson WD, Melino G, Raff M (2004) Roles for p53 and p73 during oligodendrocyte development. Development 131:1211–1220

Boyer LA, Lee TI, Cole MF, Johnstone SE, Levine SS, Zucker JP, Guenther MG, Kumar RM, Murray HL, Jenner RG, Gifford DK, Melton DA, Jaenisch R, Young RA (2005) Core transcriptional regulatory circuitry in human embryonic stem cells. Cell 122:947–956

Campeau PM, Ah Mew N, Cartier L, Mackay KL, Shaffer LG, Der Kaloustian VM, Thomas MA (2008) Prenatal diagnosis of monosomy 1p36: a focus on brain abnormalities and a review of the literature. Am J Med Genet A 146:3062–3069

Candi E, Rufini A, Terrinoni A, Dinsdale D, Ranalli M, Paradisi A, De Laurenzi V, Spagnoli LG, Catani MV, Ramadan S, Knight RA, Melino G (2006) Differential roles of p63 isoforms in epidermal development: selective genetic complementation in p63 null mice. Cell Death Differ 13:1037–1047

Candi E, Rufini A, Terrinoni A, Giamboi-Miraglia A, Lena AM, Mantovani R, Knight R, Melino G (2007) ΔNp63 regulates thymic development through enhanced expression of FgfR2 and Jag2. Proc Natl Acad Sci USA 104:11999–12004

Chambers I, Colby D, Robertson M, Nichols J, Lee S, Tweedie S, Smith A (2003) Functional expression cloning of Nanog, a pluripotency sustaining factor in embryonic stem cells. Cell 113(5):643–655

Chin MH, Mason MJ, Xie W, Volinia S, Singer M, Peterson C, Ambartsumyan G, Okita K, Ichisaka T, Yamanaka S (2007) Generation of germline-competent induced pluripotent stem cells. Nature 448:313–317

Dahéron L, Opitz SL, Zaehres H, Lensch MW, Andrews PW, Itskovitz-Eldor J, Daley GQ (2004) LIF/STAT3 signaling fails to maintain self-renewal of human embryonic stem cells. Stem Cells 22(5):770–778, Erratum in: Stem Cells. 2007 Dec;25(12):3273. Lensch, William M [corrected to Lensch, M William]

Daley GQ (2010) Stem cells: roadmap to the clinic. J Clin Invest 120:8–10

Dötsch V, Bernassola F, Coutandin D, Candi E, Melino G (2010) p63 and p73 the ancestors of p53. Cold Spring Harb Perspect Biol 2:1–14

Fujitani M, Cancino GI, Dugani CB, Weaver IC, Gauthier-Fisher A, Paquin A, Mak TW, Wojtowicz MJ, Miller FD, Kaplan DR (2010) TAp73 acts via the bHLH Hey2 to promote long-term maintenance of neural precursors. Curr Biol 20:2058–2065

González-Cano L, Herreros-Villanueva M, Fernández-Alonso R, Ayuso-Sacido Á, Meyer G, García-Verdugo JM, Silva A, Marqués MM, Marín MC (2010) p73 deficiency results in impaired self renewal and premature neuronal differentiation of mouse neural progenitors independently of p53. Cell Death Dis 1:e109. doi:10.1038/cddis.2010.87

Graham V, Khudyakov J, Ellis P, Pevny L (2003) SOX2 functions to maintain neural progenitor identity. Neuron 39(5):749–765

Hansen D, Wilson-Berry L, Dang T, Schedl T (2004) Control of the proliferation versus meiotic development decision in the C. elegans germline through regulation of GLD-1 protein accumulation. Development 131(1):93–104

Haojian Zhang, Wang ZZ (2008) Mechanisms that mediate stem cell self-renewal and differentiation. J Cell Biochem 103:709–718
Hitoshi S, Alexson T, Tropepe V, Donoviel D, Elia AJ, Nye JS, Conlon RA, Mak TW, Bernstein A, van der Kooy D (2002) Notch pathway molecules are essential for the maintenance, but not the generation, of mammalian neural stem cells. Genes Dev 16(7):846–858
Hong H, Takahashi K, Ichisaka T, Aoi T, Kanagawa O, Nakagawa M, Okita K, Yamanaka S (2009) Suppression of induced pluripotent stem cell generation by the p53-p21 pathway. Nature 460:1132–1135
Ivanova NB, Dimos JT, Schaniel C, Hackney JA, Moore KA, Lemischka IR (2002) A stem cell molecular signature. Science 298:601–604
Kaghad M, Bonnet H, Yang A, Creancier L, Biscan JC, Valent A, Minty A, Chalon P, Lelias JM, Dumont X, Ferrara P, McKeon F, Caput D (1997) Monoallelically expressed gene related to p53 at 1p36, a region frequently deleted in neuroblastoma and other human cancers. Cell 90:809–819
Kawamura T, Suzuki J, Wang YV, Menendez S, Morera LB, Raya A, Wahl GM, Belmonte JC (2009) Linking the p53 tumour suppressor pathway to somatic cell reprogramming. Nature 460:1140–1144
Lena AM, Shalom-Feuerstein R, di Val R, Cervo P, Aberdam D, Knight RA, Melino G, Candi E (2008) miR-203 represses 'stemness' by repressing ΔNp63. Cell Death Differ 15:1187–1195
Lin T, Chao C, Saito S, Mazur SJ, Murphy ME, Appella E, Xu Y (2005) p53 induces differentiation of mouse embryonic stem cells by suppressing Nanog expression. Nat Cell Biol 7:165–171
Loh YH, Wu Q, Chew JL, Vega VB, Zhang W, Chen X, Bourque G, George J, Leong B, Liu J, Wong KY, Sung KW, Lee CW, Zhao XD, Chiu KP, Lipovich L, Kuznetsov VA, Robson P, Stanton LW, Wei CL, Ruan Y, Lim B, Ng HH (2006) The Oct4 and Nanog transcription network regulates pluripotency in mouse embryonic stem cells. Nat Genet 38:431–440
Machold R, Hayashi S, Rutlin M, Muzumdar MD, Nery S, Corbin JG, Gritli-Linde A, Dellovade T, Porter JA, Rubin LL, Dudek H, McMahon AP, Fishell G (2003) Sonic hedgehog is required for progenitor cell maintenance in telencephalic stem cell niches. Neuron 39(6):937–950, Erratum in: Neuron. 2003 Sep 25;40(1):189
Marión RM, Strati K, Li H, Murga M, Blanco R, Ortega S, Fernandez-Capetillo O, Serrano M, Blasco MA (2009) A p53-mediated DNA damage response limits reprogramming to ensure iPS cell genomic integrity. Nature 460:1149–1153
Meletis K, Wirta V, Hede SM, Nistér M, Lundeberg J, Frisén J (2006) p53 suppresses the self-renewal of adult neural stem cells. Development 133:363–369
Melino G, De Laurenzi V, Vousden KH (2002) p73: friend or foe in tumorigenesis. Nat Rev Cancer 8:605–615
Mills AA, Zheng B, Wang XJ, Vogel H, Roop DR, Bradley A (1999) p63 is a p53 homologue required for limb and epidermal morphogenesis. Nature 398:708–713
Mitsui K, Tokuzawa Y, Itoh H, Segawa K, Murakami M, Takahashi K, Maruyama M, Maeda M, Yamanaka S (2003) The homeoprotein Nanog is required for maintenance of pluripotency in mouse epiblast and ES cells. Cell 113(5):631–642
Molofsky AV, Pardal R, Morrison SJ (2004) Diverse mechanisms regulate stem cell self-renewal. Curr Opin Cell Biol 16:700–707
Nichols J, Zevnik B, Anastassiadis K, Niwa H, Klewe-Nebenius D, Chambers I, Schöler H, Smith A (1998) Formation of pluripotent stem cells in the mammalian embryo depends on the POU transcription factor Oct4. Cell 95(3):379–391
Niwa H, Burdon T, Chambers I, Smith A (1998) Self-renewal of pluripotent embryonic stem cells is mediated via activation of STAT3. Genes Dev 12(13):2048–2060
Niwa H, Miyazaki J, Smith AG (2000) Quantitative expression of Oct-3/4 defines differentiation, dedifferentiation or self-renewal of ES cells. Nat Genet 24(4):372–376
Okita K, Ichisaka T, Yamanaka S (2007) Generation of germline-competent induced pluripotent stem cells. Nature 448:313–317

Pelengaris S, Khan M, Evan G (2002) c-MYC: more than just a matter of life and death. Nat Rev Cancer 2:764–776

Pellegrini G, Dellambra E, Golisano O, Martinelli E, Fantozzi I, Bondanza S, Ponzin D, McKeon F, De Luca M (2001) p63 identifies keratinocyte stem cells. Proc Natl Acad Sci USA 98: 3156–3161

Reya T, Duncan AW, Ailles L, Domen J, Scherer DC, Willert K, Hintz L, Nusse R, Weissman IL (2003) A role for Wnt signalling in self-renewal of haematopoietic stem cells. Nature 423(6938):409–414

Senoo M, Pinto F, Crum CP, McKeon F (2007) p63 is essential for the proliferative potential of stem cells in stratified epithelia. Cell 129:523–536

Shenghui H, Nakada D, Morrison SJ (2009) Mechanisms of stem cell self-renewal. Annu Rev Cell Dev Biol 25:377–406

Song H, Chung SK, Xu Y (2010) Modeling disease in human ESCs using an efficient BAC-based homologous recombination system. Cell Stem Cell 6:80–89

Su X, Paris M, Gi YJ, Tsai KY, Cho MS, Lin YL, Biernaskie JA, Sinha S, Prives C, Pevny LH, Miller FD, Flores ER (2009) TAp63 prevents premature aging by promoting adult stem cell maintenance. Cell Stem Cell 5:64–75

Suh EK, Yang A, Kettenbach A, Bamberger C, Michaelis AH, Zhu Z, Elvin JA, Bronson RT, Crum CP, McKeon F (2006) p63 protects the female germ line during meiotic arrest. Nature 444:624–628

Takahashi K, Yamanaka S (2006) Induction of pluripotent stem cells from mouse embryonic and adult fibroblast cultures by defined factors. Cell 126:663–676

Talos F, Abraham A, Holembowski L, Vaseva A, Tsirka S, Scheel A, Bode D, Dobbelstein M, Bruck W, Moll UM (2010) p73 is an essential regulator of neural stem cell maintenance in embryonal and adult CNS neurogenesis. Cell Death Differ 17:1816–1829

Utikal J, Polo JM, Stadtfeld M, Maherali N, Kulalert W, Walsh RM, Khalil A, Rheinwald JG, Hochedlinger K (2009) Immortalization eliminates a roadblock during cellular reprogramming into iPS cells. Nature 460:1145–1148

Vousden KH, Lane DP (2007) p53 in health and disease. Nat Rev Mol Cell Biol 8:275–283

Yang A, Schweitzer R, Sun D, Kaghad M, Walker N, Bronson RT, Tabin C, Sharpe A, Caput D, Crum C, McKeon F (1999) p63 is essential for regenerative proliferation in limb, craniofacial and epithelial development. Nature 398:714–718

Yang A, Walker N, Bronson R, Kaghad M, Oosterwegel M, Bonnin J, Vagner C, Bonnet H, Dikkes P, Sharpe A, McKeon F, Caput D (2000) p73-deficient mice have neurological, pheromonal and inflammatory defects but lack spontaneous tumors. Nature 404:99–103

Yang A, Kaghad M, Caput D, McKeon F (2002) On the shoulder of giants: p63, p73 and the rise of p53. Trends Genet 2:90–95

Yi R, Poy MN, Stoffel M, Fuchs E (2008) A skin microRNA promotes differentiation by repressing 'stemness'. Nature 452:225–229

Ying QL, Nichols J, Chambers I, Smith A (2003a) BMP induction of Id proteins suppresses differentiation and sustains embryonic stem cell self-renewal in collaboration with STAT3. Cell 115(3):281–292

Ying QL, Stavridis M, Griffiths D, Li M, Smith A (2003b) Conversion of embryonic stem cells into neuroectodermal precursors in adherent monoculture. Nat Biotechnol 21(2):183–186

Zhao Y, Yin X, Qin H, Zhu F, Liu H, Yang W, Zhang Q, Xiang C, Hou P, Song Z, Liu Y, Yong J, Zhang P, Cai J, Liu M, Li H, Li Y, Qu X, Cui K, Zhang W, Xiang T, Wu Y, Zhao Y, Liu C, Yu C, Yuan K, Lou J, Ding M, Deng H (2008) Two supporting factors greatly improve the efficiency of human iPSC generation. Cell Stem Cell 3:475–479

Chapter 5
Mutant p53-Driven Tumorigenesis

Tamara Terzian and Guillermina Lozano

1 Introduction

A myriad of changes occur in tumor cells some of which contribute to tumor cell behavior and many that do not (Mardis 2011; Meyerson et al. 2010; Pfeifer and Hainaut 2011). Mutations and/or alterations of multiple nodes in the p53 tumor suppressor pathway directly result in tumorigenesis (Vogelstein et al. 2000). p53 as a transcriptional regulator is key in controlling multiple signals by activating genes with roles in apoptosis, cell cycle arrest, senescence, differentiation, and survival (Riley et al. 2008). Clearly, these activities underscore the importance of p53 in regulating cell fate decisions.

The most common mechanism of inactivating the p53 pathway is by single nucleotide changes that generate missense mutations in exons encoding the DNA-binding domain (DBD) of the *TP53* gene itself. The resulting p53 mutant proteins are present at high levels in tumor cells and contribute gain-of-function (GOF) and dominant negative phenotypes. Multiple laboratories have generated mouse models with *p53* missense mutations to examine the role of these mutations in tumor development and drug resistance. These models are invaluable for testing multiple chemotherapies and for improving the efficacy of the treatment of choice, and will be the focus of this review.

T. Terzian
Department of Dermatology, University of Colorado Denver, Aurora, CO, USA

Center for Regenerative Medicine and Stem Cell Biology, University of Colorado Denver, Aurora, CO, USA

G. Lozano (✉)
Department of Genetics, University of Texas M.D. Anderson Cancer Center, Houston, TX 77030, USA
e-mail: gglozano@mdanderson.org

P. Hainaut et al. (eds.), *p53 in the Clinics*, DOI 10.1007/978-1-4614-3676-8_5,

2 *TP53* Alterations in Cancer

Most somatic mutations (82 %) in *TP53* are single nucleotide substitutions. Only 18 % are insertions/deletions that result in true loss-of-function alleles. *TP53* mutations are systematically compiled in the IARC *TP53* Database (http://www-p53.iarc.fr/). The latest release (R15, November 2010) contains 27,580 somatic and 597 germline mutations identified in human cancers. Most missense mutations are located in the DBD that encompasses *TP53* exons 5–8, and six of these (Arg175, Gly245, Arg248, Arg249, Arg273, and Arg282) are labeled "hot spot" codons because of their increased mutation frequency. Mutations at codons 248 and 273 are classified as DNA contact mutations, while alterations at codons 175, 245, 249, and 282 are considered conformational mutations (Hainaut and Hollstein 2000). Of all somatic mutations, ~20 % alter one of the three codons: 175, 248, and 273. However, no matter how these hot spot mutants are classified, all have impaired sequence-specific DNA-binding capacity resulting in a loss of transcriptional activity (Bullock et al. 2000; Epstein et al. 1998; Kern et al. 1991). In tumor cells, these p53 mutant proteins are very stable and accumulate in cell nuclei. Other alterations in the *TP53* gene include nonsense and splice site mutations. These often result in lack of accumulation of a protein product although in some cases a truncated p53 protein is made (Tsuda and Hirohashi 1994).

Germline *TP53* mutations are found in 70 % of families with *L*i–*F*raumeni *S*yndrome (LFS) (Malkin et al. 1990; Varley 2003; Varley et al. 1997a; Wu et al. 2006). LFS is a rare autosomal dominant disease that presents with an early onset of various tumor types such as soft tissue sarcomas, osteosarcomas, breast cancers, lymphomas/leukemias, brain tumors, and adrenocortical carcinomas (Li and Fraumeni 1969). LFS patients heterozygous for a mutant *TP53* allele have an increased cancer risk of 50 % by the age of 30 years old and 90 % by the age of 70 (Malkin et al. 1990; Wu et al. 2006). Additionally, females have a higher risk of developing tumors, and at a much younger age than men (Hwang et al. 2003b). Smoking, an environmental risk factor, also accelerates the onset of tumors in LFS patients (Hwang et al. 2003a). Similar to somatic mutations, the most commonly mutated codons found in LFS families are located in the central DNA-binding domain of the protein, especially at codons 248 and 273.

TP53 germline mutations were also found in children suffering from adrenocortical carcinoma with no LFS family history (Varley et al. 1999; Wagner et al. 1994). An arginine to histidine substitution at codon 337 (p53R337H) in the C-terminal oligomerization domain is frequently found in pediatric adenocarcinoma patients from southern Brazil (~35 %). This alteration was rarely observed in tumors with somatic mutations (0.1 %) (Figueiredo et al. 2006). Thus, some tissue specificity for *TP53* mutations exists.

In the light of the new era of personalized medicine, an understanding of the relationship between p53 structure, tumor-derived missense mutations, and function becomes an essential prognostic tool and predictor of the outcome for a given chemotherapeutic strategy. In 2003, Kato et al. (2003) generated 2,314 *TP53* mutants in yeast cells containing all reported single nucleotide substitutions affecting the full length of the protein to evaluate the impact on p53 structure–function and correlated their findings with the mutation frequency in tumors. The mutants were organized into three categories: nonfunctional, partially functional, and fully functional. One-third of all mutants were nonfunctional and two-thirds of these localized to the core DBD. This highlights a powerful connection between p53 activity and mutational status, and demonstrates that inactivating mutations in the sequence-specific DBD (amino acids 96–292) contribute to tumorigenesis more so than other mutations in the N-terminal transcriptional activation domain or the C-terminal oligomerization domain. The IARC *TP53* Database indicates that 73 % of germline mutations found in human cancers are non-functional, while 20 % have residual activity and only 7 % are fully functional. The age of onset of many tumor types was also significantly reduced in carriers of non functional *TP53* mutations as compared to patients with functional mutations (Petitjean et al. 2007). Together these data indicate a selective pressure to lose p53 activity in tumorigenesis and spell out the need for further characterization of individual substitutions, especially hot spot mutations, to decipher their functional impact on tumor development and resistance to therapy.

While loss of wild-type p53 function is an important step in tumorigenesis, other characteristics of mutant p53 weigh in to accelerate tumor development and metastasis. These features include (1) gain of new functions that include genetic and molecular interactions with genes or proteins that do not normally interact with p53, (2) dominant negative effects of the mutant polypeptide on the wild-type counterpart through hetero-oligomerization, and (3) mutant p53 stability.

3 Gain of New Functions

From over 1,000 core domain mutants, only a subset have been studied for GOF and these are mainly mutations occurring at high frequency (Cadwell and Zambetti 2001; Deppert et al. 2000; van Oijen and Slootweg 2000). A systematic study in yeast and mammalian cells classified 76 mutants according to their functional properties. These findings illustrate the diverse behavior of p53 mutants, and their impact on wild-type p53 and malignant transformation (Dearth et al. 2007). To examine the GOF activities of mutant p53 in context of the whole animal, several groups used mouse models to incorporate *p53* missense mutations at the *p53* locus (often referred to as knock-in alleles) (Table 5.1). The Lozano and Jacks laboratories generated knock-in mice with the hot spot *p53R172H* (175 in humans) and *p53R270H* (273 in humans) mutations on two different genetic backgrounds, C57BL/6 and 129S4/SvJae (Lang et al. 2004; Olive et al. 2004). Mice with one normal and one mutant *p53* allele are highly tumor-prone and serve as excellent models for LFS. These mice develop a variety of cancers (mainly lymphomas, sarcomas, and carcinomas) and exhibit a GOF metastatic phenotype not observed in mice with a *p53*-null allele ($p53^{+/-}$). On the C57BL/6 background, the tumor spectrum between $p53^{R172H/+}$ and $p53^{+/-}$ mice was similar although $p53^{R172H/+}$ adenocarcinomas and osteosarcomas metastasized at high frequencies.

Table 5.1 Mutant p53 tumor-prone models available for clinical testing

Type of mutant	Mouse	Characteristics	Phenotypes	References	Relevance to human disease
Deletion	$p53^{-}$	Deletion of part of intron 4 to part of exon 5. C57BL6 and 129/Sv mixed background	$p53^{+/-}$ and $p53^{-/-}$ mice are highly tumor-prone. Homozygous mice develop mainly lymphomas (70 %)	Donehower et al. (1992)	Mice have a similar but not identical tumor spectrum to human LFS patients
	$p53^{-}$	Deletion exon 2 to intron 6. C57BL6 strain	$p53^{+/-}$ mice develop sarcomas (57 %) and lymphomas (25 %). $p53^{-/-}$ mice develop mostly lymphomas (71 %)	Jacks et al. (1994)	Mice have a similar but not identical tumor spectrum to human LFS patients
	$p53^{-}$	Deletion of part of intron 4 to part of exon 5. BalbC background	$p53^{-/-}$ mice developed mainly sarcomas and lymphomas whereas 55 % of the female $p53^{+/-}$ mice developed mammary carcinomas	Kuperwasser et al. (2000)	This mouse develops mammary tumors as do patients with LFS. Also the important effect of modifiers on tumor spectrum is demonstrated
Transgenic	Tg^{A135V}	CD-1 strain	30 % tumor incidence by the age of 18 mo. High frequency of lung adenocarcinomas, osteosarcomas, and lymphomas	Lavigueur et al. (1989)	Overexpressed mutant exhibits DN functions. Tumors in these mice show genomic rearrangements of the *p53* gene as observed in human cancers
DBD KI	$p53^{R172H}$	C57BL/6 strain	$p53^{R172H/+}$ has a similar tumor spectrum and survival to $p53^{+/-}$ mice but frequent metastasis from osteosarcomas and carcinomas	Lang et al. (2004)	One of the most common hot spot mutations in human cancers. Mutant exhibits both GOF and DN functions. $p53^{R172H/+}$ mouse represents a good model of LFS
	$p53^{R172H}$	129Sv strain	$p53^{R172H/-}$ develops novel tumors compared to $p53^{-/-}$ mice. $p53^{R172H/+}$ had a similar survival curve to $p53^{+/-}$ mice but shows increase in osteosarcomas and frequent metastases	Olive et al. (2004)	$p53^{R172H/+}$ mouse is a faithful genetic model of LFS. Mutant exhibits GOF and DN phenotypes

p53^{R172P}	Mutant is cell cycle arrest competent but apoptosis defective. C57BL/6 and 129Sv mixed background	Late onset of tumors (lymphomas and sarcomas). Tumors maintain genomic stability in contrast to *p53*$^{-/-}$ mice	Liu et al. (2004)	A rare mutation found in human tumors. p53-dependent cell cycle arrest plays an important role in early tumor suppression
Hupki-TP53^{R248W}	C57BL/6 and 129Sv mixed background	Homozygous mutants have similar survival curve to *p53*$^{-/-}$ mice, and develop lymphomas and sarcomas with genomic instability. The DNA damage-response pathway is impaired	Song et al. (2007)	A common hot spot mutation in human cancers. Model demonstrates importance of GOF in tumorigenesis
p53^{R270H} (273 in humans)	129Sv strain. This codon is the most frequently mutated in p53 and the R to H substitution is the most common one	*p53*$^{R270H/-}$ developed a different tumor spectrum than *p53*$^{-/-}$ mice. *p53*$^{R270H/+}$ have a similar survival curve to *p53*$^{+/-}$ mice but display an increase in carcinomas and hemangiosarcomas and frequent metastases	Olive et al. (2004)	A common hot spot mutation in human cancers that displays GOF and DN phenotypes
Hupki-TP53^{R273H}	C57BL/6 and 129Sv mixed background	Genomic instability is observed in pre-tumor thymi. The DNA damage-response pathway is impaired	Song et al. (2007)	A common hot spot mutation in human cancers that displays GOF phenotypes

KI knock-in in the endogenous gene, *DBD* DNA binding domain, *DN* dominant negative, *GOF* gain-of-function, *HUPKI* humanized *p53* knock-in, *mo* months, *LFS* Li–Fraumeni syndrome

On the 129S4/SvJae background, metastatic osteosarcomas were more prevalent in *p53*$^{R172H/+}$ mice in comparison to *p53*$^{+/-}$ mice, whereas *p53*$^{R270H/+}$ mice displayed a different tumor spectrum (B-cell lymphomas were more frequent), a higher tumor burden and metastatic carcinomas as compared to *p53*$^{+/-}$ mice. This difference in tumor spectrum between the two mouse strains is indicative of the importance of the genetic background and the effect of modifiers on the tumor phenotype (Donehower et al. 1995; Harvey et al. 1993). This modifier effect is also seen in *p53*$^{+/-}$ mice on the BALB/c background that develop spontaneous mammary tumors (55 % with a latency of 8–14 months) in contrast to C57BL/6 and 129/Sv *p53*$^{+/-}$ mice, which rarely develop this type of cancer (Kuperwasser et al. 2000; Koch et al. 2007) . Loss of heterozygosity (LOH) through deletion of the wild-type *p53* allele was also observed in some tumors from *p53* mutant heterozygous mice regardless of background. Thus, cancers from mice with mutant *p53* alleles originate from different tissues, are more aggressive, and metastasize at a higher frequency than *p53*$^{+/-}$ tumors. In fact, patients with *TP53* missense mutations suffer a worse prognosis than those carrying *TP53* deletions (Cadwell and Zambetti 2001; Deppert et al. 2000; Roemer 1999; van Oijen and Slootweg 2000). Thus, animal models that reproduce the genetic defects in LFS patients support the GOF hypothesis.

The role of *p53* deletions/mutations in mice with cooperating oncogenic events has also been examined (Donehower and Lozano 2009; Lozano and Liu 1998). In a skin carcinogenesis model, transgenic *K-ras*^{G12D} mice carrying the *p53*^{R172H} mutation (*p53*$^{R172H/-}$) have increased tumor formation, progression, and metastasis compared to *p53-null* mice (Caulin et al. 2007). In a model of intestinal tumors, 100 % of mice that carry one inactivated allele of the *Apc* gene (*adenomatous polyposis coli*) and one mutant *p53* allele (*p53*$^{R172H/+}$) suffered from invasive intestinal adenocarcinomas as compared to 25 % of mice with one inactivated *Apc* allele and a loss of one *p53* allele (Muller et al. 2009). Another example of the GOF properties of mutant p53 is that of transgenic mice expressing *K-ras*^{G12D} specifically in the pancreas coupled with the *p53*^{R172H} mutation. This cohort of mice heterozygous for the *p53* mutation (*p53*$^{R172H/+}$) gets a high incidence of metastatic pancreatic ductal adenocarcinoma (65 %) in contrast to those with a heterozygous deletion of *p53* (0 %) (Morton et al. 2010). These models most closely mimic the kinds of cooperating mutations that occur in specific human cancers and will be valuable in evaluating drug efficacy, tumor response, and tumor relapse.

Multiple mechanisms have been identified that contribute to the GOF phenotype of mutant p53 proteins (Brosh and Rotter 2009). Mutant p53 protein–protein interactions with p63 and p73 family members was proposed as an underlying mechanism of GOF (Deyoung and Ellisen 2007; Di Como et al. 1999; Gaiddon et al. 2001; Lang et al. 2004; Strano et al. 2002). In this scenario, mutant p53 inhibits the transcriptional and tumor suppressive activities of p63 and p73. In support of this hypothesis, mutant p53 co-immunoprecipitated endogenous p63 and p73 proteins, and decreased levels of mutant p53 were accompanied by increased p63/p73 activity in tumor cells derived from *p53*^{R172H} knock-in mice (Lang et al. 2004; Olive et al. 2004). Additionally, in focus-forming assays the number of foci in *p53*-null MEFs is less than that of *p53*$^{R172H/H}$ MEFS but becomes equal to that of mutant *p53* cells with down regulation of *p63* or *p73* expression (Lang et al. 2004). Lastly, in vivo data also support the inhibition of p63 and p73 activities as a GOF mechanism. Similar to *p53* mutant mice, $p53^{+/-}$ $p63^{+/-}$ and $p53^{+/-}$ $p73^{+/-}$ compound mice developed metastatic tumors at high frequencies not observed with any of the heterozygous alleles alone (Flores et al. 2005). These data demonstrate that one GOF activity of p53 mutants occurs through inhibition of p63 and p73 activities [reviewed in Li and Prives (2007) and Lozano (2007)].

More recently, Song et al. (2007) introduced *TP53* missense mutations R248W and R273H into the *hu*manized *p*53 *k*nock-*in* (HUPKI) mouse (Luo et al. 2001). These mice have the human *TP53* sequences encompassing exons 4–9 (encoding p53 amino acids 33–332) incorporated into the murine *p53* locus. The resulting *p53* homozygous mutant mice exhibit a shift in tumor spectrum and are predisposed to metastatic cancers in comparison to $p53^{-/-}$ mice further supporting the GOF hypothesis. Additionally, precancerous thymocytes from *p53* mutant mice exhibit genetic instability observed as interchromosomal translocations rarely seen in $p53^{-/-}$ cells. Following DNA damage, these p53 mutants, but not wild-type p53, interact with the nuclease Mre11 and suppress the binding of the Mre11–Rad50–NBS1 (MRN) complex to DNA double-stranded breaks thereby impairing the activation of the DNA damage sensor ataxia-telangiectasia mutated (ATM) (Song et al. 2007). Therefore, these GOF mutant models develop tumors due to increased genetic instability by the disruption of ATM signaling (DNA damage-response pathway). More recently, Liu et al. (2010) showed that the p53R175H structural mutant shares the same GOF properties of Mre11 binding and disruption of the ATM signaling as that observed for the contact mutants p53R248H and p53R273H. These findings could explain the age-associated increase in cancer, and age-dependent decrease in ATM signaling and p53 function in LFS patients (Donehower 2009). Thus, the genomic instability associated with impairment of ATM signaling by mutant p53 is another mechanism by which mutant p53 exhibits GOF properties. According to the IARC *TP53* Database, 30 germline *TP53* mutants were identified as having GOF properties. However, these mutants have yet to be tested for GOF activities in vivo and may exhibit different GOF mechanisms depending on cellular and tissue context (Petitjean et al. 2007).

4 Dominant Negative Effects on Wild-Type p53

The p53 protein is functionally active as a tetramer that consists of four identical subunits combined as two dimers, and as such, p53 binds to a consensus DNA sequence (Jeffrey et al. 1995). In addition to GOF, mutant p53 can display dominant negative (DN) effects resulting from mutant and wild-type p53 proteins competing for binding to target gene promoters (Farmer et al. 1992; Kern et al. 1992). Additionally, mutant p53 can form mixed hetero-oligomers with wild-type p53 dampening its activity (Eliyahu et al. 1988; Finlay et al. 1989; Kraiss et al. 1988; Milner and Medcalf 1991; Rovinski and Benchimol 1988). In vitro data indicate that mutant p53 drives the wild-type protein into a mutant conformation further contributing to inactivation of wild-type p53 in the context of mutant p53 (Milner and Medcalf 1991).

Many yeast and cell culture studies have assayed for the DN function of various mutants (Brachmann et al. 1996; Dearth et al. 2007). These studies show that DN mutations cluster in the hot spot region, and that missense DN mutations are greatly over represented in human cancers. Interestingly, the age of onset of tumors carrying DN and non-DN mutants was similar but the types of tumors varied. For example, in breast cancers, DN mutations were twice more frequent and in bone sarcomas twice less frequent than non-DN mutation carriers (Petitjean et al. 2007). Studies using mice recapitulate this DN effect under some circumstances. Transgenic mice overexpressing the *p53A135V* mutant represent a good example. This mutation has demonstrated a potent oncogenic activity in biochemical and cellular assays [Harvey et al. (1995) and few references within]. The transgene accelerated tumor development in mice with one or two wild-type *p53* alleles, but not in $p53^{-/-}$ mice. Thus, this mutant allele was more tumorigenic in the presence of wild-type *p53* ($p53^{+/+}$ and $p53^{+/-}$) than in the absence of *p53* ($p53^{-/-}$) implicating a DN effect on wild-type p53 activity. These data also support a GOF activity of mutant p53, but importantly, this mutant does not display GOF activities in the absence of wild-type p53 ($p53^{-/-}$) (Harvey et al. 1995). Another example of the DN activities of mutant p53 is found in the $p53^{R172H\Delta G/+}$ mouse that carries a heterozygous *p53R172H* mutation and expresses low levels of the mutant p53 protein due to the deletion of a G nucleotide at a splice acceptor site (Liu et al. 2000). In these mice, an increased incidence of carcinomas and metastatic osteosarcomas were observed as compared to $p53^{+/-}$ mice. Also, in contrast to $p53^{+/-}$ tumors, the wild-type allele was rarely lost in $p53^{R172H\Delta G/+}$ tumors. The enhanced tumor phenotypes in these mice, despite the presence of wild-type p53, support the DN hypothesis.

Further compelling evidence for mutant p53 DN activity was provided by experiments studying the *p53R270H* or *p53P275S* mutations in mouse embryonic stem (ES) cells. The human equivalent of the *p53P275S* mutation (*p53P278S*) occurs most commonly in epithelial cancers arising in sites such as skin, vulva, oropharynx (de Vries et al. 2002; Petitjean et al. 2007). Following γ-radiation in mice, heterozygous mutant ES cells display a weaker and delayed induction of wild-type p53 target genes as compared to $p53^{+/-}$ cells. These heterozygous mutant cells show resistance to apoptosis equal to that of $p53^{-/-}$ cells upon treatment with the chemotherapeutic drug, doxorubicin. The two murine models of LFS described above also display DN properties. γ-irradiated $p53^{R172H/+}$ and $p53^{R270H/+}$ thymocytes exhibit decreased apoptosis intermediate between that of treated $p53^{+/-}$ and $p53^{-/-}$ cells (Olive et al. 2004). In the γ-irradiated $p53^{R172H/+}$ hypothalamus, apoptosis decreased similar to $p53^{-/-}$ treated cells (Lang et al. 2004). These examples demonstrate that the presence of a single p53 mutant can inhibit the function of the wild-type p53 consistent with a DN effect. Thus, a missense *p53* mutation in tumors is more detrimental than a loss-of-function allele, and advocates for a role of selective pressure in manifesting these mutations.

However, the DN properties of missense mutants are not always observed in vivo suggesting a much more complex interaction between wild-type and mutant p53 in tumorigenesis. As an example, we have observed that the presence of one copy of the mutant $p53^{R172H}$ allele cannot rescue the embryonic lethality of *Mdm2-null* mice as does the absence of *p53* (Lang et al. 2004). This finding suggests incomplete dominance of mutant over wild-type *p53* activities in $p53^{R172H/+}$ mice.

5 Mutant p53 Stability

p53 levels are tightly controlled by multiple regulators to avoid its deleterious effects on homeostasis (Levine and Oren 2009). Specifically, the ubiquitin ligase Mdm2 represents *the* major negative regulator of p53 and is able to target p53 to the proteasome for degradation (Haupt et al. 1997; Honda et al. 1997; Jones et al. 1995; Kubbutat et al. 1997; Montes de Oca Luna et al. 1995). Another important inhibitor, Mdm4, belongs to the same family and shares similarities in structure with Mdm2 but cannot degrade p53 since it lacks an ubiquitin ligase activity (Jackson and Berberich 2000; Sharp et al. 1999; Shvarts et al. 1996; Stad et al. 2001). Importantly for this discussion, Mdm2 and Mdm4 also bind mutant p53 proteins (Momand et al. 1992; Shvarts et al. 1997; Terzian et al. 2008).

Following stress signals such as inappropriate activation of an oncogene or DNA damage, p53 is post-translationally modified in cells via several mechanisms which include phosphorylation, acetylation, and methylation (Appella and Anderson 2001; Chuikov et al. 2004; Gu and Roeder 1997; Meek 1994; Sakaguchi et al. 1998). Modifications at the p53 amino terminus disrupt interaction with Mdm2, thereby releasing p53 activity. *Mdm2* is also a transcriptional target of p53 and thus mutant p53 cannot induce *Mdm2* expression. The decreased levels of *Mdm2* coupled with post-translational modifications of the amino terminus of mutant p53 allow the latter to accumulate in the nucleus. Immunohistochemistry for p53 levels is commonly used in the clinic to evaluate the *TP53* mutational state, cancer predisposition, risk, prognosis, and therapeutic choices. However, p53 positivity in tumors by immunostaining is not always an indicator of its mutational state (Petitjean et al. 2007). Thus, identifying the specific *TP53* alterations and the levels of Mdm2 and Mdm4 that occur in human tumors is likely to provide a more complete picture of how the p53 pathway is disrupted and thus serve as a better predictor of clinical outcome.

The development of genetically engineered mice carrying *p53* missense mutations led to several surprises (Poyurovsky and Prives 2006; Prives and White 2008). One such surprise indicated that p53 mutants were inherently unstable in normal tissues but were stabilized in response to DNA damage (γ-radiation) or oncogenic stress (in tumors) (Lang et al. 2004; Liu et al. 2010; Olive et al. 2004; Terzian et al. 2008). This mutant p53 stabilization uncovered its GOF capability, promoted tumorigenesis (discussed in detail below), and depending on its expression level, altered the tumor spectrum (Lang et al. 2004; Liu et al. 2000; Song et al. 2007). For example, mice with the hypomorphic *p53*$^{R172H\Delta G}$ allele (Liu et al. 2000) mainly develop sarcomas with some carcinomas and lymphomas; while mice with more moderate p53R172H protein levels, as observed in *p53*$^{R172H/+}$ mice, still develop sarcomas but now develop more lymphomas than carcinomas. In *p53*$^{R172H/H}$ mice, on the same C57/BL6 background, higher mutant p53 levels further shifted the tumor spectrum toward lymphomas, with a moderate amount of sarcomas and no carcinomas (Lang et al. 2004).

To understand the players regulating mutant p53 stability and their effect on tumorigenesis, *p53*$^{R172H/H}$ mice were crossed to *Mdm2* and *p16*^{Ink4a} knock-out mice. *p16*^{Ink4a} is a member of the Rb pathway and the loss of either *Mdm2* or *p16*^{Ink4a} stabilizes wild-type p53 (Sherr and McCormick 2002). Many but not all normal cells from *p53*$^{R172H/H}$ mice lacking *Mdm2* or *p16*^{Ink4a} display high mutant p53 levels. Similar to wild-type p53, homozygous mutant *p53* mice also stabilized mutant p53 after γ-radiation (Terzian et al. 2008). Taken together, these data demonstrate that similar signals (loss of *Mdm2* or *p16*^{Ink4a} and γ-radiation) regulate both wild-type and mutant p53 in a cell type-dependent manner. Importantly, mutant p53 stabilization in mice lacking *Mdm2* or *p16*^{Ink4a} manifests a GOF metastatic phenotype and decreases survival without a significant change in tumor spectrum as compared to mutant *p53* mice alone (Terzian et al. 2008). Thus, a *p53* missense mutation likely functions as a null allele and exhibits its GOF properties only if stabilized, and the earlier the stabilizing tumor-specific alterations occur, the greater the chance for metastasis.

6 Partial Loss of Function Alleles

As discussed, the p53R172H and p53R270H mutations lack p53 transcriptional activity and display GOF and DN properties. Partially functional alleles such as the p53R175P have been identified in human tumors in vivo. This particular mutant retained a limited transactivation capacity of the cell cycle inhibitor *p21* but failed to activate apoptotic genes or induce apoptosis (Rowan et al. 1996). These data have been recapitulated in a knock-in mouse model of the equivalent mutation (*p53*^{R172P}) (Liu et al. 2004). Following γ-radiation of the developing central nervous system and thymocytes of homozygous mutant mice (*p53*$^{R172P/P}$), *p21* expression is induced with no p53-dependent apoptosis suggesting an apoptosis-defective mutant. Interestingly, these mice demonstrate a significant delay in tumor onset (median of 11.5 months) compared to *p53-null* mice (5.5 months) highlighting the contribution of cell cycle arrest in tumor suppression. Cytogenetics data indicate that *p53*$^{R172P/P}$ tumors are diploid for the most part, whereas *p53-null* tumors are aneuploid suggesting a role of genome stability in delaying tumorigenesis in *p53*$^{R172P/P}$ mice. To further confirm the degree of involvement of p21 in this prolonged tumor latency, *p53*$^{R172P/P}$ mice lacking *p21* were generated (Barboza et al. 2006). These mice developed mainly lymphomas (31 %) and sarcomas (59 %) and like *p53*$^{-/-}$ tumors, these malignancies are aneuploid. Moreover, the overall survival of *p53*$^{R172P/P}$ *p21*$^{-/-}$ was considerably diminished (median survival ~7.5 months) compared to *p53*$^{R172P/P}$ mice but was still delayed compared to *p53*$^{-/-}$ mice. Taken together, these data emphasize the crucial in vivo role that cell cycle arrest, and specifically p21, plays in the maintenance of a diploid genome. Additionally, it highlights differences between *p53* missense mutations and their unique roles in tumor initiation and progression.

7 One or More Alterations in the p53 Pathway

Clearly a mutation in one allele of *TP53* and loss of the second allele eliminates p53 function in a tumor cell [the classic Knudsen's hypothesis (Knudson 1971)]. However, not all tumor cells with *TP53* mutations lose the wild-type *TP53* allele. A study of LFS patients with *TP53* mutations indicates that only 44 % of tumors lose the wild-type gene (Varley et al. 1997a, b). These data are supported by studies in mice. Greater than 85 % of tumors from *p53* heterozygous mice older than 18 months of age retain the wild-type *p53* allele (Venkatachalam et al. 1998). DN activities of mutant p53 (with or without stabilization) may lessen the need to lose the wild-type allele. Another possibility is that a combination of other molecular changes in the pathway could lead to the same end result: insufficient p53 to cause apoptosis, senescence, or cell cycle arrest. The presence of high levels of the p53 inhibitors Mdm2 and Mdm4, for example, could preclude loss of the remaining wild-type *TP53* allele. Thus, a cell heterozygous for *TP53* with increased p53 inhibitors may not need to lose the remaining wild-type *TP53* allele in the process of transformation. In fact, LOH is not common in $p53^{+/-}$ mice over producing Mdm4 (Xiong et al. 2010).

The discussion above suggests that the number of modifications required to inactivate the p53 pathway may vary depending on the circumstances. The nature of the *p53* mutation may also contribute to the number of changes required for transformation. Obviously, loss of *TP53* predisposes to tumor development. However, some *TP53* mutations identified in human cancers are clearly partial loss of function mutations that retain the ability to activate some but not all p53 targets (Resnick and Inga 2003). For example, the rare human $p53^{R175P}$ mutation discussed above is able to delay a tumor phenotype in mice in comparison to $p53^{R172H}$ or null alleles. Further loss of *p21* yields a tumor phenotype that now more closely resembles that of *p53*-null mice (Barboza et al. 2006). Thus, compounding weaker mutations may yield the same end result, inactivation of the p53 pathway. While *p21* mutations have not been identified in human tumors, perhaps a different cohort needs to be examined. *p21* mutations may contribute to a tumor phenotype in older patients in whom mutations with a smaller impact have accumulated. Finally, other mutations such as the GOF mutations discussed above would require even fewer changes. Splicing mutations may fall into either category depending on whether or not a truncated protein is made. Even a truncated p53 protein may have some GOF activities.

It is important to note that various cancers inactivate the p53 pathway by different mechanisms. For example, *TP53* missense mutations are the most common defect in lung tumors (50 %) (Toledo and Wahl 2006), whereas *Mdm2* and *Mdm4* amplification are prominent in osteosarcomas (30 %) and retinoblastomas (65 %), respectively (Laurie et al. 2006; Toledo and Wahl 2006). Moreover, in retinoblastomas, *TP53* mutations are rare, indicating that a single event such as amplification of *Mdm4* may be sufficient to inhibit p53 activity and allow tumor progression. Why a specific mechanism of inactivating the p53 pathway is preferred in a particular tumor type over another is not yet understood. Transgenic mouse models that overexpress *Mdm2* or *Mdm4* have been generated and may begin to address these questions (Jones et al. 1998; Xiong et al. 2010). These transgenic mice are tumor-prone and the tumors that develop do not necessarily delete or mutate the endogenous *p53* locus.

8 Restoration of p53 in Tumorigenesis

Support for p53 restoration as a tumor suppressive mechanism came from reactivation of wild-type p53 in mouse tumor models. Embryonic liver progenitor cells (hepatoblasts) infected with a retrovirus expressing the oncogenic *HrasV12* and a *p53* shRNA generated hepatocarcinomas in athymic nude mice (Xue et al. 2007). In this model, inhibition of production of the *p53* shRNA by doxycycline allowed p53 reactivation which led to tumor regression via a cellular senescence program. Additionally, an innate immune response contributed to tumor regression in this model. In another study, regression of lymphomas and sarcomas that developed in *p53*-null mice occurred upon reactivation of p53 (Ventura et al. 2007). Mechanistically, tumor cells from the lymphomas regressed by initiating apoptosis while sarcoma cells initiated numerous markers of senescence and were unable to divide. Thus, this in vivo evidence for tumor regression upon restoration of wild-type p53 provides support for a functional consequence of reactivating p53 in human cancers.

Reactivation of p53 was also performed in the *Eμ-myc* model using the $p53ER^{TAM}$ knock-in allele that is functionally null but susceptible to p53 reactivation by the addition of 4-hydroxytamoxifen (4-OHT) (Martins et al. 2006). During tumor progression (without 4-OHT), all *Eμ-myc* $p53ER^{TAM/+}$ mice either delete or functionally inactivate (via point mutations in the DBD) the wild-type *p53* gene. When p53 reactivation was examined in *Eμ-myc* $p53ER^{TAM/-}$ mice, the administration of 4-OHT enhances survival with a delay in lymphomagenesis. Upon reactivation of p53, apoptosis is observed in all tumors. However, in this model, resistance emerges as a result of loss of the $p53ER^{TAM}$ allele or the positive regulator of p53, $p19^{Arf}$ (an inhibitor of Mdm2). Transplants of primary lymphomas with an intact $p53ER^{TAM}$ gene lacking $p19^{Arf}$ show a delay of tumor progression upon p53 restoration in combination with γ-radiation which stabilizes the p53 protein. Thus, for effective p53-dependent tumor suppression, persistent p53-activating signals are required. In these experiments, inappropriate Myc activity was the initial signal via $p19^{Arf}$ activation, but upon loss of $p19^{Arf}$, γ-radiation restored that signal. These data demonstrate the value of p53 in cancer therapy and emphasize the importance of combination therapy to target p53 by multiple ways to maximize its tumor suppressive potential for an effective cure.

These examples of p53 restoration are based on p53 reactivation in tumors lacking *p53*, and disregard the fact that most cancers inactivate p53 by missense mutations. As discussed above, mutant p53 has DN effects on wild-type p53 activity and also exhibits GOF phenotypes. Another murine model that inherits a latent *p53* allele was developed to examine the effects of p53 restoration in tumors with mutant p53 (Wang et al. 2011). In agreement with the data described above, this model also showed tumor regression after p53 restoration in the absence of *p53*. However, in the presence of the *p53*^{R172H} allele, p53 reactivation led to tumor stasis not tumor regression. In these mutant *p53* mice, the transcriptional activity of the restored p53 was less than that in *p53*-null mice suggesting a DN effect of mutant p53 on wild-type p53. Additionally, senescence was the most common mechanism of p53 inhibition. Senescence and apoptosis were observed in lymphomas and sarcomas from *p53* mutant mice upon p53 restoration, but only senescence was detected in angiosarcomas. These data highlight the importance of the ratio of wild-type to mutant p53 in influencing tumor suppression. The delayed tumor growth observed in *p53*^{R172H} mice after wild-type p53 reactivation indicates the exciting therapeutic potential that p53 restoration in human tumors presents in its frequent p53 mutational setting. The genetic manipulation of *p53* in these models and the outcomes with regard to tumor response indicates that animal models are ready for clinical testing of drugs that reactivate mutant p53 or inhibit interactions with the p53 inhibitors, Mdm2 and Mdm4.

9 Conclusion

In conclusion, while several mechanisms inactivate the p53 pathway in human tumorigenesis, by far the most common mechanism is via *TP53* missense mutations, and many of these encode proteins that exhibit GOF and DN activities. Depending on the type of *TP53* mutation and other tumor specific events, p53 mutants exert different activities. Therefore, loss of the wild-type *TP53* allele is only necessary in a subset of tumors. Other alterations such as overexpression of the p53 inhibitors Mdm2 and Mdm4 also contribute to tumor growth and such changes preclude loss of wild-type *TP53*. To test novel chemotherapies targeting the p53 pathway, it is critical to generate mouse models that faithfully mimic the kinds of mutations observed in human cancers. To this end, the *p53* missense mutant mouse models described in this review are invaluable models to test the efficacy of therapeutic drugs.

Numerous factors impinge on the p53 pathway regulating its stability and tumor suppressor capacity. The mechanisms leading to the stability of wild-type p53 in tumors must be explored in greater detail as they also affect mutant p53 stability. High levels of mutant p53 reveal GOF and DN activities promoting tumor development. Thus, it is of great importance to differentiate between wild-type and mutant p53 stability in therapy, as ultimately, it is necessary to activate wild-type and not mutant p53. Unraveling these factors will evidently contribute to the success of therapeutic strategies targeting p53. Lastly and excitingly, tumor regression upon p53 restoration occurs in murine tumors lacking *p53*, whereas in mutant *p53* malignancies, this effect promotes tumor stasis. Additional studies to determine the tissue specific differences will be important along with single cell knock-outs to best model somatic events in mutant p53-driven tumorigenesis. The timing of mutations and the sequence of events must also be examined in the future. This is an exciting era for human tumor biology and for mouse models of cancer.

References

Appella E, Anderson CW (2001) Post-translational modifications and activation of p53 by genotoxic stresses. Eur J Biochem 268:2764–2772

Barboza JA, Liu G, Ju Z et al (2006) p21 delays tumor onset by preservation of chromosomal stability. Proc Natl Acad Sci USA 103:19842–19847

Brachmann RK, Vidal M, Boeke JD (1996) Dominant-negative p53 mutations selected in yeast hit cancer hot spots. Proc Natl Acad Sci USA 93:4091–4095

Brosh R, Rotter V (2009) When mutants gain new powers: news from the mutant p53 field. Nat Rev Cancer 9:701–713

Bullock AN, Henckel J, Fersht AR (2000) Quantitative analysis of residual folding and DNA binding in mutant p53 core domain: definition of mutant states for rescue in cancer therapy. Oncogene 19:1245–1256

Cadwell C, Zambetti GP (2001) The effects of wild-type p53 tumor suppressor activity and mutant p53 gain-of-function on cell growth. Gene 277:15–30

Caulin C, Nguyen T, Lang GA et al (2007) An inducible mouse model for skin cancer reveals distinct roles for gain- and loss-of-function p53 mutations. J Clin Invest 117:1893–1901

Chuikov S, Kurash JK, Wilson JR et al (2004) Regulation of p53 activity through lysine methylation. Nature 432:353–360

de Vries A, Flores ER, Miranda B et al (2002) Targeted point mutations of p53 lead to dominant-negative inhibition of wild-type p53 function. Proc Natl Acad Sci USA 99:2948–2953

Dearth LR, Qian H, Wang T et al (2007) Inactive full-length p53 mutants lacking dominant wild-type p53 inhibition highlight loss of heterozygosity as an important aspect of p53 status in human cancers. Carcinogenesis 28:289–298

Deppert W, Gohler T, Koga H et al (2000) Mutant p53: "gain of function" through perturbation of nuclear structure and function? J Cell Biochem Suppl Suppl 35:115–122

Deyoung MP, Ellisen LW (2007) p63 and p73 in human cancer: defining the network. Oncogene 26:5169–5183

Di Como CJ, Gaiddon C, Prives C (1999) p73 function is inhibited by tumor-derived p53 mutants in mammalian cells. Mol Cell Biol 19:1438–1449

Donehower LA (2009) Using mice to examine p53 functions in cancer, aging, and longevity. Cold Spring Harb Perspect Biol 1:a001081

Donehower LA, Lozano G (2009) 20 years studying p53 functions in genetically engineered mice. Nat Rev Cancer 9:831–841

Donehower LA, Harvey M, Vogel H et al (1995) Effects of genetic background on tumorigenesis in p53-deficient mice. Mol Carcinog 14:16–22

Donehower LA, Harvey M, Slagle BL et al (1992) Mice deficient for p53 are developmentally normal but susceptible to spontaneous tumours. Nature 356:215–221

Eliyahu D, Goldfinger N, Pinhasi-Kimhi O et al (1988) Meth A fibrosarcoma cells express two transforming mutant p53 species. Oncogene 3:313–321

Epstein CB, Attiyeh EF, Hobson DA et al (1998) p53 mutations isolated in yeast based on loss of transcription factor activity: similarities and differences from p53 mutations detected in human tumors. Oncogene 16:2115–2122

Farmer G, Bargonetti J, Zhu H et al (1992) Wild-type p53 activates transcription in vitro. Nature 358:83–86

Figueiredo BC, Sandrini R, Zambetti GP et al (2006) Penetrance of adrenocortical tumours associated with the germline TP53 R337H mutation. J Med Genet 43:91–96

Finlay CA, Hinds PW, Levine AJ (1989) The p53 proto-oncogene can act as a suppressor of transformation. Cell 57:1083–1093

Flores ER, Sengupta S, Miller JB et al (2005) Tumor predisposition in mice mutant for p63 and p73: evidence for broader tumor suppressor functions for the p53 family. Cancer Cell 7:363–373

Gaiddon C, Lokshin M, Ahn J et al (2001) A subset of tumor-derived mutant forms of p53 down-regulate p63 and p73 through a direct interaction with the p53 core domain. Mol Cell Biol 21:1874–1887

Gu W, Roeder RG (1997) Activation of p53 sequence-specific DNA binding by acetylation of the p53 C-terminal domain. Cell 90:595–606

Hainaut P, Hollstein M (2000) p53 and human cancer: the first ten thousand mutations. Adv Cancer Res 77:81–137

Harvey M, McArthur MJ, Montgomery CA Jr et al (1993) Genetic background alters the spectrum of tumors that develop in p53-deficient mice. FASEB J 7:938–943

Harvey M, Vogel H, Morris D et al (1995) A mutant p53 transgene accelerates tumour development in heterozygous but not nullizygous p53-deficient mice. Nat Genet 9:305–311

Haupt Y, Maya R, Kazaz A et al (1997) Mdm2 promotes the rapid degradation of p53. Nature 387:296–299

Honda R, Tanaka H, Yasuda H (1997) Oncoprotein MDM2 is a ubiquitin ligase E3 for tumor suppressor p53. FEBS Lett 420:25–27

Hwang SJ, Cheng LS, Lozano G et al (2003a) Lung cancer risk in germline p53 mutation carriers: association between an inherited cancer predisposition, cigarette smoking, and cancer risk. Hum Genet 113:238–243

Hwang SJ, Lozano G, Amos CI et al (2003b) Germline p53 mutations in a cohort with childhood sarcoma: sex differences in cancer risk. Am J Hum Genet 72:975–983

Jacks T, Remington L, Williams BO et al (1994) Tumor spectrum analysis in p53-mutant mice. Curr Biol 4:1–7

Jackson MW, Berberich SJ (2000) MdmX protects p53 from Mdm2-mediated degradation. Mol Cell Biol 20:1001–1007

Jeffrey PD, Gorina S, Pavletich NP (1995) Crystal structure of the tetramerization domain of the p53 tumor suppressor at 1.7 angstroms. Science 267:1498–1502

Jones SN, Roe AE, Donehower LA et al (1995) Rescue of embryonic lethality in Mdm2-deficient mice by absence of p53. Nature 378:206–208

Jones SN, Hancock AR, Vogel H et al (1998) Overexpression of Mdm2 in mice reveals a p53-independent role for Mdm2 in tumorigenesis. Proc Natl Acad Sci USA 95:15608–15612

Kato S, Han SY, Liu W et al (2003) Understanding the function-structure and function-mutation relationships of p53 tumor suppressor protein by high-resolution missense mutation analysis. Proc Natl Acad Sci USA 100:8424–8429

Kern SE, Kinzler KW, Baker SJ et al (1991) Mutant p53 proteins bind DNA abnormally in vitro. Oncogene 6:131–136

Kern SE, Pietenpol JA, Thiagalingam S et al (1992) Oncogenic forms of p53 inhibit p53-regulated gene expression. Science 256:827–830

Koch JG, Gu X, Han Y et al (2007) Mammary tumor modifiers in BALB/cJ mice heterozygous for p53. Mamm Genome 18:300–309

Knudson AG Jr (1971) Mutation and cancer: statistical study of retinoblastoma. Proc Natl Acad Sci USA 68:820–823

Kraiss S, Quaiser A, Oren M et al (1988) Oligomerization of oncoprotein p53. J Virol 62:4737–4744

Kubbutat MH, Jones SN, Vousden KH (1997) Regulation of p53 stability by Mdm2. Nature 387:299–303

Kuperwasser C, Hurlbut GD, Kittrell FS et al (2000) Development of spontaneous mammary tumors in BALB/c p53 heterozygous mice. A model for Li-Fraumeni syndrome. Am J Pathol 157:2151–2159

Lang GA, Iwakuma T, Suh YA et al (2004) Gain of function of a p53 hot spot mutation in a mouse model of Li-Fraumeni syndrome. Cell 119:861–872

Laurie NA, Donovan SL, Shih CS et al (2006) Inactivation of the p53 pathway in retinoblastoma. Nature 444:61–66

Lavigueur A, Maltby V, Mock D et al (1989) High incidence of lung, bone, and lymphoid tumors in transgenic mice overexpressing mutant alleles of the p53 oncogene. Mol Cell Biol 9: 3982–3991

Levine AJ, Oren M (2009) The first 30 years of p53: growing ever more complex. Nat Rev Cancer 9:749–758

Li FP, Fraumeni JF Jr (1969) Soft-tissue sarcomas, breast cancer, and other neoplasms. A familial syndrome? Ann Intern Med 71:747–752
Li Y, Prives C (2007) Are interactions with p63 and p73 involved in mutant p53 gain of oncogenic function? Oncogene 26:2220–2225
Liu G, McDonnell TJ, de Oca M, Luna R et al (2000) High metastatic potential in mice inheriting a targeted p53 missense mutation. Proc Natl Acad Sci USA 97:4174–4179
Liu G, Parant JM, Lang G et al (2004) Chromosome stability, in the absence of apoptosis, is critical for suppression of tumorigenesis in Trp53 mutant mice. Nat Genet 36:63–68
Liu DP, Song H, Xu Y (2010) A common gain of function of p53 cancer mutants in inducing genetic instability. Oncogene 29:949–956
Lozano G (2007) The oncogenic roles of p53 mutants in mouse models. Curr Opin Genet Dev 17:66–70
Lozano G, Liu G (1998) Mouse models dissect the role of p53 in cancer and development. Semin Cancer Biol 8:337–344
Luo JL, Yang Q, Tong WM et al (2001) Knock-in mice with a chimeric human/murine p53 gene develop normally and show wild-type p53 responses to DNA damaging agents: a new biomedical research tool. Oncogene 20:320–328
Malkin D, Li FP, Strong LC et al (1990) Germ line p53 mutations in a familial syndrome of breast cancer, sarcomas, and other neoplasms. Science 250:1233–1238
Mardis ER (2011) A decade's perspective on DNA sequencing technology. Nature 470:198–203
Martins CP, Brown-Swigart L, Evan GI (2006) Modeling the therapeutic efficacy of p53 restoration in tumors. Cell 127:1323–1334
Meek DW (1994) Post-translational modification of p53. Semin Cancer Biol 5:203–210
Meyerson M, Gabriel S, Getz G (2010) Advances in understanding cancer genomes through second-generation sequencing. Nat Rev Genet 11:685–696
Milner J, Medcalf EA (1991) Cotranslation of activated mutant p53 with wild type drives the wild-type p53 protein into the mutant conformation. Cell 65:765–774
Momand J, Zambetti GP, Olson DC et al (1992) The mdm-2 oncogene product forms a complex with the p53 protein and inhibits p53-mediated transactivation. Cell 69:1237–1245
Montes de Oca Luna R, Wagner DS, Lozano G (1995) Rescue of early embryonic lethality in mdm2-deficient mice by deletion of p53. Nature 378:203–206
Morton JP, Timpson P, Karim SA et al (2010) Mutant p53 drives metastasis and overcomes growth arrest/senescence in pancreatic cancer. Proc Natl Acad Sci USA 107:246–251
Muller PA, Caswell PT, Doyle B et al (2009) Mutant p53 drives invasion by promoting integrin recycling. Cell 139:1327–1341
Olive KP, Tuveson DA, Ruhe ZC et al (2004) Mutant p53 gain of function in two mouse models of Li-Fraumeni syndrome. Cell 119:847–860
Petitjean A, Mathe E, Kato S et al (2007) Impact of mutant p53 functional properties on TP53 mutation patterns and tumor phenotype: lessons from recent developments in the IARC TP53 database. Hum Mutat 28:622–629
Pfeifer GP, Hainaut P (2011) Next-generation sequencing: emerging lessons on the origins of human cancer. Curr Opin Oncol 23:62–68
Poyurovsky MV, Prives C (2006) Unleashing the power of p53: lessons from mice and men. Genes Dev 20:125–131
Prives C, White E (2008) Does control of mutant p53 by Mdm2 complicate cancer therapy? Genes Dev 22:1259–1264
Resnick MA, Inga A (2003) Functional mutants of the sequence-specific transcription factor p53 and implications for master genes of diversity. Proc Natl Acad Sci USA 100:9934–9939
Riley T, Sontag E, Chen P et al (2008) Transcriptional control of human p53-regulated genes. Nat Rev Mol Cell Biol 9:402–412
Roemer K (1999) Mutant p53: gain-of-function oncoproteins and wild-type p53 inactivators. Biol Chem 380:879–887
Rovinski B, Benchimol S (1988) Immortalization of rat embryo fibroblasts by the cellular p53 oncogene. Oncogene 2:445–452

Rowan S, Ludwig RL, Haupt Y et al (1996) Specific loss of apoptotic but not cell-cycle arrest function in a human tumor derived p53 mutant. EMBO J 15:827–838
Sakaguchi K, Herrera JE, Saito S et al (1998) DNA damage activates p53 through a phosphorylation-acetylation cascade. Genes Dev 12:2831–2841
Sharp DA, Kratowicz SA, Sank MJ et al (1999) Stabilization of the MDM2 oncoprotein by interaction with the structurally related MDMX protein. J Biol Chem 274:38189–38196
Sherr CJ, McCormick F (2002) The RB and p53 pathways in cancer. Cancer Cell 2:103–112
Shvarts A, Steegenga WT, Riteco N et al (1996) MDMX: a novel p53-binding protein with some functional properties of MDM2. EMBO J 15:5349–5357
Shvarts A, Bazuine M, Dekker P et al (1997) Isolation and identification of the human homolog of a new p53-binding protein, Mdmx. Genomics 43:34–42
Song H, Hollstein M, Xu Y (2007) p53 gain-of-function cancer mutants induce genetic instability by inactivating ATM. Nat Cell Biol 9:573–580
Stad R, Little NA, Xirodimas DP et al (2001) Mdmx stabilizes p53 and Mdm2 via two distinct mechanisms. EMBO Rep 2:1029–1034
Strano S, Fontemaggi G, Costanzo A et al (2002) Physical interaction with human tumor-derived p53 mutants inhibits p63 activities. J Biol Chem 277:18817–18826
Terzian T, Suh YA, Iwakuma T et al (2008) The inherent instability of mutant p53 is alleviated by Mdm2 or p16INK4a loss. Genes Dev 22:1337–1344
Toledo F, Wahl GM (2006) Regulating the p53 pathway: in vitro hypotheses, in vivo veritas. Nat Rev Cancer 6:909–923
Tsuda H, Hirohashi S (1994) Association among p53 gene mutation, nuclear accumulation of the p53 protein and aggressive phenotypes in breast cancer. Int J Cancer 57:498–503
van Oijen MG, Slootweg PJ (2000) Gain-of-function mutations in the tumor suppressor gene p53. Clin Cancer Res 6:2138–2145
Varley JM (2003) Germline TP53 mutations and Li-Fraumeni syndrome. Hum Mutat 21:313–320
Varley JM, McGown G, Thorncroft M et al (1997a) Germ-line mutations of TP53 in Li-Fraumeni families: an extended study of 39 families. Cancer Res 57:3245–3252
Varley JM, Thorncroft M, McGown G et al (1997b) A detailed study of loss of heterozygosity on chromosome 17 in tumours from Li-Fraumeni patients carrying a mutation to the TP53 gene. Oncogene 14:865–871
Varley JM, McGown G, Thorncroft M et al (1999) Are there low-penetrance TP53 Alleles? Evidence from childhood adrenocortical tumors. Am J Hum Genet 65:995–1006
Venkatachalam S, Shi YP, Jones SN et al (1998) Retention of wild-type p53 in tumors from p53 heterozygous mice: reduction of p53 dosage can promote cancer formation. EMBO J 17:4657–4667
Ventura A, Kirsch DG, McLaughlin ME et al (2007) Restoration of p53 function leads to tumour regression in vivo. Nature 445:661–665
Vogelstein B, Lane D, Levine AJ (2000) Surfing the p53 network. Nature 408:307–310
Wagner J, Portwine C, Rabin K et al (1994) High frequency of germline p53 mutations in childhood adrenocortical cancer. J Natl Cancer Inst 86:1707–1710
Wang Y, Suh YA, Fuller MY et al (2011) Restoring expression of wild-type p53 suppresses tumor growth but does not cause tumor regression in mice with a p53 missense mutation. J Clin Invest 121:893–904
Wu CC, Shete S, Amos CI et al (2006) Joint effects of germ-line p53 mutation and sex on cancer risk in Li-Fraumeni syndrome. Cancer Res 66:8287–8292
Xiong S, Pant V, Suh YA et al (2010) Spontaneous tumorigenesis in mice overexpressing the p53-negative regulator Mdm4. Cancer Res 70:7148–7154
Xue W, Zender L, Miething C et al (2007) Senescence and tumour clearance is triggered by p53 restoration in murine liver carcinomas. Nature 445:656–660

Chapter 6
Humanised Mouse Models: Targeting the Murine p53 Locus with Human Sequences

Monica Hollstein and Yang Xu

1 Introduction

Many of the advances in our understanding of cancer come from studies of cells in culture. However, as elegantly demonstrated by Mina Bissell, the behaviour of cancer cells grown as a monolayer in a tissue culture dish can be a very poor representation of their behaviour in an organism (Nelson and Bissell 2006). Consequently, whole animal studies are vital for the study of tumour biology. The animals of choice are usually mice and rats although clever approaches with zebra fish have recently been developed. The study of oncogenes and tumour suppressors has benefited greatly from the use of animal models, and the construction of transgenic mice in particular (Van Dyke and Jacks 2002; Attardi and Donehower 2005), although it is now recognised that manipulation of cancer-related genes in mice does not always succeed fully in recapitulating disease phenotypes observed in humans, despite considerable conservation in evolution of cell cycle regulation, apoptotic and signal transduction pathways, and ageing amongst mammalian species.

Gene-targeting technology has had a dramatic effect on the development of transgenic mouse models to study cancer. Precise modification of a mouse cancer gene, whilst maintaining endogenous regulation and microenvironment, has become increasingly easy and efficient, leading to an explosion of new mouse strains for research. The landmark studies with p53 null mice two decades ago (Donehower et al. 1992; Clarke et al. 1993; Jacks et al. 1994) have now been followed by development of numerous p53 knock-in mice designed to clarify the

M. Hollstein (✉)
Faculty of Medicine and Health, LIGHT Laboratories, University of Leeds,
Clarendon Way, Leeds LS2 9JT, UK

German Cancer Research Center, Heidelberg, Germany
e-mail: m.hollstein@leeds.ac.uk

Y. Xu
Division of Biological Sciences, University of California San Diego, La Jolla, CA, USA

P. Hainaut et al. (eds.), *p53 in the Clinics*, DOI 10.1007/978-1-4614-3676-8_6,

Table 6.1 Selected key residues modulating p53-dependent apoptosis and their flanking sequences

Site of interest (human p53/*TP53*)		Amino acid sequence
Ser46 (PTM site)	Human	asp leu met leu SER pro asp asp
	Mouse	asp leu leu leu pro gln asp val
Codon 72 polymorphism	Human	glu ala ala pro ARG/PRO val ala pro ala
	Mouse	No homologous mouse segment
Lys120 (PTM site)	Human	gly thr ala LYS ser val thr cys
	Mouse	gly thr ala LYS ser val met cys
Lys164 (PTM site)	Human	ala ile tyr LYS gln ser gln
	Mouse	ala ile tyr LYS lys ser gln
Ser269 (PTM site)	Human	gly arg asn SER phe glu val
	Mouse	gly arg asp SER phe glu val

Specific residues and flanking sequences in the polyproline and DNA binding domains of human p53 that modulate p53 control of apoptosis: comparison with aligned murine amino acid sequences. The key residue is in uppercase. Neighbouring residues that differ in mouse and man are underlined. PTM: post-translational modification

role of specific residues and their modifications in p53 function (Donehower and Lozano 2009). Several excellent reviews on the lessons learned from experiments with these mice are available that highlight the ways in which our understanding of the p53 pathway has advanced from this body of work (Kenzelmann Broz and Attardi 2010; Lozano 2010; Chap. 5).

Mouse models in which murine genes have been replaced with their human homologues, either in their unmutated or mutated forms, are more recent arrivals to the field (Freese and Tuveson 2007; Luo et al. 2001; Song et al. 2007b). They offer unique opportunities to probe various aspects of gene regulation and dysfunction in human disease. In this chapter we summarise our work with strains in which the mouse *Trp53* sequences have been replaced with their human counterpart to study the role of p53 gene regulation and p53 mutant dysfunction in an in vivo experimental setting. We also address ways in which primary cells from these mice are being used to learn how normal p53 and its variants behave under different physiological and stress conditions.

In neoplastic development, point mutations occur in *TP53* throughout the coding sequences, although they do cluster at certain hotspots (Olivier et al. 2010; Hainaut and Hollstein 2000). These mutations result either in truncation of the gene product, or more commonly, to amino acid substitutions that interfere with p53 structural stability, DNA binding properties, and interaction with target proteins. Post-translational modification (PTM) of *specific* residues is a key component in the regulation of p53 function (Hollstein and Hainaut 2010). Despite the stunning overall conservation of the p53 gene from mouse to man, some PTM sites or adjacent residues, and some mutation hotspots in the DNA sequence are not identical in mice and humans (Hergenhahn et al. 2003; Table 6.1). A downside in combining components of two species in one biological system is, of course, the potential incompatibility in molecular interactions, and species differences in molecular pathways, so interpretation of findings have to take this into account.

Replacement of the endogenous mouse p53 gene with human *TP53* sequences by knock-in gene targeting provides an experimental model that permits us to examine features of mutation spectra associated with human exposure to specific carcinogenic agents such as tobacco smoke or the plant carcinogen aristolochic acid (Liu et al. 2004, 2005; Reinbold et al. 2008; Nedelko et al. 2009). Signature mutations are clues to the aetiology of human tumours. Secondly, mice with humanised p53 alleles provide us with a unique model system to explore fundamental biological questions that continue to plague the p53 field, such as:

- Can we develop tumour immunotherapies to overcome the effects of human tumour p53 mutants?
- How do the more recently discovered PTMs of residues in the DNA binding domain of p53 affect function in vivo, or in primary cells of different tissues?
- How does sequence context affect PTM and its impact on protein function?
- Can inactivated mutants of *human* p53 be reactivated in vivo (pre-clinical model)?
- How do mutations in other genes influence p53 function (cross breeding experiments)?
- How do human p53 gene polymorphisms affect the p53 pathway in vivo or in primary cells?
- What are the cellular factors that contribute to mutation load in human cancers?

2 Human p53 Knock-in (Hupki) Mouse Strains Harbouring Normal (Wild-Type *TP53*) Sequences

In normal healthy individuals, two distinct forms of the p53 tumour suppressor protein may be present. This is due to variation in a single DNA base at codon 72 (polymorphism), either guanine or cytosine, encoding either arginine or proline. Both proteins are considered normal (wild-type) and are found in people around the world, although one or the other variant prevails, depending on the population and geographical location, leading to various theories on the possible role of selection and climate in the uneven global distribution of the variants (Whibley et al. 2009; Hu et al. 2008; Bond and Levine 2007; Chap. 2). Despite the fact that both variants are, by definition, normal, this does not preclude the possibility that the two proteins have different biological properties, some of which may affect disease susceptibility or outcome, and this has been the subject of intense research over the last two decades.

Two wild-type (prototype) Hupki strains have been generated expressing the two codon72 variants: Hupki arg72 and Hupki pro72 (Table 6.2) (Luo et al. 2001; Reinbold et al. 2008). Both strains were generated by homologous recombination in embryonic stem cells, using a gene-targeting plasmid containing human *TP53*

Table 6.2 Genetically engineered mice with human sequences embedded in the murine p53 locus

Strain[a]	Human *TP53* exons[b]	Codon 72	Mutation	References
Hupki arg72 (Trp53[tm1/Holl])	Exons 4–9	CGC (arg)	(wt)	Luo et al. (2001)
Hupki pro72 (Trp53[tm2/Holl])	Exons 4–9	CCC (pro)	(wt)	Reinbold et al. (2008)
p53 72R/72R	Exon 4	CGC (arg)	(wt)	Zhu et al. (2010)
p53 72P/72P	Exon 4	CCC (pro)	(wt)	Zhu et al. (2010)
HupkiR175H	Exons 4–9	CGC (arg)	Codon 175 tumour mutation	Liu et al. (2010)
HupkiR248W	Exons 4–9	CGC (arg)	Codon 248 tumour mutation	Song et al. (2007b)
HupkiR273H	Exons 4–9	CGC (arg)	Codon 273 tumour mutation	Song et al. (2007b)
HupkiS46A	Exons 4–9	CGC (arg)	Codon 46 PTM site	Feng et al. (2006)
HupkiA138V	Exons 4–9	CGC(pro)	Codon 138 tumour mutation	In progress

Hupki strains except for Hupki A138V (Fig. 6.1b) were constructed by homologous recombination and gene-targeting Hupki plasmids as depicted in Fig. 6.1a

Notes: (1) Johnson and colleagues (Zhu et al. 2010) constructed both transgenic and knock-in mouse models of the human p53 R72P polymorphism: the knock-in strains have a modified mouse p53 allele encoding the human amino acid sequence from residue 33 to 125 (exon 4). The transgenic mice have randomly integrated BAC sequences encompassing human p53 sequences spanning upstream sequences and the entire cds, encoding either arginine or proline at residue 72. Transgenic models generated by integration of human *TP53* sequences by random integration of exogenous DNA are not in the remit of this chapter. (2) Excision of drug selection cassette was in vitro for Hupki arg72, and in vivo (by crossing with cre-expressing strain) for Hupki pro72 and all other Hupki strains

[a]Strain names are acronyms or designations as used by the authors. Strains named according to standardised nomenclature of Jackson Laboratories are given in parentheses

[b]Intervening introns are also present in the Hupki mice (See Fig. 6.1a)

genomic sequences from exon 4 to intron 9, flanked by murine *Trp53*-derived segments encompassing exons 2 and 3 (the 5′ homologous "arm") and exon 10 (the 3′ homologous arm). This strategy is designed to replace mouse p53 sequences with the homologous human sequences, allowing the expression of a complete, chimeric mRNA that is under control of the endogenous promoter (Fig. 6.1a). The Hupki allele will thus code for a protein that is identical to the mouse p53 protein at the N- and C-termini, whereas most of the protein, including the polyproline segment and the core DNA-binding domain corresponds exactly to the p53 amino acid sequence of humans. Retention of murine sequences at the two ends of the tumour suppressor is a key feature of the model, chosen because most of the PTM sites that regulate normal cellular p53 function and degradation are positioned at the termini, thus in the Hupki prototype strains the murine kinases, acetylases, and other effectors that target these sites are presented with protein domains that are entirely familiar (i.e., murine). Indeed, experiments testing the functionality of Hupki p53 have all pointed to appropriate p53 expression, degradation, transcriptional transactivation and repression, induction of apoptosis, senescence, and tumour suppression

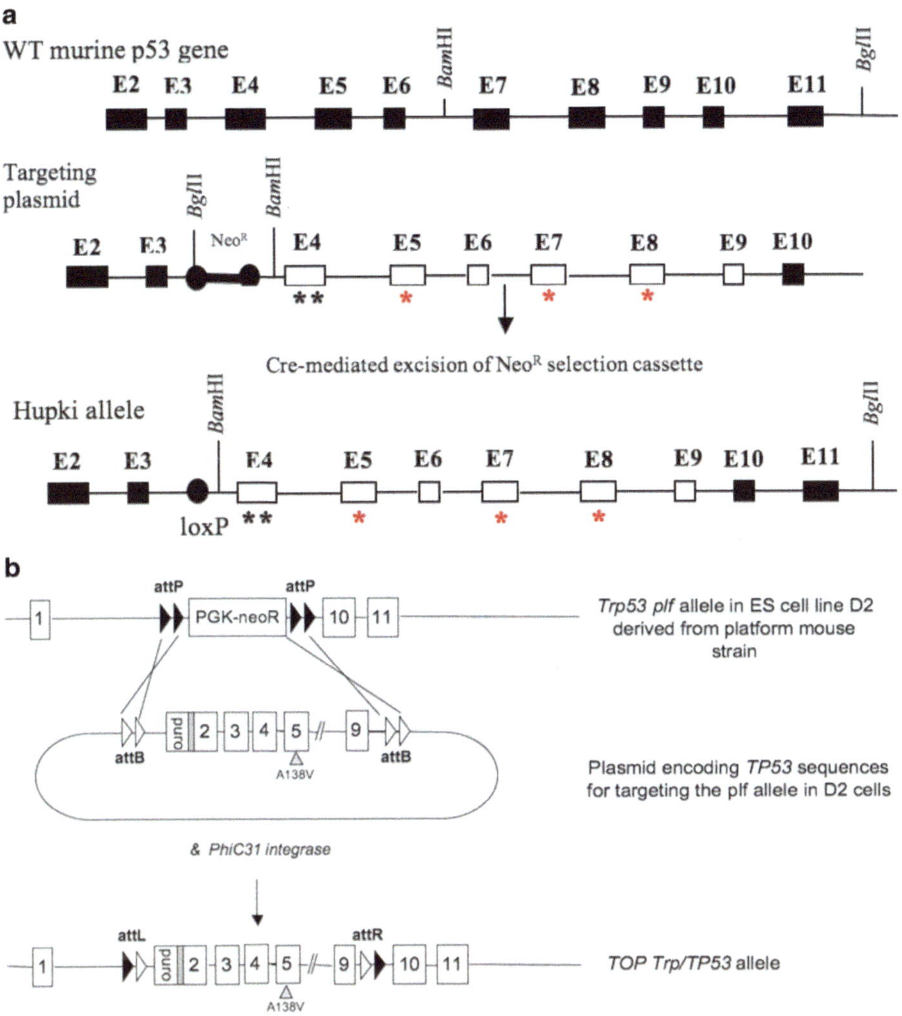

Fig. 6.1 Two approaches for incorporation of human p53 sequences at the murine p53 locus. (**a**) Homologous recombination to target the murine p53 gene. *Asterisks* show locations of mutations introduced to generate variant strains shown in Table 6.2. *Clear boxes*: human *TP53* exons. *Filled boxes*: murine *Trp53* exons (not to scale). (**b**) Site-specific recombination: integrase-mediated cassette exchange in the p53 (null) Platform mouse (see text and Wei et al. 2011)

(Luo et al. 2001; Song et al. 2007b; Frank et al. 2011; Whibley et al. 2010; unpublished observations). In contrast, a transgenic strain with randomly integrated DNA encompassing tandem copies of the full human p53 genomic sequence was reported to be defective in p53 function due to inappropriate interaction of the human p53 N-terminal sequence with the suppressor's major negative regulator, the ubiquitin ligase MDM2, which binds to the 5′ transactivation domain of p53 (Dudgeon et al. 2006). While this result underscored the prudence in keeping the N-terminal of murine p53 in humanisation of the gene in Hupki mice, a more recent report re-opens the possibility that full-length human p53 can be regulated sufficiently accurately in mice to retain major wild-type p53 properties in vivo (Zhu et al. 2010). In this new report, a mouse strain was constructed in which a large DNA segment encompassing all of the human p53 gene and upstream regulatory sequences was integrated randomly into the murine genome. A Hupki strain expressing all coding exons and intervening introns of *TP53* identical to the human gene has not been constructed to probe this issue further, but could be produced relatively easily using the new platform targeting strategy discussed below (Sect. 5).

The Hupki wild-type prototype strains have been applied to questions regarding the origins of human tumour mutations (Olivier et al. 2010; Pfeifer and Besaratinia 2009), the effect of mutation on p53 control of senescence (Whibley et al. 2010), the role of the codon72 polymorphism in NFkB signalling and susceptibility to inflammation (Frank et al. 2011), and immunotolerance to p53 vaccines (Song et al. 2007a; Ishizaki et al. 2010). Studies with Hupki cells to test hypotheses on the factors that shape mutation patterns in tumours of discrete patient groups have been reviewed in detail recently (Pfeifer and Besaratinia 2009; vom Brocke et al. 2006) and will not be discussed in detail here. The major lessons learned from this work are these:

1. Senescence bypass of primary MEFs occurs when *TP53* mutations corresponding to those frequently arising in human tumours arise in cells from Hupki mice, and this leads to outgrowth of immortalised cell lines with mutant *TP53*. This observation emphasises the importance of a normally functioning p53 in signalling cells to senesce when unduly stressed (Hahn and Weinberg 2002), and adds to the accumulated evidence that the (unmutated) Hupki p53 protein has retained wild-type p53 activity.
2. Highly mutagenic human carcinogens that certain patient cohorts have been exposed to induce a pattern of *TP53* mutations in laboratory-exposed Hupki cells that have the defining characteristics of mutations observed in the human tumours from those patients. The normal mouse *Trp53* DNA sequence is by definition essentially unequal to this task because major human tumour mutation hotspot signatures can occur at DNA sequences that are different in mouse and man.
3. Mutations selected for during senescence bypass, whether in Hupki MEFs or in MEFs from genetically unmanipulated laboratory mice present a spontaneous mutation pattern that is unexpected. G to C hotspot mutations stand out, whereas the G to A transitions that predominate across the board in spontaneous mutation spectra of various genes of numerous organisms are conspicuously rare (Whibley et al. 2010). It is unclear whether this is due to the fact that the cells are embryonic, and/or that the hotspot mutants arising spontaneously are selected for because of an exceptional ability to bypass senescence, and/or that they are induced in vitro by high oxygen levels that standard tissue culture protocols employ.

The opportunity to examine the distinctions in biological properties/impact of the codon 72 polymorphism with the Hupki model is a comparatively recent development because construction of the Hupki pro72 strain was completed in 2005 and first reported in 2008 (Reinbold et al. 2008). Attempts to understand the biological implications of the two most common forms of the human p53 protein have been frustrated by the daunting genetic heterogeneity of outbred populations, and the unavoidably anecdotal nature of in vitro experiments using a few widely used human tumour cell lines transfected with constructs expressing the different polymorphic variants. Interestingly, the polymorphism is embedded in the proline-rich domain of p53, which is uncharacteristically ill-conserved in evolution compared to the DNA binding domain (DBD). Although epidemiological studies and experimental studies with human cell lines to assess the importance of the polymorphism have been of unquestionable value in exploring this issue, investigations with inbred, gene-targeted animal strains have the potential to chart new territory due to the greatly reduced background genetic spread. A detailed comparison of p53 function in the Hupki pro72 and Hupki arg72 strains led to the discovery that pro72 mice mount a significantly more dramatic response than arg72 mice to the endotoxin LPS, a powerful stimulator of inflammatory processes (Frank et al. 2011). Consistent with this finding, thymocytes from pro72 mice showed enhanced transactivation of pro-inflammatory target genes with NF-kB binding sites. This difference is attributable to distinctions between the two variant proteins per se and not to unexpected differences in expression or stability, because levels of the two p53 variants, whether uninduced or induced by DNA damage (and thus phosphorylated at serine 15 indicating an activated state), were confirmed to be equivalent in the two strains. Another mechanistic clue to the elevated response of pro72 mice to inflammatory challenge was the finding that the NF-kB subunit p65rel binds preferentially to the p53 pro72 variant. This set of experiments provides a new focus for research on the public health impact of this notorious polymorphism. The search for its effects on disease risk and outcome in human populations, which has been fraught with numerous inconclusive or contradictory reports, should intensify in the area of inflammation-associated cancers and other pathologies associated with tissue damage by inflammation. Much of the difficulty in evaluating the literature on the biological importance of the codon72 polymorphism to human health, if any, is attributable to the emerging truth that the activity of p53, its polymorphic variants, and its multiple isoforms can be tissue and cell-type specific, highlighting the need for models in which function can be tested experimentally in intact organisms and primary cells derived from them. Studies with tumour cell lines overexpressing p53 of exogenous origin are valuable for generating hypotheses, but require corroboration with other, physiologically relevant approaches.

3 Hupki Strains to Probe the In Vivo Impact of Post-Translational Modifications of the p53 Protein

Regulation of p53 activity is extremely complex. Cellular mechanisms that fine-tune function by influencing intracellular localisation, p53 message or protein levels, degradation, interaction with nucleic acids and proteins, or relative levels of the various p53 isoforms have been discovered (Vousden and Prives 2009; Hollstein and Hainaut 2010). PTMs are considered to be primary factors affecting the ability of p53 to influence expression of a large repertoire of genes, and thereby, to suppress tumourigenesis, provoke senescence, regulate metabolism, stimulate innate immunity, and other protective tasks, some of which appear at the molecular level to be contradictory (Kruse and Gu 2009). The current working model in vogue is one which likens the different possible modifications of p53 specific residues to a bar code that allows the cell to use p53 to orchestrate a particular cellular response to a specific challenge or condition (Murray-Zmijewski et al. 2008). To fully decode this mode of p53 regulation is difficult work because there are dozens of residues known to be either phosphorylated, acetylated, methylated, sumoylated, or ubiquitinated, leading to a host of possible combinations. Added to this diversity is the complication that in vitro studies with cell lines are inadequate reporters of what may occur in different cell types in vivo in non-neoplastic cells (Toledo and Wahl 2006; Wahl 2006). Gene-targeting in mice to block or to constitutively mimic this control axis has provided valuable information on the physiological impact of these modifications. Phosphorylation of serine N-terminal residues and acetylation of C-terminal lysine residues have been the focus of this approach, and required construction of different gene-targeted strains harbouring single amino acid substitutions at key relevant PTM locations of the mouse p53 sequence (Kenzelmann Broz and Attardi 2010). Modification sites that are not identical in mice and humans, however, call for mouse strains with humanised sequences at the *Trp53* locus. In the Hupki mouse model, one variant strain has been constructed thus far to assess the impact in normal (non-neoplastic) cells in vitro and in vivo of an N-terminal phosphorylation site, Ser46 (Feng et al. 2006;Table 6.2). On the basis of in vitro work with human cell lines this modification had been judged to have a pivotal role in apoptotic functions of p53 (Oda et al. 2000). Here, use of the humanised p53 mouse model is appropriate because the PTM site is in a gene segment that is not well conserved in mammalian evolution (Soussi et al. 1990; Table 6.1). Existence of the biologically equivalent serine residue in mouse p53 is difficult to pinpoint.

The relatively modest impact of serine 46 phosphorylation on DNA damage-triggered apoptosis and stress-induced cellular senescence, assessed by comparing Hupki WT and Hupki Ser46A responses, support data emerging on other putatively key modification sites acquired with mouse strains harbouring gene-targeted murine p53: the profound effects seen in cell lines from manipulating individual modification sites are much less catastrophic in an in vivo setting (Toledo and Wahl 2006). Numerous studies argue for a model in which there are redundancies in modification sites, although this does not preclude the possibility that there are a few key modifications that individually exercise more control over p53 activity than others, such as phosphorylation of Ser315, which modulates p53 activity in Hupki embryonic stem cells undergoing differentiation (Lin et al. 2005). Also, there are three sites in the DNA-binding domain, Lys120, Lys164, and Ser269, that may have substantial effects on p53 function individually (Sykes et al. 2006; Tang et al. 2008; Fraser et al. 2010; Table 6.1), and are currently slated for study in the Hupki mouse model. The residues themselves are conserved across mammalian species, but the amino acid sequences nearby are not entirely the same in mouse and man (Table 6.1). The apparently minor differences may or may not be relevant, but are worth considering because the core domain structure of p53 is known to be extremely sensitive to single residue substitutions, and to specific combinations of two residue substitutions (Baroni et al. 2004; Tang et al. 2008; Otsuka et al. 2007).

4 Hupki Strains Harbouring Human Tumour Mutations (Fig. 6.1a; Table 6.2)

Perhaps the most striking feature of p53 tumour biology, and certainly an area that is still receiving great attention, is the significant proportion of cancer-associated dysfunctional mutations that cause high cellular levels of a mutant protein that has acquired oncogenic properties, whilst shedding wild-type suppressor activities (Brosh and Rotter 2009; Goh et al. 2010). Although some human tumours have simply lost p53 wild-type function by virtue of nonsense mutations and/or allelic loss, the vast majority express nefarious p53 variants (gain-of-function mutants) with novel activities that are not accessible for study with the p53 null mouse models to which admittedly we owe so many first insights on p53 and cancer. The oncogenic functions of mutant p53 have been examined in transgenic mice with randomly inserted, exogenously derived DNA, and more recently, in gene-targeted strains that express amino acid substitutions of residue 172 or 270 in the *mouse* p53 gene (corresponding to human p53 residue 175 or 273) (discussed in Chap. 5). These knock-in strains have provided strong evidence for an impact of common tumour mutant p53s on metastasis, and demonstrated in vivo that different mutants can have distinct biological activities, emphasising the need to examine different mutants in order to gain a more complete picture (Lang et al 2004; Olive et al 2004).

With the Hupki targeting strategy (Fig. 6.1a), three strains of mutant mice have been constructed modelling three different common human tumour mutants: R175H, R248W, and R273H (Table 6.2). These mutant knock-in alleles all encode arginine at the human polymorphic site at codon72. The most salient finding from experiments with these mice is the deleterious effect all three mutant proteins have on genomic stability in vivo. Interchromosomal translocations are exceedingly rare in normal cells, but surprisingly, given the guardian of the genome function of wild-type p53, they are also undetectable in p53 null cells. In contrast, these aberrant events were consistently observed in thymocytes from Hupki mice expressing mutant p53, a finding possibly attributable to the interference of these mutant p53 molecules with the Mre11/ATM DNA damage response pathway Song et al. (2007b; Liu et al. 2010). Intriguingly, as first observed in experiments with the knock-in mouse p53 R172H (corresponding to human R175H), mutant p53 protein does not accumulate in vivo in normal cells of the Hupki strains; overexpression appears to require a neoplastic cellular environment. The molecular basis for this observation is unclear, but is important to pursue, because new ways to cancel p53 oncogenic gain of function in tumours may then become evident. One hypothesis is that accumulation of DNA damage/oxidative stresses in cancer cells activates DNA damage response pathways, leading to hyperphosphorylation of p53 cancer mutants and their stabilisation in cancer cells (Xu 2008).

5 P53 (Null) Platform Mouse Strain to Generate P53 Knock-in Strains and Cell Lines

Given that human tumour p53 mutants fall into different structural classes and have distinct , although as yet largely uncharacterised biological effects that can be cell and tissue-type specific, there is need for additional mouse strains modelling cancer mutations. It is an arduous task, however, to generate each new p53 knock-in strain by standard gene-targeting strategies involving the mammalian cellular recombination machinery and long DNA segments of incoming targeting vectors that are homologous to sequences of the mouse endogenous p53 locus. In contrast, integrase-mediated cassette exchange (IMCE) is a powerful method for achieving reiterative gene targeting at a given locus once it has been furnished with recombinase-specific recognition sequences commandeered from lower organisms (Groth et al. 2000; Belteki et al. 2003; Chalberg et al. 2006; Capecchi 2005; Glaser et al. 2005; Raymond and Soriano 2007; Toledo et al. 2006). Such an approach can be applied to the efficient generation of a series of genetically matched Hupki mutant cell lines or mouse strains, each harbouring a different *TP53* mutation found in human tumours. This strategy requires three steps: (1) generation (by standard homologous recombination-based technology) of a mouse with a docking platform at the murine p53 locus containing integrase recognition sequences; (2) establishment of a pluripotent embryonic stem (ES) cell line from the new mouse; (3) DNA swap between the new ES cell line and an incoming vector with human p53 sequences flanked by integrase recognition sequences (Fig. 6.1b). We recently created such a mouse strain which we call the p53 Platform strain (PLF mouse) that has PhiC31 integrase-specific attP sequences flanking a PGK-neo cassette at the murine p53 locus and replacing exons 2 to 9 (Wei et al. 2011). ES cell lines generated from the mice re-express a humanised p53 allele upon co-electroporation of a targeting vector with Hupki-derived p53 sequences and an integrase-expressing plasmid. Recovery of correctly targeted ES clones is very efficient; they can be selected for with puromycin because a TOP (targeting on platform) p53 plasmid set designed for IMCE has a puromycin resistance gene ORF fused to viral Thosea asigna 2A sequences just upstream of the ATG start codon of *Trp53* exon 2. ORFs bridged by the viral segment are co-expressed as a single mRNA molecule, but the two proteins are not fused because of the ribosome-skipping effect of the intervening sequence (Tang et al. 2009). We are currently using IMCE clones, derived from the platform ES cell line D2 (plf/+) that harbour the temperature-sensitive *TP53* A138V mutation to test whether D2-derived clones are able to generate a new p53 mutant mouse strain following blastocyst injection and mating of chimeric offspring (Hollstein laboratory, unpublished). This extends the platform strategy to in vivo applications, reducing the labour involved in identifying clones with accurately targeted alleles. Approximately one clone in three recovered after nucleofection of D2 cells with the TOP-puro2AHupkiA138V plasmid was positive for IMCE at the plf locus. Even without selection against random insertions, recovery of targeted clones is efficient, raising the hope that IMCE in PLF ES cell lines may be used at some point for directly targeting the mouse genome by injection of TOP plasmids into platform mouse zygotes, which would vastly accelerate the procedure for generating new oncogenic gain-of-function p53 mutant strains.

A current application of the platform mouse model is the rapid production of Hupki PLF-derived MEF cell lines that are nearly identical in the genetic background, yet where each cell line carries a different human tumour mutation. Sets of mutant cell lines can be established with TOP plasmids from a single population of primary PLF embryonic fibroblasts within a few weeks (Wei et al. 2011).

6 Future Perspectives

There is considerable untapped potential for elucidating genetic pathways and molecular interactions by cross breeding Hupki strains with other transgenic strains. The effect of mutations and polymorphisms in various genes that control p53 function, for example variants of the p53 regulator MDM2, could be addressed with this approach. The impact of DNA repair enzymes, or the effect of DNA modifiers such as cytidine deaminases on the pattern of mutations arising in the human p53 coding sequence during conversion of normal cells to an immortal cell in vitro or a tumour cell in vivo, is another area that could be investigated by crossing Hupki strains with mice genetically engineered to overexpress or underexpress the enzymes of interest.

The opportunity to access "normal" primary cells of different tissues, stem cells, and pluripotent embryonic stem cells from the mice provides a host of possibilities for revisiting in vitro experiments but in the wider context of an organism.

Finally, carcinogen-exposed Hupki mice bearing tumours with *TP53* mutations provide a stringent tool to test the efficacy in vivo of novel drugs designed to precisely restructure the aberrant p53 molecules found in human cancers and recover their wild-type properties.

Acknowledgements The Hollstein laboratory acknowledges the financial support from Yorkshire Cancer Research, Cancer Research UK, and core support from the German Cancer Research Center. The Xu laboratory is funded by grants from NIH and California Institute for Regenerative Medicine.

References

Attardi LD, Donehower LA (2005) Probing p53 biological functions through the use of genetically engineered mouse models. Mutat Res 576:4–21

Baroni TE, Wang T, Qian H, Dearth LR, Truong LN, Zeng J et al (2004) A global suppressor motif for p53 cancer mutants. Proc Natl Acad Sci USA 101:4930–4935

Belteki G, Gertsenstein M, Ow DW, Nagy A (2003) Site-specific cassette exchange and germline transmission with mouse ES cells expressing phiC31 integrase. Nat Biotechnol 21:321–324
Bond GL, Levine AJ (2007) A single nucleotide polymorphism in the p53 pathway interacts with gender, environmental stresses and tumour genetics to influence cancer in humans. Oncogene 26:1317–1323
Brosh R, Rotter V (2009) When mutants gain new powers: news from the mutant p53 field. Nat Rev Cancer 9:701–713
Capecchi MR (2005) Gene targeting in mice: functional analysis of the mammalian genome for the twenty-first century. Nat Rev Genet 6:507–512
Chalberg TW, Portlock JL, Olivares EC, Thyagarajan B, Kirby PJ, Hillman RT et al (2006) Integration specificity of phage phiC31 integrase in the human genome. J Mol Biol 357: 28–48
Clarke AR, Purdie CA, Harrison DJ, Morris RG, Bird CC, Hooper ML, Wyllie AH (1993) Thymocyte apoptosis induced by p53-dependent and independent pathways. Nature 362:849–852
Donehower LA, Harvey M, Slagle BL, McArthur MJ, Montgomery CA Jr, Butel JS et al (1992) Mice deficient for p53 are developmentally normal but susceptible to spontaneous tumours. Nature 356:215–221
Donehower LA, Lozano G (2009) 20 years studying p53 functions in genetically engineered mice. Nat Rev Cancer 9:831–841
Dudgeon C, Kek C, Dmeidov ON, Saito S, Fernandes K, Diot A et al (2006) Tumor susceptibility and apoptosis defect in a mouse strain expressing a human p53 transgene. Cancer Res 66: 2928–2936
Feng L, Hollstein M, Xu Y (2006) Ser46 phosphorylation regulates p53-dependent apoptosis and replicative senescence. Cell Cycle 5:2812–2819
Frank A, Leu J, Zho Y, Devarajan K, Nedelko T, Klein-Szanto A, Hollstein M, Murphy ME (2011) The codon 72 polymorphism of p53 regulates interaction with NF-kB and transactivation of genes involved in immunity and inflammation. Mol Cell Biol 31:1201–1213
Fraser JA, Vojtesek B, Hupp TR (2010) A novel p53 phosphorylation site within the MDM2 ubiquitination signal. J Biol Chem 285:37762–37772
Freese KK, Tuveson DA (2007) Maximizing mouse cancer models. Nat Rev Cancer 7:645–658
Glaser S, Anastassiadis K, Stewart AF (2005) Current issues in mouse genome engineering. Nat Genet 37:1187–1193
Goh AM, Coffill CR, Lane DP (2010) The role of mutant p53 in human cancer. J Pathol 223:116–126
Groth AC, Olivares EC, Thyagarajan B, Calos MP (2000) A phage integrase directs efficient site-specific integration in human cells. Proc Natl Acad Sci USA 97:5995–6000
Hahn WC, Weinberg RA (2002) Modelling the molecular circuitry of cancer. Nat Rev Cancer 2:331–341
Hainaut P, Hollstein M (2000) The first ten thousand mutations. Adv Cancer Res 77:81–137
Hergenhahn M, Luo JL, Hollstein M (2003) p53 designer genes for the modern mouse. Cell Cycle 3:738–741
Hollstein M, Hainaut P (2010) Massively regulated genes: the example of *TP53*. J Pathol 220:164–173
Hu W, Feng Z, Atwal GS, Levine AJ (2008) p53: a new player in reproduction. Cell Cycle 7:848–852
Ishizaki H, Song GY, Srivastava T, Carroll KD, Shahabi V, Manuel ER (2010) Heterologous prime/boost immunization with p53-based vaccines combined with toll-like receptor stimulation enhances tumor regression. J Immunother 33:609–617
Jacks T, Remington L, Williams BO, Schmitt EM, Halachmi S, Bronson RT, Weinberg RA (1994) Tumor spectrum analysis in p53-mutant mice. Curr Biol 4:1–7
Kenzelmann Broz D, Attardi LD (2010) In vivo analysis of p53 tumor suppressor function using genetically engineered mouse models. Carcinogenesis 31:1311–1318
Kruse JP, Gu W (2009) Modes of p53 regulation. Cell 137:609–622

Lang GA, Iwakuma T, Suh YA, Liu G, Rao VA, Parant JM et al (2004) Gain of function of a p53 hot spot mutation in a mouse model of Li-Fraumeni syndrome. Cell 119:861–972

Lin T, Chao C, Saito S, Mazur SJ, Murphy ME, Appella E et al (2005) p53 induces differentiation of mouse embryonic stem cells by suppressing Nanog expression. Nat Cell Biol 7:165–171

Liu DP, Song H, Xu Y (2010) A common gain of function of p53 cancer mutants in inducing genetic instability. Oncogene 29:949–956

Liu Z, Hergenhahn M, Schmeiser HH, Wogan GN, Hong A, Hollstein M (2004) Human tumor p53 mutations are selected for in mouse embryonic fibroblasts harboring a humanized p53 gene. Proc Natl Acad Sci USA 101:2963–2968

Liu Z, Muehlbauer KR, Schmeiser HH, Hergenhahn M, Belharazem D, Hollstein M (2005) p53 mutations in benzo(a)pyrene-exposed human p53 knock-in murine fibroblasts correlate with p53 mutations in human lung tumors. Cancer Res 65:2583–2587

Lozano G (2010) Mouse models of p53 functions. Cold Spring Harb Perspect Biol 2:a001115

Luo JL, Yang Q, Tong WM, Hergenhahn M, Wang ZQ, Hollstein M (2001) Knock-in mice with a chimeric human/murine p53 gene develop normally and show wild-type p53 responses to DNA damaging agents: a new biomedical research tool. Oncogene 20:320–328

Murray-Zmijewski F, Slee EA, Lu X (2008) A complex barcode underlies the heterogeneous response of p53 to stress. Nature Rev Mol Cell Biol 9:702–712

Nedelko T, Arlt VM, Phillips DH, Hollstein M (2009) *TP53* mutation signature supports involvement of aristolochic acid in the aetiology of endemic nephropathy-associated tumors. Int J Cancer 124:987–990

Nelson CM, Bissell MJ (2006) Of extracellular matrix, scaffolds, and signalling: tissue architecture regulates development, homeostasis, and cancer. Annu Rev Cell Dev Biol 22:287–309

Oda K, Arakawa H, Tanaka T, Matsuda K, Tanikawa C, Mori T et al (2000) p53AIP1, a potential mediator of p53-dependent apoptosis and its regulation by Ser-46-phosphorylated p53. Cell 102:849–862

Olive KP, Tuveson DA, Ruhe ZC, Yin B, Willis NA, Bronson RT et al (2004) Mutant p53 gain of function in two mouse models of Li-Fraumeni syndrome. Cell 119:847–860

Olivier M, Hollstein M, Hainaut P (2010) *TP53* mutations in human cancers: origins, consequences, and clinical use. Cold Spring Harb Perspect Biol 2(1):a001008

Otsuka K, Kato S, Kakudo Y, Mashiko S, Shibata H, Ishioka C (2007) The screening of the second-site suppressor mutations of the common p53 mutants. Int J Cancer 121:559–561

Pfeifer GP, Besaratinia A (2009) Mutational spectra of human cancer. Hum Genet 125:493–506

Raymond CS, Soriano P (2007) High-efficiency FLP and PhiC31 site-specific recombination in mammalian cells. PLoS One 2:e162

Reinbold M, Luo JL, Nedelko T, Jerchow B, Murphy ME, Whibley C et al (2008) Common tumour p53 mutations in immortalized cells from Hupki mice heterozygous at codon 72. Oncogene 27:2788–2794

Song GY, Gibson G, Haq W, Huang EC, Srivasta T, Hollstein M, Daftarian P, Wang Z, Diamond D, Ellenhorn JDI (2007a) An MVA vaccine overcomes tolerance to human p53 in mice and humans. Cancer Immunol Immunother 56:1193–1205

Song H, Hollstein M, Xu Y (2007b) p53 gain-of-function cancer mutants induce genetic instability by inactivating ATM. Nat Cell Biol 9:573–580

Soussi T, Caron de Fromentel C, May P (1990) Structural aspects of the p53 protein in relation to gene evolution. Oncogene 5:945–952

Sykes SM, Mellert HS, Holbert MA, Li K, Marmorstein R, Lane WS, McMahon SB (2006) Acetylation of the p53 DNA-binding domain regulates apoptosis induction. Mol Cell 24: 841–851

Tang W, Ehrlich I, Wolff SB, Michalski AM, Wolf S, Hasan MT et al (2009) Faithful expression of multiple proteins via 2A-peptide self-processing: a versatile and reliable method for manipulating brain circuits. J Neurosci 29:8621–8629

Tang Y, Zhao W, Chen Y, Zhao Y, Gu W (2008) Acetylation is indispensable for p53 activation. Cell 133:612–626

Toledo F, Wahl GM (2006) Regulating the p53 pathway: in vitro hypotheses, in vivo veritas. Nature Rev Cancer 6:909–923

Toledo F, Liu CW, Lee CJ, Wahl GM (2006) RMCE-ASAP: a gene targeting method for ES and somatic cells to accelerate phenotype analyses. Nucleic Acids Res 34:e92

Van Dyke T, Jacks T (2002) Cancer modelling in the modern era: progress and challenges. Cell 108:135–144

Vom Brocke J, Schmeiser HH, Reinbold M, Hollstein M (2006) MEF immortalization to investigate the ins and outs of mutagenesis. Carcinogenesis 27:2141–2147

Vousden KH, Prives C (2009) Blinded by the light: the growing complexity of p53. Cell 137:413–431

Wahl GM (2006) Mouse bites dogma: how mouse models are changing our views of how P53 is regulated in vivo. Cell Death Differ 13:973–983

Wei QX, Odell AF, van der Hoeven F, Hollstein M (2011) Rapid derivation of genetically related mutants from embryonic cells harbouring a recombinase-specific Trp53 platform. Cell Cycle 10:1261–1270

Whibley C, Pharoah PD, Hollstein M (2009) p53 polymorphisms: cancer implications. Nature Rev Cancer 9:95–107

Whibley C, Odell AF, Nedelko T, Balaburski G, Murphy M, Liu Z et al (2010) Wild-type and Hupki (human p53 knock-in) murine embryonic fibroblasts: p53/ARF pathway disruption in spontaneous escape from senescence. J Biol Chem 285:11326–11335

Xu Y (2008) Induction of genetic instability by gain-of-function p53 cancer mutants. Oncogene 27:3501–3507

Zhu F, Dolle MET, Berton TR, Kuiper RV, Capps C, Espejo A et al (2010) Mouse models for the p53 R72P polymorphism mimic human phenotypes. Cancer Res 70:5851–5859

Chapter 7
p53 Models for Mammary Carcinogenesis

Wolfgang Deppert and Genrich Tolstonog

1 Introduction

With approximately 600,000 new cases per year, breast cancer is the most often diagnosed tumor disease of women in the United States and in Europe. Breast cancer is a very heterogeneous disease comprising various subgroups that differ in histopathological appearance, molecular profile, and clinical response (Weigelt and Reis-Filho 2009). At least part of the heterogeneity of breast cancer is due to its complex biogenesis. The mammalian breast harbors a variety of cell types which develop from progenitors over different stages of differentiation. As it is now commonly accepted that tumors arise from stem and progenitor cells, each of these progenitors might serve as cell of tumor origin (Visvader 2009). Combined with the well-known fact that there are many molecular pathways that drive tumorigenesis, it at first glance appears that understanding breast cancer is an unmanageable task. Fortunately, however, it seems that the development of various breast cancer subtypes follows specific molecular traits, which can be imitated in model systems.

The probably most famous tumor-associated factor is the *TP53* gene. Identified in 1979 and initially dubbed an oncogene, it was classified as a tumor suppressor gene in 1989. By now it is evident that p53, especially its functional inactivation, plays a pivotal role in tumorigenesis. A lot of our knowledge on the functions of p53 as a tumor suppressor has been obtained from mouse models (for reviews

W. Deppert (✉) • G. Tolstonog
Heinrich-Pette-Institute, Leibniz Institute for Experimental Virology,
Martinistr. 52, 20251 Hamburg, Germany
e-mail: wolfgang.deppert@hpi.uni-hamburg.de; genrich.tolstonog@hpi.uni-hamburg.de

P. Hainaut et al. (eds.), *p53 in the Clinics*, DOI 10.1007/978-1-4614-3676-8_7,

see Donehower and Lozano 2009; Lozano 2010). The role of p53 specifically in mammary carcinogenesis as defined by analyses of genetically modified mice has been reviewed by Blackburn and Jerry (2002). As a tumor suppressor, loss of p53 functions accelerates tumorigenesis compared to mice expressing at least one wild-type *Trp53* allele. Moreover, in addition to loss of function, certain mutant p53 molecules exert "gain of function" (GOF) properties, as had already been demonstrated years ago in cell culture (Blandino et al. 1999). Meanwhile a number of molecular pathways have been identified which could explain the oncogenic functions of mutant p53 in vitro (Brosh and Rotter 2009; Oren and Rotter 2010), while the data on mutant p53 GOF in human tumors still are ambiguous (Olivier et al. 2010). Mouse models then were first to demonstrate certain GOF properties of mutant p53 in vivo (Adhikari and Iwakuma 2009; Lozano 2007).

About 20–40 % of breast cancers harbor *TP53* mutations, which mostly are associated with the "basal-like" and "ERBB2+" subtypes, and are a strong prognostic marker for poor prognosis (Langerod et al. 2007). Loss of transactivation activities, and to a lesser extent, dominant-negative activities are the main driving forces that determine *TP53* mutation patterns and influence tumor phenotype (Petitjean et al. 2007). Current data on the acquisition of oncogenic activities (GOF) by p53 mutants in human breast cancers are too scarce and heterogeneous to assess whether this property has an impact on tumor development and prognosis (Olivier et al. 2006).

2 Model Systems (General)

2.1 Cell Culture Studies

Cell lines established from human breast cancer, or transformed mammary epithelial cells (MECs) obtained by in vitro transformation of isolated primary human MECs by various oncogenes (Dimri et al. 2005), constitute the most simple model systems, and are widely used in breast cancer research, either in cell culture or in xenograft studies. While analysis in cell culture allows easy access to study certain biochemical and molecular aspects of breast cancer cells, transfer of the obtained knowledge to breast cancer biology is limited. An important limit for instance is that cells in 2D culture do not exhibit the same properties as cells in a spatial arrangement. While this problem can be partially resolved by culturing such cells under 3D conditions (Weigelt and Bissell 2008), this approach is also limited, because even in 3D culture tumor cells lack the dynamic interactions with the other cells that build up the tumor. Thereby such studies do not take into account that cancers are heterogeneous cellular "ecosystems" whose composition and expansion are dependent upon mutual interactions between epithelial tumor cells in various states of differentiation and their microenvironment (Tlsty and Coussens 2006). Despite these limitations, the heterogeneous molecular and cellular properties of commonly used human breast cancer epithelial cell lines together represent the heterogeneity of human breast cancers, thereby providing suitable models for in vitro studies (Keller et al. 2010).

2.2 *Xenograft Studies*

Xenograft studies, i.e., the transplantation of primary or established breast cancer cells into immune-compromised mice, partially resolve the limitations of cell culture. Such studies are widely used for the identification of the tumor-initiating ("cancer stem cell") compartment within a primary tumor or a tumor cell line, for identifying genes involved, e.g., in metastasis, and, last but not least in preclinical studies. Although xenograft models are an approximation to endogenous human tumor growth, they cannot recapitulate early steps in tumor development. Another caveat in xenograft studies is that the human epithelial tumor compartment has to interact with the mouse microenvironment, inevitably leading to selection not primarily according to their stemness or metastatic potential, but rather according to their ability to grow in a mouse microenvironment (Kelly et al. 2007). To circumvent this problem the human-in-mouse (HIM) model has been developed (Proia and Kuperwasser 2006). The HIM protocol creates functionally normal breast tissues by orthotopic xenograft implantation of co-mixed human epithelial and stromal cells into a cleared mouse mammary fat pad that has been "humanized" by primary fibroblasts. By lentiviral transduction of human breast epithelial organoids prior to transplantation, Wu et al. (2009) were able to successfully reconstitute preneoplastic lesions and tumors that closely mimicked individual steps of tumorigenesis in patients, and reflected the chain of underlying genetic events, i.e., disruption of tumor-suppressive pathways, e.g., the p53 pathway, and enforced expression of oncogenes, e.g., ERBB2/HER2. The HIM protocol thus provides a feasible tool to analyze human breast cancer development, with emphasis on the discovery and validation of new therapies.

3 Role of p53 in Mammary Carcinogenesis as Deduced from Mouse Models

The development of techniques that allow manipulation of the mouse germ line has initiated the development of mouse models for human cancer. Meanwhile, a large number of tumor models have been generated to follow the effects of gain or loss of a single tumor-associated factor or their combinations over time, monitoring the chain of tumorigenic events, from hyperplasia to invasive carcinoma and metastasis. However, it deems fair to say that despite the undisputable advantages of mouse models the problem of cross-validating molecular mouse data with the corresponding human disease has not yet been satisfactorily resolved for many models.

3.1 *p53 in Mammary Gland Development and Its Relation to Tumorigenesis*

For deciphering the role of p53 in mammary carcinogenesis it would be desirable to understand its functions in the normal gland. Evidence is accumulating that p53 regulates self-renewal of stem cells (Zhao and Xu 2010) and differentiation in a cell type specific and cell fate dependent manner (Molchadsky et al. 2010). In mammary glands p53 restricts the pool of epithelial stem/progenitor cells (Tao et al. 2011), indicating an important role for p53 in regulating MEC hierarchy. However, mammary gland development in C57BL/6x129/Sv (Donehower et al. 1992) and BALB/c (Jerry et al. 1998) *Trp53−/−* mice appears to be normal. Transplantation of cells from *Trp53+/+* and *Trp53−/−* mice into cleared fat pads did not reveal morphological differences in the outgrowths, although their number was higher for transplanted *Trp53−/−* cells (Tao et al. 2011). This is consistent with the observation that ductal morphogenesis, which requires a high level of programmed cell death in the terminal end buds, is not affected in *Trp53−/−* mice, thereby suggesting that apoptosis in the early development of the mammary gland is p53 independent (Humphreys et al. 1996). Also lactation was not affected in both *Trp53−/−* mice, although in wild-type mice *Trp53* transcription is strongly upregulated by nuclear factor 1-C2 (Nfic) at midpregnancy, i.e., at a time when expression of milk genes, e.g., *Wap*, is dramatically induced (Johansson et al. 2003), and is expected to be functional. Indeed, removal of pups from the lactating mammary glands induced the p53 target gene *p21/WAF1* within 24–48 h in wild-type, but not in BALB/c *Trp53−/−* mice (Jerry et al. 1998). This correlated with a delayed involution (postlactational tissue remodeling) in BALB/c *Trp53−/−* mice (Jerry et al. 1998), a phenotype that is aggravated by the combined loss of p53 and Stat3 (Matthews and Clarke 2005).

In the mostly quiescent cells of the ductal epithelium of nulliparous mice, p53 is sequestered into the cytoplasm, but translocated into the nucleus and became transcriptionally active when the mice were transiently treated with ovarian steroids to mimic hormonal stimulation that precedes ovulation (Kuperwasser et al. 2000b). In line with this observation, the proliferation and differentiation associated with pregnancy renders the mammary epithelium resistant to the development of tumors, most likely because hormonal stimulation during pregnancy activates p53. Indeed, in p53-null mice the protective effect of hormones against carcinogen-induced mammary carcinogenesis is abrogated, pointing to the necessity of the p53 pathway for tumor protection (Medina and Kittrell 2003). However, if the duration of hormonal stimulation is extended, as e.g., in luteinizing hormone overexpressing transgenic mice, a model for ovarian hyperstimulation (Milliken et al. 2008), the effect of hormones became carcinogenic, as spontaneous mammary tumors appeared despite wild-type p53 expression. In summary, the function of p53 in the normal mammary gland is influenced by steroid hormones in a complex manner which is not yet fully understood.

3.2 p53 Loss as an Initiating Event in Mammary Carcinogenesis

Targeted germ line disruption of the *Trp53* gene results in the absence of p53 in all tissues of the mouse throughout development and in adulthood. Adult *Trp53−/−* mice succumb to cancer death mostly by developing lymphomas at an early age (between 4 and 6 months, depending on mouse strain). Heterozygous *Trp53+/–* mice have an increased life span, and in addition to lymphomas also develop sarcomas, and to a much lesser extent carcinomas. In the *Trp53+/–* C57BL/6 × 129/Sv mouse strain mammary tumors account for 1 % of the tumors (Donehower et al. 1995b), while back-crossing the *Trp53*-minus allele onto the BALB/c background increased the frequency of mammary carcinomas to about 55 % in *Trp53+/–* BALB/c female mice (Kuperwasser et al. 2000a). This implies that in BALB/c mice genetic modifiers that promote mammary carcinogenesis cooperate with p53 deficiency (Blackburn et al. 2003). *Trp53+/–* BALB/c mice thus constitute a suitable model for the human Li–Fraumeni syndrome (LFS), which is characterized by the development of multiple tumors in patients, with the highest relative frequency of breast carcinomas at the age 15–29 (Birch et al. 2001). Importantly, like in breast carcinomas of LFS patients, mammary carcinomas that had developed in *Trp53+/–* BALB/c mice had frequently lost the wild-type *Trp53* allele (Blackburn et al. 2004; Yan et al. 2010). The data support the conclusion that loss of p53 is an important collaborating genetic event in mammary carcinogenesis.

A similar conclusion was reached from transplanting BALB/c *Trp53–/–* (Jerry et al. 2000) and *Trp53+/–* (Yan et al. 2010) MECs into the cleared fat pad of BALB/c wild-type mice. This transplant model circumvents the problem that mammary carcinogenesis cannot be studied in BALB/c *Trp53–/–* mice due to their early death from lymphomas. In this model, mammary carcinomas developed in about 60 % of the transplants after about ten consecutive transplantations with a latency of about 50 weeks (Jerry et al. 2000). The rather long latency period, as well as the incomplete penetrance of the phenotype indicates that germ line p53 deficiency as such is not sufficient for mammary tumor initiation, but requires either inactivation of p53 in a functional context or cooperating genetic events. The first possibility is supported by the observation that conditional MMTV-Cre- and WAP-Cre-directed deletion of the floxed *Trp53* alleles in MECs leads to the development of a broad spectrum of mammary tumors associated with pulmonary metastasis (Lin et al. 2004). The importance of a functional context is strengthened by the finding that conditional deletion of the *Trp53* gene using the Cre-recombinase under the K14 promoter, restricted to the basal mammary compartment (Jonkers et al. 2001), does not efficiently induce mammary tumorigenesis unless the mice were treated with estrogen (Shai et al. 2008). The data indicate that the autonomous tumor-suppressive functions of p53 are limited to the fraction of hormone-responsive mammary cells. Conditional loss of p53 in K14-expressing cells in combination with estrogen stimulation resulted not only in an early tumor onset but also in a frequent estrogen receptor a (ERα) positivity and an occurrence of an epithelial to mesenchymal transition (EMT) phenotype characterized by loss of E-cadherin expression (Shai et al. 2008). By introducing conditional E-cadherin mutations into the *K14Cre*p53$^{f/f}$ mouse tumor model, the synergism of E-cadherin and p53 loss resulted in accelerated development of invasive mammary carcinomas that are richly vascularized and significantly metastasize (Derksen et al. 2006), thus providing a model for human invasive lobular carcinomas (ILC) that constitute about 10–15 % of all human breast carcinomas.

Spontaneous mammary carcinogenesis in *Trp53+/−* and *Trp53−/−* mammary tissue is greatly facilitated by the carcinogen DMBA (7,12-dimethylbent[a]anthracene) and by γ-irradiation (Blackburn and Jerry 2002), supporting the conclusion that p53 deficiency acts rather synergistically with other oncogenic stimuli. The probably best studied genetic factors cooperating with loss of p53 in mammary carcinogenesis are the two breast cancer susceptibility genes *Brca1* and *Brca2*. Conditional deletion of exon 11 in a single *Brca1* allele in mammary glands after a long latency period occasionally leads to tumor formation. However, shorter latency and higher tumor incidence were observed on a *Trp53+/−* background (Xu et al. 1999), thereby demonstrating the necessity of p53 inactivation in Brca1-mediated tumorigenesis. The conclusion was supported by introducing a *Brca1* deletion into the germ line and crossing with *Trp53+/−* mice (Xu et al. 2001). In this study most female $Brca1^{\Delta 11/\Delta 11}$ *x Trp53+/−* mice developed mammary tumors with loss of the remaining *Trp53* allele within 6–12 months. The synergistic effect of Brca1 and p53 deficiency in mammary tumorigenesis was further confirmed by conditional inactivation of their alleles in K14-expressing (Liu et al. 2007) and MMTV-expressing (Shafee et al. 2008) MECs.

Like inheritance of a defective *BRCA1* allele, also the inheritance of a defective *BRCA2* allele predisposes women to breast tumors, and, like for *Brca1*, heterozygous deficiency of Brca2 and p53 cooperates in the induction of mammary carcinomas in genetically engineered mice (Cheung et al. 2004; Jonkers et al. 2001). In addition to deficiency in the *Brca1/Brca2* susceptibility genes, heterozygous deficiency in a major DNA repair gene, *Atm*, contributes to mammary tumorigenesis on a *Trp53+/−* background. An increased frequency of spontaneous and γ-irradiation-induced mammary carcinomas and their number per mouse was observed in *Trp53+/− x Atm+/−* mice than in the *Atm+/−* mice, demonstrating genetic interaction of heterozygous *Atm* and *Trp53* alleles (Umesako et al. 2005). Taken together, germ line and mammary cell restricted p53 deficiency contributes to initiation of mammary carcinogenesis, but p53 deficiency is not always sufficient as it requires additional genetic lesion.

3.3 Cooperation of p53 Loss with Oncogene-Induced Mammary Carcinogenesis

P53 deficiency not only cooperates with deficiency in breast cancer susceptibility genes, but also with oncogenic pathways activated in sporadic breast cancer. The respective studies were mostly performed by transgenic expression of various oncogenes under control of the mouse mammary tumor virus (MMTV) promoter. Use of this promoter has the advantage that the mammary glands can develop normally up to puberty, i.e., in the absence of oncogene expression. MMTV-driven expression of oncogenes, e.g., *Wnt-1* (Donehower et al. 1995a) and v-Ha-*ras* (Hundley et al. 1997), in transgenic mice led to mammary carcinomas whose onset was accelerated when the oncogene was expressed in a *Trp53+/−* or a *Trp53−/−* background. While MMTV-*Wnt1* mammary carcinomas induced in a *Trp53+/+* background were usually diploid, karyotype instability, enhanced proliferation, and reduced fibrosis in mammary tumors were observed in a *Trp53–/–* background, combined with an early onset of tumor growth. An accelerated onset of mammary tumors with a more aggressive phenotype was also observed in MMTV-*ras*/*Trp53+/–* mice, but in a p53-null background tumor occurrence was shifted from mammary to salivary glands. The data imply that tissue-specific tumorigenesis induced by MMTV-driven oncogenes is dependent on the *Trp53* gene status, and that p53 deficiency synergistically contributes to mammary tumorigenesis. This conclusion is further supported by crossing of MMTV-*myc* with p53-deficient mice. On a *Trp53+/–* background, MMTV-*myc* induced much more hyperplastic lesions in the mammary gland than on the *Trp53+/+* background, but mammary carcinomas were considerably fewer due to early death of the animals by lymphomas (McCormack et al. 1998). While the data clearly show that Myc-induced mammary carcinogenesis does not require loss of p53, the early onset of lymphomas in MMTV-*myc* transgenic mice with a *Trp53+/–* or *Trp53–/–* background might have obscured a cooperative effect between loss of p53 and Myc in mammary carcinogenesis.

In summary, the data reveal a complex picture regarding the cooperative effect of p53 loss on mammary carcinogenesis. As oncogene-induced tissue-specific tumorigenesis seems to be differently affected by p53 loss, tumorigenesis in other organs can obscure the analysis of mammary carcinogenesis. Thus more sophisticated models with conditional tissue-specific deletion of the *Trp53* gene and tissue-specific expression of the respective oncogenes are required. Alternative to conditional ablation of *Trp53* alleles, organ-specific expression of the SV40 large T antigen (LT) creates the organ-specific absence of a functional p53, as LT efficiently complexes and functionally inactivates p53. While most SV40-based models for mammary carcinogenesis make use of the WAP-promoter (see Sect. 3.4), the 5′ flanking region of the C3(1) component of the rat prostate steroid binding protein (PSBP) has been successfully used to target LT expression to the epithelium of both the mammary and the prostate glands, resulting in models of mammary and prostate cancers which histologically resemble the respective human diseases (Green et al. 2000). In contrast to oncogenes driven by the MMTV and the WAP-promoter, which require hormonal stimuli for their activation, C3(1)Tag mice develop mammary tumors in virgin animals without hormone supplementation.

3.4 WAP-SV40 Transgenic Models for Mammary Carcinogenesis: pRB and p53 Inactivation as Major Causes for Tumor Initiation

Mammary carcinogenesis in WAP-SV40 mice is determined by the activity of the WAP-promoter, which is regulated similar to the endogenous *Wap*-gene coding for the whey acid protein, a milk protein in rodents. The *Wap* gene is transiently induced by hormones during estrus, stably during midpregnancy, and expressed highest during lactation in differentiated MECs that disappear during involution (Robinson et al. 1995). Tumorigenic events are initiated by the expression of two synergistically acting SV40 early region encoded proteins, the large T (LT) and the small t (sT) antigens, produced by alternative splicing. While sT displaces the regulatory B-subunit from the PP2A complex and inhibits dephosphorylation of PP2A targets, the transforming activity of LT is mainly mediated by direct binding of LT to the p53 and pRb proteins (Pipas 2009). The effect of the functional inactivation of pocket proteins by LT closely mimics the consequences of *RB1* LOH in human tumors. *RB1* LOH is a frequent event in basal-like and luminal B human mammary carcinomas and is associated with a high proliferative activity (Herschkowitz et al. 2008). The group of basal-like breast carcinomas is also characterized by a high frequency of mutations in the *TP53* gene (Manie et al. 2009), possibly indicating a cooperative effect of the concomitant inactivation of the pRB and the p53 pathways.

While expression of LT and sT is absolutely required to initiate tumor development, this is not sufficient for malignant cell transformation, as additional genetic alterations, which arise from the induction of genomic instability by LT, are required for tumor growth. Tumor development in WAP-T mouse lines (WAP-SV40 transgenic mice generated in a BALB/c genetic background (Schulze-Garg et al. 2000)) is characterized by a moderately long latency time of up to 8 months after the first pregnancy (Heinlein et al. 2008; Schulze-Garg et al. 2000). About 60 days after involution, virtually each terminal end bud in each mammary gland has developed lesions which histologically resemble human ductal carcinomas in situ (DCIS). The process of selection of tumor-permissive genotypes within the multiple DCIS that develop after first pregnancy is stochastic and may take several months (Jannasch et al. 2009; Santarelli et al. 1996; Schulze-Garg et al. 2000) (Fig. 7.1). Similar kinetics of pregnancy-induced tumor development were observed in transgenic mice generated by introduction of the same WAP-SV40 transgene construct by pronuclear injection into zygotes of NMRI outbred mice (Santarelli et al. 1996). However, in the C57BL/6J inbred background, a low tumor incidence and a long latency after multiple pregnancies were observed despite strong LT/sT-expression in MECs of the lactating gland (Husler et al. 1998), pointing to epistatic effects of modifier genes. Indeed, as already outlined before, in the BALB/c genetic background genetic modifiers, e.g., mutations in the *Cdkn2a* and *Prkdc* genes (Blackburn et al. 2003), as well as *Trp53* gene haploinsufficiency or complete loss of p53 expression (Kuperwasser et al. 2000a) favor the development of mammary carcinomas.

Given the regulatory functions of p53 in the physiological context of the mammary gland, inhibition of p53 by SV40 LT in MECs expressing nuclear p53 is a prerequisite for tumor formation. Hormonal stimuli in the mammary gland induce mitotic activity in MECs, as well as differentiation of epithelial progenitors along their developmental routes. Upon LT expression the concomitant de-repression of E2F-regulated promoters and the deregulation of p53 target genes create genetically instable cells in a proliferative state. These cells most likely originate from the mammary stem/progenitor cell compartment, where p53 exerts its regulatory function through the Notch pathway (Tao et al. 2011) controlling polarity of stem cell divisions (Cicalese et al. 2009). The observation that tumors in WAP-T mice comprise differentiated and undifferentiated ductal adenocarcinomas (Fig. 7.2) and that their phenotype can be reproduced upon transplantation also points to a stem/progenitor cell origin of WAP-T tumors (Wegwitz et al. 2010).

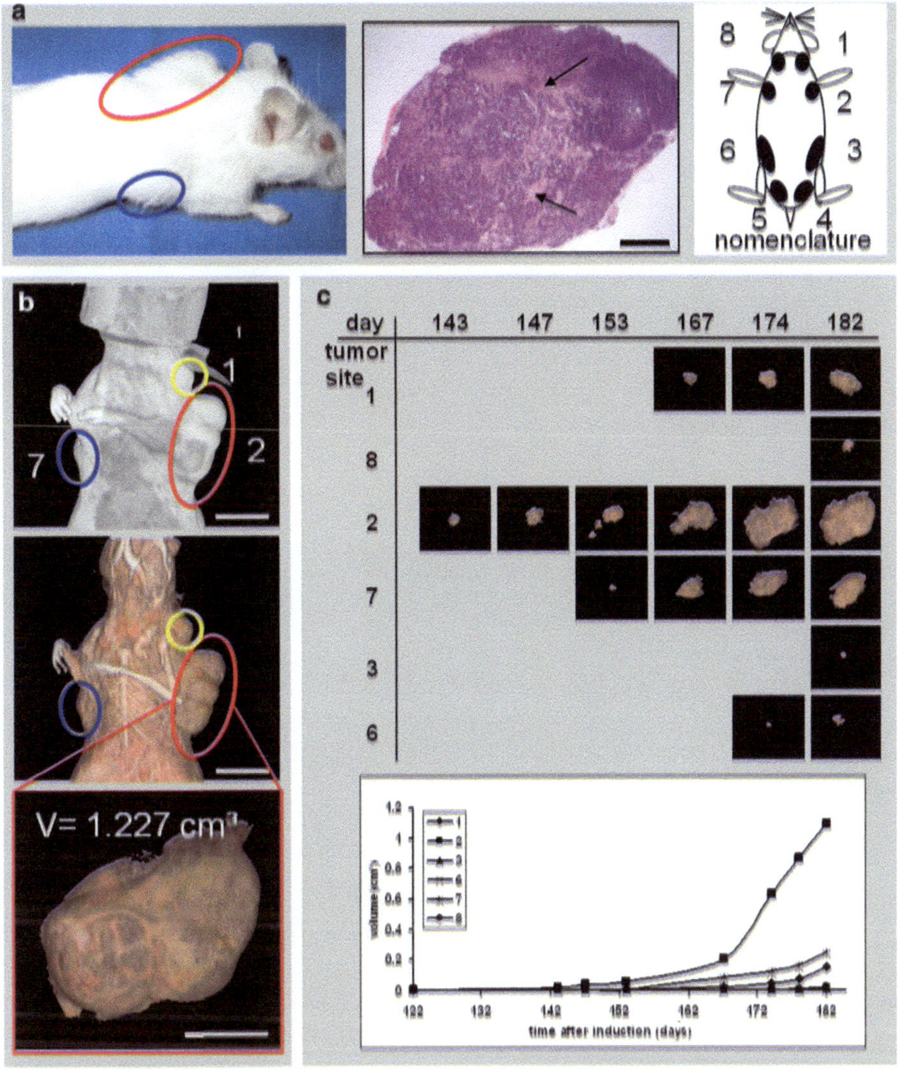

Fig. 7.1 Growth rate of multiple mammary tumors within one mouse depicted in vivo by 3D reconstructions of fpVCT data sets. (**a**) Representation of a tumor-bearing WAP-T transgenic mouse. Visible mammary carcinomas that developed at different sites are indicated by *circles*. The nomenclature of the tumors depending on their site of origin is displayed on the *right side*. Section of a representative mammary tumor stained with H&E is depicted (*middle*). Note that the tumor contains several small areas of necrosis within intact tumor tissue (*arrows*). (**b**) Shows a 3D fpVCT image of a mammary tumor 182 days after induction in situ (*top*), followed by another visualization protocol (*middle*), displaying particularly soft tissue and vessels. After segmentation only the tumor is depicted (*bottom*). (**c**) *Top panel*: Visualization of the enlargement of the virtually isolated mammary carcinomas by repeated fpVCT scans in combination with contrast agent performed between day 143 and 182 after induction. *Bottom panel*: Resulting tumor growth curves are displayed. Note that each mammary carcinoma is depicted at first by fpVCT at different time points but all tumors display similar growth kinetics up to a volume of 0.2 cm^3. Scale bars: (**a**) 2 mm, (**b**) 10 mm. Reprinted with permission from Jannasch et al. (2009)

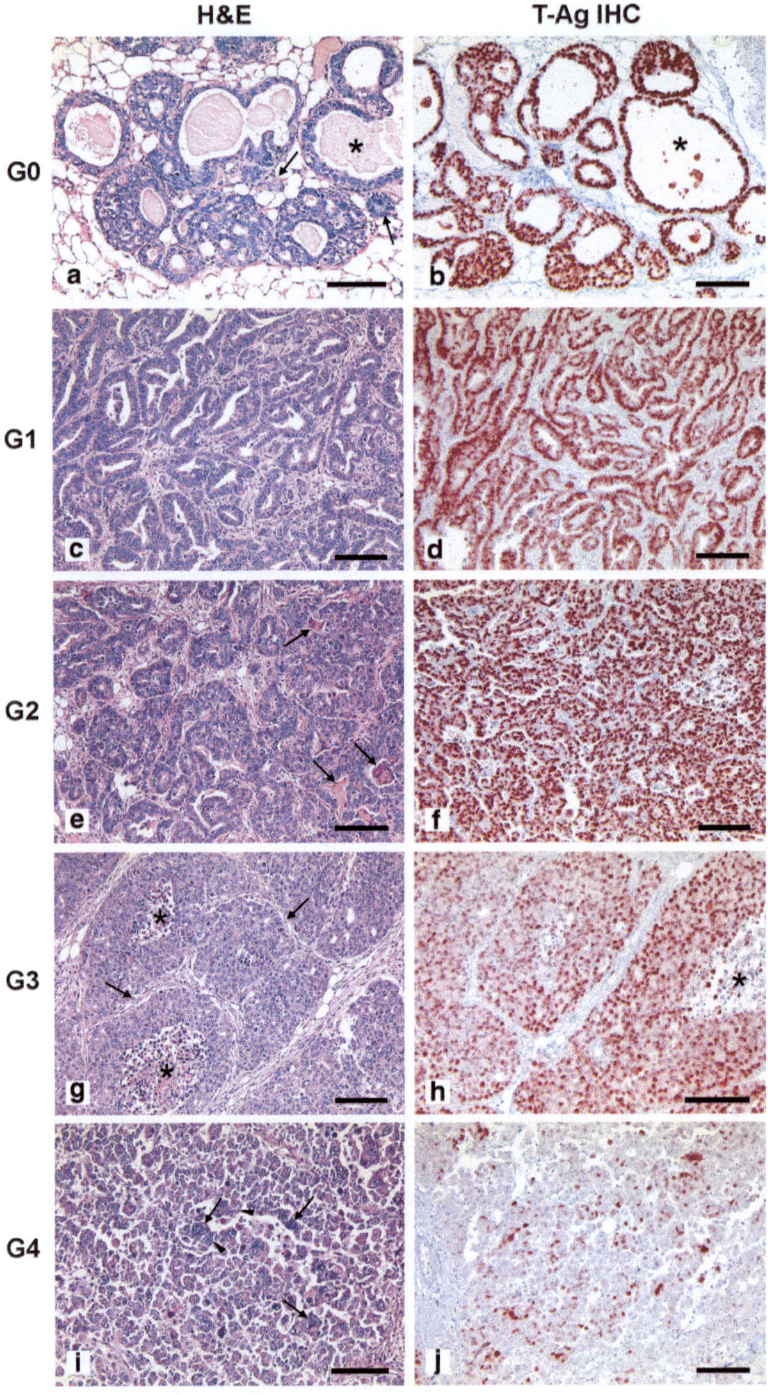

Fig. 7.2 Morphology and T-Ag immunostaining of mammary carcinoma of different histological grades in WAP-T mice. Sequential sections of pathologically altered mammary glands were stained

Further work is needed to understand the function of p53 in the MEC hierarchy and to identify there the p53-controlled transcriptional network, the deregulation of which contributes to mammary tumorigenesis.

3.5 Models for Analyzing mutp53 GOF in Mammary Carcinogenesis

The first in vivo demonstration of mutant p53 GOF was obtained in mice by comparing *Trp53+/−* mice with *Trp53+/R172H* mice, expressing mutant $p53^{R172H}$, an ortholog of the human "hot spot" mutant $p53^{R175H}$. In two genetic backgrounds, C57BL/6 (Lang et al. 2004) and $129S_4$/SvJae (Olive et al. 2004), mice with the mutant *Trp53*^{R172H} allele exhibited more aggressive and metastatic tumor phenotypes than *Trp53+/−* mice. On the $129S_4$/SvJae background, *Trp53+/R172H* mice exhibited increased numbers of carcinomas and B-cell lymphomas compared to *Trp53+/−* mice, which were not observed in the C57BL/6 background. These results indicate that the in vivo GOF phenotypes by $p53^{R172H}$ are affected by the genetic background of the mice. An interesting observation was made in these mice regarding mutant p53 protein stability. While it so far had been assumed that high metabolic stability is an intrinsic property of mutant p53 proteins, immunohistochemistry for p53 in tumors and tumor-free organs revealed that the $p53^{R172H}$ protein was strongly detectable only in tumors, whereas all normal tissues stained negative for p53 (Terzian et al. 2008). The data indicate that also mutant p53 must become activated during tumorigenesis, e.g., by oncogenic stress, and that only activated mutant p53 exerts oncogenic properties.

Fig. 7.2 (continued) with H&E (**a**, **c**, **e**, **g**, **i**) and immunostained with polyclonal rabbit antibodies against T-Ag (**b**, **d**, **f**, **h**, **j**). (**a**, **b**): Grade 0. (**a**) End-stage of fully developed intraepithelial neoplasia with typical DCIS-like structures caricaturing the lobular architecture of an expanded TDLU-like murine LA unit (lobulo-alveolar unit). *Arrows* indicate beginning invasive growth (microinvasive carcinoma). *Asterisks* mark the greatly dilated lumen of the terminal ductule. (**b**) Nuclear T-Ag staining in the pathologically altered TDLU-like "lobules". (**c**, **d**) Grade 1. (**c**) Well-differentiated, invasive tubulo-papillary adenocarcinoma. (**d**) Uniform antibody decoration of tumor cell nuclei. (**e**, **f**): Grade 2. (**e**) Moderately differentiated, invasive tubulo-papillary adenocarcinoma. *Arrows* indicate comedo necroses. (**f**) Uniform antibody decoration of tumor cell nuclei. (**g**, **h**) Grade 3. (**g**) Relatively monomorphous tumor cells of an invasive adenocarcinoma are arranged in roundish tracts separated by thin septa of stroma (*arrows*). Centrally located comedo necroses are indicated by *asterisks*. (**h**) Typically, grade 3 carcinoma show varying degrees of T-Ag expression. (**i**, **j**) Grade 4. (**i**) Invasive mammary carcinoma consisting of discohesively growing, pleomorphous tumor cells with obvious scarcity of stroma. *Arrows* indicate examples of multinuclear tumor cells, arrow heads nuclear atypias. (**j**) The reduced to absent T-Ag immunoreactivity. Scale bars (**a**)–(**j**) represent: 100 μm. Reprinted with permission from Heinlein et al. (2008)

So far, mouse models that specifically address the role of mutant p53 in mammary carcinogenesis are scarce. This is somewhat surprising, considering the importance of human breast cancer and the finding that mutations in the *TP53* gene have been established as an independent factor for bad prognosis (Langerod et al. 2007; Olivier et al. 2006). The first mouse models for analyzing the effect of mutant p53 on mammary carcinogenesis used transgenic WAP-promoter-driven expression of $p53^{R172H}$ in a FVB *Trp53+/+* genetic background (Li et al. 1998). Transgene expression alone had no apparent effect on normal mammary development, and spontaneous mammary tumor development occurred at a rather low incidence and with a long latency. However, treatment of transgenic mice with the chemical carcinogen DMBA significantly shortened latency of tumor development compared to control littermates and led to a greater tumor burden. Tumors arising in transgenic mice displayed increased genomic instability compared to spontaneously arising tumors in control littermates. However, due to the presence of wild-type p53 in these mice, it is difficult to discern between a dominant-negative and a true oncogenic effect of the transgenic mutant p53. The dominant-negative effect of $p53^{R270H}$, an equivalent to the human "hot-spot" mutant $p53^{R273H}$, seems to be responsible for spontaneous, mostly ERα-positive mammary tumors in $p53^{R270H/+}$WAPCre mice with mammary gland-specific expression of the $p53^{R270H}$ mutation (Wijnhoven et al. 2005). The authors show that heterozygous expression of $p53^{R270H}$ is not equivalent to a *Trp53+/−*, but rather to a *Trp53−/−* situation, also upon exposure to DMBA which accelerates mammary carcinogenesis.

Like *Trp53+/−* mice, the WAP-p53R172H mouse has been used to study cooperation with other oncogenes in mammary carcinogenesis, e.g., *Erbb2/neu*, encoding a receptor tyrosine kinase (RTK) and des(1–3)hIGF-I, an IGF1 analog with reduced affinity for IGFBP proteins. On a *Trp53+/+* background, mammary tumors in MMTV-*neu* mice develop with a long latency, which is reduced in mice carrying the $p53^{R172H}$ and Erbb2 transgenes, suggesting cooperative events and dominant-oncogenic properties of this mutant p53 protein during tumor formation (Li et al. 1997). However, in a follow-up study the effect of $p53^{R172H}$ expression was rather attributed to a p53-null situation (Zelazny et al. 2001).

Transgenic expression of the des(1–3)hIGF-I in WAP-DES mice is associated with hyperplastic lesions; however, mammary adenocarcinomas grow in about half of the mice after a long latency period and multiple pregnancies (Hadsell et al. 2000). WAP-DES × WAP-p53R172H bi-transgenic mice had a tumor incidence like WAP-DES mice, but tumor latency was reduced by 8 months. Aneuploidy was frequently observed in tumors from bi-transgenic, but not from mice expressing only the des(1–3)hIGF-I transgene. The increase in genomic instability in bi-transgenic tumors was attributed to oncogenic functions of $p53^{R172H}$, but the mechanism remains to be elucidated.

The only currently available mouse model addressing mutant p53 ($p53^{R270H}$, equivalent to human $p53^{R273H}$, and $p53^{R245W}$, equivalent to human $p53^{R248W}$) GOF in mammary carcinogenesis in the context of functional inactivation of wild-type p53 are WAP-T mice crossed with WAP-mutp53 mice (Heinlein et al. 2008; Krepulat et al. 2005). In contrast to $p53^{R270H/+}$WAPCre mice (Wijnhoven et al. 2005), even multiparous WAP-mutp53 females show no mammary gland anomalies, indicating that mutant p53 in the presence of both wild-type p53 alleles and in the absence of an oncogenic stimulus is inert, arguing against a solely dominant-negative effect. However, in the context of SV40 LT-mediated inactivation of wild-type p53 the GOF phenotype of mutant p53 in mammary carcinogenesis is manifested by an increased proportion of undifferentiated tumors and more frequent metastasis (Heinlein et al. 2008).

The molecular basis of the mutant p53 GOF phenotype in WAP-T/WAP-mutp53 mice is still under investigation. The finding that even a small fraction (2 %) of highly mutant p53 expressing cells suffices for conferring the mutant p53 GOF tumor phenotype (Heinlein et al. 2008) points to the possibility that the mutant p53 function is associated with tumor initiating events in the mammary stem/progenitor compartment where wild-type p53 function is compromised by LT. Given the pronounced effect of mutant p53 on reprogramming of mouse embryo fibroblasts (Sarig et al. 2010), the assumption that mutant p53 supports expansion of cancer stem cells seems to be reasonable. Indirect support for this possibility came from cell culture studies. While we consistently failed to establish cell lines from WAP-T tumors we repeatedly succeeded in establishing such cells from WAP-T/WAP-mutp53 tumors. The characteristic features of one of these cell lines (e.g., the G-2 cell line) have been described in detail (Wegwitz et al. 2010). G-2 cells display stem-like traits, and like their clonal derivates, and like tumors derived from these cells after syngeneic orthotopic transplantation, are closely related to primary tumors regarding histology and gene expression profiles, and reflect heterogeneity regarding their differentiation states. G-2 cultures comprise cell populations in distinct differentiation states, which, when isolated separately, are competent to reconstitute the initial cellular composition. The availability of G-2 cells and additional cell lines with similar properties will greatly facilitate molecular analyses of various aspects of mammary tumorigenesis and progression, as these cells are amenable to genetic manipulation, and the effect of such manipulations can be analyzed in vitro as well as in vivo after orthotopic transplantation. Furthermore, this model system will be useful for preclinical studies, e.g., in testing novel drugs and their influence on tumor growth, tumor cell dissemination, and metastasis.

4 Mouse Models in Preclinical Studies

The use of transgenic mouse models will not only shed light on mechanistic aspects of mammary carcinogenesis, but should be useful also for preclinical studies. In line, mouse models have been used in preclinical settings to study the effects of known or novel experimental drugs on mammary carcinogenesis, e.g., tumor-suppressive effects of chemopreventive drugs in the p53-null transplantation model (Medina et al. 2009). The following example shows that transgenic mouse models allow a thorough dissection of the parameters by which drugs influence mammary carcinogenesis and progression, as well as focusing on molecular and cellular mechanisms of drug resistance and tumor relapse in a defined biological context.

Mammary adenocarcinomas grown in *K14cre*; *Brca1*$^{F/F}$; *p53*$^{F/F}$ mice recapitulate features of BRCA1- and p53-deficient human basal-like tumors, including a triple-negative receptor status and significant genomic instability (Liu et al. 2007). Therefore this mouse provides a model to test therapeutic approaches against this type of tumors. Application of the topoisomerase I inhibitor topotecan alone or together with the poly(ADP-ribose) polymerase inhibitor olaparib showed a heterogeneous response, but always tumor relapse and acquired resistance, which could be attributed to overexpression of the ATP binding cassette half-transporter Abcg2/Bcrp (Zander et al. 2010). Ablation of the *Abcg2* alleles in this tumor model demonstrated, however, a subordinate function of this drug transporter in the chemoresistance of tumor-initiating cells (TICs) (Zander et al. 2010). The rather dormant state of these TICs is attributed to tumor re-growth after eradication of tumor cells with cisplatin (Pajic et al. 2010).

In conclusion, a wider use of mouse models in preclinical studies should be strongly encouraged with the aim to finally substitute xenograft models.

5 Concluding Remarks

The model systems reviewed here have substantially expanded our knowledge on mammary carcinogenesis, in general, and of the role of p53 in this process. As expected, functional elimination of p53 strongly contributes to mammary carcinogenesis, but the molecular pathways initiated by p53 loss in this process are far from being understood. Although loss of the established tumor suppressor functions of p53 is part of this process, as demonstrated by the cooperativity of p53 loss and DNA damage, additional functions of p53, e.g., in mammary gland differentiation and in the maintenance of the stem/progenitor cell hierarchy seem to play an important, but so far scarcely understood, role. Similarly, a mutant p53 GOF in mammary carcinogenesis has been shown, but not yet been explained at the molecular level. Further analysis of existing models as well as the development of new models specifically addressing these aspects thus is required. Furthermore, molecular data obtained in mouse models will have to be cross-validated with the corresponding human diseases. The knowledge developing from such analyses not only will further our understanding of the multiple roles of p53 in mammary carcinogenesis, but will improve the value of such models for preclinical studies.

References

Adhikari AS, Iwakuma T (2009) Mutant p53 gain of oncogenic function: in vivo evidence, mechanism of action and its clinical implications. Fukuoka Igaku Zasshi 100:217–228

Birch JM, Alston RD, McNally RJ, Evans DG, Kelsey AM, Harris M, Eden OB, Varley JM (2001) Relative frequency and morphology of cancers in carriers of germline TP53 mutations. Oncogene 20:4621–4628

Blackburn AC, Brown JS, Naber SP, Otis CN, Wood JT, Jerry DJ (2003) BALB/c alleles for Prkdc and Cdkn2a interact to modify tumor susceptibility in Trp53+/– mice. Cancer Res 63: 2364–2368

Blackburn AC, Jerry DJ (2002) Knockout and transgenic mice of Trp53: what have we learned about p53 in breast cancer? Breast Cancer Res 4:101–111

Blackburn AC, McLary SC, Naeem R, Luszcz J, Stockton DW, Donehower LA, Mohammed M, Mailhes JB, Soferr T, Naber SP et al (2004) Loss of heterozygosity occurs via mitotic recombination in Trp53+/– mice and associates with mammary tumor susceptibility of the BALB/c strain. Cancer Res 64:5140–5147

Blandino G, Levine AJ, Oren M (1999) Mutant p53 gain of function: differential effects of different p53 mutants on resistance of cultured cells to chemotherapy. Oncogene 18:477–485

Brosh R, Rotter V (2009) When mutants gain new powers: news from the mutant p53 field. Nat Rev Cancer 9:701–713

Cheung AM, Elia A, Tsao MS, Done S, Wagner KU, Hennighausen L, Hakem R, Mak TW (2004) Brca2 deficiency does not impair mammary epithelium development but promotes mammary adenocarcinoma formation in p53(+/–) mutant mice. Cancer Res 64:1959–1965

Cicalese A, Bonizzi G, Pasi CE, Faretta M, Ronzoni S, Giulini B, Brisken C, Minucci S, Di Fiore PP, Pelicci PG (2009) The tumor suppressor p53 regulates polarity of self-renewing divisions in mammary stem cells. Cell 138:1083–1095

Derksen PW, Liu X, Saridin F, van der Gulden H, Zevenhoven J, Evers B, van Beijnum JR, Griffioen AW, Vink J, Krimpenfort P et al (2006) Somatic inactivation of E-cadherin and p53 in mice leads to metastatic lobular mammary carcinoma through induction of anoikis resistance and angiogenesis. Cancer Cell 10:437–449

Dimri G, Band H, Band V (2005) Mammary epithelial cell transformation: insights from cell culture and mouse models. Breast Cancer Res 7:171–179

Donehower LA, Godley LA, Aldaz CM, Pyle R, Shi YP, Pinkel D, Gray J, Bradley A, Medina D, Varmus HE (1995a) Deficiency of p53 accelerates mammary tumorigenesis in Wnt-1 transgenic mice and promotes chromosomal instability. Genes Dev 9:882–895

Donehower LA, Harvey M, Slagle BL, McArthur MJ, Montgomery CA Jr, Butel JS, Bradley A (1992) Mice deficient for p53 are developmentally normal but susceptible to spontaneous tumours. Nature 356:215–221

Donehower LA, Harvey M, Vogel H, McArthur MJ, Montgomery CA Jr, Park SH, Thompson T, Ford RJ, Bradley A (1995b) Effects of genetic background on tumorigenesis in p53-deficient mice. Mol Carcinog 14:16–22

Donehower LA, Lozano G (2009) 20 years studying p53 functions in genetically engineered mice. Nat Rev Cancer 9:831–841

Green JE, Shibata MA, Yoshidome K, Liu ML, Jorcyk C, Anver MR, Wigginton J, Wiltrout R, Shibata E, Kaczmarczyk S et al (2000) The C3(1)/SV40 T-antigen transgenic mouse model of mammary cancer: ductal epithelial cell targeting with multistage progression to carcinoma. Oncogene 19:1020–1027

Hadsell DL, Murphy KL, Bonnette SG, Reece N, Laucirica R, Rosen JM (2000) Cooperative interaction between mutant p53 and des(1–3)IGF-I accelerates mammary tumorigenesis. Oncogene 19:889–898

Heinlein C, Krepulat F, Lohler J, Speidel D, Deppert W, Tolstonog GV (2008) Mutant p53(R270H) gain of function phenotype in a mouse model for oncogene-induced mammary carcinogenesis. Int J Cancer 122:1701–1709

Herschkowitz JI, He X, Fan C, Perou CM (2008) The functional loss of the retinoblastoma tumour suppressor is a common event in basal-like and luminal B breast carcinomas. Breast Cancer Res 10:R75

Humphreys RC, Krajewska M, Krnacik S, Jaeger R, Weiher H, Krajewski S, Reed JC, Rosen JM (1996) Apoptosis in the terminal endbud of the murine mammary gland: a mechanism of ductal morphogenesis. Development 122:4013–4022

Hundley JE, Koester SK, Troyer DA, Hilsenbeck SG, Subler MA, Windle JJ (1997) Increased tumor proliferation and genomic instability without decreased apoptosis in MMTV-ras mice deficient in p53. Mol Cell Biol 17:723–731

Husler MR, Kotopoulis KA, Sundberg JP, Tennent BJ, Kunig SV, Knowles BB (1998) Lactation-induced WAP-SV40 Tag transgene expression in C57BL/6J mice leads to mammary carcinoma. Transgenic Res 7:253–263

Jannasch K, Dullin C, Heinlein C, Krepulat F, Wegwitz F, Deppert W, Alves F (2009) Detection of different tumor growth kinetics in single transgenic mice with oncogene-induced mammary carcinomas by flat-panel volume computed tomography. Int J Cancer 125:62–70

Jerry DJ, Kittrell FS, Kuperwasser C, Laucirica R, Dickinson ES, Bonilla PJ, Butel JS, Medina D (2000) A mammary-specific model demonstrates the role of the p53 tumor suppressor gene in tumor development. Oncogene 19:1052–1058

Jerry DJ, Kuperwasser C, Downing SR, Pinkas J, He C, Dickinson E, Marconi S, Naber SP (1998) Delayed involution of the mammary epithelium in BALB/c-p53null mice. Oncogene 17: 2305–2312

Johansson EM, Kannius-Janson M, Bjursell G, Nilsson J (2003) The p53 tumor suppressor gene is regulated in vivo by nuclear factor 1-C2 in the mouse mammary gland during pregnancy. Oncogene 22:6061–6070

Jonkers J, Meuwissen R, van der Gulden H, Peterse H, van der Valk M, Berns A (2001) Synergistic tumor suppressor activity of BRCA2 and p53 in a conditional mouse model for breast cancer. Nat Genet 29:418–425

Keller PJ, Lin AF, Arendt LM, Klebba I, Jones AD, Rudnick JA, Dimeo TA, Gilmore H, Jefferson DM, Graham RA et al (2010) Mapping the cellular and molecular heterogeneity of normal and malignant breast tissues and cultured cell lines. Breast Cancer Res 12:R87

Kelly PN, Dakic A, Adams JM, Nutt SL, Strasser A (2007) Tumor growth need not be driven by rare cancer stem cells. Science 317:337

Krepulat F, Lohler J, Heinlein C, Hermannstadter A, Tolstonog GV, Deppert W (2005) Epigenetic mechanisms affect mutant p53 transgene expression in WAP-mutp53 transgenic mice. Oncogene 24:4645–4659

Kuperwasser C, Hurlbut GD, Kittrell FS, Dickinson ES, Laucirica R, Medina D, Naber SP, Jerry DJ (2000a) Development of spontaneous mammary tumors in BALB/c p53 heterozygous mice. A model for Li-Fraumeni syndrome. Am J Pathol 157:2151–2159

Kuperwasser C, Pinkas J, Hurlbut GD, Naber SP, Jerry DJ (2000b) Cytoplasmic sequestration and functional repression of p53 in the mammary epithelium is reversed by hormonal treatment. Cancer Res 60:2723–2729

Lang GA, Iwakuma T, Suh YA, Liu G, Rao VA, Parant JM, Valentin-Vega YA, Terzian T, Caldwell LC, Strong LC et al (2004) Gain of function of a p53 hot spot mutation in a mouse model of Li-Fraumeni syndrome. Cell 119:861–872

Langerod A, Zhao H, Borgan O, Nesland JM, Bukholm IR, Ikdahl T, Karesen R, Borresen-Dale AL, Jeffrey SS (2007) TP53 mutation status and gene expression profiles are powerful prognostic markers of breast cancer. Breast Cancer Res 9:R30

Li B, Murphy KL, Laucirica R, Kittrell F, Medina D, Rosen JM (1998) A transgenic mouse model for mammary carcinogenesis. Oncogene 16:997–1007

Li B, Rosen JM, McMenamin-Balano J, Muller WJ, Perkins AS (1997) neu/ERBB2 cooperates with p53-172H during mammary tumorigenesis in transgenic mice. Mol Cell Biol 17: 3155–3163

Lin SC, Lee KF, Nikitin AY, Hilsenbeck SG, Cardiff RD, Li A, Kang KW, Frank SA, Lee WH, Lee EY (2004) Somatic mutation of p53 leads to estrogen receptor alpha-positive and -negative mouse mammary tumors with high frequency of metastasis. Cancer Res 64:3525–3532

Liu X, Holstege H, van der Gulden H, Treur-Mulder M, Zevenhoven J, Velds A, Kerkhoven RM, van Vliet MH, Wessels LF, Peterse JL et al (2007) Somatic loss of BRCA1 and p53 in mice induces mammary tumors with features of human BRCA1-mutated basal-like breast cancer. Proc Natl Acad Sci USA 104:12111–12116

Lozano G (2007) The oncogenic roles of p53 mutants in mouse models. Curr Opin Genet Dev 17:66–70

Lozano G (2010) Mouse models of p53 functions. Cold Spring Harb Perspect Biol 2:a001115

Manie E, Vincent-Salomon A, Lehmann-Che J, Pierron G, Turpin E, Warcoin M, Gruel N, Lebigot I, Sastre-Garau X, Lidereau R et al (2009) High frequency of TP53 mutation in BRCA1 and sporadic basal-like carcinomas but not in BRCA1 luminal breast tumors. Cancer Res 69:663–671

Matthews JR, Clarke AR (2005) p53 mediates a default programme of mammary gland involution in the absence of STAT3. Oncogene 24:3083–3090

McCormack SJ, Weaver Z, Deming S, Natarajan G, Torri J, Johnson MD, Liyanage M, Ried T, Dickson RB (1998) Myc/p53 interactions in transgenic mouse mammary development, tumorigenesis and chromosomal instability. Oncogene 16:2755–2766

Medina D, Kittrell F, Hill J, Zhang Y, Hilsenbeck SG, Bissonette R, Brown PH (2009) Prevention of tumorigenesis in p53-null mammary epithelium by rexinoid bexarotene, tyrosine kinase inhibitor gefitinib, and celecoxib. Cancer Prev Res (Phila Pa) 2:168–174

Medina D, Kittrell FS (2003) p53 function is required for hormone-mediated protection of mouse mammary tumorigenesis. Cancer Res 63:6140–6143

Milliken EL, Lozada KL, Johnson E, Landis MD, Seachrist DD, Whitten I, Sutton AL, Abdul-Karim FW, Keri RA (2008) Ovarian hyperstimulation induces centrosome amplification and aneuploid mammary tumors independently of alterations in p53 in a transgenic mouse model of breast cancer. Oncogene 27:1759–1766

Molchadsky A, Rivlin N, Brosh R, Rotter V, Sarig R (2010) p53 is balancing development, differentiation and de-differentiation to assure cancer prevention. Carcinogenesis 31:1501–1508

Olive KP, Tuveson DA, Ruhe ZC, Yin B, Willis NA, Bronson RT, Crowley D, Jacks T (2004) Mutant p53 gain of function in two mouse models of Li-Fraumeni syndrome. Cell 119:847–860

Olivier M, Hollstein M, Hainaut P (2010) TP53 mutations in human cancers: origins, consequences, and clinical use. Cold Spring Harb Perspect Biol 2:a001008

Olivier M, Langerod A, Carrieri P, Bergh J, Klaar S, Eyfjord J, Theillet C, Rodriguez C, Lidereau R, Bieche I et al (2006) The clinical value of somatic TP53 gene mutations in 1,794 patients with breast cancer. Clin Cancer Res 12:1157–1167

Oren M, Rotter V (2010) Mutant p53 gain-of-function in cancer. Cold Spring Harb Perspect Biol 2:a001107

Pajic M, Kersbergen A, van Diepen F, Pfauth A, Jonkers J, Borst P, Rottenberg S (2010) Tumor-initiating cells are not enriched in cisplatin-surviving BRCA1;p53-deficient mammary tumor cells in vivo. Cell Cycle 9:3780–3791

Petitjean A, Mathe E, Kato S, Ishioka C, Tavtigian SV, Hainaut P, Olivier M (2007) Impact of mutant p53 functional properties on TP53 mutation patterns and tumor phenotype: lessons from recent developments in the IARC TP53 database. Hum Mutat 28:622–629

Pipas JM (2009) SV40: cell transformation and tumorigenesis. Virology 384:294–303

Proia DA, Kuperwasser C (2006) Reconstruction of human mammary tissues in a mouse model. Nat Protoc 1:206–214

Robinson GW, McKnight RA, Smith GH, Hennighausen L (1995) Mammary epithelial cells undergo secretory differentiation in cycling virgins but require pregnancy for the establishment of terminal differentiation. Development 121:2079–2090

Santarelli R, Tzeng YJ, Zimmermann C, Guhl E, Graessmann A (1996) SV40 T-antigen induces breast cancer formation with a high efficiency in lactating and virgin WAP-SV-T transgenic animals but with a low efficiency in ovariectomized animals. Oncogene 12:495–505

Sarig R, Rivlin N, Brosh R, Bornstein C, Kamer I, Ezra O, Molchadsky A, Goldfinger N, Brenner O, Rotter V (2010) Mutant p53 facilitates somatic cell reprogramming and augments the malignant potential of reprogrammed cells. J Exp Med 207:2127–2140

Schulze-Garg C, Lohler J, Gocht A, Deppert W (2000) A transgenic mouse model for the ductal carcinoma in situ (DCIS) of the mammary gland. Oncogene 19:1028–1037

Shafee N, Smith CR, Wei S, Kim Y, Mills GB, Hortobagyi GN, Stanbridge EJ, Lee EY (2008) Cancer stem cells contribute to cisplatin resistance in Brca1/p53-mediated mouse mammary tumors. Cancer Res 68:3243–3250

Shai A, Pitot HC, Lambert PF (2008) p53 Loss synergizes with estrogen and papillomaviral oncogenes to induce cervical and breast cancers. Cancer Res 68:2622–2631

Tao L, Roberts AL, Dunphy KA, Bigelow C, Yan H, Jerry DJ (2011) Repression of mammary stem/progenitor cells by p53 is mediated by notch and separable from apoptotic activity. Stem Cells 29:119–127

Terzian T, Suh YA, Iwakuma T, Post SM, Neumann M, Lang GA, Van Pelt CS, Lozano G (2008) The inherent instability of mutant p53 is alleviated by Mdm2 or p16INK4a loss. Genes Dev 22:1337–1344

Tlsty TD, Coussens LM (2006) Tumor stroma and regulation of cancer development. Annu Rev Pathol 1:119–150

Umesako S, Fujisawa K, Iiga S, Mori N, Takahashi M, Hong DP, Song CW, Haga S, Imai S, Niwa O et al (2005) Atm heterozygous deficiency enhances development of mammary carcinomas in p53 heterozygous knockout mice. Breast Cancer Res 7:R164–R170

Visvader JE (2009) Keeping abreast of the mammary epithelial hierarchy and breast tumorigenesis. Genes Dev 23:2563–2577

Wegwitz F, Kluth MA, Manz C, Otto B, Gruner K, Heinlein C, Kuhl M, Warnecke G, Schumacher U, Deppert W et al (2010) Tumorigenic WAP-T mouse mammary carcinoma cells: a model for a self-reproducing homeostatic cancer cell system. PLoS One 5:e12103

Weigelt B, Bissell MJ (2008) Unraveling the microenvironmental influences on the normal mammary gland and breast cancer. Semin Cancer Biol 18:311–321

Weigelt B, Reis-Filho JS (2009) Histological and molecular types of breast cancer: is there a unifying taxonomy? Nat Rev Clin Oncol 6:718–730

Wijnhoven SW, Zwart E, Speksnijder EN, Beems RB, Olive KP, Tuveson DA, Jonkers J, Schaap MM, van den Berg J, Jacks T et al (2005) Mice expressing a mammary gland-specific R270H mutation in the p53 tumor suppressor gene mimic human breast cancer development. Cancer Res 65:8166–8173

Wu M, Jung L, Cooper AB, Fleet C, Chen L, Breault L, Clark K, Cai Z, Vincent S, Bottega S et al (2009) Dissecting genetic requirements of human breast tumorigenesis in a tissue transgenic model of human breast cancer in mice. Proc Natl Acad Sci USA 106:7022–7027

Xu X, Qiao W, Linke SP, Cao L, Li WM, Furth PA, Harris CC, Deng CX (2001) Genetic interactions between tumor suppressors Brca1 and p53 in apoptosis, cell cycle and tumorigenesis. Nat Genet 28:266–271

Xu X, Wagner KU, Larson D, Weaver Z, Li C, Ried T, Hennighausen L, Wynshaw-Boris A, Deng CX (1999) Conditional mutation of Brca1 in mammary epithelial cells results in blunted ductal morphogenesis and tumour formation. Nat Genet 22:37–43

Yan H, Blackburn AC, McLary SC, Tao L, Roberts AL, Xavier EA, Dickinson ES, Seo JH, Arenas RB, Otis CN et al (2010) Pathways contributing to development of spontaneous mammary tumors in BALB/c-Trp53+/− mice. Am J Pathol 176:1421–1432

Zander SA, Kersbergen A, van der Burg E, de Water N, van Tellingen O, Gunnarsdottir S, Jaspers JE, Pajic M, Nygren AO, Jonkers J et al (2010) Sensitivity and acquired resistance of BRCA1;p53-deficient mouse mammary tumors to the topoisomerase I inhibitor topotecan. Cancer Res 70:1700–1710

Zelazny E, Li B, Anagnostopoulos AM, Coleman A, Perkins AS (2001) Cooperating oncogenic events in murine mammary tumorigenesis: assessment of ErbB2, mutant p53, and mouse mammary tumor virus. Exp Mol Pathol 70:183–193

Zhao T, Xu Y (2010) p53 and stem cells: new developments and new concerns. Trends Cell Biol 20:170–175

Chapter 8
TP53 Somatic Mutations: Prognostic and Predictive Value in Human Cancers

Magali Olivier

1 Introduction

Somatic inactivation of the tumor suppressor gene *TP53* (OMIM #191117) contributes to a large proportion of human cancers (Olivier et al. 2010). The most frequent mode of inactivation is mutation of the gene, although mechanisms such as loss of allele or inactivation of the protein through sequestration by viral or cellular proteins may occur in specific types of cancers. Recent data produced by genome-wide sequencing have confirmed that mutation patterns in many cancer genomes are dominated by *TP53* mutations. Because p53 protein plays a central anti-proliferative role through induction of apoptosis, senescence and cell-cycle arrest in response to various types of stresses, including DNA damage, the *TP53* status of a tumor should have deep implications for tumor growth potential as well as tumor response to various genotoxic chemotherapeutic agents. Indeed, preclinical studies have shown that p53 is an important factor in the apoptotic and cytotoxic activity of many DNA damaging chemotherapies (Lowe et al. 1994; Wahl et al. 1996; O'Connor et al. 1997). However, despite extensive investigation in various types of cancer, the prognostic and predictive values of *TP53* status in patient cohorts remain controversial in many cancer types and the routine clinical use of *TP53* status remains limited (see Chap. 17).

There are several reasons that may explain the controversial results obtained for the prognostic and predictive value of *TP53* status: (1) heterogeneity in the methods used to assess *TP53* status, (2) the types of *TP53* mutations considered, (3) heterogeneity of tumor types studied, (4) the large range of drugs used, and (5) heterogeneity in study design.

M. Olivier (✉)
Group of Molecular Mechanisms and Biomarkers, International Agency for Research on Cancer (CIRC/IARC), 150 Cours Albert Thomas, 69372 Lyon, France
e-mail: olivierm@iarc.fr

P. Hainaut et al. (eds.), *p53 in the Clinics*, DOI 10.1007/978-1-4614-3676-8_8,

Among various methods that have been developed to assess *TP53* status (see Chap. 9), massive translational efforts have been made using immunohistochemistry (IHC) as the sole assessment of p53 functional status. Indeed, more than 70% of *TP53* mutations observed in cancers are missense mutations producing mutant proteins that often accumulate in the nucleus of tumors cells and can thus be detected by IHC. Although some studies have shown a prognostic or predictive value of p53 expression by immunostaining, the overall results from these studies are not consistent. In fact, up to 25% of mutations are truncating mutations (nonsense, small insertions or deletions that produce frameshift) that stain negative by IHC and are more often located outside exons 5–8. Moreover, although wild-type p53 protein is usually not detectable in normal conditions because of permanent degradation through MDM2-dependent ubiquitination, wild-type p53 may lead to positive IHC if activated by cellular stress. IHC is thus a poor surrogate marker for assessing *TP53* gene status as it produces a high rate of false-positive (overexpression of wild-type p53 protein) and false-negative cases (truncating mutations that stain negative by IHC).

Studies that have used gene resequencing to assess *TP53* status have produced more consistent results, at least for certain types of cancer such as breast, head and neck, or hematological cancers. However, because most mutations are missense substitutions, another level of heterogeneity comes from the impact of these substitutions on protein functions. Indeed, functional assays in yeast (FASAY) developed to test the transactivation capacity of p53 mutant proteins on different p53 target sequences have shown a great variability of activities between mutant proteins and between target sequences (Inga et al. 2002; Kato et al. 2003). Whereas hotspot missense mutations in DNA-binding motifs (DBM, including L1, L2, H1, L3, L1, S2, S2′, H2 structural motifs) lead to a general loss of specific transactivation capacity on all wild-type target sequences, missense mutations outside DBM more often retain transcriptional activity on some promoters. Mutant proteins may also exert dominant-negative effects (DNE) over wild-type p53, an activity that may influence tumor phenotype (Petitjean et al. 2007a). More missense mutations in DBM have been shown to have DNE compared to missense mutations outside these motifs. A number of resequencing studies performed in different types of cancer have reported that mutations within DBM (L2–L3 loops of the DNA-binding domain in particular) were associated with poorer prognosis compared to mutations outside DBM (Borresen et al. 1995; Alsner et al. 2000; Skaug et al. 2000; Kihara et al. 2000; Temam et al. 2000). Thus, variations in the proportion of mutations outside DBM may account for some of the differences in the prognostic impact of the overall p53 status reported in different studies.

As discussed in detail in Chap. 16, it is important to distinguish prognostic and predictive factors. A prognostic marker is predicting risk of relapse and thus able to identify patients who are at risk for relapse. A predictive marker is predicting response to a certain therapy and thus able to identify patients who will respond to a certain therapy. Study design to identify a prognostic marker will thus differ from the one of a predictive marker with in particular different endpoints. While a marker may be prognostic and predictive, it may in fact predict a poor prognosis due to a yet undetected interaction between the marker and the standard treatment. For *TP53*, most studies have investigated its prognostic value in patients receiving treatment without addressing the effect of treatment. While *TP53* gene mutations have been consistently associated with poor prognosis in many cancer types, recent studies point to the fact that the prognostic value of *TP53* mutations depends on the type of treatment regimen. In breast cancer for example, while *TP53* mutations are usually associated with poor survival, a recent study has shown that the presence of a *TP53* mutation was associated with good response to treatment. In fact, in this study, patients were treated with a dose-dense epirubicin–cyclophosphamide regimen that targets highly proliferating tumors, a hallmark of *TP53*-mutated tumors (Bertheau et al. 2007). In other types of cancer such as brain and pancreas tumors, mutations were found to be associated with both poor and good prognosis depending on the study. The tissue-specific effects of wild-type p53 may thus be another crucial determinant of the ultimate outcome of any anticancer treatment that activates p53.

In this chapter, we discuss *TP53* mutation occurrence in human cancers and review studies that have investigated the prognostic and predictive value of *TP53* gene status (studies using IHC are not reviewed), with an emphasis on the ones that have screened at least exons 4–10 of *TP53* gene and have been done in the context of specific and controlled treatment regimens.

2 Diversity of *TP53* Somatic Mutations in Human Cancers

Somatic *TP53* gene mutations are found in almost every types of cancer at various frequencies (Fig. 8.1). Cancers with high mutation frequencies (35–50%) include ovarian, esophageal, colorectal, head and neck, larynx, and lung cancers. Tumors of the brain, breast, stomach, and liver show intermediate mutation rates (20–35%). Cancers with low mutation frequency include leukemia, sarcoma, testicular cancer, malignant melanoma, and cervical cancer. While it is well known that p53 is inactivated at the protein level by E6 viral protein in cervical cancers, how p53 may be inactivated in other cancers with low *TP53* mutation frequency is not known. Mutations are usually more frequent in advanced stage of the disease or in cancer subtypes of more aggressive behavior (such as triple-negative or HER2-amplified breast cancers and stage III ovarian cancers) (Sorlie et al. 2003; Langerod et al. 2007). In most cancers, multiple mutations in one tumor sample are infrequent (less than 5% of tumor samples in cancer such as breast, lung, colorectum, ovary, esophagus, or liver). However, some tumor types may present multiple mutations in 5–15% of samples. This is the case for skin, bladder, head and neck, prostate, brain, nasal cavity, or soft tissue cancers. These multiple mutations may correspond to a high mutation load due to carcinogen exposures and/or defects in repair pathways (Olivier et al. 2004). The increase in sensitivity of more recent mutation detection methods may provide further insight into the occurrence of multiple *TP53* mutations in different types of cancers.

A large majority of *TP53* mutations are missense substitutions (70%) that cluster in the DNA-binding domain of the protein encoded by exons 5–8 (codons 125–300)

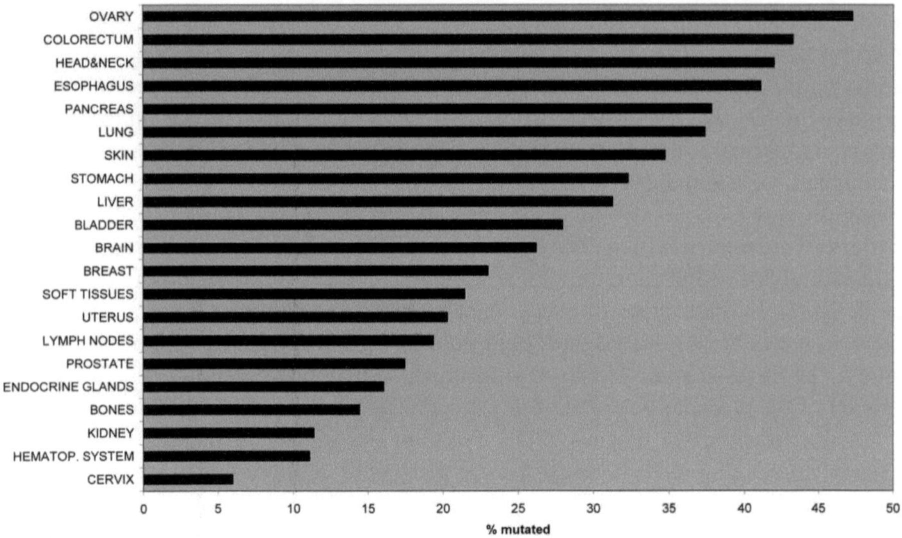

Fig. 8.1 Prevalence of *TP53* somatic mutations in various cancer types. Data from the IARC *TP53* Database (R15, November 2010)

(Fig. 8.2). The rest of the mutations are truncating mutations, including frameshift insertions and deletions (10%), nonsense mutations (10%), and mutations at splice sites (4%). The proportion of truncating mutations is higher outside the DNA-binding domain (Fig. 8.2b). About 25% of all mutations correspond to G:C>A:T nucleotide substitutions at CpG sites within the coding sequence. These sites are all methylated in the *TP53* sequence and thus highly mutable due to deamination of 5-methylcytosine leading to the emergence of thymine base-paired with guanine at CpG sites (Pfeifer 2000). Three hotspots occur at these CpG sites, codons 175, 248, and 273 which represent 60% of the mutated residues at CpG sites. Another five residues at these sites (196, 213, 245, 282, and 306) represent 26% of all CpG mutations. Apart from these hotspots, the spectrum of mutation is very diverse with all residues within the DNA-binding domain having reported to be mutated in cancer.

Some cancers have very characteristic mutation patterns. This is the case in particular for bladder carcinoma, non-small cell lung carcinoma, non-melanoma skin cancer, or hepatocellular carcinoma. These differences have been linked to differences in tumor etiology where the involvement of a specific mutagen has been identified or is suspected (Olivier et al. 2004, 2010). Geographic differences have been reported that are thought to be due to differences in environmental exposures. In addition, genetic background may also strongly influence *TP53* mutation patterns as exemplified by breast tumors arising in BRCA1 germline mutation carriers or skin tumors arising in XP patients. Indeed, BRCA1-related breast cancers were reported to carry a high incidence of protein-truncating *TP53* mutations that stained negative by IHC (Holstege et al. 2009). In fact the majority of BRCA1-related tumors that stained negative for p53 accumulation had protein-truncating *TP53*

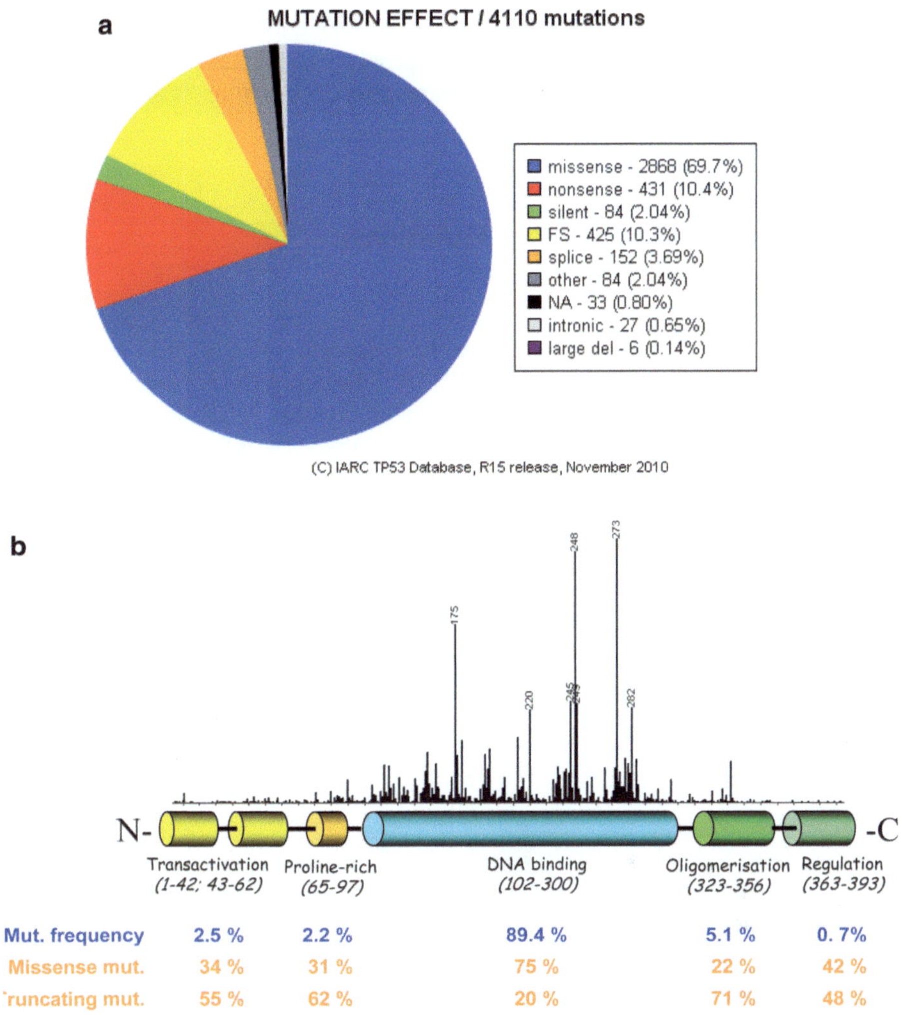

Fig. 8.2 Characteristics of somatic *TP53* mutations found in human cancers. (**a**) Type of mutations according to their impact on the coding sequence. (**b**) Distribution of mutations according to their location in the protein sequence. Data from the IARC *TP53* Database (R15, November 2010) taking into account only studies that have screened the full coding sequence

mutations. In XP skin tumors, mutations are located at PY–PY sites and more than 50% are tandem CC to TT transitions considered as signatures of UV-induced lesions (Dumaz et al. 1993). These mutations, although far more frequent in skin tumors compared to other tumor types, were only found in 14% of skin tumors from normal individuals. These results show the influence of DNA repair enzyme activity on *TP53* mutation spectra.

This diversity of *TP53* mutation patterns results in a functional diversity that raises several questions regarding the classification of patient according to *TP53* status. Are truncating mutations similar to loss-of-function missense mutations in terms of prognostication? Which biological property of mutant proteins should be taken into account to derive clinically meaningful classification systems based on *TP53* status?

3 Prognostic Value of *TP53* Mutations: Mutation Type Does Matter

Around 200 studies that have investigated the prognostic value of *TP53* status using gene sequencing are reported in the IARC *TP53* Database (R15, November 2010). A majority of studies have reported an association between *TP53* mutation and poor outcome, but consistency between studies varies between cancer types. Breast, colorectal, hematological, and head and neck cancers show more consistent results (Fig. 8.3).

In several studies, results suggest that different types of mutations have different prognostic value, although differences in mutation grouping and tumor types analyzed make it difficult to draw general conclusions. In a study performed on colorectal hepatic metastases, while p53 protein immunostaining did not show correlation with

TUMOR SITE	Number of studies* reporting that TP53 mutations are : * Only studies with > 50 cases were included		
	Related to bad prognosis	Related to good prognosis	Not related to prognosis
Bladder	4	-	3
Bones	1	-	1
Brain	3	2	4
Breast	28	1	5
Colorectum	16	-	8
Esophagus	2	-	2
Head & Neck	7	-	2
Hematol.	12	-	-
Larynx	-	-	1
Liver	3	-	-
Lung	8	-	6
Ovary	7	1	2
Pancreas	1	1	1
Prostate	1	-	1
Soft tissues	2	-	-
Stomach	1	-	2
Renal pelvis	1	-	-

Fig. 8.3 Number of studies on the prognostic and predictive value of somatic *TP53* mutations in human cancers and their overall results. Studies compiled in the IARC *TP53* Database (R15, November 2010)

survival, the presence of a *TP53* gene mutation was associated with poor prognosis in univariate and multivariate analyses (Mollevi et al. 2007). These results were valid both in patients undergoing radical resection only or in patients receiving adjuvant chemotherapy (Mollevi et al. 2007). In these series, 40% of mutations were truncating mutations, which is high compared to the overall 20% reported in the IARC *TP53* Database. Such a high level of truncating mutations was reported in another study on colorectal cancers (Miyaki et al. 2002). This study showed an excess of truncating mutations in hepatic metastases compared to primary carcinomas without liver metastasis. Two studies of unselected lung cancers and one study of unselected ovarian cancers reported a worse prognosis associated with truncating mutations compared to missense mutations (de Anta et al. 1997; Hashimoto et al. 1999; Rose et al. 2003). These results suggest that protein-truncating mutations may be more relevant prognostic factors than missense mutations in certain types of cancer.

Some studies have reported that missense mutations located in L2/L3 loops or at codons involved in zinc binding (also located within L2/L3 loops) were associated with a worse outcome compared to mutations outside these motifs (Kucera et al. 1999; Skaug et al. 2000; Kihara et al. 2000; Geisler et al. 2001, 2003; Poeta et al. 2007; Chrisanthar et al. 2008). In a large series of unselected breast cancers, we also found that missense mutations in DBM were associated with poorer survival than missense mutations outside DBM (Olivier et al. 2006). However, we found that truncating and missense mutations in DBM had similar prognostic value. In fact, in Geisler et al. (2001, 2003), about half (47% and 55%) of mutations that were classified as L2/L3 mutations were truncating mutations. Thus in these studies, both truncating and missense mutations in DBM were associated with poor survival. In lung, esophagus and head and neck studies, most L2/L3 mutations were missense substitutions, confirming that missense mutations in DBM have poor prognostic value in these types of cancer.

The proportions of missense mutations in DBM and truncating mutations vary according to cancer type. Truncating mutations are more prevalent in breast and skin cancers, while colorectal, esophagus, stomach, and brain cancers show the highest rates of missense mutations in DBM (Table 8.1). It is of note that these proportions show inter-study variability that may be due to differences in screening method, tumor subtypes, or tumor etiology. To overcome these variability and because truncating and missense mutations may carry different prognostic value in some cancer types, some classification systems are needed that take into account the predicted impact of mutations on protein functions. The IARC *TP53* Database (http://www-p53.iarc.fr/) provides annotations on various properties of missense mutations, including transactivation (TA) and dominant-negative (DN) activities of mutant proteins, and structural features (Petitjean et al. 2007b). These annotations are derived from both experimental assessment and in silico prediction of mutant protein activities. According to these annotations, a higher proportion of missense mutations in DBM, compared to mutations outside DBM, displays complete loss of TA capacity (45% vs 14%), have DN activity (65% vs 20%) and fall at conserved residues (83% vs 42%). Both DN and TA activities may thus be important determinant of the prognostic value of missense mutations in DBM. Functional

Table 8.1 Prevalence and type of *TP53* gene mutations by cancer type (data from IARC *TP53* Database, R15, including only studies that have screened the entire coding sequence of *TP53* and excluding cell lines)

Cancer type	Mutation prevalence (%)	% Truncating mutations[a]	% Missense mutations in DBM[b]
Bladder	34.4	22.8	50.0
Breast	26.0	30.1	39.2
Brain	23.4	16.2	52.4
CRC	46.9	19.8	53.0
Esophagus	33.8	4.8[c]	51.6[c]
Head and neck	46.5	27.1	45.9
Hematol	12.6	13.9[c]	44.2[c]
Kidney	6.1	27.8[c]	33.3[c]
Lung	47.1	25.8	42.1
Ovary	51.4	25.2	44.1
Skin	30.9	29.9	45.4
Stomach	31.1	22.0[c]	61.0[c]

[a]Proportion of mutations relative to the total number of mutations
[b]Mutations located in L2 and L3 loops (codons 164–194 and 237–250, respectively) that are involved in DNA contacts in the minor groove, and in the LSH motif (codons 119–135 and codons 272–287) that are involved in DNA contacts in the major groove of DNA
[c]These figures may not be representative as the total number of mutations is <100

annotations from the IARC *TP53* Database may thus be useful tools to better classify missense mutations and should be tested on different series of patients to determine their clinical relevance.

4 Prognostic and Predictive Value of *TP53* Mutations: Treatment Does Matter

Survival in patients receiving anticancer treatments is highly dependent on treatment response. While experimental models have identified p53 as an important determinant of the cytostatic or cytotoxic effects of various types of chemotherapeutic agents (Lowe et al. 1994; Cimoli et al. 2004), results obtained in clinical studies have shown inconsistencies and whether p53 status may be a useful predictive marker is still a matter of debate. In fact, as discussed in details in Chap. 16, none of the published *TP53* studies was performed in the context of an adequately designed clinical trial for the evaluation of the predictive value of *TP53* mutations. Most studies retrospectively assessed the prognostic value of *TP53* mutations in patients treated with specific regimens. These studies show heterogeneity in *TP53* status assessment and type of treatment regimens. Indeed, when compiling studies that have analyzed at least exons 4–10 of *TP53* gene (where more than 95% of mutations occur) and assessed the predictive value of *TP53* status in the context of a specific type of treatment (Table 8.2), one can see that each study has been done in a specific treatment setting, with different doses and different drug combinations. No real replication has been done for any treatment setting. Moreover, only a few studies have used neoadjuvant setting which allow for a better assessment of treatment response. It is thus difficult to draw any strong conclusion regarding the predictive value of *TP53* even for a class of drug.

For example, experimental models have shown p53-dependent cytotoxic activity of anthracyclines and p53-independent activity of taxanes. These treatments are widely used in breast cancers and some studies did found an association between mutant p53 status and poorer response to anthracyclines or better response to taxanes when used as single agents (Table 8.2). A large recent study confirmed the resistance of *TP53*-mutated patients to anthracyclines but found no association between *TP53* status and response to taxanes (Chrisanthar et al. 2011). Results from the first phase III trial designed to address the predictive value of *TP53* status regarding a taxane versus non-taxane-based regimen showed that *TP53* status, assessed by FASAY, was not predictive of preferential sensitivity to taxanes (Bonnefoi et al. 2011). These studies suggest that *TP53* status may be a good marker for response to anthracyclines but not taxanes.

However, most current therapeutic regimens are using combination of different agents. In particular, anthracyclines are combined with widely different doses of alkylating agents such as cyclophosphamide. It was recently shown that cyclophosphamide dose intensification may circumvent anthracycline resistance in *TP53*-mutated breast cancers (Lehmann-Che et al. 2010). High response rates were observed in *TP53*-mutated breast cancers, especially ER-negative cases, in patients treated with a dose-dense epirubicin and cyclophosphamide regimen. Based on these results, *TP53* mutations may be critical biomarkers for sensitivity to high-dose cyclophosphamide. The predictive value of *TP53* mutations may thus depend on the dosage combination of cyclophosphamide/anthracycline regimens.

Recent developments in cancer therapies have introduced molecules that target specific oncogenes such as HER2, EGFR, or VEGF (Zhang et al. 2009). These treatments are highly selective and effective in selective types of cancers that carry specific genetic alterations. However, as for other types of treatment, de novo or secondary treatment resistance is a major issue and finding additional markers that can predict treatment efficacy is an intense field of research. Because p53 is functionally connected with some of the targeted oncogenes, *TP53* mutations may potentially affect responses to some of these treatments. One study in chemorefractory metastatic CRC treated with cetuximab-based chemotherapy showed that *TP53* mutations were associated with response to this anti-EGFR drug (Oden-Gangloff 2009). In this series, both *TP53* and KRAS mutations were independent predictive markers. It remains to be determined if similar results may be obtained with other anti-EGFR drugs or with first-line cetuximab.

Knowing p53 protein status may be most useful for one type of treatment, namely p53 gene therapy. As discussed in details in Chap. 11, p53 gene therapies have shown efficacy in several types of cancers and one of them, Gendicine, has been approved in China for the treatment of head and neck squamous cell carcinoma. Because p53 mutant proteins may exert DN activity, it is predicted that tumors that express high levels of dominant-negative mutant proteins will be less likely to respond to p53 gene therapy. This was confirmed in different clinical trials using

Table 8.2 Studies having evaluated the prognostic/predictive value of *TP53* gene mutations in the context of a specific primary therapeutic regimen and have resequenced at least exons 4–10 of *TP53* gene

References	Mutation detection method	Treatment	Response assessment	Mutated samples	Response associated with mutations	Impact on survival in univariate analysis
Breast cancer						
Bergh et al. (1995)	cDNA sequencing (exons 2–11)	Adjuvant tamoxifen+RX	n.d.	25% (12/48)	n.d.	Short DFS and OS in LN+ patients ($P>0.05$)
		No adjuvant therapy	n.d.	18% (13/72)	n.d.	Short DFS ($P<0.0002$) and OS in LN+ patients ($P>0.05$)
Berns et al. (2000)	cDNA sequencing (exons 2–11)	First-line tamoxifen (40 mg)	Pathological tumor size, UICC criteria	32.2% (65/202)	Poor response ($P<0.0001$)	Short OS and DFS ($P<0.001$)
Kandioler-Eckersberger et al. (2000)	Direct sequencing (exons 2–11)	Neoadjuvant 5-FU (600 mg), epirubicin (60 mg), cyclophosphamide (600 mg); four cycles	Pathological tumor size, UICC criteria	22.8% (8/35)	Treatment failure ($P=0.0029$)	n.d.
		Neoadjuvant paclitaxel (215–300 mg); four cycles	Pathological tumor size, UICC criteria	18.7% (6/32)	Treatment response ($P=0.07$)	n.d.
Takahashi et al. (2000), Kandioler-Eckersberger et al. (2000)	Yeast assays confirmed by sequencing (exons 4–10)	Adjuvant 5-FU, tamoxifen in ER+	n.d.	39.5% (30/76)	n.d.	Short OS ($P=0.03$); no association with DFS ($P=0.25$)
Geisler et al. (2001)	TTGE-sequencing (exons 2–11)	Neoadjuvant doxorubicin (14 mg)	Tumor size, UICC criteria	28.9% (26/90)	Treatment failure ($P=0.063$ for all mutations; $P=0.008$ for L2/L3 mutations)	Poor DFS and OS associated with mutations (all and L2/L3)

Geisler et al. (2003)	TTGE-sequencing (exons 2–11)	Neoadjuvant 5-FU (1000 mg), mitomycin (6 mg) (FUMI); four cycles	Tumor size, UICC criteria	51.4% (18/35)	Treatment failure ($P=0.177$ for all mutations; $P=0.006$ for L2/L3 mutations)	n.d.
Andersson et al. (2005)	cDNA sequencing (exons 2–11)	Adjuvant CMF	n.d.	37% (46/125)	n.d.	Poor OS ($P=0.001$)
		Adjuvant tamoxifen (10–40 mg)	n.d.	22% (40/183)		No association ($P=0.1$)
		Adjuvant CMF+tamoxifen	n.d.	28% (19/68)		No association ($P=0.27$)
Bertheau et al. (2007)	Yeast assays confirmed by sequencing (exons 4–10)	Neoadjuvant epirubicin (75 mg), cyclophosphamide (1200 mg); six cycles	Pathological tumor size, UICC criteria	35% (28/80)	Treatment response ($P<0.001$)	WT or MUT with CR were associated with better OS than MUT with IR
Lehmann-Che et al. (2010)	Yeast assays confirmed by sequencing (exons 4–10)	Dose-dense anthracycline-cyclophosphamide	Pathologic response	40.6% (99/244)	Treatment response	n.d.
Bonnefoi et al. (2011)	Yeast assays confirmed by sequencing (exons 4–10)	Neoadjuvant epirubicin, CMF, ±docetaxel	Pathologic response		No association	Poor survival
Chrisanthar et al. (2011)	Direct sequencing (exons 2–11)	Epirubicin (90 mg); four cycles	UICC criteria	21% (23/109)	Treatment failure ($P<0.05$); more significant for L2/L3 mutations	No association
		Paclitaxel (200 mg); four cycles	UICC criteria	22% (25/114)	No association	Poor DSS
Gluck et al. (2011)	AmpliChip *TP53* assay (exons 2–11)	Neoadjuvant capecitabine, docetaxel, ±trastuzumab (based on HER2 status)	Pathologic response	53.1% (78/147)	Treatment response ($P=0.0032$)	n.d.

(continued)

Table 8.2 (continued)

References	Mutation detection method	Treatment	Response assessment	Mutated samples	Response associated with mutations	Impact on survival in univariate analysis
Lung cancer (NSCC)						
Kandioler-Eckersberger et al. (1999)	Direct sequencing (exons 2–11)	Neoadjuvant cisplatin (30 mg)+ifosfamide (1 g): three cycles	Pathological tumor size, UICC criteria	33% (8/24)	Treatment failure ($P<0.001$)	Poor survival ($P=0.021$)
Tomizawa et al. (1999)	SSCP and sequencing (exons 2–11)	Surgery only	n.d.	47.5% (49/103)	n.d.	Missense mutation associated with poor survival ($P<0.001$)
Tomizawa et al. (2002)	SSCP and sequencing (exons 2–11)	Surgery only	n.d.	33.6% (37/110)	n.d.	Not related to survival ($P=0.1630$)
Kandioler et al. (2008)	Direct sequencing (exons 2–11)	Neoadjuvant: three cycles of cisplatin (60 mg, day 1, 7)+etoposide (150 mg days 3–5) followed by one cycle of simultaneous cisplatin (50 mg, day 2, 9)+etoposide (100 mg, days 4–6)+irradiation (45Gy)	Pathological tumor size, RECIST	31% (11/35)	Treatment failure ($P=0.002$)	Poor survival ($P=0.02$)
Head and neck cancer						
Obata et al. (2000)	Yeast assay and sequencing (exons 2–11)	Rx (45 Gy)	Tumor size	50% (9/18)	Treatment failure ($P<0.01$)	Poor OS ($P=0.035$)
Temam et al. (2000)	Direct sequencing (exons 2–11)	Simultaneous cisplatin (20 mg) and 5-FU (1,000 mg)	Tumor size	37% (39/105)	Treatment failure ($P=0.002$)	n.d.
Nemunaitis et al. (2009)	GeneChip and sequencing (exons 2–11)	p53 gene therapy (Advexin)	n.d.	30.5% (11/36)	Shorter DFS	Mutation+positive IHC associated with poor survival ($P<0.0001$)
		Methotrexate	n.d.	30.5% (11/36)	No association	No association

Ovarian cancer						
Smith-Sorensen et al. (1998)	TTGE sequencing (exons 2–11)	Adjuvant paclitaxel (175 mg), cisplatin (75 mg); six cycles	WHO criteria	70% (14/20)	No association	No association
		Adjuvant cyclophosphamide (750 mg), cisplatin (75 mg); six cycles	WHO criteria	77.8% (14/18)	No association	Shorter DFS
Reles et al. (2001)	SSCP sequencing (exons 2–11)	Adjuvant cisplatin-cyclophosphamide or carboplatin-cyclophosphamide	Clinical FU	54% (40/74)	Missense mutations associated with resistance ($P=0.008$)	Shorter DFS and OS ($P=0.014$)
Havrilesky et al. (2003)	Direct sequencing (exons 2–11)	Platinium-based adjuvant treatment with or without paclitaxel	n.d.	77.6% (73/94)	n.d.	Improved short term DFS
		Adjuvant paclitaxel	n.d.	73.3% (11/15)	n.d.	No association
Concin et al. (2005)	cDNA sequencing (exon 2–11)	Adjuvant platinium-based ± paclitaxel	Clinical FU	65.6% (80/122)	Associated with treatment failure	Shorter DFS ($P<0.0001$) and OS ($P=0.003$)
Colorectal cancer						
Kandioler et al. (2002)	Direct sequencing (exons 2–10)	Preoperative RX (25 Gy)	Tumor size	45.3% (29/64)	Associated with treatment failure	Poor survival ($P=0.049$)
Rebischung et al. (2002)	DGGE and sequencing (exons 2–11)	Preoperative RX (25 Gy)	Tumor size			
Katkoori et al. (2009)	Direct sequencing (exons 4–9)	Surgery only	n.d.	52.3% (195/373)	n.d.	Poor survival ($P<0.05$)
Brain tumors (GBM)						
Chen et al. (2006)	SSCP and sequencing (exons 2–11)	Surgery + RX (60 Gy)	n.d.	26.8% (26/97)	n.d.	Longer OS ($P=0.003$)

(continued)

Table 8.2 (continued)

References	Mutation detection method	Treatment	Response assessment	Mutated samples	Response associated with mutations	Impact on survival in univariate analysis
Felsberg et al. (2009)	SSCP and sequencing (exons 4–10)	Surgery, RX, +2 cycles temozolomide (150–200 mg)	Time to recurrence, Macdonald criteria	20% (13/65)	No association with response ($P=0.061$)	No association with OS ($P=0.397$)
Hematol cancers						
Forconi et al. (2009)	Direct sequencing (exons 2–10)	Cladribine (0.1 mg)	Consensus resolution criteria	3.4% (2/58)	Poor response ($P=0.006$)	n.d.
Esophagus						
Casson et al. (2003)	SSCP and sequencing (exons 4–10)	Surgery only	n.d.	50.5% (46/91)	n.d.	Shorter OS ($P=0.01$)
Goan et al. (2005)	Direct sequencing (exons 2–11)	Surgery only	n.d.	52.5% (21/40)	n.d.	No association
Yamasaki et al. (2010)	SSCP and sequencing (exons 4–9)	Neoadjuvant 5-FU (700 mg), adriamycin (35 mg), cisplatin (70 mg); two cycles FAP	Tumor size	46.7% (36/77)	Poor response ($P<0.0001$)	Shorter OS ($P=0.0007$)

Advexin in recurrent head and neck squamous cell carcinomas, where it was found that tumors carrying missense mutations and staining positive for p53 were less likely to respond to p53 gene-therapy combined with irradiation than tumors expressing a non-mutated p53 (Nemunaitis et al. 2009).

The heterogeneity of studies and results presented in Table 8.2 points to the need for large clinical trial studies to determine in which context (type of cancer and type of regimen) *TP53* mutations may be a useful predictive marker.

5 Polymorphisms in *TP53* Gene and Prognosis

Several polymorphisms in *TP53* gene have been identified and validated in human populations (IARC *TP53* Database, R15) (Whibley et al. 2009). One coding SNP, P72R, has been extensively studied for its association with cancer risk but with limited results. A few studies have analyzed its impact on prognosis or response to treatment. Here again conflicting results have been obtained. In general, evaluation of survival studies is affected by problems similar to those described for cancer risk association studies and, importantly, most of these studies did not take into account *TP53* somatic mutations that are more likely to affect prognosis or response to treatment. In fact, studies that have analyzed both P72R SNP and *TP53* somatic mutations suggest that this SNP may have modifier effects on *TP53* mutations. Indeed, several studies have found a bias towards the expression of a mutated R72 allele in tumors of Pro/Arg heterozygous patients, with LOH affecting the Pro allele (Langerod et al. 2002; Furihata et al. 2002; Schneider-Stock et al. 2004; Nelson et al. 2005; Lind et al. 2007). In brain tumors, it was found that *TP53* P72 allele was significantly associated with shorter survival among patients with glioblastomas carrying a *TP53* mutation, and among those treated with surgery plus radiotherapy (Zawlik et al. 2009). Furthermore, experimental studies showed that p53 mutant proteins with an arginine at codon 72 were more potent in cooperating with EJ-Ras to transform cells and in neutralizing p73-induced activity (Marin et al. 2000). Neutralization of p73-dependent apoptosis correlated well with the response to chemo-radiotherapy treatment of HNSCC patients, carriers of a 72Pro mutant showing longer overall and progression-free survivals (Bergamaschi et al. 2003). This modifier effect has also been reported in the context of *TP53* germline mutations in LFS, where R72 was associated with earlier onset of cancer in *TP53* germline mutation carriers (Bougeard et al. 2006). The haplotype context of *TP53* mutations may thus be another parameter to take into account in future studies on the prognostic and predictive value of *TP53* somatic mutations.

6 Conclusion

Among hundreds of biomarkers that have been investigated for cancer prognostication, few have reached clinical practice. *TP53*, although promising in pioneering studies, is among the ones that has reached the pathologist's hands for only a limited number of applications (see Chap. 17). Lack of easy methodologies for assessing *TP53* gene status in routine practice has been a major drawback. With easier access to sequencing technologies and mutation analysis close to become standard practice in molecular pathology, *TP53* gene status is expected to come back as one of the candidate markers to be tested in clinical trials.

Recent whole genome expression analyses have described gene signatures reflecting *TP53* status (Miller et al. 2005; Troester et al. 2006; Langerod et al. 2007). These so-called "p53 signatures" may be better predictors of "functional" loss of wild-type p53 and of cancer outcome compared to *TP53* gene status (see Chap. 9). Downstream functional assessment of p53 may capture the impact of other *TP53* features that may influence tumor outcome or treatment sensitivity. Further studies using both downstream functional assessment of *TP53* status and *TP53* gene status itself are thus needed to better understand the impact of *TP53* mutations on tumor phenotype and behavior. To perform these studies in the context of clinical trials with specific treatment regimens would be essential to avoid previous pitfalls.

References

Alsner J, Yilmaz M, Guldberg P, Hansen LL, Overgaard J (2000) Heterogeneity in the clinical phenotype of *TP53* mutations in breast cancer patients. Clin Cancer Res 6:3923–3931

Andersson J, Larsson L, Klaar S, Holmberg L, Nilsson J, Inganas M, Carlsson G, Ohd J, Rudenstam CM, Gustavsson B, Bergh J (2005) Worse survival for *TP53* (p53)-mutated breast cancer patients receiving adjuvant CMF. Ann Oncol 16:743–748

Bergamaschi D, Gasco M, Hiller L, Sullivan A, Syed N, Trigiante G, Yulug I, Merlano M, Numico G, Comino A, Attard M, Reelfs O, Gusterson B, Bell AK, Heath V, Tavassoli M, Farrell PJ, Smith P, Lu X, Crook T (2003) p53 polymorphism influences response in cancer chemotherapy via modulation of p73-dependent apoptosis. Cancer Cell 3:387–402

Bergh J, Norberg T, Sjogren S, Lindgren A, Holmberg L (1995) Complete sequencing of the p53 gene provides prognostic information in breast cancer patients, particularly in relation to adjuvant systemic therapy and radiotherapy. Nat Med 1:1029–1034

Berns EM, Foekens JA, Vossen R, Look MP, Devilee P, Henzen-Logmans SC, van Staveren IL, van Putten WL, Inganas M, Meijer-van Gelder ME, Cornelisse C, Claassen CJ, Portengen H, Bakker B, Klijn JG (2000) Complete sequencing of *TP53* predicts poor response to systemic therapy of advanced breast cancer. Cancer Res 60:2155–2162

Bertheau P, Turpin E, Rickman DS, Espie M, de RA, Feugeas JP, Plassa LF, Soliman H, Varna M, de RA, Lehmann-Che J, Beuzard Y, Marty M, Misset JL, Janin A, de TH (2007) Exquisite sensitivity of *TP53* mutant and basal breast cancers to a dose-dense epirubicin-cyclophosphamide regimen. PLoS Med 4:90

Bonnefoi H, Piccart M, Bogaerts J, Mauriac L, Fumoleau P, Brain E, Petit T, Rouanet P, Jassem J, Blot E, Zaman K, Cufer T, Lortholary A, Lidbrink E, Andre S, Litiere S, Lago LD, Becette V, Cameron DA, Bergh J, Iggo R (2011) *TP53* status for prediction of sensitivity to taxane versus non-taxane neoadjuvant chemotherapy in breast cancer (EORTC 10994/BIG 1–00): a randomised phase 3 trial. Lancet Oncol 12:527–539

Borresen AL, Andersen TI, Eyfjord JE, Cornelis RS, Thorlacius S, Borg A, Johansson U, Theillet C, Scherneck S, Hartman S (1995) *TP53* mutations and breast cancer prognosis: particularly poor survival rates for cases with mutations in the zinc-binding domains. Genes Chromosomes Cancer 14:71–75

Bougeard G, Baert-Desurmont S, Tournier I, Vasseur S, Martin C, Brugieres L, Chompret A, Bressac-de Paillerets B, Stoppa-Lyonnet D, Bonaiti-Pellie C, Frebourg T (2006) Impact of the MDM2 SNP309 and p53 Arg72Pro polymorphism on age of tumour onset in Li-Fraumeni syndrome. J Med Genet 43:531–533

Casson AG, Evans SC, Gillis A, Porter GA, Veugelers P, Darnton SJ, Guernsey DL, Hainaut P (2003) Clinical implications of p53 tumor suppressor gene mutation and protein expression in esophageal adenocarcinomas: results of a ten-year prospective study. J Thorac Cardiovasc Surg 125:1121–1131

Chen YJ, Hakin-Smith V, Teo M, Xinarianos GE, Jellinek DA, Carroll T, McDowell D, MacFarlane MR, Boet R, Baguley BC, Braithwaite AW, Reddel RR, Royds JA (2006) Association of mutant *TP53* with alternative lengthening of telomeres and favorable prognosis in glioma. Cancer Res 66:6473–6476

Chrisanthar R, Knappskog S, Lokkevik E, Anker G, Ostenstad B, Lundgren S, Berge EO, Risberg T, Mjaaland I, Maehle L, Engebretsen LF, Lillehaug JR, Lonning PE (2008) CHEK2 mutations affecting kinase activity together with mutations in *TP53* indicate a functional pathway associated with resistance to epirubicin in primary breast cancer. PLoS One 3:e3062

Chrisanthar R, Knappskog S, Lokkevik E, Anker G, Ostenstad B, Lundgren S, Risberg T, Mjaaland I, Skjonsberg G, Aas T, Schlichting E, Fjosne HE, Nysted A, Lillehaug JR, Lonning PE (2011) Predictive and prognostic impact of *TP53* mutations and MDM2 promoter genotype in primary breast cancer patients treated with epirubicin or paclitaxel. PLoS One 6:e19249

Cimoli G, Malacarne D, Ponassi R, Valenti M, Alberti S, Parodi S (2004) Meta-analysis of the role of p53 status in isogenic systems tested for sensitivity to cytotoxic antineoplastic drugs1. Biochim Biophys Acta 1705:103–120

Concin N, Hofstetter G, Berger A, Gehmacher A, Reimer D, Watrowski R, Tong D, Schuster E, Hefler L, Heim K, Mueller-Holzner E, Marth C, Moll UM, Zeimet AG, Zeillinger R (2005) Clinical relevance of dominant-negative p73 isoforms for responsiveness to chemotherapy and survival in ovarian cancer: evidence for a crucial p53-p73 cross-talk in vivo. Clin Cancer Res 11:8372–8383

de Anta JM, Jassem E, Rosell R, Martinez-Roca M, Jassem J, Martinez-Lopez E, Monzo M, Sanchez-Hernandez JJ, Moreno I, Sanchez-Cespedes M (1997) *TP53* mutational pattern in Spanish and Polish non-small cell lung cancer patients: null mutations are associated with poor prognosis. Oncogene 15:2951–2958

Dumaz N, Brougard C, Sarasin A, Daya-Grosjean L (1993) Specific UV-induced mutation spectrum in the p53 gene of skin tumors from DNA-repair-defiicent xeroderma pigmentosum patients. Proc Natl Acad Sci USA 90:10529–10533

Felsberg J, Rapp M, Loeser S, Fimmers R, Stummer W, Goeppert M, Steiger HJ, Friedensdorf B, Reifenberger G, Sabel MC (2009) Prognostic significance of molecular markers and extent of resection in primary glioblastoma patients. Clin Cancer Res 15:6683–6693

Forconi F, Sozzi E, Cencini E, Zaja F, Intermesoli T, Stelitano C, Rigacci L, Gherlinzoni F, Cantaffa R, Baraldi A, Gallamini A, Zaccaria A, Pulsoni A, Gobbi M, Tassi M, Raspadori D, Leoncini L, Rinaldi A, Sabattini E, Bertoni F, Pileri SA, Lauria F (2009) Hairy cell leukemias with unmutated IGHV genes define the minor subset refractory to single-agent cladribine and with more aggressive behavior. Blood 114:4696–4702

Furihata M, Takeuchi T, Matsumoto M, Kurabayashi A, Ohtsuki Y, Terao N, Kuwahara M, Shuin T (2002) p53 mutation arising in Arg72 allele in the tumorigenesis and development of carcinoma of the urinary tract. Clin Cancer Res 8:1192–1195

Geisler S, Borresen-Dale AL, Johnsen H, Aas T, Geisler J, Akslen LA, Anker G, Lonning PE (2003) *TP53* gene mutations predict the response to neoadjuvant treatment with 5-fluorouracil and mitomycin in locally advanced breast cancer. Clin Cancer Res 9:5582–5588

Geisler S, Lonning PE, Aas T, Johnsen H, Fluge O, Haugen DF, Lillehaug JR, Akslen LA, Borresen-Dale AL (2001) Influence of *TP53* gene alterations and c-erbB-2 expression on the response to treatment with doxorubicin in locally advanced breast cancer. Cancer Res 61:2505–2512

Gluck S, Ross JS, Royce M, McKenna EF, Jr., Perou CM, Avisar E, Wu L (2011) *TP53* genomics predict higher clinical and pathologic tumor response in operable early-stage breast cancer treated with docetaxel-capecitabine +/– trastuzumab. Breast Cancer Res Treat, Epub ahead of print

Goan YG, Hsu HK, Chang HC, Chou YP, Chiang KH, Cheng JT (2005) Deregulated p21(WAF1) overexpression impacts survival of surgically resected esophageal squamous cell carcinoma patients. Ann Thorac Surg 80:1007–1016

Hashimoto T, Tokuchi Y, Hayashi M, Kobayashi Y, Nishida K, Hayashi S, Ishikawa Y, Tsuchiya S, Nakagawa K, Hayashi J, Tsuchiya E (1999) p53 null mutations undetected by immunohistochemical staining predict a poor outcome with early-stage non-small cell lung carcinomas. Cancer Res 59:5572–5577

Havrilesky L, Darcy M, Hamdan H, Priore RL, Leon J, Bell J, Berchuck A (2003) Prognostic significance of p53 mutation and p53 overexpression in advanced epithelial ovarian cancer: a Gynecologic Oncology Group Study. J Clin Oncol 21:3814–3825

Holstege H, Joosse SA, van Oostrom CT, Nederlof PM, de VA, Jonkers J (2009) High incidence of protein-truncating *TP53* mutations in BRCA1-related breast cancer 1. Cancer Res 69:3625–3633

Inga A, Storici F, Darden TA, Resnick MA (2002) Differential transactivation by the p53 transcription factor is highly dependent on p53 level and promoter target sequence. Mol Cell Biol 22:8612–8625

Kandioler D, Stamatis G, Eberhardt W, Kappel S, Zochbauer-Muller S, Kuhrer I, Mittlbock M, Zwrtek R, Aigner C, Bichler C, Tichy V, Hudec M, Bachleitner T, End A, Muller MR, Roth E, Klepetko W (2008) Growing clinical evidence for the interaction of the p53 genotype and response to induction chemotherapy in advanced non-small cell lung cancer. J Thorac Cardiovasc Surg 135:1036–1041

Kandioler D, Zwrtek R, Ludwig C, Janschek E, Ploner M, Hofbauer F, Kuhrer I, Kappel S, Wrba F, Horvath M, Karner J, Renner K, Bergmann M, Karner-Hanusch J, Potter R, Jakesz R, Teleky B, Herbst F (2002) *TP53* genotype but not p53 immunohistochemical result predicts response to preoperative short-term radiotherapy in rectal cancer. Ann Surg 235:493–498

Kandioler-Eckersberger D, Kappel S, Mittlbock M, Dekan G, Ludwig C, Janschek E, Pirker R, Wolner E, Eckersberger F (1999) The *TP53* genotype but not immunohistochemical result is predictive of response to cisplatin-based neoadjuvant therapy in stage III non-small cell lung cancer. J Thorac Cardiovasc Surg 117:744–750

Kandioler-Eckersberger D, Ludwig C, Rudas M, Kappel S, Janschek E, Wenzel C, Schlagbauer-Wadl H, Mittlbock M, Gnant M, Steger G, Jakesz R (2000) *TP53* mutation and p53 overexpression for prediction of response to neoadjuvant treatment in breast cancer patients. Clin Cancer Res 6:50–56

Katkoori VR, Jia X, Shanmugam C, Wan W, Meleth S, Bumpers H, Grizzle WE, Manne U (2009) Prognostic significance of p53 codon 72 polymorphism differs with race in colorectal adenocarcinoma. Clin Cancer Res 15:2406–2416

Kato S, Han SY, Liu W, Otsuka K, Shibata H, Kanamaru R, Ishioka C (2003) Understanding the function-structure and function-mutation relationships of p53 tumor suppressor protein by high-resolution missense mutation analysis. Proc Natl Acad Sci USA 100:8424–8429

Kihara C, Seki T, Furukawa Y, Yamana H, Kimura Y, van Schaardenburgh P, Hirata K, Nakamura Y (2000) Mutations in zinc-binding domains of p53 as a prognostic marker of esophageal-cancer patients. Jpn J Cancer Res 91:190–198

Kucera E, Speiser P, Gnant M, Szabo L, Samonigg H, Hausmaninger H, Mittlbock M, Fridrik M, Seifert M, Kubista E, Reiner A, Zeillinger R, Jakesz R (1999) Prognostic significance of mutations in the p53 gene, particularly in the zinc-binding domains, in lymph node- and steroid receptor positive breast cancer patients. Austrian Breast Cancer Study Group. Eur J Cancer 35:398–405

Langerod A, Bukholm IR, Bregard A, Lonning PE, Andersen TI, Rognum TO, Meling GI, Lothe RA, Borresen-Dale AL (2002) The *TP53* codon 72 polymorphism may affect the function of *TP53* mutations in breast carcinomas but not in colorectal carcinomas. Cancer Epidemiol Biomarkers Prev 11:1684–1688

Langerod A, Zhao H, Borgan O, Nesland JM, Bukholm IR, Ikdahl T, Karesen R, Borresen-Dale AL, Jeffrey SS (2007) *TP53* mutation status and gene expression profiles are powerful prognostic markers of breast cancer. Breast Cancer Res 9:R30

Lehmann-Che J, Andre F, Desmedt C, Mazouni C, Giacchetti S, Turpin E, Espie M, Plassa LF, Marty M, Bertheau P, Sotiriou C, Piccart M, Symmans WF, Pusztai L, de TH (2010)

Cyclophosphamide dose intensification may circumvent anthracycline resistance of p53 mutant breast cancers1. Oncologist 15:246–252

Lind H, Ekstrom PO, Ryberg D, Skaug V, Andreassen T, Stangeland L, Haugen A, Zienolddiny S (2007) Frequency of *TP53* mutations in relation to Arg72Pro genotypes in non small cell lung cancer. Cancer Epidemiol Biomarkers Prev 16:2077–2081

Lowe SW, Bodis S, McClatchey A, Remington L, Ruley HE, Fisher DE, Housman DE, Jacks T (1994) p53 status and the efficacy of cancer therapy in vivo. Science 266:807–810

Marin MC, Jost CA, Brooks LA, Irwin MS, O'Nions J, Tidy JA, James N, McGregor JM, Harwood CA, Yulug IG, Vousden KH, Allday MJ, Gusterson B, Ikawa S, Hinds PW, Crook T, Kaelin WG Jr (2000) A common polymorphism acts as an intragenic modifier of mutant p53 behaviour. Nat Genet 25:47–54

Miller LD, Smeds J, George J, Vega VB, Vergara L, Ploner A, Pawitan Y, Hall P, Klaar S, Liu ET, Bergh J (2005) An expression signature for p53 status in human breast cancer predicts mutation status, transcriptional effects, and patient survival. Proc Natl Acad Sci USA 102:13550–13555

Miyaki M, Iijima T, Yasuno M, Kita Y, Hishima T, Kuroki T, Mori T (2002) High incidence of protein-truncating mutations of the p53 gene in liver metastases of colorectal carcinomas1. Oncogene 21:6689–6693

Mollevi DG, Serrano T, Ginesta MM, Valls J, Torras J, Navarro M, Ramos E, Germa JR, Jaurrieta E, Moreno V, Figueras J, Capella G, Villanueva A (2007) Mutations in *TP53* are a prognostic factor in colorectal hepatic metastases undergoing surgical resection1. Carcinogenesis 28:1241–1246

Nelson HH, Wilkojmen M, Marsit CJ, Kelsey KT (2005) *TP53* mutation, allelism and survival in non-small cell lung cancer. Carcinogenesis 26:1770–1773

Nemunaitis J, Clayman G, Agarwala SS, Hrushesky W, Wells JR, Moore C, Hamm J, Yoo G, Baselga J, Murphy BA, Menander KA, Licato LL, Chada S, Gibbons RD, Olivier M, Hainaut P, Roth JA, Sobol RE, Goodwin WJ (2009) Biomarkers predict p53 gene therapy efficacy in recurrent squamous cell carcinoma of the head and neck. Clin Cancer Res 15:7719–7725

O'Connor PM, Jackman J, Bae I, Myers TG, Fan S, Mutoh M, Scudiero DA, Monks A, Sausville EA, Weinstein JN, Friend S, Fornace AJ Jr, Kohn KW (1997) Characterization of the p53 tumor suppressor pathway in cell lines of the National Cancer Institute anticancer drug screen and correlations with the growth-inhibitory potency of 123 anticancer agents. Cancer Res 57:4285–4300

Obata A, Eura M, Sasaki J, Saya H, Chikamatsu K, Tada M, Iggo RD, Yumoto E (2000) Clinical significance of p53 functional loss in squamous cell carcinoma of the oropharynx. Int J Cancer 89:187–193

Oden-Gangloff A, Di Fiore F, Bibeau F, Lamy A, Bougeard G, Charbonnier F, Blanchard F, Tougeron D, Ychou M, Boissière F, Le Pessot F, Sabourin JC, Tuech JJ, Michel P, Frebourg T (2009) *TP53* mutations predict disease control in metastatic colorectal cancer treated with cetuximab-based chemotherapy. Br J Cancer 100:1330–1335

Olivier M, Hollstein M, Hainaut P (2010) *TP53* mutations in human cancers: origins, consequences, and clinical use1. Cold Spring Harb Perspect Biol 2:a001008

Olivier M, Hussain SP, Caron dF, Hainaut P, Harris CC (2004) *TP53* mutation spectra and load: a tool for generating hypotheses on the etiology of cancer. IARC Sci Publ 157:247–270

Olivier M, Langerod A, Carrieri P, Bergh J, Klaar S, Eyfjord J, Theillet C, Rodriguez C, Lidereau R, Bieche I, Varley J, Bignon Y, Uhrhammer N, Winqvist R, Jukkola-Vuorinen A, Niederacher D, Kato S, Ishioka C, Hainaut P, Borresen-Dale AL (2006) The clinical value of somatic *TP53* gene mutations in 1,794 patients with breast cancer. Clin Cancer Res 12:1157–1167

Petitjean A, Achatz MI, Borresen-Dale AL, Hainaut P, Olivier M (2007a) *TP53* mutations in human cancers: functional selection and impact on cancer prognosis and outcomes. Oncogene 26:2157–2165

Petitjean A, Mathe E, Kato S, Ishioka C, Tavtigian SV, Hainaut P, Olivier M (2007b) Impact of mutant p53 functional properties on *TP53* mutation patterns and tumor phenotype: lessons from recent developments in the IARC *TP53* database. Hum Mutat 28:622–629

Pfeifer GP (2000) p53 mutational spectra and the role of methylated CpG sequences. Mutat Res 450:155–166

Poeta ML, Manola J, Goldwasser MA, Forastiere A, Benoit N, Califano JA, Ridge JA, Goodwin J, Kenady D, Saunders J, Westra W, Sidransky D, Koch WM (2007) *TP53* mutations and survival in squamous-cell carcinoma of the head and neck1. N Engl J Med 357:2552–2561

Rebischung C, Gerard JP, Gayet J, Thomas G, Hamelin R, Laurent-Puig P (2002) Prognostic value of P53 mutations in rectal carcinoma. Int J Cancer 100:131–135

Reles A, Wen WH, Schmider A, Gee C, Runnebaum IB, Kilian U, Jones LA, El-Naggar A, Minguillon C, Schönborn I, Reich O, Kreienberg R, Lichtenegger W, Press MF (2001) Correlation of p53 mutations with resistance to platinum-based chemotherapy and shortened survival in ovarian cancer. Clin Cancer Res 7:2984–2997

Rose SL, Robertson AD, Goodheart MJ, Smith BJ, DeYoung BR, Buller RE (2003) The impact of p53 protein core domain structural alteration on ovarian cancer survival. Clin Cancer Res 9:4139–4144

Schneider-Stock R, Boltze C, Peters B, Szibor R, Landt O, Meyer F, Roessner A (2004) Selective loss of codon 72 proline p53 and frequent mutational inactivation of the retained arginine allele in colorectal cancer. Neoplasia 6:529–535

Skaug V, Ryberg D, Kure EH, Arab MO, Stangeland L, Myking AO, Haugen A (2000) p53 mutations in defined structural and functional domains are related to poor clinical outcome in non-small cell lung cancer patients. Clin Cancer Res 6:1031–1037

Smith-Sorensen B, Kaern J, Holm R, Dorum A, Trope C, Borresen-Dale AL (1998) Therapy effect of either paclitaxel or cyclophosphamide combination treatment in patients with epithelial ovarian cancer and relation to *TP53* gene status. Br J Cancer 78:375–381

Sorlie T, Tibshirani R, Parker J, Hastie T, Marron JS, Nobel A, Deng S, Johnsen H, Pesich R, Geisler S, Demeter J, Perou CM, Lonning PE, Brown PO, Borresen-Dale AL, Botstein D (2003) Repeated observation of breast tumor subtypes in independent gene expression data sets. Proc Natl Acad Sci USA 100:8418–8423

Takahashi M, Tonoki H, Tada M, Kashiwazaki H, Furuuchi K, Hamada J, Fujioka Y, Sato Y, Takahashi H, Todo S, Sakuragi N, Moriuchi T (2000) Distinct prognostic values of p53 mutations and loss of estrogen receptor and their cumulative effect in primary breast cancers. Int J Cancer 89:92–99

Temam S, Flahault A, Perie S, Monceaux G, Coulet F, Callard P, Bernaudin JF, St Guily JL, Fouret P (2000) p53 gene status as a predictor of tumor response to induction chemotherapy of patients with locoregionally advanced squamous cell carcinomas of the head and neck. J Clin Oncol 18:385–394

Tomizawa Y, Kohno T, Fujita T, Kiyama M, Saito R, Noguchi M, Matsuno Y, Hirohashi S, Yamaguchi N, Nakajima T, Yokota J (1999) Correlation between the status of the p53 gene and survival in patients with stage I non-small cell lung carcinoma [In Process Citation]. Oncogene 18:1007–1014

Tomizawa Y, Kohno T, Kondo H, Otsuka A, Nishioka M, Niki T, Yamada T, Maeshima A, Yoshimura K, Saito R, Minna JD, Yokota J (2002) Clinicopathological significance of epigenetic inactivation of RASSF1A at 3p21.3 in stage I lung adenocarcinoma. Clin Cancer Res 8:2362–2368

Troester MA, Herschkowitz JI, Oh DS, He X, Hoadley KA, Barbier CS, Perou CM (2006) Gene expression patterns associated with p53 status in breast cancer. BMC Cancer 6:276

Wahl AF, Donaldson KL, Fairchild C, Lee FY, Foster SA, Demers GW, Galloway DA (1996) Loss of normal p53 function confers sensitization to Taxol by increasing G2/M arrest and apoptosis. Nat Med 2:72–79

Whibley C, Pharoah PD, Hollstein M (2009) p53 polymorphisms: cancer implications. Nat Rev Cancer 9:95–107

Yamasaki M, Miyata H, Fujiwara Y, Takiguchi S, Nakajima K, Nishida T, Yasuda T, Matsuyama J, Mori M, Doki Y (2010) p53 genotype predicts response to chemotherapy in patients with squamous cell carcinoma of the esophagus. Ann Surg Oncol 17:634–642

Zawlik I, Kita D, Vaccarella S, Mittelbronn M, Franceschi S, Ohgaki H (2009) Common polymorphisms in the MDM2 and *TP53* genes and the relationship between *TP53* mutations and patient outcomes in glioblastomas. Brain Pathol 19:188–194

Zhang J, Yang PL, Gray NS (2009) Targeting cancer with small molecule kinase inhibitors. Nat Rev Cancer 9:28–39

Chapter 9
Assessing *TP53* Status in Human Tumors: Lessons from Breast Cancer

Anita Langerød, Magali Olivier, and Anne-Lise Børresen-Dale

1 Introduction

Since the discovery of mutations in the *TP53* gene as a frequent event in cancer in the beginning of the 1990th, various approaches have been used to assess *TP53* mutation status in human tumors. The main challenges of *TP53* mutation screening in solid tumors are the difficulties in detecting the large spectrum of mutation types occurring in these tumors with adequate sensitivity and specificity. Methods have been developed to better meet these challenges, and some of the most commonly used techniques will be presented to show the advancement and current trends.

In the present era of whole genome analysis, mutation analysis in a single gene such as *TP53* may be replaced by surrogate markers or molecular profiles that may be more applicable in different aspects of clinical decision making. It is important to understand *TP53* mutations in the context of the whole range of aberrations seen in tumors, both to better understand the underlying biology and the clinical aspects of cancer. Here we will discuss what we have learned from studies of breast cancer

A. Langerød, Ph.D. (✉)
Department of Genetics, Institute for Cancer Research, The Norwegian Radium Hospital, Oslo University Hospital, Ullernchausseen 70, 0310 Oslo, Norway
e-mail: anita.langerod@rr-research.no

M. Olivier, Ph.D.
Group of Molecular Mechanisms and Biomarkers, International Agency for Research on Cancer (IARC/CIRC), 150 Cours Albert Thomas, 69372 Lyon CEDEX 08, France
e-mail: olivierm@iarc.fr

A.-L. Børresen-Dale, Ph.D., M.D. (h.c.)
Department of Genetics, Institute for Cancer Research, The Norwegian Radium Hospital, Oslo University Hospital, Ullernchausseen 70, 0310 Oslo, Norway

The K.G. Jebsen Center for Breast Cancer Research, Institute for Clinical Medicine Faculty of Medicine University of Oslo, University of Oslo, Pb 1078 Blindern, 0316 Oslo, Norway
e-mail: a.l.borresen-dale@medisin.uio.no

P. Hainaut et al. (eds.), *p53 in the Clinics*, DOI 10.1007/978-1-4614-3676-8_9,

combining *TP53* mutation status and different levels of whole genome analyses. Since breast cancer is one of the cancer diseases where *TP53* mutation status shows the most significant prognostic impact and clinical potential, this chapter will be anchored by experiences from research on this tumor type, although it may also be relevant to several other cancer forms. We will discuss some aspects of using *TP53* as a biomarker for breast cancer in clinical practice.

2 Requirements and Challenges in Mutation Detection Analysis

During the last two decades different approaches have been used to obtain *TP53* mutation status, a consequence of technological development and increased knowledge about the distribution and character of *TP53* mutations in different cancer diseases. A selection of these methods will be described in brief below, with focus on the challenges of screening for somatic mutations (point mutations, as well as small insertions and deletions) in DNA from solid tumors.

The main challenge in the analysis of solid tumors, such as those arising in the breast, is to find a method with sufficient sensitivity to deal with an occasionally low fraction of mutated cells in the biopsy. The fraction of cells with a *TP53* mutation may vary between types of breast cancer, and may in some heterogeneous tumors only be present in a small sub-fraction. Another substantial challenge is the huge spectrum of mutations, both with regard to the character of mutations and the wide distribution along the gene. It is not sufficient to analyze the 5–10 hotspot mutations found in *TP53*, since less frequent mutations also show severe effects on the protein and furthermore on patient outcome. Many studies have not analyzed the whole coding gene (exons 2–11; 393 amino acids), but only focused on the conserved regions of the gene (exons 5–8) that includes the DNA binding domain where mutations tend to cluster. This may seem like a cost- and time-efficient choice, but inadequate by potentially leaving as much as 10–20 % of the mutations undetected (own published and unpublished data; IARC TP53 database), resulting in a bias both in individual studies and in the *TP53* mutation database (Petitjean et al. 2007). Because *TP53* mutations are very diverse and are spread along the entire gene, it is recommended to analyze all exons (exons 2–11), both for research and clinical purposes.

For clinical use, the mutation detection method needs to be easy to perform, cost-efficient, rapid, and with high sensitivity and specificity. Also, it is essential that the amount of tumor material needed is feasible for the clinic to provide. Formalin-fixed paraffin-embedded (FFPE) tissue may be used, but DNA or RNA from fresh frozen tissue is for most methods preferred to get a high-quality analysis result. General guidelines need to be established when the analysis is introduced for clinical use, and the approaches may vary according to cancer type and treatment options. In summary, the most important criteria that need to be considered for the appropriate choice of method is the following: Type of specimen (peripheral blood, tissue, etc.) and starting material available (DNA/RNA), previous knowledge about the mutations (known/unknown, character, frequency), sensitivity and specificity requirements, ratio between mutant and wild-type allele, reliability of method and its possible standardization for clinical routine analysis (Nollau and Wagener 1997).

3 Approaches for Detection of *TP53* Mutation Status in Solid Tumors

Below we describe methods for mutation detection focusing on those that have been most commonly used for the analysis of *TP53* mutations in solid tumors. It is not a complete list of mutation detection techniques available. We discuss advantages and disadvantages of these methods regarding their sensitivity and specificity and how this may affect current knowledge on *TP53* mutations in human cancers.

3.1 Immunohistochemistry

Immunohistochemistry (IHC) analysis, measuring the expression level of p53 protein, has been used in many studies and in the clinic as a surrogate marker for *TP53* gene mutation. A substantial fraction of mutant p53 proteins found in tumors show a high protein level compared to wild type, due to a combination of decreased MDM2-mediated degradation and sustained stabilization, and partly explained by the fact that a high fraction of mutations are missense with minor effects on the protein structure, in contrast to the less frequent truncating mutations that usually are prone to a faster degradation. Studies comparing *TP53* mutation status assessed by IHC and sequencing have shown that IHC is a less sensitive method and that many of the mutations found by sequencing are not detected by IHC (Alsner et al. 2008; Langerød et al. 2007). In particular, *TP53* truncating mutations and splice mutations may not be scored. The p53 IHC staining is also dependent on the antibody used, as the location of the mutation may affect epitope recognition. As discussed in another chapter of this book, staining could be used as an indicator of *TP53* mutation if no other options are available and taking into account the limitation of this method. IHC is certainly useful in research to explore the subcellular location and protein levels of wild-type and different types of mutated p53 in different tissues.

3.2 Single-Strand Conformation Polymorphism

Single-strand conformation polymorphism (SSCP) (Orita et al. 1989) was developed more than two decades ago and has been widely used. It is a mutation detection method, based on the principle that single-nucleotide changes in the DNA sequence alter the structure and hence the electrophoretic mobility of single-stranded DNA. The advantages of the traditional SSCP method imply that it is technically easy and can be used for screening large numbers of samples. A significant disadvantage of this procedure is that not all types of potential mutations are detectable even under optimal conditions (Hayashi and Yandell 1993), and therefore several modifications of the SSCP methods have been developed over the years to increase the number of mutations detected, e.g., combining SSCP with restriction endonuclease fingerprinting (Liu and Sommer 1995).

3.3 Denaturing Gel Electrophoresis

Several mutation detection methods able to identify single-base mutations in a gene of interest are based on the melting behavior of the DNA double helix of a given fragment (Fischer and Lerman 1983). This melting behavior is sequence-dependent and generally occurs in discrete domains, rather than at single bases. DNA are denatured when being exposed to a chemical denaturant (formamide and urea) or a high temperature (all DNA fragments are considered melted at around 95 °C), and both denaturing methods have been used in different combination in the techniques described hereafter. Using *denaturing gradient gel electrophoresis (DGGE)* (Lerman and Silverstein 1987; Sheffield et al. 1989), the melting is achieved by electrophoresis of DNA fragments in a polyacrylamide gel containing a chemical denaturant gradient. Strand separation of a domain in a DNA fragment causes a change in conformation, and consequently, its electrophoretic mobility is reduced. Four bands may be observed on the gel representing two heteroduplexes (wt/mut, mut/wt) and two homoduplexes (wt/wt, mut/mut) that are partly melted double strands. *Constant denaturant gel electrophoresis (CDGE)* was developed as an improvement of DGGE (Børresen et al. 1991; Hovig et al. 1991), as choice of the exact running time and conditions to achieve the optimal separation was difficult using DDGE. CDGE applies one single denaturant concentration determined to be optimal for the analysis of a given fragment. In contrast to the parallel DGGE the separation in CDGE is time dependent only, meaning that a separation pattern once achieved will not disappear but will continue to increase the longer the gel is run. The procedure of making gels for CDGE is simplified compared to DGGE; it is faster and more reproducible. *Temporal temperature gel electrophoresis (TTGE)* was first introduced by Yoshino et al. (1991), and was later optimized for *TP53* mutation analysis (Børresen-Dale et al. 1997; Sørlie et al. 2004). It is based on the sequence-specific melting behavior of wild-type and mutant DNA in a temporal temperature gradient that increases gradually in a linear fashion over the time the electrophoresis is run. TTGE combines some of the advantages of CDGE and DGGE while eliminating some of their disadvantages. In TTGE a single concentration of urea and formamide is used, as in CDGE, but the temperature during the run is gradually increased. The focusing of bands obtained in DGGE seems to be partly retained, and the requirement for a chemical denaturing gradient is avoided as in CDGE. The challenge of determining the exact running time and denaturing conditions is strongly diminished, and since the denaturing conditions span a wider range, several fragments with different melting behavior can be analyzed on the same gel. TTGE differs from a method with a somewhat similar name; *temperature gradient gel electrophoresis (TGGE)* that has a fixed temperature gradient from top to bottom of the gel. In TTGE however, the temperature at any location of the entire gel is the same at any given time but changes with respect to time (temporal temperature). Hence it is easier to modulate the temperature over time, and thus TTGE provides a wider separation range that increases sensitivity.

3.4 Denaturing High-Performance Liquid Chromatography

Denaturing high-performance liquid chromatography (DHPLC) is a technique, which uses heteroduplex formation between wild-type and mutated DNA strands to identify mutations. Heteroduplex molecules are separated from homoduplex molecules by reverse-phase liquid chromatography on a special column matrix with temperature-based denaturation of the DNA strands (O'Donovan et al. 1998). DHPLC is easily automated and suitable for mutation screening of a large volume of samples. A major drawback of DHPLC in order to detect somatic mutations is that both wild-type and mutant DNA have to be present to form heteroduplexes. In germline DNA the variants are present in equal amounts, whereas in tumor specimens this is not always the situation due to low tumor cell content with mutation or the absence of wild-type cells (Keller et al. 2001).

3.5 Functional Yeast Assay

Functional assay in yeast (FASAY) is a functional assay for uncovering mutations in the *TP53* gene that alters the transactivation capability of the protein (Ishioka et al. 1993; Flaman et al. 1995). A vector with a *TP53* insert (exons 4–10) from the sample of interest is introduced into the yeast, and results in red colonies when mutant p53 is expressed and white colonies when wild-type p53 is expressed. Advantages of this method are that it is based on measuring the main function of p53 as a transcription factor. Disadvantages are that the whole gene is not analyzed (only a region between exons 4 and 10), RNA is needed that is not always easily accessible, and to determine the exact nature of the mutations further characterization is subsequently necessary.

3.6 Prescreening Methods

All methods listed above may be regarded as -prescreening methods, since they need sequencing to uncover the character of mutations as well as to validate them. Several -prescreening methods have been frequently used for mutation detection, mainly in order to increase the sample throughput. A number of studies, however, have not performed subsequent sequencing and the results from those may likely be biased based on low sensitivity and specificity.

3.7 Gene Sequencing

DNA sequencing is the most common method currently used to identify mutations in the *TP53* gene and is considered to be the gold standard (an example protocol is provided at http://www-p53.iarc.fr/p53sequencing.html). DNA sequencing includes several methods and technologies that are used for determining the order of the nucleotide bases (adenine, guanine, cytosine, and thymine) in a molecule of DNA. To search for mutations in single genes, the *Sanger sequencing method* (Sanger et al. 1977) is most commonly used. The Sanger method is based on DNA synthesis and the use of dideoxynucleotide triphosphates (ddNTPs) as DNA chain terminators. Sequencing platforms including capillary electrophoresis in a 96/384-well plate format has generated a more cost- and time-efficient method with an increased sensitivity. A drawback of sequencing solid tumors is the low sensitivity that may occur as a result of samples with small fraction of mutated cells, with most studies reporting a minimum of 10–20 % mutated alleles to be present in order to detect the mutation. Another challenge is when only DNA from FFPE (formalin-fixed paraffin-embedded) tissue is available, since the DNA from such samples usually is heavily fragmented and chemically modified. Software available to analyze raw data and further compare the sample sequence to a reference sequence is very useful in mutation scoring. However, fully automated scoring still has improvement potential due to scoring failure in poor quality sequences, caused by, e.g., tumor percentage or fragmentation. *Next-generation sequencing (NGS)* technologies are likely to become the new gold standard when the costs are being further reduced, and sensitivity/specificity issues of the methods have been properly addressed. NGS technologies are high-throughput methods that parallelize the sequencing process, producing thousands or millions of sequences at once [e.g., 454 pyrosequencing, Illumina (Solexa) sequencing, SOLiD sequencing, and Ion semiconductor sequencing] (Glenn 2011). NGS may be used for targeted -resequencing of *TP53* gene with unprecedented high throughput. Whether targeted -resequencing of a limited number of cancer genes such as *TP53* or whole genome mutation pattern will be used in the future for clinical care remains to be seen.

3.8 Microarrays for TP53 Mutation Detection

Several array-based platforms have been developed to increase the throughput and reduce the costs of *TP53* mutation analysis. *Polymerase chain reaction/ligase detection reaction (PCR/LDR)* is a platform developed for high-throughput detection of specific mutations in various genes involved in human neoplasia (Barany and Gelfand 1991). Oligonucleotide primers are annealed adjacently to target DNA and joined by a ligase to form an LDR product if the target nucleotide at mutation site is complementary to the discriminating primer. The method is sensitive and detects low abundance mutations, but it does only allow detection of the selected mutations it is designed for. *Arrayed primer extension reaction (APEX)* (Asper Biotech) is based on the incorporation of four dye terminators into oligonucleotide primers that each identifies a base in the target sequence. Several APEX -resequencing assays including one or several genes are being applied in diagnostics of a number of diseases caused by germline variation. A *TP53* array was also developed, and although the method was promising for analyzing somatic mutations (Kringen et al. 2005; Tonisson et al. 2002), the *TP53* APEX array has drawbacks such as low sensitivity and lack of robustness. *AmpliChip p53 test* is a microarray-based -resequencing test developed by Roche Molecular Diagnostics (previously Affymetrix) and is the most promising array-based *TP53* mutation detection technology available (Chiaretti et al. 2011; Grollman et al. 2007). The method is based on the hybridization of labeled target nucleic acid to a chip followed by fluorescence detection. The probe design and tiling arrangements as well as the scoring algorithms are critical for the detection of mutations. The design of the test allows detection of single-nucleotide substitutions and 1 bp deletions in the entire coding region and the flanking splice sites of exons 2–11 of the *TP53* gene. The predictive value of AmpliChip p53 is investigated in several ongoing clinical trials in collaboration with Merck (Merck Sharp & Dohme Corp.). The advantage of the AmpliChip p53 test is that it provides a rapid analysis of *TP53* mutation status and is able to detect a low abundance of mutated cells in FFPE or fresh frozen tissues. A disadvantage is that AmpliChip p53 is not designed to detect insertion mutations or >1 bp deletion mutations that on average may represent up to 12 % of the mutations in breast cancers (data from IARC TP53 database, R15 release), depending on the subtype (Holstege et al. 2010). A number of other existing technology platforms such as molecular inversion probe (MIP) arrays (Affymetrix) (Hardenbol et al. 2003) used for genotyping and copy number variation may have the potential of being developed and optimized for *TP53* -resequencing. For all the array-based methods described, a clear advantage is that subsequent sequencing is not required.

3.9 Gene Profiles as Surrogate Measures of Mutant TP53

In the past decade, whole genome analysis using microarrays has revealed gene profiles on different levels (DNA, mRNA, miRNA, and methylation) associated with *TP53* mutations status (Bergamaschi et al. 2006; Enerly et al. 2011; Miller et al. 2005). The profiles are typically defined by the action of p53 as a transcription factor regulating downstream pathways. Although no such test is currently offered, these gene patterns may have the potential of being used as a substitute assessment of *TP53* status, possibly through an integrated approach, as discussed in the next section.

3.10 Concluding Remarks on TP53 Mutation Detection Methods

Data extracted from the IARC TP53 database show that the different technologies described above have been used to a very different extent (Fig. 9.1). SSCP was by far the most popular -prescreening methods in the 1990s and early years of 2000 as

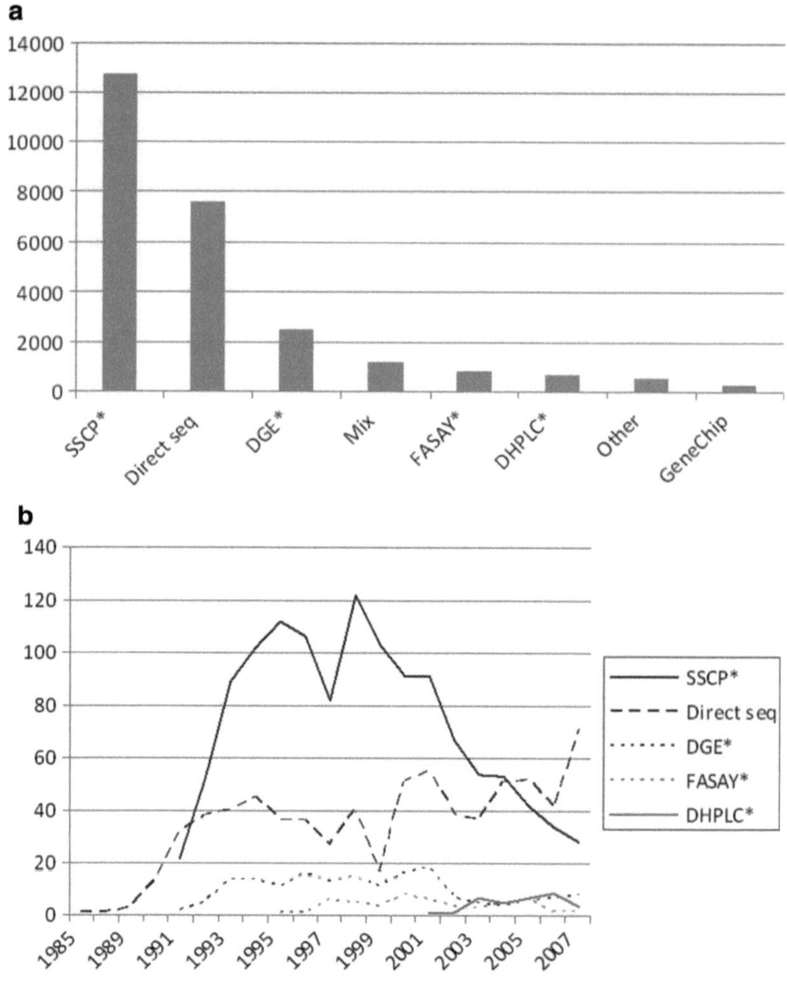

Fig. 9.1 Methods used for *TP53* mutation analysis (IARC p53 database, R15 release) (Petitjean et al. 2007). An inclusion criterion for the IARC p53 database is that mutations have been sequenced. In all studies using prescreening methods (*) sequencing was applied to find the nature of the mutations. NGS (next generation sequencing) is not yet included in the database (first reports from 2008). (**a**) Distribution of methods used to detect 26,409 *TP53* mutations registered in the IARC p53 database. (**b**) Timeline showing methods used from 1985 to 2007 in 2425 publications to detect *TP53* mutations in cancer (IARC p53 database). [DGE (denaturing gel electrophoresis), DHPLC (denaturing high performance liquid chromatography), Direct sequencing, FASAY (functional assay in yeast), Gene Chip, Mix (combination of different techniques), Other (various infrequently used methods), and SSCP (single stranded confirmation analysis).]

it was easy to perform. DGE methods (DGGE, CDGE, TTGE, etc.) were developed to reach better sensitivity but were only used in the 1990th by a limited number of labs. Direct sequencing, which was introduced in the early 1990s, has increasingly been used based on reduced time and cost, as well as improved sensitivity.

The domination of SSCP, which has low sensitivity and may detect different types of mutations with different efficiency, may have caused biases in mutation patterns and frequencies. Studies that used IHC or a combination of IHC and SSCP for -prescreening (included in the "other" and "mixed" categories, respectively, in Fig. 9.1) are even more likely biased based on low sensitivity and specificity, but they account for a very small fraction of studies compiled in the database. Gene chips show an infrequent use, but the increasing curve suggests that such methods may be more common in the future, especially in routine analysis in a clinical setting. The more common use of direct sequencing nowadays and of NGS in the years to come may reveal *TP53* mutation patterns that slightly differ from current knowledge as their sensitivity and chemistry differ from previous methods. It is thus important to critically consider the methodology used in every study and take into account these methodological biases when analyzing data from mutation databases.

4 *TP53* Status in the Context of Whole Genome Analyses

4.1 TP53 is Associated with Subtypes of Breast Cancer

A decade ago the development of the microarray technology allowed us to perform whole genome analysis of tumor transcriptomes. The early microarray analyses of breast tumors identified five novel subtypes based on their gene expression profiles that also showed difference in clinical outcomes (Perou et al. 2000; Sørlie et al. 2001). Interestingly, the distribution of *TP53* mutations differed between the subtypes. Overall, 20–30 % of breast tumors have a mutation in the *TP53* gene, but the subtypes have mutation rates varying from <5 to >95 %. The basal-like and HER2-enriched subtypes have the highest frequency of *TP53* mutations, luminal B an intermediate frequency, while luminal A and normal-like breast tumors are only occasionally mutated. The subgroups with the highest frequency of *TP53* mutations also have a gene expression profile representing high proliferation and cell division, and these groups have the poorest prognosis (Langerød et al. 2007; Naume et al. 2007; Sørlie et al. 2001).

The basal-like subtype in which *TP53* mutation frequency is around 70–90 % has shown a gene expression pattern with similarities to that of stem cells, and has been suggested to originate from breast stem cells (Ben-Porath et al. 2008; Liu et al. 2007). As *TP53* has been reported to be involved in the regulation of stem cells (Bonizzi et al. 2011; Mizuno et al. 2010), and has been shown to be a target of early mutational events (Chitemerere et al. 1996; Zhou et al. 2009), p53 may be involved in the initiation and development of basal-like breast cancer. However, the role of p53 in different developmental stages of the different subtypes of breast cancer remains to be determined.

Not only mutation frequency, but also *TP53* mutation types may differ between breast cancer subtypes, with basal-like and *BRCA1*-mutated breast cancers having a high incidence of protein-truncating mutations (Holstege et al. 2010). This finding supports diverse effects and functions of the various *TP53* mutations in different cells of origin and conditions of cancer development. Since p53 is a transcription factor regulating a vast network of downstream targets, gene expression profiles reflecting *TP53* status have been determined as they may provide a better assessment of p53 functionality in tumors. Such *TP53* expression signatures have been derived from different series of breast cancer as discussed below.

4.2 *TP53* Gene Expression Signatures in Human Breast Cancer

Among methods described previously in this chapter to assess *TP53* status, only FASAY was an attempt to directly measure the functional effects of *TP53* mutations (Ishioka et al. 1993). However, this test only measures the capacity of mutant proteins to transactivate one specific p53 response element at a time, such as p21, GADD45, BAX, or MDM2. It is well known that p53 uses response elements that differ in sequences and affinity in each target gene and that it also controls gene expression through repression. In addition, p53 may be inactivated by other mechanisms than gene mutations (e.g., MDM2 amplification, viral protein expression, other mutations in the p53 pathway and associated networks). A measure of downstream effects of p53 inactivation, such as the one given by whole transcriptional profiling, is thus expected to provide a better assessment of p53 functional status.

Despite an accelerating amount of whole genome data, only a few datasets have adequate *TP53* mutation data and gene expression data from breast tumors available to derive a p53 signature. So far, five gene expression signatures of *TP53* mutation status have been reported from four different series of breast cancer. The most comprehensive study has analyzed the transcription profile of 251 primary breast tumors (58/251 with *TP53* mutation) and identified a 32-gene signature that distinguishes *TP53* mutant and wild-type breast tumors (Miller et al. 2005). This *TP53* signature was shown to outperform sequence-based assessments of *TP53* status in predicting prognosis and therapeutic response. Sørlie et al. obtained a gene expression profile associated with *TP53* mutation status in a cohort of 122 primary breast tumors (51/106 with *TP53* mutation) previously used to define the five molecular subtypes (Sørlie 2004). In a set of 80 breast tumors (20/80 with *TP53* mutation), Langerød et al. identified a list of genes that were significantly associated with *TP53* mutation status (Fig. 9.2) (Langerød et al. 2007). Perou and colleagues identified a set of genes associated with *TP53* status based on a combined approach using breast cancer cell lines (isogenic cell line pairs with and without p53 function engineered using siRNA and examined for doxorubicin-induced p53 response) and breast tumor samples. The expression pattern of this gene set could predict *TP53* mutation status and had prognostic value in different clinical datasets (Troester et al. 2006). Another *TP53* signature was identified by Takahashi et al. based on a smaller sample set of 26 breast tumors (of which 12 had mutation) and a validation set of 12 samples (6 with mutation) (Takahashi et al. 2008). This signature had prognostic significance in two previously published datasets. Oshima et al. (2011) showed that the same gene signature could be used to predict response to neoadjuvant treatment P-FEC

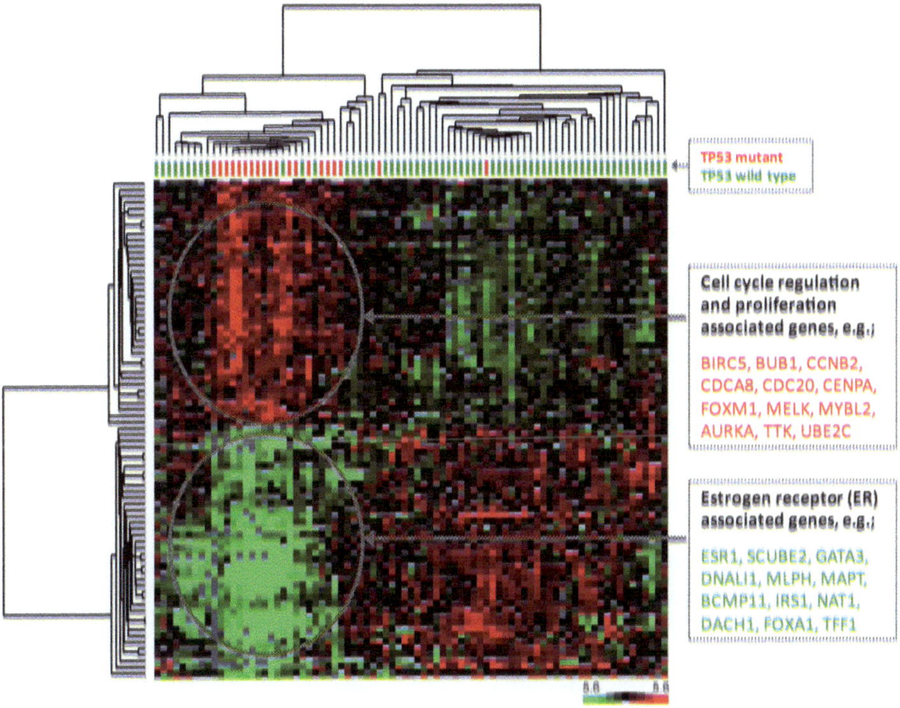

Fig. 9.2 Hierarchical clustering of 80 genes (*vertical* dendrogram) that are associated with *TP53* mutation status in 80 breast cancer patients (*horizontal* dendrogram) (Langerød et al. 2007). Genes highly expressed (*gray*) in tumors with *TP53* mutation are associated with cell cycle regulation and proliferation and genes downregulated (*white*) in tumors with *TP53* mutation are estrogen receptor (ER) associated genes

(paclitaxel followed by 5-FU/epirubicin/cyclophosphamide) in a cohort of 72 breast tumors of which 16 had a *TP53* mutation.

From the studies described above, it seems that a small number of genes (~30–50) may be sufficient to predict *TP53* mutations status, patient outcome, and treatment response. Surprisingly, the genes differentially expressed according to *TP53* mutation status do not represent known transcriptional targets of p53; instead genes upregulated in *TP53* mutant tumors are mainly associated with cell cycle and proliferation, while genes upregulated in *TP53* wild-type tumors are associated with ER status. These transcriptomics data thus confirm the strong association between *TP53* status, ER status, and grade that has previously been recognized by other methods. It is of note that the degree of overlap between these different *TP53* gene signatures is minimal, which may reflect different study design, sample sets included (number and selection of patients), the analysis platform used (Affymetrix, Agilent, Stanford custom made, etc.), and the type of *TP53* mutations present (Ein-Dor et al. 2005). These studies did not address the potential diverse effects of different types of *TP53* mutations on gene expression. Further investigation is thus needed to determine the biological and clinical implications of the partly functional or gain of function mutants.

The overlap between *TP53* signatures and breast cancer subtype signatures points to the fact that it is difficult to distinguish the components in a gene signature that represents the *TP53* mutation status of a tumor from those representing the subtype of breast cancer (Langerød et al. 2007; Sørlie et al. 2001). The complexity of both the disease and the *TP53* network complicates such interpretation. Miller et al. addressed the question by controlling for ER status and grade to derive a p53 signature. Troester et al. combined data from primary tumors with data from isogenic cell line pairs (with and without p53 function), and the resulting signature excluded genes that were breast cancer subtype-associated, but not downstream of p53 (Troester et al. 2006). Coutant et al. reassessed the Miller dataset and derived a gene expression-based predictor of *TP53* status separately for ER-positive and ER-negative tumors (Coutant et al. 2011). The ER+ *TP53* signature was predictive of poor prognosis in the ER+ cancers but not in the ER− cancers in a validation set of 98 breast cancers. The ER− *TP53* signature showed no prognostic or predictive values in ER+ cancers, but it was associated with prognosis in ER− cancers. These data support the hypothesis that p53 dysfunction has partly different consequences in ER− and ER+ cancers.

A recent study by Prives and colleagues illustrates how downstream effects of p53 inactivation measured by transcriptional profiling not only retrieves a *TP53* signature with prognostic and/or predictive power, but also reveals how mutant *TP53* may contribute to mammary carcinogenesis. Mutant p53 was, in a three-dimensional cell culture model, found to affect mammary tissue architecture by upregulation of the sterol biosynthesis genes (mevalonate pathway), which also were found highly expressed in a large series of human breast tumors with mutant p53 (Freed-Pastor et al. 2012). Statins, the drugs extensively used to reduce cholesterol levels, are blocking the mevalonate pathway and may have potential in cancer therapy. Similar approaches using transcriptional profiling of cells/tissue with *TP53* mutant vs. wild-type status may reveal novel effects of mutant p53 and drugable targets in cancer.

4.3 *TP53* in the Landscape of Mutational Processes and Somatic Rearrangements

Although it has been claimed for years (Hollstein et al. 1991), we can now, based on whole genome sequencing of tumors by the NGS technology and data from global initiatives (such as ICGC and TCGA) (Hudson et al. 2010; The Cancer Genome Atlas Research Network 2011), state as a fact that *TP53* is the most frequently mutated gene in human cancers. It is also clear that the mutation frequency differs considerably between tumors in different organs as well as between tumor subtypes within an organ (Chap. 8). Upcoming whole genome sequencing studies of larger series of breast carcinomas will validate the currently recognized 20–30 % *TP53* mutation prevalence (based on preliminary data). However, in some cancer types NGS data are revealing different pictures. In ovarian cancer in particular, the prevalence of *TP53* mutations is known to vary widely between studies, ranging from 5 to 80 % (IARC p53 database). Based on recent NGS data, *TP53* mutations may be present in up to 96 % of tumors in high-grade serous ovarian cancer (ICGC consortium 2010). Whether these differences are due to differences in methodologies (better sensitivity of NGS), tumor subtypes, or tumor stage remain to be determined.

Whole genome sequencing of breast tumors shows a diversity of mutational events and rearrangements, both in quantity and in quality (Stephens et al. 2009). So far only a handful of genes, such as *TP53*, *PIK3CA*, and *GATA3*, have been found fairly frequent somatically mutated in breast cancer. An increasing number of affected pathways will probably be uncovered when the volume of sequencing data piles up. It is challenging to understand the significance of the huge spectrum of somatic mutations found in human tumors, and in particular to identify and distinguish driver from passenger mutations. *TP53* mutations are believed to be strong promoters of breast carcinogenesis, but whether a *TP53* mutation always is a "driver" and the main cause of cancer, one of many crucial factors, or sometimes just a "passenger" mutation is unclear. It has been claimed that *TP53* or its pathway somehow are affected in every cancer and thus being one of the hallmarks of cancer, whereas others may claim that *TP53* seems to be involved only because it is the central unit of the cell's large molecular network. A systems biology approach is needed to fully understand the effect the various *TP53* mutations have on the perturbations of all its downstream targets in concert with other alterations in that specific tumor.

In the large-scale "-omics" analyses, *TP53* mutations have also been found to be highly associated with patterns of DNA aberrations and rearrangements (genomics) (Bergamaschi et al. 2006; Jain et al. 2001), miRNA expression patterns (transcriptomics) (Brosh et al. 2008; Enerly et al. 2011), and recently with methylation patterns (epigenomics) (Rønneberg et al. 2011) of breast cancer. The data point to *TP53* as a key node in numerous cellular processes. Data produced at all these levels in large series of breast tumors give us a unique opportunity to understand processes related to the *TP53* network in breast cancer that previously were inaccessible to us.

5 Clinical Relevance of *TP53* Status in Breast Cancer

The impact of the *TP53* tumor suppressor gene in cancer development has been demonstrated through various approaches, in particular by the important observation of *TP53* knockout mice developing various tumors (Donehower et al. 1992). Li–Fraumeni Syndrome (LFS) patients carrying a germline mutation in *TP53* are frequently hit by cancer at young age, and one of the organs most commonly affected is the breast, suggesting a crucial role for *TP53* in breast cancer development. Overall, one in four breast cancer patients has a somatic mutation in the *TP53* gene, and this is significantly associated with an aggressive cancer with a poor prognosis and development of metastasis. *TP53* mutations also have been observed as an early event in tumor development, found in premalignant hyperplasia and in situ stages of breast cancer (Chitemerere et al. 1996; Zhou et al. 2009). *TP53* may play an important role in the origin of certain breast cancer subtypes, such as the basal-like and

HER2-enriched groups described previously in this chapter. A very intriguing finding is that the majority of breast cancers arising in LFS patients are *HER2+* (83 %) (Wilson et al. 2010), suggesting that a *TP53* mutation may be an event prior to *HER2* amplification/overexpression in sporadic breast tumors.

Altogether, there is little doubt that *TP53* is very important in many aspects concerning breast cancer. However, a clinically clear application and benefit is required for a biomarker to be introduced in the clinic, and the clinical value of *TP53* as a prognostic or predictive biomarker of breast cancer is still under debate. Measuring a pure prognostic or predictive effect of tumor markers is difficult in an ethical perspective. The original definition of a *prognostic* biomarker is a biomarker that provides information on the likely course of the cancer disease in an untreated individual. Since the majority of breast cancer patients receive adjuvant treatment, most studies of prognostic markers will today include patients who received systemic treatment, influencing the natural course of the disease. A prognostic marker may, however, be extremely valuable in selecting early stage patients for the appropriate adjuvant systemic treatment. The term *predictive* biomarker is defined as a marker which can be used to identify subpopulations of patients who are most likely to respond to a given therapy. Predictive markers are the basis of personalized medicine, and in breast cancer patients, estrogen and progesterone receptors are used to predict sensitivity to endocrine therapy, whereas *HER2* is used to predict sensitivity to Herceptin treatment.

The data supporting a prognostic power of *TP53* mutation status in breast cancer are rather convincing, as the majority of studies show association between mutant *TP53* and poor prognosis (Petitjean et al. 2007). As discussed in Chap. 8, the predictive value of *TP53* status remains uncertain because of the lack of study replication. Indeed, available studies are heterogeneous in the type, dosage and combination of drugs used, the methods used to assess *TP53* status and treatment response, and the type of tumor included. *TP53* mutations seems to confer resistance to single-agent anthracyclines (Aas et al. 1996; Chrisanthar et al. 2011), but to predict *response* to dose intense combined anthracycline/cyclophosphamide treatment (Bertheau et al. 2007) or to a regimen of docetaxel-capecitabine ± trastuzumab (Glück et al. 2011). A recent phase III clinical trial showed no impact of *TP53* status on response to taxanes (Bonnefoi et al. 2011). Many large clinical trials have unfortunately not included *TP53* mutation analysis in their study design, but based on promising results as prognostic and predictive marker, we encourage including *TP53* gene mutation analysis in all relevant clinical trials to resolve these important issues. Breast cancer on the molecular bases appears to be several different diseases with seemingly different origin, progression, and outcome, and the significance of *TP53* as a marker should be evaluated in the different subtypes of breast cancer.

5.1 Current Guidelines and Future Perspectives

Despite many reports on somatic *TP53* mutation status as a strong prognostic and also predictive marker of breast cancer, analysis of *TP53* mutations in tumor tissue is generally not routinely performed in clinical practice for breast cancer patients. In a guideline from the American Society of Clinical Oncology (ASCO) concerning the use of tumor markers in breast cancer, they concluded that the present data in 2007 were insufficient to recommend use of *TP53* measurements for management of patients with breast cancer (Harris et al. 2007), referring to diagnosis, staging, prognosis, surveillance, or monitoring treatment of patients with breast cancer. They acknowledge *TP53* gene mutations as associated with poor prognosis, as shown, e.g., in a study of almost 1,800 breast cancer patients (Olivier et al. 2006). Based on this study they also suggested that *TP53* status may, if confirmed, be used to select patients that will benefit from systemic adjuvant therapy in node-negative, ER-positive patients.

Breast cancer is in rare cases a clinical indication for *germline TP53* mutation analysis, when the presence of BRCA1/2 mutations is excluded from cancer prone families or if the breast cancer patient is affected by multiple cancers. *TP53* mutation screening is then recommended in order to reveal whether the patient has LFS (http://www.nbcg.no/index.html; national guidelines, Norway). Germline *TP53* mutations in cancer families can be considered a biomarker of increased risk of cancer and may also have clinical implications, such as avoiding therapy where the effects are dependent on normal p53 function (http://www.cancer.gov/cancertopics/genetics/breast/predictive-testing-p53-mutations).

6 Conclusions

There are many levels of complexity in assessment, interpretation, and utilization of the *TP53* mutation status and many questions remain to be addressed. Mutation detection approaches may be improved, but the current task is rather how to best define *TP53* status and how to utilize this knowledge in a clinical scenery. The success of a tumor marker in clinical practice is dependent on several aspects. An easily accessible sample from the patient is crucial. Currently a tumor biopsy/surgical specimen is used, but biomarkers in blood are usually preferred. Efforts have been made to measure *TP53* mutation status in tumor DNA from circulating tumor cells in blood or as fragments in plasma (Chen et al. 2009), and also to measure circulating autoantibodies to tumor-associated antigens such as p53 (Piura and Piura 2010); however, no useful clinical information has yet been provided. Generally a sensitive, specific, robust, cost- and time-efficient method is important. Whether NGS will be the answer remains to be seen. Understanding the effect of *TP53* mutations may require more complex levels of analysis such as expression level of p53, regulation through miRNA, protein modifications, and haplotype and isoform expression. The challenge is still to integrate these levels of molecular knowledge in order to understand the role of p53 in the biology of breast cancer. It is also essential that a biomarker contributes significantly to the choice of treatment. *TP53* mutation status in breast cancer is recommendable based on its prognostic power, but its predictive value is less clear and needs to be further explored in clinical trials testing different treatment regimens.

References

Aas T, Børresen AL, Geisler S, Smith-Sørensen B, Johnsen H, Varhaug JE, Akslen LA, Lønning PE (1996) Specific P53 mutations are associated with de novo resistance to doxorubicin in breast cancer patients. Nat Med 2:811–814

Alsner J, Jensen V, Kyndi M, Offersen BV, Vu P, Børresen-Dale AL, Overgaard J (2008) A comparison between p53 accumulation determined by immunohistochemistry and TP53 mutations as prognostic variables in tumours from breast cancer patients. Acta Oncol 47:600–607

Barany F, Gelfand DH (1991) Cloning, overexpression and nucleotide sequence of a thermostable DNA ligase-encoding gene. Gene 109:1–11

Ben-Porath I, Thomson MW, Carey VJ, Ge R, Bell GW, Regev A, Weinberg RA (2008) An embryonic stem cell-like gene expression signature in poorly differentiated aggressive human tumors. Nat Genet 40:499–507

Bergamaschi A, Kim YH, Wang P, Sørlie T, Hernandez-Boussard T, Lønning PE, Tibshirani R, Børresen-Dale AL, Pollack JR (2006) Distinct patterns of DNA copy number alteration are associated with different clinicopathological features and gene-expression subtypes of breast cancer. Genes Chromosomes Cancer 45:1033–1040

Bertheau P, Turpin E, Rickman DS, Espié M, de Reyniès A, Feugeas JP, Plassa LF, Soliman H, Varna M, de Roquancourt A, Lehmann-Che J, Beuzard Y, Marty M, Misset JL, Janin A, de Thé H (2007) Exquisite sensitivity of TP53 mutant and basal breast cancers to a dose-dense epirubicin-cyclophosphamide regimen. PLoS Med 2007;4(3):e90. PMID:17388661

Bonizzi G, Cicalese A, Insinga A, Pelicci PG (2011) The emerging role of p53 in stem cells. Trends Mol Med 18(1):6–12

Bonnefoi H, Piccart M, Bogaerts J, Mauriac L, Fumoleau P, Brain E, Petit T, Rouanet P, Jassem J et al (2011) TP53 status for prediction of sensitivity to taxane versus non-taxane neoadjuvant chemotherapy in breast cancer (EORTC 10994/BIG 1–00): a randomised phase 3 trial. Lancet Oncol 12:527–539

Børresen AL, Hovig E, Smith-Sørensen B, Malkin D, Lystad S, Andersen TI, Nesland JM, Isselbacher KJ, Friend SH (1991) Constant denaturant gel electrophoresis as a rapid screening technique for p53 mutations. Proc Natl Acad Sci USA 88:8405–8409

Børresen-Dale AL, Lystad S, Langerød A (1997) Temporal temperature gradient gel electrophoresis on the DCode system. Bio Rad Bulletin 2133:12–13

Brosh R, Shalgi R, Liran A, Landan G, Korotayev K, Nguyen GH, Enerly E, Johnsen H, Buganim Y et al (2008) p53-Repressed miRNAs are involved with E2F in a feed-forward loop promoting proliferation. Mol Syst Biol 4:229

Chen Z, Feng J, Buzin CH, Liu Q, Weiss L, Kernstine K, Somlo G, Sommer SS (2009) Analysis of cancer mutation signatures in blood by a novel ultra-sensitive assay: monitoring of therapy or recurrence in non-metastatic breast cancer. PLoS One 4:e7220

Chiaretti S, Tavolaro S, Marinelli M, Messina M, Del GI, Mauro FR, Santangelo S, Piciocchi A, Peragine N et al (2011) Evaluation of TP53 mutations with the AmpliChip p53 research test in chronic lymphocytic leukemia: correlation with clinical outcome and gene expression profiling. Genes Chromosomes Cancer 50:263–274

Chitemerere M, Andersen TI, Holm R, Karlsen F, Børresen AL, Nesland JM (1996) TP53 alterations in atypical ductal hyperplasia and ductal carcinoma in situ of the breast. Breast Cancer Res Treat 41:103–109

Chrisanthar R, Knappskog S, Lokkevik E, Anker G, Ostenstad B, Lundgren S, Risberg T, Mjaaland I, Skjonsberg G et al (2011) Predictive and prognostic impact of TP53 mutations and MDM2 promoter genotype in primary breast cancer patients treated with epirubicin or paclitaxel. PLoS One 6:e19249

Coutant C, Rouzier R, Qi Y, Lehmann-Che J, Bianchini G, Iwamoto T, Hortobagyi GN, Symmans WF, Uzan S et al (2011) Distinct p53 gene signatures are needed to predict prognosis and response to chemotherapy in ER-positive and ER-negative breast cancers. Clin Cancer Res 17:2591–2601

Donehower LA, Harvey M, Slagle BL, McArthur MJ, Montgomery CAJ, Butel JS, Bradley A (1992) Mice deficient for p53 are developmentally normal but susceptible to spontaneous tumours. Nature 356:215–221

Ein-Dor L, Kela I, Getz G, Givol D, Domany E (2005) Outcome signature genes in breast cancer: is there a unique set? Bioinformatics 21:171–178

Enerly E, Steinfeld I, Kleivi K, Leivonen SK, Aure MR, Russnes HG, Rønneberg JA, Johnsen H, Navon R et al (2011) miRNA-mRNA integrated analysis reveals roles for miRNAs in primary breast tumors. PLoS One 6:e16915

Fischer SG, Lerman LS (1983) DNA fragments differing by single base-pair substitutions are separated in denaturing gradient gels: correspondence with melting theory. Proc Natl Acad Sci USA 80:1579–1583

Flaman JM, Frebourg T, Moreau V, Charbonnier F, Martin C, Chappuis P, Sappino AP, Limacher IM, Bron L et al (1995) A simple p53 functional assay for screening cell lines, blood, and tumors. Proc Natl Acad Sci USA 92:3963–3967

Freed-Pastor WA, Mizuno H, Zhao X, Langerød A, Moon SH, Rodriguez-Barrueco R, Barsotti A, Chicas A, Li W et al (2012) Mutant p53 disrupts mammary tissue architecture via the mevalonate pathway. Cell 148:244–258

Glenn TC (2011) Field guide to next-generation DNA sequencers. Mol Ecol Resour 11:759–769

Glück S, Ross JS, Royce M, McKenna EF Jr, Perou CM, Avisar E, Wu L (2012) TP53 genomics predict higher clinical and pathologic tumor response in operable early-stage breast cancer treated with docetaxel-capecitabine ± trastuzumab. Breast Cancer Res Treat, Epub 2012;132(3):781–791. PMID:21373875

Grollman AP, Shibutani S, Moriya M, Miller F, Wu L, Moll U, Suzuki N, Fernandes A, Rosenquist T et al (2007) Aristolochic acid and the etiology of endemic (Balkan) nephropathy. Proc Natl Acad Sci USA 104:12129–12134

Hardenbol P, Baner J, Jain M, Nilsson M, Namsaraev EA, Karlin-Neumann GA, Fakhrai-Rad H, Ronaghi M, Willis TD et al (2003) Multiplexed genotyping with sequence-tagged molecular inversion probes. Nat Biotechnol 21:673–678

Harris L, Fritsche H, Mennel R, Norton L, Ravdin P, Taube S, Somerfield MR, Hayes DF, Bast RC Jr (2007) American Society of Clinical Oncology 2007 update of recommendations for the use of tumor markers in breast cancer. J Clin Oncol 25:5287–5312

Hayashi K, Yandell DW (1993) How sensitive is PCR-SSCP? Hum Mutat 2:338–346

Hollstein M, Sidransky D, Vogelstein B, Harris CC (1991) p53 mutations in human cancers. Science 253:49–53

Holstege H, Horlings HM, Velds A, Langerød A, Børresen-Dale AL, van de Vijver MJ, Nederlof PM, Jonkers J (2010) BRCA1-mutated and basal-like breast cancers have similar aCGH profiles and a high incidence of protein truncating TP53 mutations. BMC Cancer 10:654

Hovig E, Smith-Sørensen B, Brøgger A, Børresen AL (1991) Constant denaturant gel electrophoresis, a modification of denaturing gradient gel electrophoresis, in mutation detection [published erratum appears in Mutat Res 1991 May;263(1):61]. Mutat Res 262:63–71

Hudson TJ, Anderson W, Artez A, Barker AD, Bell C, Bernabe RR, Bhan MK, Calvo F, Eerola I et al (2010) International network of cancer genome projects. Nature 464:993–998

ICGC consortium (2010) International network of cancer genome projects. Nature 464:993–998

Ishioka C, Frebourg T, Yan YX, Vidal M, Friend SH, Schmidt S, Iggo R (1993) Screening patients for heterozygous p53 mutations using a functional assay in yeast. Nat Genet 5:124–129

Jain AN, Chin K, Børresen-Dale AL, Erikstein BK, Eynstein LP, Kaaresen R, Gray JW (2001) Quantitative analysis of chromosomal CGH in human breast tumors associates copy number abnormalities with p53 status and patient survival. Proc Natl Acad Sci USA 98:7952–7957

Keller G, Hartmann A, Mueller J, Hofler H (2001) Denaturing high pressure liquid chromatography (DHPLC) for the analysis of somatic p53 mutations. Lab Invest 81:1735–1737

Kringen P, Bergamaschi A, Due EU, Wang Y, Tagliabue E, Nesland JM, Nehman A, Tonisson N, Børresen-Dale AL (2005) Evaluation of arrayed primer extension for TP53 mutation detection in breast and ovarian carcinomas. Biotechniques 39:755–761

Langerød A, Zhao H, Borgan O, Nesland J, Bukholm I, Ikdahl T, Kaaresen R, Børresen-Dale AL, Jeffrey SS (2007) TP53 mutation status and gene expression profiles are powerful prognostic markers of breast cancer. Breast Cancer Res 9:R30

Lerman LS, Silverstein K (1987) Computational simulation of DNA melting and its application to denaturing gradient gel electrophoresis. Methods Enzymol 155:482–501

Liu Q, Sommer SS (1995) Restriction endonuclease fingerprinting (REF): a sensitive method for screening mutations in long, contiguous segments of DNA. Biotechniques 18:470–477

Liu R, Wang X, Chen GY, Dalerba P, Gurney A, Hoey T, Sherlock G, Lewicki J, Shedden K et al (2007) The prognostic role of a gene signature from tumorigenic breast-cancer cells. N Engl J Med 356:217–226

Miller LD, Smeds J, George J, Vega VB, Vergara L, Ploner A, Pawitan Y, Hall P, Klaar S et al (2005) An expression signature for p53 status in human breast cancer predicts mutation status, transcriptional effects, and patient survival. Proc Natl Acad Sci USA 102:13550–13555

Mizuno H, Spike BT, Wahl GM, Levine AJ (2010) Inactivation of p53 in breast cancers correlates with stem cell transcriptional signatures. Proc Natl Acad Sci USA 107:22745–22750

Naume B, Zhao X, Synnestvedt M, Borgen E, Russnes HG, Lingjaerde OC, Stromberg M, Wiedswang G, Kvalheim G et al (2007) Presence of bone marrow micrometastasis is associated with different recurrence risk within molecular subtypes of breast cancer. Mol Oncol 1:160–171

Nollau P, Wagener C (1997) Methods for detection of point mutations: performance and quality assessment. IFCC Scientific Division, Committee on Molecular Biology Techniques. Clin Chem 43:1114–1128

O'Donovan MC, Oefner PJ, Roberts SC, Austin J, Hoogendoorn B, Guy C, Speight G, Upadhyaya M, Sommer SS et al (1998) Blind analysis of denaturing high-performance liquid chromatography as a tool for mutation detection. Genomics 52:44–49

Olivier M, Langerød A, Carrieri P, Bergh J, Klaar S, Eyfjord J, Theillet C, Rodriguez C, Lidereau R et al (2006) The clinical value of somatic TP53 gene mutations in 1,794 patients with breast cancer. Clin Cancer Res 12:1157–1167

Orita M, Iwahana H, Kanazawa H, Hayashi K, Sekiya T (1989) Detection of polymorphisms of human DNA by gel electrophoresis as single-strand conformation polymorphisms. Proc Natl Acad Sci USA 86:2766–2770

Oshima K, Naoi Y, Kishi K, Nakamura Y, Iwamoto T, Shimazu K, Nakayama T, Kim SJ, Baba Y et al (2011) Gene expression signature of TP53 but not its mutation status predicts response to sequential paclitaxel and 5-FU/epirubicin/cyclophosphamide in human breast cancer. Cancer Lett 307:149–157

Perou CM, Sørlie T, Eisen MB, van de Rijn M, Jeffrey SS, Rees CA, Pollack JR, Ross DT, Johnsen H et al (2000) Molecular portraits of human breast tumours. Nature 406:747–752

Petitjean A, Mathe E, Kato S, Ishioka C, Tavtigian SV, Hainaut P, Olivier M (2007) Impact of mutant p53 functional properties on TP53 mutation patterns and tumor phenotype: lessons from recent developments in the IARC TP53 database. Hum Mutat 28:622–629

Piura E, Piura B (2010) Autoantibodies to tumor-associated antigens in breast carcinoma. J Oncol 2010:264926

Rønneberg JA, Fleischer T, Solvang HK, Nordgard SH, Edvardsen H, Potapenko I, Nebdal D, Daviaud C, Gut I et al (2011) Methylation profiling with a panel of cancer related genes: association with estrogen receptor, TP53 mutation status and expression subtypes in sporadic breast cancer. Mol Oncol 5:61–76

Sanger F, Nicklen S, Coulson AR (1977) DNA sequencing with chain-terminating inhibitors. Proc Natl Acad Sci USA 74:5463–5467

Sheffield VC, Cox DR, Lerman LS, Myers RM (1989) Attachment of a 40-base-pair G + C-rich sequence (GC-clamp) to genomic DNA fragments by the polymerase chain reaction results in improved detection of single-base changes. Proc Natl Acad Sci USA 86:232–236

Sørlie T (2004) Molecular portraits of breast cancer: tumour subtypes as distinct disease entities. Eur J Cancer 40:2667–2675

Sørlie T, Perou CM, Tibshirani R, Aas T, Geisler S, Johnsen H, Hastie T, Eisen MB, van de Rijn M et al (2001) Gene expression patterns of breast carcinomas distinguish tumor subclasses with clinical implications. Proc Natl Acad Sci USA 98:10869–10874

Sørlie T, Johnsen H, Vu P, Lind GE, Lothe R, Børresen-Dale AL (2004) Mutation screening of the TP53 gene by temporal temperature gradient gel electrophoresis. methods mol biol 291:207–216

Stephens PJ, McBride DJ, Lin ML, Varela I, Pleasance ED, Simpson JT, Stebbings LA, Leroy C, Edkins S et al (2009) Complex landscapes of somatic rearrangement in human breast cancer genomes. Nature 462:1005–1010

Takahashi S, Moriya T, Ishida T, Shibata H, Sasano H, Ohuchi N, Ishioka C (2008) Prediction of breast cancer prognosis by gene expression profile of TP53 status. Cancer Sci 99:324–332

The Cancer Genome Atlas Research Network (2011) Integrated genomic analyses of ovarian carcinoma. Nature 474:609–615

Tonisson N, Zernant J, Kurg A, Pavel H, Slavin G, Roomere H, Meiel A, Hainaut P, Metspalu A (2002) Evaluating the arrayed primer extension -resequencing assay of TP53 tumor suppressor gene. Proc Natl Acad Sci USA 99:5503–5508

Troester MA, Herschkowitz JI, Oh DS, He X, Hoadley KA, Barbier CS, Perou CM (2006) Gene expression patterns associated with p53 status in breast cancer. BMC Cancer 6:276

Wilson JR, Bateman AC, Hanson H, An Q, Evans G, Rahman N, Jones JL, Eccles DM (2010) A novel HER2-positive breast cancer phenotype arising from germline TP53 mutations. J Med Genet 47:771–774

Yoshino K, Nishigaki K, Husimi Y (1991) Temperature sweep gel electrophoresis: a simple method to detect point mutations. Nucleic Acids Res 19:3153

Zhou W, Muggerud AA, Vu P, Due EU, Sørlie T, Børresen-Dale AL, Warnberg F, Langerød A (2009) Full sequencing of TP53 identifies identical mutations within in situ and invasive components in breast cancer suggesting clonal evolution. Mol Oncol 3(3):214–219

Chapter 10
TP53 Germline Mutations: Genetics of Li–Fraumeni Syndrome

Doua Bakry and David Malkin

1 Introduction

It is estimated that 5–10 % of human malignancies occur in individuals with an inborn cancer susceptibility. Since the late 1980s, several cancer susceptibility genes have been cloned. Detection of heterozygous constitutional mutations in these genes permits the identification of families with inherited susceptibilities to cancer. It also facilitates the identification of individuals who though they do not belong to families with extended cancer pedigrees are still a "high cancer risk" because they carry germline alterations of cancer susceptibility genes. Although oncogenes have been occasionally implicated in cancer predisposition disorders, genes that have been most strongly associated with familial tumors are tumor suppressor genes.

Somatic mutations of the *TP53* tumor suppressor gene are frequently detected in most types of human cancers (Nigro et al. 1989; Royds and Iacopetta 2006). These are described more thoroughly in other chapters in this book. Germline mutations of the *TP53* tumor suppressor gene are associated with the complex cancer syndrome, Li–Fraumeni syndrome (LFS); LFS is a clinically and genetically uncommon heterogeneous autosomal dominantly inherited disorder caused primarily by germline *TP53* mutations. LFS patients and families exhibit a remarkable phenotype of early onset cancers (Li et al. 1995) and diverse neoplasms in both children and young adults. The predominant tumors are soft-tissue sarcoma, osteosarcoma, breast cancer, brain tumors, leukemia, and adrenocortical carcinoma, as well as a predilection to development of multiple primary neoplasms. A wide spectrum of other cancers has also been frequently reported.

D. Bakry, M.D. • D. Malkin, M.D. (✉)
Genetics and Genome Biology Program, Division of Hematology/Oncology, The Hospital for Sick Children, University of Toronto, 555 University Avenue, Toronto, ON, Canada M5G 1X8
e-mail: david.malkin@sickkids.ca

P. Hainaut et al. (eds.), *p53 in the Clinics*, DOI 10.1007/978-1-4614-3676-8_10,

The relative rarity of "classic" LFS families together with the high mortality rate within these families precluded conventional linkage approaches to identify the likely chromosomal location of the causative genetic defect. Therefore, a candidate gene approach was taken that ultimately focused on *TP53* (Malkin et al. 1990).

Surprisingly, the tumor spectrum seen in this familial cancer syndrome is quite different from that associated with somatic *TP53* mutations. Thus instead of the epithelial cancers where somatic mutations are prevalent, LFS patients present more commonly with tumors of embryonal, neuroectodermal, and mesenchymal cell lineage, with the striking and overrepresented exceptions of carcinomas of the breast, choroid plexus epithelium, and adrenal gland. This chapter will describe the phenotype and genotype characteristics of LFS and place this intriguing syndrome in the context of clinical relevance from both diagnostic and therapeutic perspectives.

2 *TP53*: Basis of LFS

The complex biochemical features of the p53 protein are described in depth elsewhere in this book. Here, we will highlight those properties most relevant to the following discussion of its relevance to LFS. p53 mediates cellular stress responses and initiates DNA repair, cell-cycle arrest, senescence, and apoptosis. The critical biochemical function of p53, which is intimately linked to its tumor suppressor activity, is the ability to activate transcription (Pietenpol et al. 1994). Surprisingly, whereas mutant p53 is stabilized in tumor cells, p53 is not stabilized in nontumor tissue of patients with LFS or mice bearing a *TP53* mutation. These observations suggest that expression and function of p53 is context dependent and particularly associated with the process of malignant transformation and tumor development. Thus, it is perhaps not surprising that while wild-type protein accumulation may be observed in response to an inflammatory microenvironment (Hofseth et al. 2003) or even senescence (Collado et al. 2005), there is no evidence to date of aberrant inflammatory or senescence response in LFS patients. Close to 4,000 different *TP53* mutations have been reported, with the large majority localized to the DNA-binding domain (Petitjean et al. 2007). In general, clinical correlative data do not implicate which properties of particular mutants are selected for during tumorigenesis in a cell-specific context. This suggests that mutant p53 loss-of-function, dominant-negative, and gain-of-function properties are all important for tumorigenesis in humans (Farnebo et al. 2010) with gain-of-function activities being particularly relevant (Brosh and Rotter 2009).

The great majority of the mutant *TP53* alleles found in human cancer genomes encode amino acid substitutions in the DNA-binding domain of p53. The resulting defective p53 proteins, being unable to bind to the promoters of downstream target genes, have therefore lost the ability to mediate most if not all of p53's multiple functions. Unlike most other tumor suppressor genes that are inactivated by frame shift or nonsense mutations leading to disappearance or aberrant synthesis of a gene product, the overwhelming majority (>80 %) of *TP53* mutations are missense (Soussi and

Lozano 2005) that lead to the synthesis of stable full length protein (Soussi and Beroud 2001) with a mutated DNA-binding domain core. The most common *TP53* germline missense mutations are found in the DNA-binding domain at codons 175, 248, 249, 273, and 281 (Royds and Iacopetta 2006). Two classes of mutations have been distinguished on the basis of various in vitro assays and the quaternary structure of the protein (Cho et al. 1994). Class I mutations, exemplified by mutants at codon 248, affect amino acids directly involved in the protein–DNA interaction. They have a wild-type conformation and they do not bind to the hsp70 chaperone (Hinds et al. 1990; Ory et al. 1994). Class II mutations, exemplified by the R175H mutant (4.9 % in the database), have an altered conformation with intense binding to hsp70. The amino acids altered in this class are involved in stabilizing the tertiary structure of the protein. Class II mutations are associated with a more severe phenotype in vitro than class I mutations (Ory et al. 1994). Some mutant p53 display only partial loss of their DNA-binding activity allowing the mutant to bind only to a subset of p53 response elements (Friedlander et al. 1996; Rowan et al. 1996). Biologically, these mutants have lost their apoptotic properties, but their cell cycle arrest activity remains similar to that of the wild-type protein. Several studies have revealed that specific *TP53* mutations are associated with either a poorer prognosis or a poor response to treatment (Borresen et al. 1995; Goh et al. 1995; Berns et al. 1998; Borresen-Dale et al. 1998; Kucera et al. 1999)—an observation that is likely based in part on the disparate properties of the different mutants.

However, the frequency and distribution of LFS-associated mutations does otherwise differ somewhat from the somatic pattern in that splice-acceptor mutations are more frequent in the germline, with as many as 15 % reported in one study (Varley et al. 2001). "Hot-spot" mutants at codons 175, 248, and 273 represent almost 20 % of all somatic *TP53* mutations found in human tumors (Soussi and Wiman 2007).

Germline *TP53* gene deletions are extremely rare. Shlein et al. report a large deletion at chromosome locus 17q13.1 in eight families, four of whom exhibited a phenotype consistent with LFS and four of whom presented with a unique phenotype that included a spectrum of congenital anomalies including developmental delay, hypotonia, and hand and foot abnormalities. Where the deletion breakpoints at 17p13.1 were intragenic to *TP53*, only the LFS cancer phenotype was observed; however, where the breakpoints fell outside the *TP53* open reading frame, no cancers were described and the phenotype was restricted to the congenital anomalies despite the germline heterozygous deletion of most of the *TP53* gene (Shlien 2010) (Fig. 10.1). These observations suggest that while p53-mediated DNA-binding activity (and therefore transcriptional activity) is the major defect selected during cell transformation, the possibility that other rare alterations could also be selected cannot be excluded. Such defects could be more specific to particular cancer subtypes or other diseases. In another context, the c.1010G4G (p.R337H) mutation exemplifies this possibility. This alteration was initially found as a germline mutation specifically associated with pediatric adrenal cortical carcinoma in southern Brazil in several independent families that were not predisposed to other tumors, a feature which is quite different from LFS (Ribeiro et al. 2001). In conventional

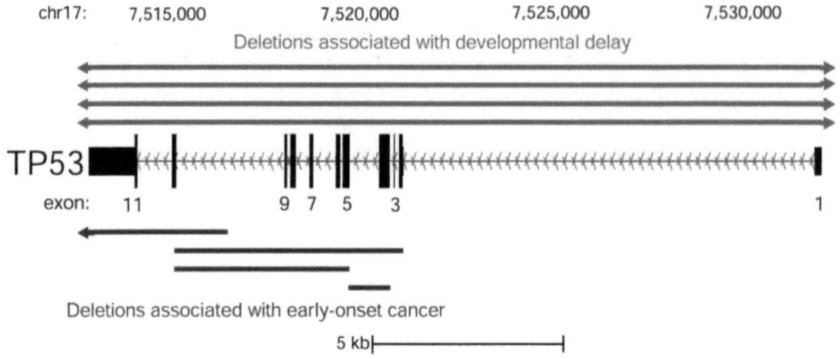

Fig. 10.1 Differential phenotypic impact of large genic vs. small intragenic deletions across *TP53*. High-resolution single nucleotide polymorphism (SNP) array and direct sequencing analysis demonstrate that deletions that encompass the entire open reading frame of *TP53* are associated with a phenotype of congenital anomalies and cognitive developmental delay (denoted by *gray bidirectional arrows*), whereas small intragenic deletions (denoted below the *TP53* exon map) are associated with early onset cancer in the context of LFS

transactivation assays, this mutant demonstrated wild-type behavior. Precise biochemical analysis revealed that this mutant is highly sensitive to pH in the physiological range, leading to folding changes dependent on the protonated state of the protein (DiGiammarino et al. 2001).

TP53 harbors several biologically intriguing polymorphisms that do not appear to be phenotypically silent (Grochola et al. 2009; Whibley et al. 2009). Ninety percent of *TP53* polymorphisms occur in the noncoding sequences (Lazar et al. 1993; Pleasants and Hansen 1994). Three common polymorphisms occur upstream of the DNB domain; two of these are intronic (PIN2 in intron 2 and PIN3, a 16 base pair insertion in intron 3) (Lazar et al. 1993). To date, the latter is the only intronic polymorphism that has been associated with an increase in the risk of several types of cancer (Whibley et al. 2009). This allele has been associated with lower levels of *TP53* transcripts, suggesting that the polymorphism causes an alteration in mRNA processing, providing a possible molecular basis for the associated increase in risk of developing cancer (Gemignani et al. 2004; Boldrini et al. 2008; Costa et al. 2008). Less frequent polymorphisms occur in the DBD domain of p53 such as V217M and G360A that are in the linker region adjacent to the tetramerization domain (Whibley et al. 2009). The third widely studied polymorphism in this region is a single nucleotide substitution [single nucleotide polymorphism (SNP)] at codon 72 in exon 4. In humans, the most common polymorphism involves the substitution of an arginine for a proline base. The current consensus from a large number of studies is that R72 is more effective in inducing apoptosis than P72 (Vazquez et al. 2008; Olivier et al. 2010). There are insufficient experimental data to establish consistent differences in biological activities between p53–R72 and p53–P72 in normal primary human cells or in tumors (Whibley et al. 2009). However, the allelic frequencies vary significantly in different ethnic and racial populations. For example, the *TP53* codon 72 C-allele frequency is 33 % in the sub-Sahara African population, but over 50 % in Asian populations and 77 % in Europeans.

A second exonic polymorphism at codon 47 changing a proline to serine has also been described. The frequency is very low, ranging from 0.5 to 5 % in various studies. It has only been reported in populations of African origin, in which it is found at an allele frequency of ~5 % (Felley-Bosco et al. 1993). These polymorphisms are adjacent to serine 46, phosphorylation of which is a key event for the apoptotic function of p53. In vitro studies have shown that the ability of p53 to suppress cell growth is not affected by the p53–S47; however, Li et al. observed that the S47 variant is a poorer substrate for S46 phosphorylation and has an impaired proapoptotic ability (Li et al. 2005). The clinical significance of this p53 variant is not known.

The last polymorphic site worth commenting on in the p53 pathway is SNP309, rs2279744 T/G in the *MDM2* gene (Harris et al. 1986; Matlashewski et al. 1987; Bond et al. 2004). In a fashion similar to the *TP53* codon 72 SNP, the G-allele frequency of MDM2 SNP309 appears to be demographically variable: 10 % in African Americans (Atwal et al. 2007), 33 % in Northern Europeans (Bond et al. 2006), 45 % in Asians (Hong et al. 2005), and 50 % in Ashkenazi Jews (Bond et al. 2006). Allele-specific selective advantages and disadvantages for evolution and/or reproduction have been proposed to explain these differences (Atwal et al. 2007).

Recently a natural antisense transcript, Wrap53 (*W*D40 encoding *R*NA *a*ntisense to *p53*), was discovered, which plays a key role in regulating the steady-state level of p53 transcripts by interacting with the p53 5'UTR (Farnebo 2009; Mahmoudi et al. 2009). The *Wrap53* gene overlaps with the *TP53* gene in a head-to-head fashion. As a result, their respective RNAs show perfect sequence complementarity in the first exons, suggesting that they interact and influence each other's expression. Indeed, Wrap53 regulates p53 at the posttranscriptional level. Wrap53-mediated regulation of p53 is critical for the p53 response to DNA damage. Steady-state levels of both p53 and Wrap53 mRNA increase upon DNA damage, indicating not only that it maintains basal p53 mRNA levels but also plays a role in stabilizing p53 mRNA response to cellular stress. In addition, Wrap53 antisense transcripts sensitize cells to p53-induced apoptosis (Mahmoudi et al. 2009). The potential role of Wrap53 in cancer susceptibility is not known. While it appears to be deleted together with p53 in the 17p13.1 microdeletion syndrome, its specific effect on the resulting cancer (or noncancer phenotype) is not clear.

3 LFS: The Clinical Story

In 1969 Li and Fraumeni described four families in whom soft-tissue sarcoma was associated with early onset premenopausal breast cancer in close female relatives. The term LFS was initially coined in 1982 (Pearson et al. 1982). LFS is a clinically and genetically heterogeneous autosomal dominant disorder, which gives rise to a wide spectrum of early onset cancers (Li and Fraumeni 1969; Birch et al. 1994; Li et al. 1995; Nichols et al. 2001). Li and Fraumeni proposed that the observed clustering of cancers in these families was most likely the result of an inherited predisposition (Li and Fraumeni 1969; Birch 1994). In 1990 germline *TP53* mutations were discovered to be the underlying cause of the majority of LFS cases (Malkin et al. 1990). Approximately 80 % of LFS families that fulfill the classical clinical criteria have mutations in *TP53*, with loss of p53 pathway function probably accounting for the rest (Royds and Iacopetta 2006). Although some studies have suggested that LFS could also result from *CHEK2* mutations (Bell et al. 1999; Sodha et al. 2000; Vahteristo et al. 2001) or be linked to an as yet unidentified gene cluster on chromosome 1q23 in some families (Bachinski et al. 2005), there are at present no definitive data supporting the involvement of a gene other than *TP53* in LFS (Kleihues et al. 1997; Sodha et al. 2002; Varley 2003).

LFS, although rare, accounts for 17 % of all genetically defined familial cancer cases. Over 500 families have been reported worldwide with complete or partial [Li–Fraumeni like (LFS)] phenotypes, and many more families have been identified and not reported. The *TP53* carrier rate may be more common than previously recognized, occurring in about 1 in 5,000 to 1 in 20,000 birth (Lalloo et al. 2006; Gonzalez et al. 2009). LFS is a highly penetrant disorder with the lifetime risk of developing cancer being over 90 % (Royds and Iacopetta 2006). The risk is more pronounced in women than in men (lifetime risk of 93 and 68 %, respectively) (Gonzalez et al. 2009). While initially this sex difference was thought to be primarily due to the high incidence of female breast cancer (Nagy et al. 2004), Hwang et al. (2003) and Chompret et al. (2004) suggest that the explanation may be more complicated as they showed the cumulative risks estimated by ages 20, 30, 40, and 50 years to be 18, 49, 77, and 93 %, respectively, in female carriers, and 10, 21, 33, and 68 %, respectively, in male carriers. Hwang shows that the difference between males and females cannot be explained by breast or sex-related cancers, since the difference remained after exclusion of breast, ovary, and prostate cancer. Females also exhibit an earlier average age of onset (29 vs. 40 years in men) (Hwang et al. 2003).

In all reported studies, the age distribution at cancer diagnosis was strikingly young. In at least one report, 56 % of the cancer cases were diagnosed under 30 years of age and 100 % under 50 years; in LFL families the figures were 44 and 78 %, respectively. These findings are in stark contrast to the general population in which only 2 % of cancer cases occur under the age of 30 years and 11 % under the age of 50 years (Varley et al. 1997).

It has been suggested that the specific *TP53* mutant genotype may influence age of onset and tumor spectrum. Birch et al. (1998) reported a significantly higher cancer incidence and earlier age at cancer diagnosis in families with LFS who carry missense mutations in the DNA-binding domain than in families that harbor protein inactivating mutations. In addition, a significantly higher risk for breast and brain cancers was observed in families with mutations in the DNA-binding domain. Conversely, while families in whom adrenocortical carcinomas (ACC) occur together with a wider spectrum of cancers harbor the usual spectrum of germline mutations, those with isolated ACC or apparently de novo mutations are quite (though not universally) consistently found to occur outside the DNA-binding loops (Tabori and Malkin 2008). Furthermore, nonsense, frame shift, and splice mutations, which are likely to result in loss of expression of the protein or in nonfunctional proteins, seem

Table 10.1 Clinical criteria for classic Li–Fraumeni syndrome (LFS), LFS-like (LFL) criteria, and Chompret criteria

Classic LFS criteria (Li et al. 1988)
Proband diagnosed with a sarcoma before age 45; AND
A first-degree[a] relative with cancer diagnosed before age 45; AND
Another first- or second-degree[b] relative on the same side of the family with cancer diagnosed before age 45 or a sarcoma at any age
LFS-like (LFL) criteria Birch (1994)
Proband with any childhood cancer or sarcoma, brain tumor, or adrenocortical carcinoma diagnosed before age 45; AND
First- or second-degree relative with a typical LFS cancer (sarcoma, breast cancer, brain tumor, leukemia, or adrenocortical carcinoma) diagnosed at any age; AND
A first- or second-degree relative on the same side of the family with any cancer diagnosed under age 60
Eeles definition of LFL (Eeles 1995*)*
Two first- or second-degree relatives with LFS-related malignancies (sarcoma, breast cancer, brain tumor, leukemia, adrenocortical tumor, melanoma, prostate cancer, pancreatic cancer) at any age
Chompret criteria for TP53 germline mutation screening (Tinat et al. 2009)
Proband diagnosed with a tumor belonging to the LFS tumor spectrum (sarcoma, brain tumor, premenopausal breast cancer, adrenocortical carcinoma, leukemia, lung bronchoalveolar cancer) before age 46, and at least one first- or second-degree relative affected with a LFS tumor (other than breast cancer if the proband is affected by breast cancer) before age 56, or a relative with multiple primary tumors at any age; OR
A proband with multiple primary tumors (except multiple breast tumors), two of which belong to the LFS tumor spectrum and the first of which occurred before age 46, regardless of family history; OR
A proband with adrenocortical carcinoma or choroid plexus tumor, regardless of family history

[a]First-degree relative is defined as a parent, sibling, or child
[b]Second-degree relative is defined as a grandparent, aunt, uncle, niece, nephew, or grandchild

to be related to early onset of cancer, especially brain tumors (Olivier et al. 2003). However, these attempts at genotype:phenotype correlations are difficult to pin down, as numerous families carrying the same mutation express widely divergent clinical manifestations of age of onset and cancer type (Fiszer-Maliszewska et al. 2000).

The difficulty in clarifying an absolute clinical definition for LFS has likely led to underreporting of cases. Furthermore, the high frequency of new mutations also results in under-recognition of the syndrome, for the diagnosis of LFS requires a detailed family history that confirms occurrence of cancers in close family relatives. These issues have resulted in various clinical definitions of LFS and as well as incomplete phenotypes: LF-like (LFL) and incomplete LFS (LFSI) (Birch et al. 1994; Eeles 1995; Chompret 2002). Table 10.1 highlights the comparative phenotypic characteristics of the classical LFS definition and its related disorders. The sensitivity and specificity of the Chompret criteria was determined to be 82 and 58 %, respectively (Bougeard et al. 2008), making it perhaps the most rigorous and relevant definition to justify *TP53* mutation analysis.

Remarkably, ~20–30 % of the tumors in *TP53* mutation positive families do not belong to the classical LFS tumor spectrum (Ruijs et al. 2010); in fact, alterations of this gene may be present in the germlines of families without histories of excess cancer (Malkin et al. 1992). Birch et al. found Wilms tumor and phyllodes tumors of the breast to be strongly associated, pancreatic cancer moderately associated, and neuroblastoma weakly associated with *TP53* mutation carrier status (Birch et al. 2001; Gonzalez et al. 2009). Hwang et al. observed significantly earlier ages at onset of gastrointestinal and lung cancers for the mutation carriers (Hwang et al. 2003), supporting a previous suggestion to include these cancer types as possible LFS components (Varley et al. 1997). Similarly, carcinomas of the lung and gastrointestinal tract, lymphomas, and other neoplasms have been shown to occur in *TP53* mutation carriers or first-degree relatives of carriers at much earlier ages than expected in the general population (Nichols et al. 2001).

It is not clear what specific cancer history in a family might almost universally point to the presence of a germline *TP53* mutation. However it has been proposed that occurrence of childhood ACC in a family in whom a member was diagnosed with one of the "core" cancers (sarcoma, brain tumor, breast cancer, ACC) is a strong indicator of the presence of a germline *TP53* mutation that should prompt a request for diagnostic testing by molecular screening (Gonzalez et al. 2009).

Occurrence of multiple metachronous or synchronous primary cancers is one of the phenotypes observed in LFS (Malkin et al. 1992; Hisada et al. 1998; Potzsch et al. 2002). Age and tumor diagnosis influence the risk of second malignant neoplasms, with the highest risk being in those who developed a first tumor in the first two decades of life or those diagnosed with rhabdomyosarcoma (RMS) (Hisada et al. 1998). Hisada evaluated the second and third cancer among 200 individuals. Overall, there was a fivefold relative risk of developing a second cancer among 30 individuals resulting in a cumulative probability of a second cancer of 57 % (±10 %) at 30 years. There was a strikingly high risk of second cancer among individuals whose first cancer was diagnosed before 20 years of age (RR = 83, 95 % 37–187); the risk decreased to RR = 9.7 (95 % CI 4.9–20) for ages 20–44, and to RR = 1.5 (95 % CI 0.5–4.2) for age 45 years and older at first cancer diagnosis.

4 The LFS Classic Component Tumors

As noted earlier, the age of diagnosis for all tumors is early compared with patients without LFS, but this varies depending on the cancer type. The earliest reported cancers are ACC and choroid plexus carcinomas (CPC), while melanoma and prostate cancer were the latest (Nichols et al. 2001; Gonzalez et al. 2009; Tabori et al. 2010). Detection of *TP53* mutations in families in whom no "core" LFS cancer has been diagnosed before age 50 years is extremely rare (Gonzalez et al. 2009).

Breast cancer is the most common tumor in *TP53* mutation carriers (24–31.2 %), followed by soft-tissue sarcomas (11.6–17.8 %), brain tumors (3.5–14 %), osteosarcomas (12.6–13.4 %), and adrenocortical tumors (6.5–9.9 %) (Kleihues et al. 1997; Olivier et al. 2003). Data regarding common histologies of LFS component tumors suggest that breast cancers are most commonly invasive ductal carcinomas (Li et al. 1988). RMS account for 55 % of soft-tissue sarcomas, followed by fibrosarcomas (13 %) and malignant fibrous histiocytomas. Sixty-nine percent of LFS-associated brain tumors are astrocytic (astrocytoma or glioblastoma), followed by medulloblastoma/primitive neuroectodermal tumors (17 %) (Kleihues et al. 1997).

Breast cancer, soft-tissue sarcoma, and bone sarcoma account for more than 50 % of tumors in *TP53* mutation carriers, followed by ACC and brain tumors. The four core cancers (breast, sarcoma, brain, and ACC) account for 77 % of all associated cancers (Gonzalez et al. 2009). Others include hematological malignancies, gastric, colorectal, and ovarian cancers, occurring at earlier ages than in the general population (Nichols et al. 2001; Olivier et al. 2003). Approximately 1 % of heritable breast cancer cases arise due to a germline *TP553* mutation (Borresen et al. 1992; Sidransky et al. 1992). Women who develop breast cancer under 30 years of age and have no first- or second-degree relatives with cancer have a 3–7 % risk to have a *TP53* mutation; however, presence of first- or second-degree relatives with cancer raises this likelihood to well over 75 % (http://www-p53.iarc.fr/GermMutPrev.html).

The initial families reported by Li and Fraumeni revealed a particularly increased risk of breast cancer in the relatives of patients with RMS (Li and Fraumeni 1969). Subsequent studies have revealed an increased relative risk of early onset breast cancer in mothers of patients with RMS (Hartley et al. 1991). This association has not been seen consistently in mothers of other childhood cancer patients.

RMS is the most frequent soft-tissue sarcoma of childhood, accounting for 10–15 % of all childhood sarcomas. Sarcomas represent the primary tumor criterion in classical LFS, and 5–9 % of childhood RMS is diagnosed in the context of LFS. A diagnosis of LFS should be considered in any child, diagnosed <3 years of age with RMS. There appears to be no correlation between anatomic site, histological subtype or clinical grouping, and the presence of a *TP53* mutation (Diller et al. 1995). RMS therapy influences second malignant neoplasms risk with the highest incidence among survivors treated with chemoradiotherapy, particularly cyclophosphamide, epipodophyllotoxins, and radiotherapy (Rich et al. 1997; Cohen et al. 2005). The question of modifying therapeutic interventions in patients with *TP53* mutations is extremely important. To date, no definitive studies have been undertaken to address this concern or to evaluate whether risk reduction can be effectively attained. With this in mind, Trahair et al. advocate that treatment approaches for LFS patients with RMS be based on standard risk stratification, but recommend that treatment alternatives be considered, particularly for patients with locally advanced or unfavorable histology tumors. In these instances, definitive surgical resection to avoid radiotherapy for local control should be considered (Trahair et al. 2007).

Although the histopathologic characteristics of tumors associated with *TP53* germline mutations are very similar to their sporadic counterparts, their ages at diagnosis show marked organ-specific differences. ACC represents 14 % of all cancers in LFS and occurs primarily in children. Childhood ACC is a hallmark for the presence of a *TP53* germline mutation. The likelihood of detecting germline *TP53* mutations in adults with ACC is much less common. Bone sarcomas occur mainly in adolescents, although with a tendency shifted to the latter part of the first decade of life rather than the more typical age range of late adolescence seen in sporadic osteosarcoma. Brain tumors and soft-tissue sarcomas exhibit a biphasic age distribution, with a first peak in very early childhood (<5 years) and a second peak between 20 and 40 years. About 10 % of patients with LFS develop gliomas, typically before the age of 45 years. Up to 5 % of patients develop supratentorial primary neuroectodermal tumors, CPCs, and medulloblastomas (Taylor et al. 2000). Almost 50 % of children with CPC harbor germline *TP53* mutations even in the absence of a typical LFS family history. Therefore, as with childhood ACC, all children with CPC should be considered for *TP53* mutation analysis as suggested by Tinat et al. (2009) and Gonzalez et al. (2009).

Breast carcinomas are the most frequent neoplasm in *TP53* mutation carriers (25.5 % of all tumors). They occur almost exclusively in women (rare male *TP53* mutation carriers have been reported with breast cancer) and are more prevalent in the third and fourth decades of life. The gender distribution of other tumors shows an excess of men for brain tumors, hematopoietic cancers, and gastric cancer, whereas an excess of women was observed for ACC and nonmelanoma skin cancer (Palmero et al. 2010).

While the unique R337H mutation is described in more detail elsewhere in this book, it is worthwhile to discuss it briefly in the context of tumor phenotype as compared with the other more ubiquitous mutations. The R337H mutation predominantly found in LFS families from southeastern Brazil is a risk allele for pediatric ACC and, at least until recently, was thought to represent a unique example of a tissue-specific predisposing mutation. The carrier frequency in southeastern Brazil is 0.3 % (Palmero et al. 2008), the annual incidence ranging from 3.4 to 4.2 cases per million children per year (Sandrini et al. 1997), compared with a worldwide incidence at ages 0–14 years of only 0.3 per million children per year. This mutation was believed to be tissue specific and to predispose only to ACC developing usually in the first 2–3 years of life with a penetrance of 10 % by age 12 (Ribeiro et al. 2001; Figueiredo et al. 2006) However, a more thorough ascertainment of many families with the R337H mutation shows that the cancer risk is much broader and can encompass the full spectrum of tumors associated with LFS and its variants, in particular early breast cancer (mean age at diagnosis below 40 years) (Achatz et al. 2007; Assumpcao et al. 2008). Coustodio et al. found the mutation in 63 % of Brazilian patients with CPC (Custodio et al. 2011). Additionally, current data on overall cancer penetrance suggest that the age-related cancer risk is somewhat lower than in LFS associated with other *TP53* mutations, with tumors detected in about 25 % of carriers at the age of 30 years and a lifetime risk of about 80 % (Achatz et al. 2007).

5 What Phenotype Should We Expect from a Germline Loss of *TP53*?

The nature of the mutation, the interplay between the p53 family members, and the genetic background of the host are all important factors in determining the final phenotype. Thus, identifying the type of *TP53* mutation may be important in the management of LFS families, as not all mutations are functionally equivalent (Bougeard et al. 2003; Soussi and Wiman 2007). Evidence from murine knock-in models supports this premise in that whereas the tumor spectrum in mice with a heterozygous R273H mutation reveals up to 50 % carcinomas with a high metastatic potential, expression of the R175H in two different genetic backgrounds results in a more diverse tumor spectrum (Lang et al. 2004; Olive et al. 2004).

It is also reasonable to expect that individuals who inherit a functionally deficient copy of *TP53* may be prone to tumorigenesis through interaction with environmental factors. To date, however, *TP53* germline mutation carriers do not show a significant increase in the risk of tobacco-associated cancers or in occupational cancers (Palmero et al. 2010). Whether other environmental factors such as exposure to tumor viruses play any role in modifying the cancer phenotype in LFS patients has not been explored (Malkin 2001).

Finally, in tumors with a high frequency of *TP53* mutation, the R72 variant can be associated with a poor prognosis, whereas in tumors with a low frequency of *TP53* mutation, the low apoptotic variant P72 could be associated with a poorer outcome. The role of this intragenic SNP as a prognostic variable in the setting of LFS has not been clearly demonstrated (Soussi and Wiman 2007).

6 Genetic Anticipation and Modifiers of *TP53* Genotype

Genetic anticipation is characterized by decreasing age at onset and/or increasing severity of symptoms of a disease in successive generations within a pedigree. In LFS, decrease in the age at cancer onset and increase in cancer incidence in successive generations have both been observed (Trkova et al. 2002; Tabori and Malkin 2008). Interestingly, telomeres in the peripheral blood leukocytes of *TP53* mutation carriers are shorter than in normal individuals of corresponding age. This difference was more pronounced in children (34 % decrease) than in adults (19 % decrease) (Trkova et al. 2007). The accelerated telomere attrition in LFS individuals compared to controls is postulated to lead to greater genomic instability and earlier age of cancer onset in successive generations (Tabori et al. 2007) and may in part explain the observed pattern of anticipation in many families. Copy number variation (CNV) has also been explored in LFS. Shlien et al. observed that while global CNV frequency is remarkably similar among healthy individuals, it is significantly increased in individuals with germline *TP53* mutations. Furthermore, total structural variation as measured by the product of the frequency and size of CNVs is also strikingly increased in mutation carriers. In addition, among families with a history of cancer, offspring were significantly more likely to have an increase in CNVs when compared with their mutation carrier parent (Shlien et al. 2008). These findings, conceptually similar to the accelerated telomere attrition data, support the notion that patients who are *TP53* mutation carriers have inherently unstable genomes and harbor other genetic and genomic alterations that can directly modify the age phenotype. What the other genetic events are that modify the "driver" genotype conferred by the germline *TP53* mutation are being actively explored and may ultimately lead to the development of more precise predictive algorithms of cancer phenotype and disease risk.

7 Treatment and Prognosis

Development of a rational approach to clinical management of LFS is difficult owing to the range of sites, types of cancer, and ages of presentation. Tumors in LFS patients are often treated with radiotherapy and chemotherapy, which have carcinogenic potential. Hisada et al. (1998) reported that multiple primary cancers occurred with high frequency within LFS cancers. These observations have been validated more recently (Cavalier et al. 2005) with further evidence to suggest that radiation-associated sarcomas were particularly common among the subsequent cancers. However, in several attempts to more accurately define links between treatment and second tumor development through comprehensive reviews, treatment data were too scanty for analyses of their carcinogenic influences (Kony et al. 1997; Nichols et al. 2001; Pierce and Haffty 2011).

In a prospective study, Heymann et al. (2010) assessed the incidence and outcome of secondary radiation-induced malignancies in patient with LFS with breast cancer as the first cancer. They recommend avoiding radiation therapy if conservative treatment is an available option.

The risks of second and third cancers many years later, primarily in the radiation treatment field, are consistent with a radiation effect. Patients with LFS may develop MDS, which is most likely chemotherapy related and is often associated with cytogenetic markers of poor prognosis (Talwalkar et al. 2010). Further evaluation of these secondary events is necessary to more accurately define the particular agents and interactions that may predispose to these second cancers. Clinicians following patients with inherited cancer syndromes should be aware of their increased susceptibility to second and third cancers that is enhanced by sensitivity to radiation (Kleinerman 2009).

TP53 mutations are reported to predict resistance to anthracyclines (Aas et al. 1996), doxorubicin (Geisler et al. 2001), and 5-fluorouracil/mitomycin (Geisler et al. 2003) in breast cancer. Mutations affecting the L2 and L3 loop structures in particular were associated with lack of response to chemotherapy. These results support the analysis of *TP53* mutation in prospective trials of new chemotherapeutic agents, with special consideration of the structural and functional consequences of mutation. A large, ongoing phase III clinical trial (EORTC 10994/BIG 00-1) of neoadjuvant chemotherapy given before breast cancer surgery has in fact incorporated *TP53* mutation analysis into its study design. However, whether extrapolation can be made from these data to the chemotherapy effect in the setting of germline *TP53* mutations is not known (Bonnefoi et al. 2011).

Recently, several discoveries have reenergized p53 as a therapeutic target as it has been shown that reintroduction of functional p53 into tumors has a therapeutic benefit.

Additionally, some of the small molecules identified that are shown to modulate p53 function may also help reduce the side effects of commonly used cancer therapeutics with relatively fewer off-target effects (Staples et al. 2008). For example, by binding directly into the p53-binding groove on mdm2, Nutlins are able to impair the interaction between p53 and mdm2 allowing p53 to function as a transcription factor (Vassilev et al. 2004). Another p53 small molecule activator, RITA, binds to the N terminus of p53 preventing mdm2 binding (Issaeva et al. 2004). In vivo reconstitution of wild-type p53 reactivates proapoptotic and senescence pathways which impede tumor development. For example, the compound PRIMA-1 (p53 reactivation and induction of massive apoptosis) binds to the p53 core and reactivates wild-type function to mutant p53 (Lambert et al. 2010). These and similarly functioning molecules have been shown to work in vivo in mouse xenograft models and data on their clinical efficacy are eagerly awaited. Whether they would have any degree of feasibility in the context of germline *TP53* mutations is entirely unexplored.

8 Screening and Surveillance

The general consensus in the LFS field is to consider testing for patients who meet the criteria of classic LFS or LFL or for children and young adults who have had two primary tumors despite a negative family history. In addition, very young women with breast cancer who are shown to be BRCA1/2 negative, and children diagnosed with adrenocortical tumors, CPC, RMSs, particularly when diagnosed <3 years, childhood brain tumors (10 %), and childhood osteosarcomas, particularly when diagnosed <10 years of age, may all be considered as candidates for testing.

A diagnosis of LFS raises the issue of preventive clinical surveillance for mutation carriers who have a high lifetime risk of malignancy. There appear to be no universally agreed rules for managing LFS, but there are some published guidelines (Varley et al. 1997). Until recently, it was widely accepted that testing of children should not be offered routinely; however, parents may exert considerable pressure to have a child tested. Further, there may be medical benefits related to early cancer detection. Each case should be treated on its own merits. If a case can be made that the child is likely to benefit from the test, then after very careful counseling that may include the child or children, proceeding to testing may be appropriate in some instances. Such an approach should be taken in the setting of a multidisciplinary team with expertise in clinical genetics, oncology and pediatric oncology, and genetic counseling, with extensive psychology and social work support.

For at-risk adults, the annual complete physical examination, complete blood count, and urinalysis should be supplemented with a yearly dermatologic examination and, for women, annual mammograms and/or breast magnetic resonance imaging (MRI) examinations should be offered starting in their early 1920s. The use of mammography continues to be under some debate due to the concern raised by exposure to the inherent radiation of the test. LFS patients should practice strict avoidance of sun and artificial tanning beds to limit UV-induced melanoma

(D'Orazio 2010). Lifestyle counseling on avoiding ionizing radiation (IR) and other DNA-damaging agents should also be given. For female *TP53* mutation carriers, the role of prophylactic mastectomy has not been carefully evaluated. Prenatal or preimplantation genetic testing can be offered to fertile couples affected by LFS. Moreover, as mentioned earlier, determining the exact nature of the *TP53* mutation may be of prognostic significance and therefore important for the clinical management of these patients (Royds and Iacopetta 2006). Most importantly, a high index of suspicion should be maintained for LFS carriers who complain of persisting but unexplained symptoms (Trahair et al. 2007).

The diversity of the tumors occurring in LFS families has prompted investigators to evaluate the use of biochemical and imaging modalities to identify tumors. One study evaluated the use of F18-fluorodeoxyglucose-positron emission tomography CT (FDG-PET/CT) to screen for new primary cancers in 15 asymptomatic members of LFS families (Masciari et al. 2008). Three tumors were detected (two thyroid and one esophageal cancer) using this surveillance method. The authors estimated the effective dose of the total FDG-PET/CT scan to be 2.4 rem and concluded that these preliminary data provide evidence for a potential cancer screening method for patients at high risk of multiple tumors.

While these results support the value of whole body imaging for tumor detection, the radiation exposure from repeated FDG-PET/CT scans may be of concern in *TP53* mutation carriers. The use of MRI is receiving greater attention (Leach et al. 2005). MRI has a higher sensitivity and specificity than mammography (Warner et al. 2001) in individuals with BRCA1/2 mutations and does not involve exposure to radiation (Warner 2011). Therefore, its application to LFS seems appropriate.

As colon cancer has been described in LFS children beginning in their school-age years, regular screening colonoscopy is probably warranted beginning sometime in childhood (Wong et al. 2006). Sarcomas and brain tumors are more difficult to identify. However, a recent study by Villani et al. suggests that an aggressive multimodality approach to clinical surveillance using a combination of biochemical markers of disease (particularly for adrenocortical carcinoma) and rapid sequence whole body MRI and dedicated brain MRI in both children and adults does identify presymptomatic cancers. Many of these are relatively low grade or low stage. Their early detection appears to be correlated with improved outcome when compared against *TP53* mutation carriers who did not undergo surveillance (Villani et al. 2011).

Because of the rarity of the syndrome, many individuals with LFS often feel isolated. Support groups for patients and families with inherited syndromes can help people cope when diagnosed with the disease. For LFS patients and family members, an online support group is available at http://www.mdjunction.com/li-fraumeni-syndrome. Members of this group include patients with LFS, healthcare providers, and spouses/friends of individuals with LFS (Vogel and Shannon 2009); several studies have explored the psychological impact of LFS in carriers and family members. While a detailed evaluation of this literature is beyond the scope of this chapter, one recent study bears particular notice. This group evaluated the uptake of genetic testing and the psychosocial impact of undergoing or not undergoing a genetic test for LFS. In total, 18 families with a *TP53* germline mutation were identified. Of the total group, 23 % reported clinically relevant levels of LFS-related distress. Carriers were not significantly more distressed than noncarriers or than those with a 50 % risk who did not undergo genetic testing. Those with a lack of social support were more prone to report clinically relevant levels of distress. Despite the limited treatment options for LFS, more than half of the individuals with a 50 % a prior risk choose to undergo presymptomatic testing. These findings point to the important fact that, in general, LFS family members, irrespective of their test result, appear to cope well (Lammens et al. 2010). As the interface between molecular diagnostics and clinical translation into treatment modification, accuracy of prognosis, and ability to predict tumor onset becomes more clear, it is reassuring to know that, for the most part, informed LFS patients will be able to effectively make rational decisions with respect to their health and that of their families.

9 Conclusions and Perspectives

Although a relatively uncommon disorder, LFS represents a remarkable paradigm of the biology and clinical complexity of cancer susceptibility. The diverse cancer phenotype has informed numerous studies that have led to great advances in our understanding of the fundamental genetic and biological properties of the p53 tumor suppressor. Conversely, technological advances in interrogating p53 structure and function have translated into improvements in gene testing, refinements in prediction of phenotype, and implementation of early detection strategies for this devastating human condition. With emergence of a multidisciplinary, international LFS consortium, it is anticipated that research efforts will be accelerated and translation of findings into clinical practice will become standard practice.

References

Aas T, Borresen AL, Geisler S, Smith-Sorensen B, Johnsen H, Varhaug JE, Akslen LA, Lonning PE (1996) Specific P53 mutations are associated with de novo resistance to doxorubicin in breast cancer patients. Nat Med 2(7):811–814

Achatz MI, Olivier M, Le Calvez F, Martel-Planche G, Lopes A, Rossi BM, Ashton-Prolla P, Giugliani R, Palmero EI, Vargas FR, Da Rocha JC, Vettore AL, Hainaut P (2007) The TP53 mutation, R337H, is associated with Li-Fraumeni and Li-Fraumeni-like syndromes in Brazilian families. Cancer Lett 245(1–2):96–102

Assumpcao JG, Seidinger AL, Mastellaro MJ, Ribeiro RC, Zambetti GP, Ganti R, Srivastava K, Shurtleff S, Pei D, Zeferino LC, Dufloth RM, Brandalise SR, Yunes JA (2008) Association of the germline TP53 R337H mutation with breast cancer in southern Brazil. BMC Cancer 8:357

Atwal GS, Bond GL, Metsuyanim S, Papa M, Friedman E, Distelman-Menachem T, Ben Asher E, Lancet D, Ross DA, Sninsky J, White TJ, Levine AJ, Yarden R (2007) Haplotype structure and selection of the MDM2 oncogene in humans. Proc Natl Acad Sci USA 104(11):4524–4529

Bachinski LL, Olufemi SE, Zhou X, Wu CC, Yip L, Shete S, Lozano G, Amos CI, Strong LC, Krahe R (2005) Genetic mapping of a third Li-Fraumeni syndrome predisposition locus to human chromosome 1q23. Cancer Res 65(2):427–431

Bell DW, Varley JM, Szydlo TE, Kang DH, Wahrer DC, Shannon KE, Lubratovich M, Verselis SJ, Isselbacher KJ, Fraumeni JF, Birch JM, Li FP, Garber JE, Haber DA (1999) Heterozygous germ line hCHK2 mutations in Li-Fraumeni syndrome. Science 286(5449):2528–2531

Berns EM, van Staveren IL, Look MP, Smid M, Klijn JG, Foekens JA (1998) Mutations in residues of TP53 that directly contact DNA predict poor outcome in human primary breast cancer. Br J Cancer 77(7):1130–1136

Birch JM (1994) Li-Fraumeni syndrome. Eur J Cancer 30A(13):1935–1941

Birch JM, Heighway J, Teare MD, Kelsey AM, Hartley AL, Tricker KJ, Crowther D, Lane DP, Santibanez-Koref MF (1994) Linkage studies in a Li-Fraumeni family with increased expression of p53 protein but no germline mutation in p53. Br J Cancer 70(6):1176–1181

Birch JM, Blair V, Kelsey AM, Evans DG, Harris M, Tricker KJ, Varley JM (1998) Cancer phenotype correlates with constitutional TP53 genotype in families with the Li-Fraumeni syndrome. Oncogene 17(9):1061–1068

Birch JM, Alston RD, McNally RJ, Evans DG, Kelsey AM, Harris M, Eden OB, Varley JM (2001) Relative frequency and morphology of cancers in carriers of germline TP53 mutations. Oncogene 20(34):4621–4628

Boldrini L, Gisfredi S, Ursino S, Lucchi M, Greco G, Mussi A, Donati V, Fontanini G (2008) Effect of the p53 codon 72 and intron 3 polymorphisms on non-small cell lung cancer (NSCLC) prognosis. Cancer Invest 26(2):168–172

Bond GL, Hu W, Bond EE, Robins H, Lutzker SG, Arva NC, Bargonetti J, Bartel F, Taubert H, Wuerl P, Onel K, Yip L, Hwang SJ, Strong LC, Lozano G, Levine AJ (2004) A single nucleotide polymorphism in the MDM2 promoter attenuates the p53 tumor suppressor pathway and accelerates tumor formation in humans. Cell 119(5):591–602

Bond GL, Hirshfield KM, Kirchhoff T, Alexe G, Bond EE, Robins H, Bartel F, Taubert H, Wuerl P, Hait W, Toppmeyer D, Offit K, Levine AJ (2006) MDM2 SNP309 accelerates tumor formation in a gender-specific and hormone-dependent manner. Cancer Res 66(10):5104–5110

Bonnefoi H, Piccart M, Bogaerts J, Mauriac L, Fumoleau P, Brain E, Petit T, Rouanet P, Jassem J, Blot E, Zaman K, Cufer T, Lortholary A, Lidbrink E, Andre S, Litiere S, Lago LD, Becette V, Cameron DA, Bergh J, Iggo R (2011) TP53 status for prediction of sensitivity to taxane versus non-taxane neoadjuvant chemotherapy in breast cancer (EORTC 10994/BIG 1–00): a randomised phase 3 trial. Lancet Oncol 12(6):527–539

Borresen AL, Andersen TI, Garber J, Barbier-Piraux N, Thorlacius S, Eyfjord J, Ottestad L, Smith-Sorensen B, Hovig E, Malkin D et al (1992) Screening for germ line TP53 mutations in breast cancer patients. Cancer Res 52(11):3234–3236

Borresen AL, Andersen TI, Eyfjord JE, Cornelis RS, Thorlacius S, Borg A, Johansson U, Theillet C, Scherneck S, Hartman S et al (1995) TP53 mutations and breast cancer prognosis: particularly poor survival rates for cases with mutations in the zinc-binding domains. Genes Chromosomes Cancer 14(1):71–75

Borresen-Dale AL, Lothe RA, Meling GI, Hainaut P, Rognum TO, Skovlund E (1998) TP53 and long-term prognosis in colorectal cancer: mutations in the L3 zinc-binding domain predict poor survival. Clin Cancer Res 4(1):203–210

Bougeard G, Brugieres L, Chompret A, Gesta P, Charbonnier F, Valent A, Martin C, Raux G, Feunteun J, Bressac-de Paillerets B, Frebourg T (2003) Screening for TP53 rearrangements in families with the Li-Fraumeni syndrome reveals a complete deletion of the TP53 gene. Oncogene 22(6):840–846

Bougeard G, Sesboue R, Baert-Desurmont S, Vasseur S, Martin C, Tinat J, Brugieres L, Chompret A, de Paillerets BB, Stoppa-Lyonnet D, Bonaiti-Pellie C, Frebourg T (2008) Molecular basis of the Li-Fraumeni syndrome: an update from the French LFS families. J Med Genet 45(8):535–538

Brosh R, Rotter V (2009) When mutants gain new powers: news from the mutant p53 field. Nat Rev Cancer 9(10):701–713

Cavalier ME, Davis MM, Croop JM (2005) Germline p53 mutation presenting as synchronous tumors. J Pediatr Hematol Oncol 27(8):441–443

Cho Y, Gorina S, Jeffrey PD, Pavletich NP (1994) Crystal structure of a p53 tumor suppressor-DNA complex: understanding tumorigenic mutations. Science 265(5170):346–355
Chompret A (2002) The Li-Fraumeni syndrome. Biochimie 84(1):75–82
Chompret A, Brugieres L, Bonaiti-Pellie C, Feunteun J (2004) Sex differences in cancer risk among germline p53 mutation carriers. Br J Cancer 91(3):603–604
Cohen CJ, Zheng Z, Bray R, Zhao Y, Sherman LA, Rosenberg SA, Morgan RA (2005) Recognition of fresh human tumor by human peripheral blood lymphocytes transduced with a bicistronic retroviral vector encoding a murine anti-p53 TCR. J Immunol 175(9):5799–5808
Collado M, Gil J, Efeyan A, Guerra C, Schuhmacher AJ, Barradas M, Benguria A, Zaballos A, Flores JM, Barbacid M, Beach D, Serrano M (2005) Tumour biology: senescence in premalignant tumours. Nature 436(7051):642
Costa S, Pinto D, Pereira D, Rodrigues H, Cameselle-Teijeiro J, Medeiros R, Schmitt F (2008) Importance of TP53 codon 72 and intron 3 duplication 16 bp polymorphisms in prediction of susceptibility on breast cancer. BMC Cancer 8:32
Custodio G, Taques GR, Figueiredo BC, Gugelmin ES, Oliveira Figueiredo MM, Watanabe F, Pontarolo R, Lalli E, Torres LF (2011) Increased incidence of choroid plexus carcinoma due to the germline TP53 R337H mutation in southern Brazil. PLoS One 6(3):e18015
D'Orazio JA (2010) Inherited cancer syndromes in children and young adults. J Pediatr Hematol Oncol 32(3):195–228
DiGiammarino EL, Filippov I, Weber JD, Bothner B, Kriwacki RW (2001) Solution structure of the p53 regulatory domain of the p19Arf tumor suppressor protein. Biochemistry 40(8): 2379–2386
Diller L, Sexsmith E, Gottlieb A, Li FP, Malkin D (1995) Germline p53 mutations are frequently detected in young children with rhabdomyosarcoma. J Clin Invest 95(4):1606–1611
Eeles RA (1995) Germline mutations in the TP53 gene. Cancer Surv 25:101–124
Farnebo M (2009) Wrap53, a novel regulator of p53. Cell Cycle 8(15):2343–2346
Farnebo M, Bykov VJ, Wiman KG (2010) The p53 tumor suppressor: a master regulator of diverse cellular processes and therapeutic target in cancer. Biochem Biophys Res Commun 396(1):85–89
Felley-Bosco E, Weston A, Cawley HM, Bennett WP, Harris CC (1993) Functional studies of a germ-line polymorphism at codon 47 within the p53 gene. Am J Hum Genet 53(3):752–759
Figueiredo BC, Sandrini R, Zambetti GP, Pereira RM, Cheng C, Liu W, Lacerda L, Pianovski MA, Michalkiewicz E, Jenkins J, Rodriguez-Galindo C, Mastellaro MJ, Vianna S, Watanabe F, Sandrini F, Arram SB, Boffetta P, Ribeiro RC (2006) Penetrance of adrenocortical tumours associated with the germline TP53 R337H mutation. J Med Genet 43(1):91–96
Fiszer-Maliszewska L, Czernik J, Sawicz-Birkowska K, Perek D, Kozera M, Wojciechowska B, Kazanowska B, Hudziec P, Kilar E (2000) Screening for germline p53 mutations in pediatric and adult patients of high-risk groups in Poland. Arch Immunol Ther Exp (Warsz) 48(4):309–315
Friedlander P, Haupt Y, Prives C, Oren M (1996) A mutant p53 that discriminates between p53-responsive genes cannot induce apoptosis. Mol Cell Biol 16(9):4961–4971
Geisler S, Lonning PE, Aas T, Johnsen H, Fluge O, Haugen DF, Lillehaug JR, Akslen LA, Borresen-Dale AL (2001) Influence of TP53 gene alterations and c-erbB-2 expression on the response to treatment with doxorubicin in locally advanced breast cancer. Cancer Res 61(6):2505–2512
Geisler S, Borresen-Dale AL, Johnsen H, Aas T, Geisler J, Akslen LA, Anker G, Lonning PE (2003) TP53 gene mutations predict the response to neoadjuvant treatment with 5-fluorouracil and mitomycin in locally advanced breast cancer. Clin Cancer Res 9(15):5582–5588
Gemignani F, Moreno V, Landi S, Moullan N, Chabrier A, Gutierrez-Enriquez S, Hall J, Guino E, Peinado MA, Capella G, Canzian F (2004) A TP53 polymorphism is associated with increased risk of colorectal cancer and with reduced levels of TP53 mRNA. Oncogene 23(10): 1954–1956

Goh HS, Yao J, Smith DR (1995) p53 point mutation and survival in colorectal cancer patients. Cancer Res 55(22):5217–5221

Gonzalez KD, Noltner KA, Buzin CH, Gu D, Wen-Fong CY, Nguyen VQ, Han JH, Lowstuter K, Longmate J, Sommer SS, Weitzel JN (2009) Beyond Li Fraumeni Syndrome: clinical characteristics of families with p53 germline mutations. J Clin Oncol 27(8):1250–1256

Grochola LF, Vazquez A, Bond EE, Wurl P, Taubert H, Muller TH, Levine AJ, Bond GL (2009) Recent natural selection identifies a genetic variant in a regulatory subunit of protein phosphatase 2A that associates with altered cancer risk and survival. Clin Cancer Res 15(19): 6301–6308

Harris N, Brill E, Shohat O, Prokocimer M, Wolf D, Arai N, Rotter V (1986) Molecular basis for heterogeneity of the human p53 protein. Mol Cell Biol 6(12):4650–4656

Hartley AL, Birch JM, Blair V (1991) Malignant disease in the mothers of a population-based series of young adults with bone and soft tissue sarcomas. Br J Cancer 63(3):416–419

Heymann S, Delaloge S, Rahal A, Caron O, Frebourg T, Barreau L, Pachet C, Mathieu MC, Marsiglia H, Bourgier C (2010) Radio-induced malignancies after breast cancer postoperative radiotherapy in patients with Li-Fraumeni syndrome. Radiat Oncol 5:104

Hinds PW, Finlay CA, Quartin RS, Baker SJ, Fearon ER, Vogelstein B, Levine AJ (1990) Mutant p53 DNA clones from human colon carcinomas cooperate with ras in transforming primary rat cells: a comparison of the "hot spot" mutant phenotypes. Cell Growth Differ 1(12):571–580

Hisada M, Garber JE, Fung CY, Fraumeni JF Jr, Li FP (1998) Multiple primary cancers in families with Li-Fraumeni syndrome. J Natl Cancer Inst 90(8):606–611

Hofseth LJ, Saito S, Hussain SP, Espey MG, Miranda KM, Araki Y, Jhappan C, Higashimoto Y, He P, Linke SP, Quezado MM, Zurer I, Rotter V, Wink DA, Appella E, Harris CC (2003) Nitric oxide-induced cellular stress and p53 activation in chronic inflammation. Proc Natl Acad Sci USA 100(1):143–148

Hong Y, Miao X, Zhang X, Ding F, Luo A, Guo Y, Tan W, Liu Z, Lin D (2005) The role of P53 and MDM2 polymorphisms in the risk of esophageal squamous cell carcinoma. Cancer Res 65(20):9582–9587

Hwang SJ, Lozano G, Amos CI, Strong LC (2003) Germline p53 mutations in a cohort with childhood sarcoma: sex differences in cancer risk. Am J Hum Genet 72(4):975–983

Issaeva N, Bozko P, Enge M, Protopopova M, Verhoef LG, Masucci M, Pramanik A, Selivanova G (2004) Small molecule RITA binds to p53, blocks p53-HDM-2 interaction and activates p53 function in tumors. Nat Med 10(12):1321–1328

Kleihues P, Schauble B, zur Hausen A, Esteve J, Ohgaki H (1997) Tumors associated with p53 germline mutations: a synopsis of 91 families. Am J Pathol 150(1):1–13

Kleinerman RA (2009) Radiation-sensitive genetically susceptible pediatric sub-populations. Pediatr Radiol 39(Suppl 1):S27–S31

Kony SJ, de Vathaire F, Chompret A, Shamsaldim A, Grimaud E, Raquin MA, Oberlin O, Brugieres L, Feunteun J, Eschwege F, Chavaudra J, Lemerle J, Bonaiti-Pellie C (1997) Radiation and genetic factors in the risk of second malignant neoplasms after a first cancer in childhood. Lancet 350(9071):91–95

Kucera E, Speiser P, Gnant M, Szabo L, Samonigg H, Hausmaninger H, Mittlbock M, Fridrik M, Seifert M, Kubista E, Reiner A, Zeillinger R, Jakesz R (1999) Prognostic significance of mutations in the p53 gene, particularly in the zinc-binding domains, in lymph node- and steroid receptor positive breast cancer patients. Austrian Breast Cancer Study Group. Eur J Cancer 35(3):398–405

Lalloo F, Varley J, Moran A, Ellis D, O'Dair L, Pharoah P, Antoniou A, Hartley R, Shenton A, Seal S, Bulman B, Howell A, Evans DG (2006) BRCA1, BRCA2 and TP53 mutations in very early-onset breast cancer with associated risks to relatives. Eur J Cancer 42(8):1143–1150

Lambert JM, Moshfegh A, Hainaut P, Wiman KG, Bykov VJ (2010) Mutant p53 reactivation by PRIMA-1MET induces multiple signaling pathways converging on apoptosis. Oncogene 29(9):1329–1338

Lammens CR, Aaronson NK, Wagner A, Sijmons RH, Ausems MG, Vriends AH, Ruijs MW, van Os TA, Spruijt L, Gomez Garcia EB, Kluijt I, Nagtegaal T, Verhoef S, Bleiker EM (2010)

Genetic testing in Li-Fraumeni syndrome: uptake and psychosocial consequences. J Clin Oncol 28(18):3008–3014

Lang GA, Iwakuma T, Suh YA, Liu G, Rao VA, Parant JM, Valentin-Vega YA, Terzian T, Caldwell LC, Strong LC, El-Naggar AK, Lozano G (2004) Gain of function of a p53 hot spot mutation in a mouse model of Li-Fraumeni syndrome. Cell 119(6):861–872

Lazar V, Hazard F, Bertin F, Janin N, Bellet D, Bressac B (1993) Simple sequence repeat polymorphism within the p53 gene. Oncogene 8(6):1703–1705

Leach MO, Boggis CR, Dixon AK, Easton DF, Eeles RA, Evans DG, Gilbert FJ, Griebsch I, Hoff RJ, Kessar P, Lakhani SR, Moss SM, Nerurkar A, Padhani AR, Pointon LJ, Thompson D, Warren RM (2005) Screening with magnetic resonance imaging and mammography of a UK population at high familial risk of breast cancer: a prospective multicentre cohort study (MARIBS). Lancet 365(9473):1769–1778

Li FP, Fraumeni JF Jr (1969) Soft-tissue sarcomas, breast cancer, and other neoplasms. A familial syndrome? Ann Intern Med 71(4):747–752

Li FP, Fraumeni JF Jr, Mulvihill JJ, Blattner WA, Dreyfus MG, Tucker MA, Miller RW (1988) A cancer family syndrome in twenty-four kindreds. Cancer Res 48(18):5358–5362

Li YJ, Sanson M, Hoang-Xuan K, Delattre JY, Poisson M, Thomas G, Hamelin R (1995) Incidence of germ-line p53 mutations in patients with gliomas. Int J Cancer 64(6):383–387

Li X, Dumont P, Della Pietra A, Shetler C, Murphy ME (2005) The codon 47 polymorphism in p53 is functionally significant. J Biol Chem 280(25):24245–24251

Mahmoudi S, Henriksson S, Corcoran M, Mendez-Vidal C, Wiman KG, Farnebo M (2009) Wrap53, a natural p53 antisense transcript required for p53 induction upon DNA damage. Mol Cell 33(4):462–471

Malkin D (2001) The role of p53 in human cancer. J Neurooncol 51(3):231–243

Malkin D, Li FP, Strong LC, Fraumeni JF Jr, Nelson CE, Kim DH, Kassel J, Gryka MA, Bischoff FZ, Tainsky MA et al (1990) Germ line p53 mutations in a familial syndrome of breast cancer, sarcomas, and other neoplasms. Science 250(4985):1233–1238

Malkin D, Jolly KW, Barbier N, Look AT, Friend SH, Gebhardt MC, Andersen TI, Borresen AL, Li FP, Garber J et al (1992) Germline mutations of the p53 tumor-suppressor gene in children and young adults with second malignant neoplasms. N Engl J Med 326(20): 1309–1315

Masciari S, Van den Abbeele AD, Diller LR, Rastarhuyeva I, Yap J, Schneider K, Digianni L, Li FP, Fraumeni JF Jr, Syngal S, Garber JE (2008) F18-fluorodeoxyglucose-positron emission tomography/computed tomography screening in Li-Fraumeni syndrome. JAMA 299(11): 1315–1319

Matlashewski GJ, Tuck S, Pim D, Lamb P, Schneider J, Crawford LV (1987) Primary structure polymorphism at amino acid residue 72 of human p53. Mol Cell Biol 7(2):961–963

Nagy R, Sweet K, Eng C (2004) Highly penetrant hereditary cancer syndromes. Oncogene 23(38):6445–6470

Nichols KE, Malkin D, Garber JE, Fraumeni JF Jr, Li FP (2001) Germ-line p53 mutations predispose to a wide spectrum of early-onset cancers. Cancer Epidemiol Biomarkers Prev 10(2):83–87

Nigro JM, Baker SJ, Preisinger AC, Jessup JM, Hostetter R, Cleary K, Bigner SH, Davidson N, Baylin S, Devilee P et al (1989) Mutations in the p53 gene occur in diverse human tumour types. Nature 342(6250):705–708

Olive KP, Tuveson DA, Ruhe ZC, Yin B, Willis NA, Bronson RT, Crowley D, Jacks T (2004) Mutant p53 gain of function in two mouse models of Li-Fraumeni syndrome. Cell 119(6):847–860

Olivier M, Goldgar DE, Sodha N, Ohgaki H, Kleihues P, Hainaut P, Eeles RA (2003) Li-Fraumeni and related syndromes: correlation between tumor type, family structure, and TP53 genotype. Cancer Res 63(20):6643–6650

Olivier M, Hollstein M, Hainaut P (2010) TP53 mutations in human cancers: origins, consequences, and clinical use. Cold Spring Harb Perspect Biol 2(1):a001008

Ory K, Legros Y, Auguin C, Soussi T (1994) Analysis of the most representative tumour-derived p53 mutants reveals that changes in protein conformation are not correlated with loss of transactivation or inhibition of cell proliferation. EMBO J 13(15):3496–3504

Palmero EI, Schuler-Faccini L, Caleffi M, Achatz MI, Olivier M, Martel-Planche G, Marcel V, Aguiar E, Giacomazzi J, Ewald IP, Giugliani R, Hainaut P, Ashton-Prolla P (2008) Detection of R337H, a germline TP53 mutation predisposing to multiple cancers, in asymptomatic women participating in a breast cancer screening program in Southern Brazil. Cancer Lett 261(1):21–25

Palmero EI, Achatz MI, Ashton-Prolla P, Olivier M, Hainaut P (2010) Tumor protein 53 mutations and inherited cancer: beyond Li-Fraumeni syndrome. Curr Opin Oncol 22(1):64–69

Pearson AD, Craft AW, Ratcliffe JM, Birch JM, Morris-Jones P, Roberts DF (1982) Two families with the Li-Fraumeni cancer family syndrome. J Med Genet 19(5):362–365

Petitjean A, Achatz MI, Borresen-Dale AL, Hainaut P, Olivier M (2007) TP53 mutations in human cancers: functional selection and impact on cancer prognosis and outcomes. Oncogene 26(15):2157–2165

Pierce LJ, Haffty BG (2011) Radiotherapy in the treatment of hereditary breast cancer. Semin Radiat Oncol 21(1):43–50

Pietenpol JA, Tokino T, Thiagalingam S, el-Deiry WS, Kinzler KW, Vogelstein B (1994) Sequence-specific transcriptional activation is essential for growth suppression by p53. Proc Natl Acad Sci USA 91(6):1998–2002

Pleasants LM, Hansen MF (1994) Identification of a polymorphism in intron 2 of the p53 gene. Hum Genet 93(5):607–608

Potzsch C, Voigtlander T, Lubbert M (2002) p53 Germline mutation in a patient with Li-Fraumeni Syndrome and three metachronous malignancies. J Cancer Res Clin Oncol 128(8):456–460

Ribeiro RC, Sandrini F, Figueiredo B, Zambetti GP, Michalkiewicz E, Lafferty AR, DeLacerda L, Rabin M, Cadwell C, Sampaio G, Cat I, Stratakis CA, Sandrini R (2001) An inherited p53 mutation that contributes in a tissue-specific manner to pediatric adrenal cortical carcinoma. Proc Natl Acad Sci USA 98(16):9330–9335

Rich DC, Corpron CA, Smith MB, Black CT, Lally KP, Andrassy RJ (1997) Second malignant neoplasms in children after treatment of soft tissue sarcoma. J Pediatr Surg 32(2):369–372

Rowan S, Ludwig RL, Haupt Y, Bates S, Lu X, Oren M, Vousden KH (1996) Specific loss of apoptotic but not cell-cycle arrest function in a human tumor derived p53 mutant. EMBO J 15(4):827–838

Royds JA, Iacopetta B (2006) p53 and disease: when the guardian angel fails. Cell Death Differ 13(6):1017–1026

Ruijs MW, Verhoef S, Rookus MA, Pruntel R, van der Hout AH, Hogervorst FB, Kluijt I, Sijmons RH, Aalfs CM, Wagner A, Ausems MG, Hoogerbrugge N, van Asperen CJ, Gomez Garcia EB, Meijers-Heijboer H, Ten Kate LP, Menko FH, van 't Veer LJ (2010) TP53 germline mutation testing in 180 families suspected of Li-Fraumeni syndrome: mutation detection rate and relative frequency of cancers in different familial phenotypes. J Med Genet 47(6):421–428

Sandrini R, Ribeiro RC, DeLacerda L (1997) Childhood adrenocortical tumors. J Clin Endocrinol Metab 82(7):2027–2031

Shlien A (2010) Differential phenotypic impact of large genic versus small intragenic deletions across TP53. Amer J Human Genet 87(5):631–642

Shlien A, Tabori U, Marshall CR, Pienkowska M, Feuk L, Novokmet A, Nanda S, Druker H, Scherer SW, Malkin D (2008) Excessive genomic DNA copy number variation in the Li-Fraumeni cancer predisposition syndrome. Proc Natl Acad Sci USA 105(32):11264–11269

Sidransky D, Tokino T, Helzlsouer K, Zehnbauer B, Rausch G, Shelton B, Prestigiacomo L, Vogelstein B, Davidson N (1992) Inherited p53 gene mutations in breast cancer. Cancer Res 52(10):2984–2986

Sodha N, Williams R, Mangion J, Bullock SL, Yuille MR, Eeles RA (2000) Screening hCHK2 for mutations. Science 289(5478):359

Sodha N, Houlston RS, Bullock S, Yuille MA, Chu C, Turner G, Eeles RA (2002) Increasing evidence that germline mutations in CHEK2 do not cause Li-Fraumeni syndrome. Hum Mutat 20(6):460–462

Soussi T, Beroud C (2001) Assessing TP53 status in human tumours to evaluate clinical outcome. Nat Rev Cancer 1(3):233–240

Soussi T, Lozano G (2005) p53 mutation heterogeneity in cancer. Biochem Biophys Res Commun 331(3):834–842

Soussi T, Wiman KG (2007) Shaping genetic alterations in human cancer: the p53 mutation paradigm. Cancer Cell 12(4):303–312

Staples OD, Steele RJ, Lain S (2008) p53 as a therapeutic target. Surgeon 6(4):240–243

Tabori U, Malkin D (2008) Risk stratification in cancer predisposition syndromes: lessons learned from novel molecular developments in Li-Fraumeni syndrome. Cancer Res 68(7):2053–2057

Tabori U, Nanda S, Druker H, Lees J, Malkin D (2007) Younger age of cancer initiation is associated with shorter telomere length in Li-Fraumeni syndrome. Cancer Res 67(4):1415–1418

Tabori U, Shlien A, Baskin B, Levitt S, Ray P, Alon N, Hawkins C, Bouffet E, Pienkowska M, Lafay-Cousin L, Gozali A, Zhukova N, Shane L, Gonzalez I, Finlay J, Malkin D (2010) TP53 alterations determine clinical subgroups and survival of patients with choroid plexus tumors. J Clin Oncol 28(12):1995–2001

Talwalkar SS, Yin CC, Naeem RC, Hicks MJ, Strong LC, Abruzzo LV (2010) Myelodysplastic syndromes arising in patients with germline TP53 mutation and Li-Fraumeni syndrome. Arch Pathol Lab Med 134(7):1010–1015

Taylor MD, Mainprize TG, Rutka JT (2000) Molecular insight into medulloblastoma and central nervous system primitive neuroectodermal tumor biology from hereditary syndromes: a review. Neurosurgery 47(4):888–901

Tinat J, Bougeard G, Baert-Desurmont S, Vasseur S, Martin C, Bouvignies E, Caron O, Bressac-de Paillerets B, Berthet P, Dugast C, Bonaiti-Pellie C, Stoppa-Lyonnet D, Frebourg T (2009) 2009 version of the Chompret criteria for Li Fraumeni syndrome. J Clin Oncol 27(26):e108–e109, author reply e110

Trahair T, Andrews L, Cohn RJ (2007) Recognition of Li Fraumeni syndrome at diagnosis of a locally advanced extremity rhabdomyosarcoma. Pediatr Blood Cancer 48(3):345–348

Trkova M, Hladikova M, Kasal P, Goetz P, Sedlacek Z (2002) Is there anticipation in the age at onset of cancer in families with Li-Fraumeni syndrome? J Hum Genet 47(8):381–386

Trkova M, Prochazkova K, Krutilkova V, Sumerauer D, Sedlacek Z (2007) Telomere length in peripheral blood cells of germline TP53 mutation carriers is shorter than that of normal individuals of corresponding age. Cancer 110(3):694–702

Vahteristo P, Tamminen A, Karvinen P, Eerola H, Eklund C, Aaltonen LA, Blomqvist C, Aittomaki K, Nevanlinna H (2001) p53, CHK2, and CHK1 genes in Finnish families with Li-Fraumeni syndrome: further evidence of CHK2 in inherited cancer predisposition. Cancer Res 61(15):5718–5722

Varley JM (2003) Germline TP53 mutations and Li-Fraumeni syndrome. Hum Mutat 21(3):313–320

Varley JM, Evans DG, Birch JM (1997) Li-Fraumeni syndrome – a molecular and clinical review. Br J Cancer 76(1):1–14

Varley JM, McGown G, Thorncroft M, Kelsey AM, Birch JM (2001) Significance of intron 6 sequence variations in the TP53 gene in Li-Fraumeni syndrome. Cancer Genet Cytogenet 129(1):85–87

Vassilev LT, Vu BT, Graves B, Carvajal D, Podlaski F, Filipovic Z, Kong N, Kammlott U, Lukacs C, Klein C, Fotouhi N, Liu EA (2004) In vivo activation of the p53 pathway by small-molecule antagonists of MDM2. Science 303(5659):844–848

Vazquez A, Bond EE, Levine AJ, Bond GL (2008) The genetics of the p53 pathway, apoptosis and cancer therapy. Nat Rev Drug Discov 7(12):979–987

Villani A, Tabori U, Schiffman J, Shlien A, Beyene J, Druker H, Novokmet A, Finlay J, Malkin D (2011) Biochemical and imaging surveillance in germline TP53 mutation carriers with Li-Fraumeni syndrome: a prospective observational study. Lancet Oncol 12(6):559–567

Vogel JM, Shannon KM (2009) Genetic counseling and testing for Li-Fraumeni syndrome: the medical and psychosocial implications. Commun Oncol 6:507–514

Warner E (2011) Clinical practice. Breast-cancer screening. N Engl J Med 365(11):1025–1032

Warner E, Plewes DB, Shumak RS, Catzavelos GC, Di Prospero LS, Yaffe MJ, Goel V, Ramsay E, Chart PL, Cole DE, Taylor GA, Cutrara M, Samuels TH, Murphy JP, Murphy JM, Narod SA (2001) Comparison of breast magnetic resonance imaging, mammography, and ultrasound for surveillance of women at high risk for hereditary breast cancer. J Clin Oncol 19(15):3524–3531

Whibley C, Pharoah PD, Hollstein M (2009) p53 polymorphisms: cancer implications. Nat Rev Cancer 9(2):95–107

Wong P, Verselis SJ, Garber JE, Schneider K, DiGianni L, Stockwell DH, Li FP, Syngal S (2006) Prevalence of early onset colorectal cancer in 397 patients with classic Li-Fraumeni syndrome. Gastroenterology 130(1):73–79

Chapter 11
Tp53 Gene Therapy for Cancer Treatment and Prevention

Robert E. Sobol, Yong-Song Guan, Long-Jiang Li, Wei-Wei Zhang, Zhaohui Peng, Kerstin B. Menander, Sunil Chada, Daniel C. Maneval, Jane A. Horowitz, Robert Warren, Gary L. Clayman, Stephen G. Swisher, W. Jarrard Goodman, John Nemunaitis, and Jack A. Roth

1 Overview

Abnormalities in the tumor suppressor *TP53* are among the most common mechanisms of cancer pathogenesis (Lane and Levine 2010) and formed the rationale for *TP53* gene therapy to restore normal p53 function in cancer treatment. Several important principles were elucidated in preclinical tumor models which predicted the outcomes of subsequent clinical trials including synergistic antitumor activity for combined *TP53* gene therapy plus DNA damaging chemotherapy and radiation (Zhang and Roth 1994; Gjerset and Sobol 1997; Nielsen and Maneval 1998).

R.E. Sobol (✉) • D.C. Maneval
p53 Inc., P.O. Box 1574, Rancho Santa Fe, CA, USA
e-mail: rsobol@p53inc.com

K.B. Menander • S. Chada
p53 Inc., Houston, USA

Y.-S. Guan • L.-J. Li
Sichuan University, Chengdu, People's Republic of China

W.-W. Zhang • Z. Peng
SiBiono Biotech LTD., Shenzen, People's Republic of China

J.A. Horowitz
Merck & Co., Whitehouse Station, USA

R. Warren
University of California, San Francisco, USA

G.L. Clayman • S.G. Swisher • J.A. Roth
MD Anderson Cancer Center, Houston, USA

W.J. Goodman
University of Miami, Miami, USA

J. Nemunaitis
Mary Crowley Cancer Center, Dallas, USA

P. Hainaut et al. (eds.), *p53 in the Clinics*, DOI 10.1007/978-1-4614-3676-8_11,

The first clinical *TP53* cancer gene therapy trial employed a retroviral vector resulting in tumor regressions in advanced lung cancer patients (Roth et al. 1996). Three similar adenoviral vectors expressing wild-type p53 were then developed and employed in clinical trials internationally: Advexin (Zhang et al. 1994), Gendicine (Peng 2005), and SCH58500 (Wills et al. 1994). In 2003, through the efforts of Peng and his colleagues, Gendicine was approved by the State Food and Drug Administration of China (SFDA) for the treatment of head and neck squamous cell carcinoma by intra-tumoral injection in combination with radiation.

Subsequently, biomarkers predictive of *TP53* gene therapy efficacy were identified by Nemunaitis et al. (2009). Molecular profiles favorable for *TP53* gene therapy efficacy had either normal *TP53* gene sequences or low-level p53 protein expression, whereas unfavorable p53 molecular profiles had high-level expression of mutated p53 that inhibits normal p53 protein function.

Impressive efficacy of *TP53* gene therapy has been demonstrated with intra-arterial *TP53* gene delivery. The Phase I safety data of intra-arterial *TP53* gene therapy with the adenoviral vector (SCH 58500) was reported by Warren and his colleagues (Reid et al. 2002; Atencio et al. 2006). Subsequently, the efficacy of intra-arterial *TP53* gene therapy in combination with chemotherapy was demonstrated in head and neck cancer by Li et al. (2006) and in advanced lung cancer by Guan et al. (2009).

A different type of *TP53* targeted adenoviral treatment involves the administration of a genetically engineered virus with its *E1B* 55-kDa gene deleted. As E1B inhibits normal p53 suppression of adenovirus cell lysis, the *E1B* deleted adenovirus termed Onyx-015 was designed to preferentially replicate in and lyse p53 deficient cancer cells while sparing normal cells (Bischoff et al. 1996). In a Phase II clinical study of ONYX-015 in combination with cisplatin and 5-fluorouracil in recurrent head and neck cancer (Khuri et al. 2000), the time to tumor progression was increased for ONYX-015 injected tumors compared to non-injected lesions ($p=0.006$, log-rank test). In 2005, the Chinese State Food and Drug Administration approved H101, an oncolytic adenovirus similar to Onyx-015 (*E1B*-55K/*E3B*-deleted), for use in combination with chemotherapy for the treatment of late-stage, refractory nasopharyngeal cancer. The results of clinical trials performed with H101 are reviewed in detail elsewhere by Yu and Fang (2007).

Overall, the results of preclinical and clinical trials performed to date demonstrate the efficacy of *TP53* gene therapy in combination with radiation and chemotherapy. Additional clinical experience has indicated that future applications of *TP53* gene therapy may be improved by intra-arterial administration and the use of predictive efficacy biomarkers. The key results of these clinical studies are summarized in this review.

2 Background/Introduction

TP53 is the prototypical tumor suppressor gene and plays a critical role in regulating tumor proliferation and death (apoptosis) (Lane and Levine 2010). In the vast majority of cancers, abnormal p53 function results in uncontrolled cellular proliferation through loss of cell cycle regulation and may contribute to treatment resistance from impaired apoptosis in response to therapy. These p53 pathways are aberrant in virtually all cancer cells, either by mutation/deletion in the *TP53* gene or by abnormal regulation of p53 expression or function in the absence of *TP53* gene mutations (Lane and Levine 2010). The frequency of p53 abnormalities and their fundamental bearing on cancer pathogenesis provided the rationale to develop *TP53* gene therapy for cancer treatment and prevention.

This review summarizes the results of *TP53* gene therapy clinical trials performed in a variety of prevention and treatment settings demonstrating broad utility in treating multiple tumor types as monotherapy and in combination with chemotherapy and radiation. These studies support applications of *TP53* gene therapy for cancer prevention, initial cancer treatment, therapy of recurrent tumors, and the treatment of disease refractory to standard therapies.

2.1 Key Treatment Principles from Preclinical Studies

Several important principles regarding p53 function impacting the potential efficacy of *TP53* gene therapy have been elucidated in preclinical tumor models. *TP53* gene therapy is one of the first anticancer agents that induces cellular senescence as a key mechanism of action (Xue et al. 2007). Importantly, only transient restoration of p53 function is required for the induction of cellular senescence which results in "permanent cell cycle arrest" (Xue et al. 2007). p53 also activates anti-angiogenic mechanisms which are also associated with the stabilization of tumor growth (Nishizaki et al. 1999). Apoptosis or "programmed cell death" is another key pathway activated by p53 restoration (Zambetti and Levine 1993; Zhang and Roth 1994; Gjerset et al. 1995; Gjerset and Sobol 1997; Nielsen and Maneval 1998).

Hence, these and many other preclinical studies indicated that transient expression of wild-type p53 mediates therapeutic benefits through the induction of either cellular senescence (permanent cell cycle arrest) or apoptosis (programmed cell death).

These observations led to additional preclinical studies demonstrating the ability of wild-type *TP53* gene transfer to sensitize tumors to the effects of DNA damaging chemotherapy or radiation (Gjerset et al. 1995; Nishizaki et al. 2001). A representative example of the preclinical studies showing the synergistic antitumor effects of combination adenoviral *TP53*, chemotherapy, and radiation is depicted in Fig. 11.1 from Nishizaki et al. (2001). These preclinical data documented improved therapeutic benefits for combined *TP53* treatment plus radiation or chemotherapy compared to standard treatment alone and equivalent but not superior antitumor effects for *TP53* monotherapy compared to single agent chemotherapy or radiation. The results of these preclinical studies predicted the outcomes of subsequent *TP53* gene therapy clinical trials as summarized below.

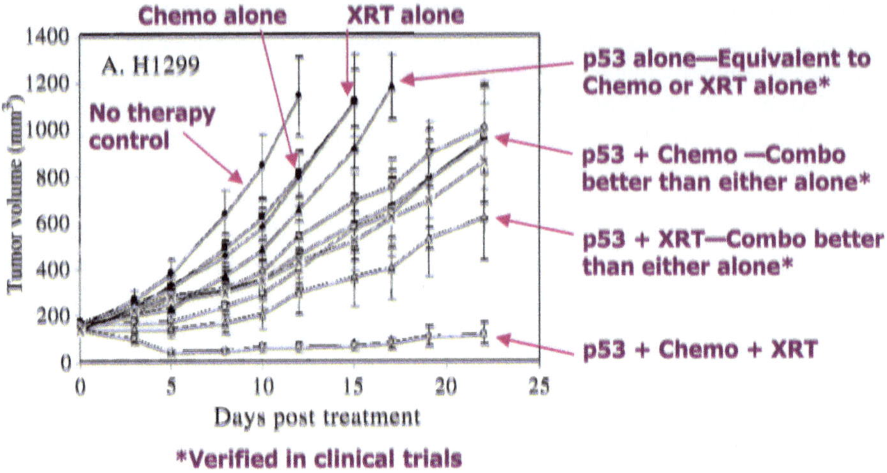

Fig. 11.1 Representative preclinical data demonstrating the synergistic antitumor effects of combination adenoviral *TP53*, chemotherapy, and radiation are shown in a human lung tumor xenograft animal model [adapted from Nishizaki et al. (2001)]. Improved therapeutic benefits for combined *TP53* treatment plus radiation or chemotherapy were observed compared to either treatment alone while equivalent but not superior antitumor effects were demonstrated for *TP53* monotherapy compared to single agent chemotherapy. These results are consistent with the outcomes of *TP53* gene therapy clinical trials

3 *TP53* Gene Therapy Clinical Trials for Cancer Treatment

3.1 Retroviral TP53 Gene Therapy

The first *TP53* cancer gene therapy employed a retroviral vector which was administered intra-tumorally to advanced lung cancer patients by Roth et al. (1996). A retroviral vector containing the wild-type *TP53* gene under control of a beta-actin promoter was delivered by direct intra-tumoral injection in nine patients with tumors that had become refractory to standard therapy. There were no clinically significant treatment-related adverse effects. In situ hybridization and DNA polymerase chain reaction demonstrated vector—*TP53* sequences in posttreatment tumor biopsies. Tumor regression was noted in three patients and tumor growth stabilized in three other patients. Apoptosis (programmed cell death) was more frequent in posttreatment biopsies than in pretreatment biopsies.

3.2 Adenoviral TP53 Gene Therapy Key Clinical Trials

3.2.1 Adenoviral *TP53* Gene Therapy Vectors

Three similar adenoviral vectors expressing wild-type p53 were developed and employed in clinical trials. Each of these vectors carried the wild-type *TP53* gene in a genetically modified *E1* deleted serotype 5 adenovirus. Advexin (Zhang et al. 1994) and SCH 58500 (Wills et al. 1994) employed a CMV promoter to drive *TP53* gene expression while Gendicine utilized an RSV promoter (Peng 2005).

Advexin, SCH 58500, and Gendicine share many features common to "first-generation" recombinant adenoviral gene delivery systems. Each adenoviral *TP53* construct encodes the full-length human wild-type p53 protein. The viral delivery systems are based on human adenovirus serotype 5 with deletions in the *E1* region that render the adenoviral *TP53* constructs replication deficient in normal cells. Manufacturing systems for each product are similar. Production in 293 human embryonic kidney (HEK-293) cells is followed by column chromatography for adenoviral *TP53* purification. These characteristics imply that Advexin, SCH 58500, and Gendicine are likely to behave similarly as therapeutic agents.

3.2.2 Adenoviral *TP53* Gene Therapy by Intra-tumoral Administration

Adenoviral *TP53* Gene Therapy as Monotherapy

Head and Neck Cancer

One of the earliest studies to evaluate adenoviral *TP53* gene therapy in human subjects was performed by Clayman et al. (1998). Advexin was injected into the tumors of patients with incurable, recurrent head and neck cancers. Groups of patients were treated with increasing doses of Advexin to identify a safe dose for future clinical development. No dose-limiting toxicity or serious adverse events were noted. Objective antitumor activity was detected in several patients despite many patients receiving subtherapeutic Advexin dosages. Clinical efficacy was evaluable in 17 patients with non-resectable tumors. One patient had a complete response with no evidence of residual tumor, two patients showed objective tumor regressions of greater than 50%, and six patients had stable disease. Similar objective tumor responses were also observed in subsequent head and neck cancer clinical trials evaluating adenoviral *TP53* as monotherapy (Nemunaitis et al. 2009) (see Fig. 11.2).

Li–Fraumeni Syndrome

Li–Fraumeni syndrome is an inherited disorder that greatly increases the risk of developing multiple types of cancer. The majority of Li–Fraumeni syndrome families inherit a mutated *TP53* tumor suppressor gene (Malkin et al. 1990). Hence, *TP53* gene therapy is prototypical molecular treatment targeting the pathogenic molecular defect. Senzer et al. (2007) reported intra-tumoral Advexin *TP53* monotherapy of a Li–Fraumeni patient with refractory, progressive embryonal carcinoma. With respect to molecular markers, the patient's tumor had mutated *TP53* and expressed coxsackie adenovirus receptors conducive for adenoviral *TP53* activity. *TP53* treatment resulted in the induction of cell cycle arrest and apoptosis documented by increased p21 and cleaved caspase-3 expression, respectively. Elevated adenoviral antibody titers after repeated therapy did not inhibit adenoviral *TP53* activity or result in pathologic side effects.

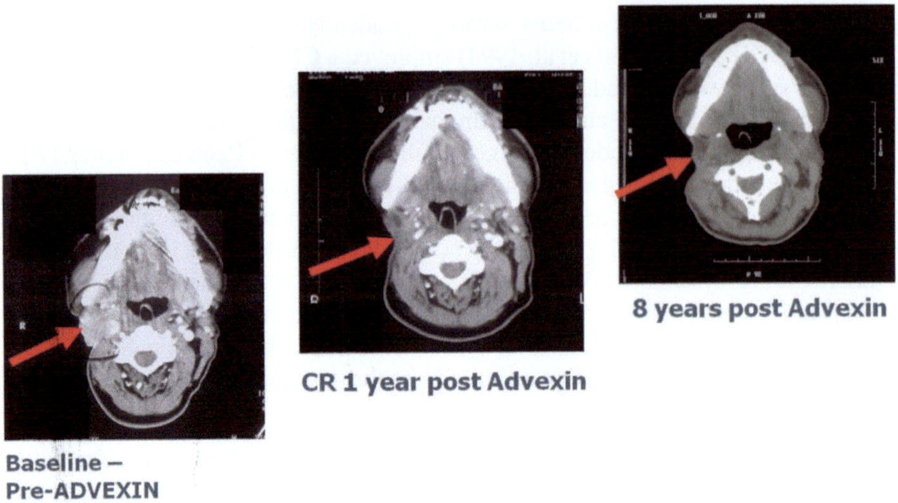

Fig. 11.2 Clinical, intra-tumoral adenoviral *TP53* monotherapy of squamous cell carcinoma of the head and neck. CT images before and after monthly doses of Advexin monotherapy (2 x 10^{11} viral particles/dose) depicting a complete tumor response (*arrows*) in a patient with recurrent squamous cell carcinoma of the head and neck

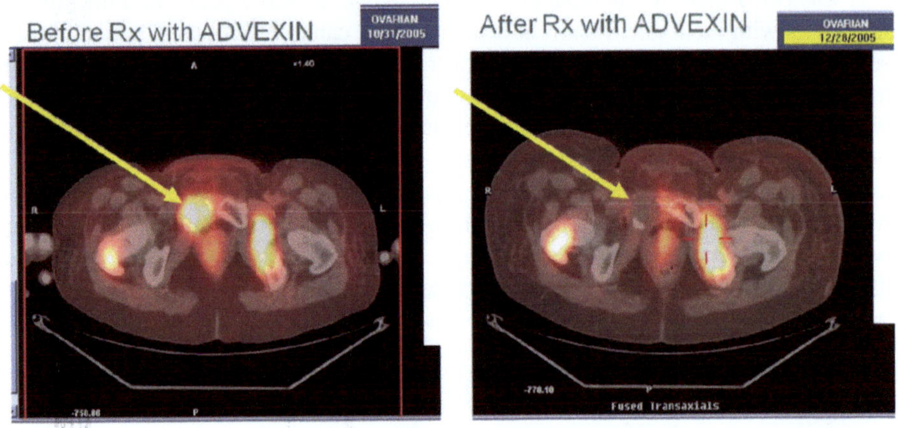

Fig. 11.3 Clinical, intra-tumoral adenoviral *TP53* monotherapy of Li–Fraumeni Syndrome. Fusion PET/CT scan of the treated pelvic tumor (*arrow*) before (*left*) and after four intra-tumoral injections of Advexin depicting complete resolution of 2,3-FDG uptake at 2 months (*right*) and progression of the untreated lesion (cross-hairs)

As demonstrated in Fig. 11.3, *TP53* treatment resulted in complete and durable remission of the injected lesion by fluorodeoxyglucose-positron emission tomography (PET) scans. The PET scan findings are consistent with the induction of p21 expression and cellular senescence. The tumor response was associated with improvement of tumor-related symptoms (decreased pelvic pain and improvement in lower extremity edema).

Adenoviral *TP53* Gene Therapy in Combination with Radiation

The results of adenoviral *TP53* gene therapy clinical trials in China led to Gendicine becoming the world's first gene therapy product approved by a regulatory agency, as reviewed in Peng (2005). After extensive multiyear and multicenter clinical studies, intra-tumoral adenovirus *TP53* injection (trademarked as Gendicine) was approved by the State Food and Drug Administration of China (SFDA) on October 16, 2003, for the treatment of head and neck squamous cell carcinoma in combination with radiation.

A multicenter, controlled, randomized clinical trial involving 135 patients with squamous cell carcinoma of the head and neck compared the efficacy of Gendicine in combination with radiation to radiation treatment alone. Of the enrolled patients, 77% had extensive disease (Stage III or IV) and had prior radiation or chemotherapy or were not eligible for surgery. The majority of the patients (85%) had nasopharyngeal cancer. Objective tumor response was evaluated by computed tomography (CT) or magnetic resonance imaging (MRI) according to World Health Organization (WHO) response criteria. The response rate in the combined *TP53* gene therapy + radiation group was 93%, with 64% showing complete response (CR) and 29% partial response (PR). The response rate in the radiation alone group was 79%, with only 19% of the patients showing CR and 60% PR. There was a significant difference ($p < 0.01$) between the two groups in terms of the CR rate (Fig. 11.4). These clinical results are consistent with preclinical studies indicating the synergistic effects of *TP53* gene therapy and radiation treatment (see Fig. 11.1).

Swisher et al. (2003) performed a prospective single arm Phase II study to evaluate the feasibility and mechanisms of apoptosis induction after adenoviral *TP53* gene transfer and radiation therapy in patients with non-small cell lung cancer. Nineteen patients with non-metastatic non-small cell lung cancer who were not eligible for chemoradiation or surgery were treated as outpatients with radiation therapy to 60 Gy over 6 weeks in conjunction with three intra-tumoral injections of Advexin on days 1, 18, and 32. Seventeen of the 19 patients completed all planned radiation and adenoviral *TP53* gene therapy as outpatients. The most common adverse events were grade 1 or 2 fevers (79%) and chills (53%). Three months after completion of therapy, biopsies of the primary tumor revealed no viable tumor in 12/19 patients, 63%; viable tumor in 3 of 19 patients, 16%; and not evaluable in 4 of 19 patients, 21%. Computed tomography and bronchoscopic findings at the primary injected tumor revealed complete response (1 of 19 patients, 5%), partial response (11 of 19 patients, 58%), stable disease (3 of 19 patients, 16%), progressive disease (2 of 19 patients, 11%), and not evaluable (2 of 19 patients, 11%). Quantitative reverse transcription-PCR analysis of four p53 related genes (*p21—CDKN1A*, *FAS*, *BAK*, and *MDM2*) revealed that *BAK* expression was increased significantly 24 h after Advexin injection and that levels of *CDKN1A* and *MDM2* expression were increased over the course of treatment. The authors concluded that intra-tumoral injection of adenoviral *TP53* in combination with radiation therapy was well tolerated and demonstrated regression of the primary injected tumor.

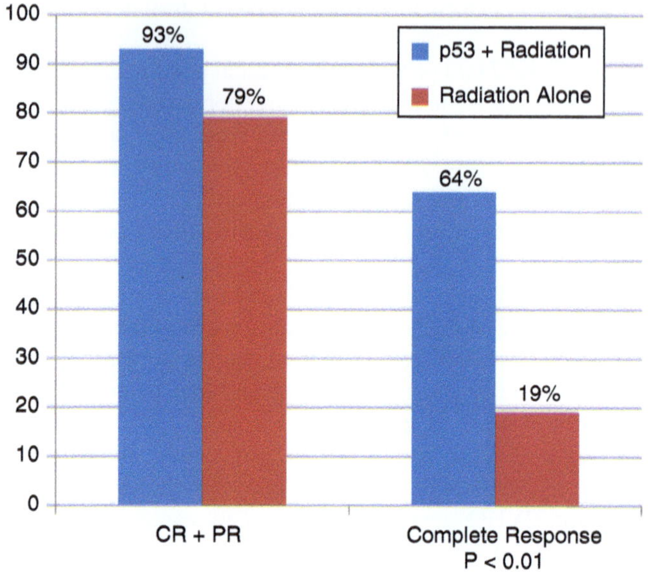

Fig. 11.4 Clinical, intra-tumoral adenoviral *TP53* gene therapy in combination with radiation for the treatment of squamous cell carcinoma of the head and neck. Statistically superior complete response rates were observed for *TP53* gene therapy in combination with radiation compared to radiation treatment alone in a multicenter, controlled, randomized clinical trial involving 135 patients [Adapted from Peng (2005)]. These results led to the world's first approved gene therapy product (Gendicine) by the State Food and Drug Administration of China

Hence, these studies indicate the safety and clinical activity of intra-tumoral adenoviral *TP53* in combination with radiotherapy.

Efficacy of Intra-tumor Adenoviral *TP53* Combined with Chemoembolization in Advanced Hepatocellular Carcinoma

Guan et al. (2011) reported positive results of intra-tumoral administration of adenoviral *TP53* (Gendicine) in combination with hepatic transcatheter arterial chemoembolization (TACE) to patients with advanced hepatocellular carcinoma (HCC). A total of 150 patients were enrolled in this two-arm controlled clinical trial. Sixty-eight patients were treated with the Gendicine–TACE combination by intra-tumoral Gendicine injection under CT guidance at a dose of 1–4 x 10^{12} viral particles, once per week for 4 weeks, 2–5 days after TACE was administered. In the control arm, 82 patients were treated with TACE alone. In both treatment groups, TACE therapy comprised intra-arterial administration of 5-fluorouracil, hydroxy-camptothecin, doxorubicin, and arterial embolization with iodized oil. Patients were followed for

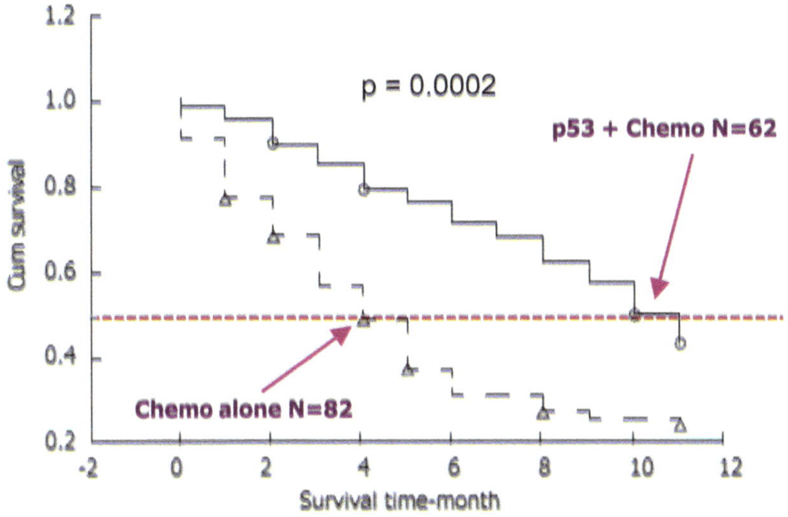

Fig. 11.5 Clinical, intra-tumoral adenoviral *TP53* gene therapy combined with chemoembolization for the treatment of hepatocellular carcinoma. Statistically significant increased survival by log-rank test ($p = 0.0002$) was observed for *TP53* gene therapy + chemoembolization compared to chemoembolization alone [Adapted from Guan et al. (2011)]

12 months with safety and efficacy evaluations. The total objective response rate by RECIST criteria was superior for the *TP53* treatment group (67.6%) compared to the TACE only control patients (51.2%) ($p < 0.05$). The 3-, 6-, and 12-mo survival rates were significantly higher for the *TP53* treatment group than for the control group (*TP53* treatment + TACE group 89.71%, 76.13%, and 43.30% compared to 68.15%, 36.98%, and 24.02% for the TACE only control group). Comparison of survival by the log-rank test demonstrated significantly increased survival for the *TP53* treatment group ($p = 0.0002$, see Fig. 11.5).

Consistent with many other safety studies, adenoviral *TP53* treatment was well tolerated with flu-like symptoms (fever, muscle and joint pain) being the most common adverse events that were significantly higher than for the TACE only group. Except for these minor adverse effects, no severe vector-related complications were identified. The data from this trial support adenoviral *TP53* gene therapy in combination with TACE as a safe and more effective treatment modality for advanced HCC compared to TACE alone.

3.2.3 Adenoviral *TP53* Gene Therapy by Intra-arterial Administration

The feasibility and impressive clinical outcomes of intra-arterial adenoviral *TP53* gene therapy have been demonstrated in several studies (Reid et al. 2002; Atencio et al. 2006; Li et al. 2006; Guan et al. 2009).

Safety and Feasibility of Repeated Intra-vascular Adenoviral *TP53* Administration

Warren and his colleagues (Reid et al. 2002; Atencio et al. 2006) were among the first to administer adenoviral *TP53* intra-arterially employing the SCH58500 vector. Following hepatic artery infusion, the majority of tumor samples evaluated after treatment (days 2–7) were positive for *TP53* gene expression from the adenoviral vector, especially at doses exceeding 2.5 x 10^{11} viral particles. Neutralizing antibody titers to the Ad5 protein coat increased following intravascular administration, but these antibodies did not adversely affect p53 expression.

Adenoviral *TP53* was well tolerated following hepatic artery infusion at doses of up to 2.5 x 10^{13} particles. At a dose of 7.5 x 10^{13} particles, adenoviral *TP53* was associated with dose-limiting cardiac output suppression. Mild to moderate fever, rigors, and transaminitis were the most common adverse events. Transient, dose-related suppression of lymphocytes, monocytes, eosinophils, and platelets were also observed generally within the first 24–48 h. These decreases resolved in less than 7 days and were not clinically significant. Associated increases in fibrin degradation products and increases in PT/PTT were demonstrated within a similar timeframe but were not clinically significant and fibrinogen levels actually increased. Acute inflammatory cytokine induction occurred within 3–6 h post-infusion in a dose-dependent fashion including interleukin (IL)-6, interferon gamma, and tumor necrosis factor. IL-10 induction generally occurred after IL-6 induction by approximately 18–24 h. However, the levels and patterns of cytokine induction varied greatly between patients. In summary, adenoviral *TP53* was well tolerated by intra-arterial infusion, both with single and repeated administrations, at doses of up to 2.5 x 10^{13} particles. Importantly, the feasibility of repeated intra-arterial delivery of adenoviral *TP53* was demonstrated and the development of neutralizing antibodies did not prevent *TP53* transgene expression.

Similar safety and feasibility data were reported by Tolcher et al. (2006) following intravenous adenoviral *TP53* when administered for three consecutive days every 28 days. Seventeen patients received 36 courses of adenoviral *TP53* at doses ranging from 3 x 10^{10} to 3 x 10^{12} virus particles (vp). Flu-like symptoms were commonly observed but rarely severe. Subclinical abnormalities of coagulation parameters, including decreases in fibrinogen and increases in fibrin degradation products, were noted in some patients at 3 x 10^{12} vp dose level. The authors concluded that adenoviral *TP53* can be safely and repetitively administered at doses up to 1 x 10^{12} vp intravenously.

In summary, adenoviral *TP53* was well tolerated by intravascular infusion, both with single and repeated administrations. The most common adverse events were very similar to those reported with intravascular administrations of adenoviral vectors carrying different transgenes and these tolerable toxicities appear to be due to adenovirus itself rather than the *TP53* transgene.

Efficacy of Intra-arterial Adenoviral *TP53* Combined with Chemotherapy for Locally Advanced Head and Neck Carcinoma

Li and colleagues (2006) pioneered the application of intra-arterial *TP53* gene therapy for the treatment of head and neck cancer. In a randomized, controlled clinical trial, patients with locoregional advanced squamous or adenocarcinomas were subdivided into three treatment groups. Group I received adenoviral *TP53* (Gendicine) combined with chemotherapeutic agents; Group II was treated with adenoviral *TP53* alone; and Group III received the chemotherapeutic agents alone. Chemotherapeutic agents employed for the treatment of squamous cell carcinoma were carboplatin (CP), bleomycin (BLM), and methotrexate (MTX), and for adenocarcinoma (including adenoid cystic carcinoma and mucoepidermoid carcinoma) were CP, 5-fluorouracil (5-FU), and cyclophosphamide (CTX).

Catheters for intra-arterial therapy were advanced in retrograde fashion into the superficial temporal artery until the tip reached the opening of the main feeder artery arising from the external carotid artery. Catheterization was confirmed with tracer labeling after infusion of methylene blue. When the catheter position was confirmed, an indwelling arterial pump was implanted subcutaneously in the temporal region. For patients with advanced primary tumors and contralateral metastases, bilateral catheterization was performed. Patients in Groups I and II received 10 cycles of adenoviral *TP53* every 4 days over 6 weeks at doses ranging from 1×10^{12} to 2×10^{12} vp per patient with unilateral or bilateral catheters, respectively. The first combination chemotherapy began 2 days after the first adenoviral *TP53* infusion. Between the treatment periods there was a 7-day break.

Preliminary results of the study have been published (Li et al. 2006). The complete response (CR) rate of the Group I patients receiving adenoviral *TP53* + chemotherapy (28.57%; $n=7$) was higher than that of both Group II patients treated with adenoviral *TP53* alone (16.67%; $n=6$ patients), and Group III patients treated with chemotherapy alone (10%; $n=10$ patients). Consistent with preclinical studies, *TP53* treatment resulted in the induction of pro-apoptotic Bax and inactivation of anti-apoptotic Bcl-2 which were only observed in Group I and II patients treated with adenoviral *TP53*. The trial has now accrued 116 patients and results demonstrating statistically significant increased survival for patients treated with the combination of adenoviral *TP53* and chemotherapy are being prepared for publication and will be reported separately (Li, personal communication).

Efficacy of Intra-arterial Adenoviral *TP53* Combined with Chemotherapy in Advanced Non-small Cell Lung Carcinoma

In addition to their pioneering adenoviral *TP53* gene therapy for the treatment of hepatocellular carcinoma, Guan et al. (2009) have also established the safety and efficacy of intra-arterial *TP53* gene therapy in combination with chemotherapy for the treatment of advanced non-small cell lung cancer (NSCLC). In their study, a total of 58 patients with advanced NSCLC (Stage IIIA, IIIB, and IV) were enrolled in a two-armed controlled clinical trial. Nineteen patients received adenoviral *TP53* treatment (10 patients intra-arterially and 9 patients intra-tumorally) combined with intra-arterial chemotherapy (cisplatin, fluorouracil, and nevelbine) while the control group of 39 patients were treated with chemotherapy alone.

The patients in the combination *TP53* treatment groups received adenoviral *TP53* gene transfer via intra-tumoral or bronchial arterial access on Days 1, 8, 15, and 22. Intra-arterial chemotherapy was administered on Day 5, with at least 1-month interval between treatment cycles. Patients in the control chemotherapy only group received chemotherapy at the same time intervals as those in the combination *TP53* gene therapy groups. Adenoviral *TP53* gene therapy intra-tumoral administration was performed in patients with easily accessible lesions by percutaneous injection under computed tomography (CT) guidance while the other group of combination *TP53* therapy patients received adenoviral *TP53* by bronchial artery administration. After the tip of the catheter was confirmed within the bronchial artery, adenoviral *TP53* was injected at doses ranging from 1 to 4 x 10^{12} viral particles according to tumor size. Patients were followed for 12 months with safety evaluated by the National Cancer Institute's Common Toxicity Criteria and local response by RECIST criteria. Time to progression (TTP) and survival were analyzed by the Kaplan-Meier method and log-rank test.

Consistent with earlier adenoviral *TP53* treatment experience, the adenoviral *TP53*+chemo treatment arm was well tolerated and these patients had more fever, arthralgia, and myalgia influenza-like symptoms compared to the control chemotherapy alone group ($p<0.05$). The complete response rate of 30% (3/10 patients) for patients receiving intra-arterial *TP53* was significantly higher statistically ($p=0.0065$ by two-sided Fisher Exact Test) compared to that of the control group receiving chemotherapy alone (0%, 0/39 patients). There were no complete remissions in the patients treated by direct intra-tumoral *TP53* injection. Overall, patients in the combination adenoviral *TP53*+chemotherapy group had a significantly longer TTP than those in the chemotherapy alone group (a median 7.75 vs 5.5 months, $p=0.018$ by log-rank test) (see Fig. 11.6a). Patients treated with intra-arterial adenoviral *TP53*+chemotherapy had significantly longer survival compared to patients treated with chemotherapy alone or with adenoviral *TP53* administered intra-tumorally in combination with chemotherapy (Fig. 11.6b) (median survival >12 months for intra-arterial *TP53*+chemotherapy vs. 9 months for intra-tumoral *TP53*+chemotherapy vs. 8 months chemotherapy alone; $p=0.0486$ log-rank test) (Fig. 11.6b). These results demonstrate that the combination of bronchial intra-arterial adenoviral *TP53*+chemotherapy was well tolerated in patients with NSCLC and resulted in statistically significant increases in complete remissions, time to progression, and survival compared to treatment with chemotherapy alone. The improved survival of the intra-arterial *TP53* treatment group compared to intra-tumoral *TP53* injection implies that intra-arterial *TP53* gene therapy may be a more efficacious administration method.

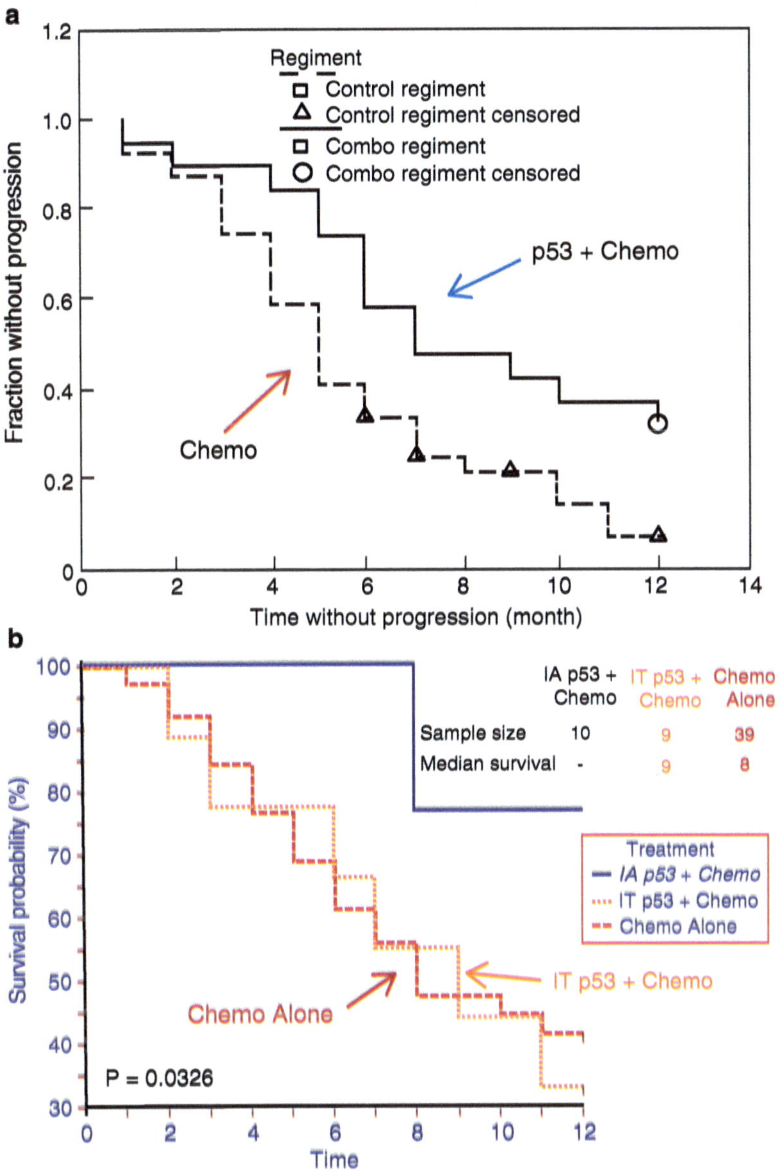

Fig. 11.6 Clinical, intra-arterial and intra-tumoral adenoviral *TP53* gene therapy combined with chemotherapy for the treatment of advanced non-small cell lung carcinoma. Figure 11.6a depicts the statistically significant (log-rank test $p = 0.018$) increased time to progression (TTP) for adenoviral *TP53* gene therapy (intra-arterial or intra-tumoral administration) + chemotherapy compared to chemotherapy alone. Figure 6b shows the statistically significant (log rank test $p = 0.0326$) increased survival for intra-arterial adenoviral *TP53* gene therapy + chemotherapy compared to intra-tumoral adenoviral *TP53* gene therapy + chemotherapy and chemotherapy alone

4 Biomarkers Predictive of *TP53* Gene Therapy Efficacy

In addition to the improved outcomes observed with intra-arterial administration, another important advance in *TP53* adenoviral treatment is the identification of biomarkers predictive of *TP53* gene therapy efficacy. Nemunaitis et al. (2009) described novel biomarkers predictive of *TP53* gene therapy efficacy based upon molecular mechanisms of tumor inactivation of wild-type p53. Profiles unfavorable for *TP53* gene therapy efficacy had high-level protein expression of mutated p53 that can inhibit normal p53 protein function whereas favorable profiles had either normal *TP53* gene sequences or low-level mutated p53 protein expression by immunohistochemistry (<20% positive cells) (see Table 11.1 and Fig. 11.7).

These tumor p53 biomarkers were evaluated by *TP53* gene sequencing and p53 immunohistochemistry evaluations in 116 patients with recurrent squamous cell carcinoma of the head and neck (SCCHN) treated intra-tumorally with adenoviral *TP53* as monotherapy. Statistically significant increases in tumor growth control responses (CR + PR + SD) following intra-tumoral *TP53* adenoviral treatment were observed for patients with favorable p53 efficacy profiles compared with those with unfavorable p53 inhibitor profiles in multiple clinical trials (Nemunaitis et al. 2009) (summarized in Table 11.2). In the Phase I/II trials of recurrent SCCHN, the tumor growth control rate was significantly higher for patients with favorable vs. unfavorable tumor p53 profiles (74%; $n=34/46$ vs. 20%; $n=1/5$; Fisher's exact test

Table 11.1 Tumor p53 Biomarker Profiles Predictive of *TP53* Gene Therapy Efficacy

p53 Biomarker Profiles			
TP53 Gene Therapy Efficacy	Tumor *TP53* Gene Sequence	Tumor p53 Protein Level (Immunohistology)	Mechanisms
Favorable	Wild type	Low <20% positive nuclear staining cells	Advexin reverses p53 hdm2 and hdm4 inhibition
Favorable	Wild type	High ≥ 20% positive nuclear staining cells	
Favorable	Mutated	Low	Advexin restores mutated p53 function
Unfavorable	Mutated	High	Advexin inhibited by dominant negative effects

The tumor p53 biomarker profiles predictive of TP53 gene therapy efficacy are consistent with the methods tumors utilize to block p53 function and the ability of TP53 gene therapy to reverse certain mechanisms of p53 inactivation. The biomarker profiles predictive of TP53 gene therapy efficacy are determined by the level of mutated p53 protein expressed by the tumor. Tumor p53 biomarker profiles favorable for gene therapy efficacy have wild-type p53 sequences where TP53 gene therapy is able to reverse p53 inhibition typically caused by up regulation of the p53 protein inhibitors hdm2 and hdm4 (also known as mdm2 and mdm4) (Zambetti and Levine 1993). Tumor p53 biomarker profiles unfavorable for gene therapy efficacy have mutated p53 sequences which are expressed at high levels (>20% positive p53 nuclear staining by immunohistochemistry). When unfavorable tumor profiles for TP53 gene therapy efficacy are present, the high levels of mutated p53 protein with dominant negative effects inhibit the normal p53 protein delivered by gene therapy.

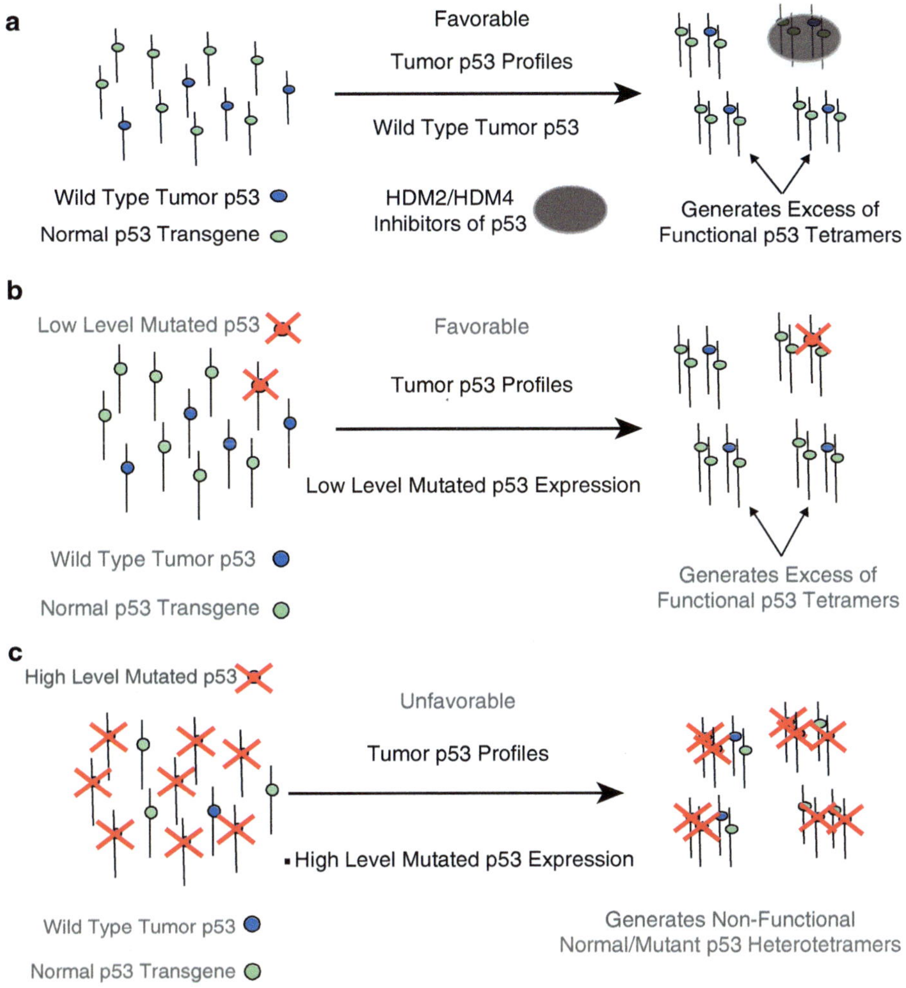

Fig. 11.7 The principles of tumor p53 biomarker profiles predictive of *TP53* gene therapy efficacy are depicted. These predictive biomarker profiles are determined by the level of mutated p53 protein expressed by the tumor. Tumor p53 biomarker profiles favorable for gene therapy efficacy have either wild-type p53 sequences or low levels of mutated p53 expression. Under these conditions, additional normal p53 provided by gene therapy results in an excess of functional p53 tetramers as shown in Fig. 11.7a and b. As shown in Fig. 11.7a, for favorable wild-type p53 tumor profiles, additional normal p53 provided by gene therapy overcomes p53 inhibition typically due to upregulation of the p53 inhibitors hdm2 or hdm4. Similarly in Fig. 11.7b, normal p53 provided by gene therapy results in an excess of functional p53 tetramers when the expression level of mutated p53 is low. In contrast as shown in Fig. 11.7c, tumor p53 biomarker profiles unfavorable for gene therapy efficacy have high levels of mutated p53 protein expression. The high levels of mutated p53 protein have dominant negative effects and inhibit the normal p53 protein delivered by gene therapy generating non-functional normal-mutant p53 heterotetramers

Table 11.2 Tumor p53 Biomarker Profiles Predict Tumor Growth Control Responses of *TP53* Gene Therapy

Clinical Trial	Tumor p53 Efficacy Profile	% Tumor Growth Control Response	Fisher's Exact Test
Phase I/II	Favorable	74% (34/46)	p = 0.0290
TP53 Gene Therapy	Unfavorable	20% (1/5)	
Phase III	Favorable	71% (17/24)	p = 0.0088
TP53 Gene Therapy	Unfavorable	18% (2/11)	
Phase III	Favorable	50% (11/22)	p = 0.1965
Methotrexate	Unfavorable	83% (5/6)	(Not Significant)

Statistically significant increases in tumor growth control responses (CR+PR+SD) following intratumoral TP53 adenoviral treatment in recurrent squamous cell carcinoma of the head and neck were observed for patients with favorable p53 efficacy profiles compared with those with unfavorable p53 inhibitor profiles in multiple clinical trials. In contrast, the biomarker profiles predictive of TP53 gene therapy efficacy did not predict tumor growth control outcomes in the Phase III trial control group treated with methotrexate (Nemunaitis 2009).

$p=0.0290$). Similar statistically increased tumor growth control rates for favorable vs. unfavorable tumor p53 profiles were also observed in a Phase III recurrent SCCHN trial (71%; $n=17/24$ vs. 18%; $n=2/11$; Fisher's exact test $p=0.0088$).

Consistent with the tumor response data, statistically significant increases in time to progression (TTP) and survival following *TP53* gene therapy were also observed in patients with favorable tumor p53 profiles compared to unfavorable p53 inhibitor profiles (Nemunaitis et al. 2009). In the Phase Ill recurrent SCCHN trial, there was a statistically significant increased TTP following *TP53* gene therapy for patients with favorable tumor p53 profiles compared to those with unfavorable profiles (median TTP, 2.7 months vs. 1.4 months, $p=0.0121$) (see Fig. 11.8a). There was a statistically significant increased survival following *TP53* gene therapy for patients with favorable tumor p53 profiles compared with those with unfavorable profiles (median survival, 7.2 months vs. 2.7 months; $p<0.0001$, log-rank test) (see Fig. 11.8b). In contrast, the biomarker profiles predictive of *TP53* gene therapy efficacy in the Phase III trial did not predict tumor response, TTP, or survival outcomes in the control group treated with methotrexate (Nemunaitis et al. 2009). The findings indicate that these tumor p53 biomarker profiles are predictive of *TP53* gene therapy efficacy and that they are not general prognostic markers for other treatments.

5 Cancer Prevention

The fundamental role of p53 dysfunction in cancer pathogenesis (Lane and Levine 2010) and the exemplary safety profile of *TP53* gene therapy support its application to prevent cancer in patients with premalignant lesions or with genetic risk factors for malignancies. In this regard, loss of p53 function is known to be a critical

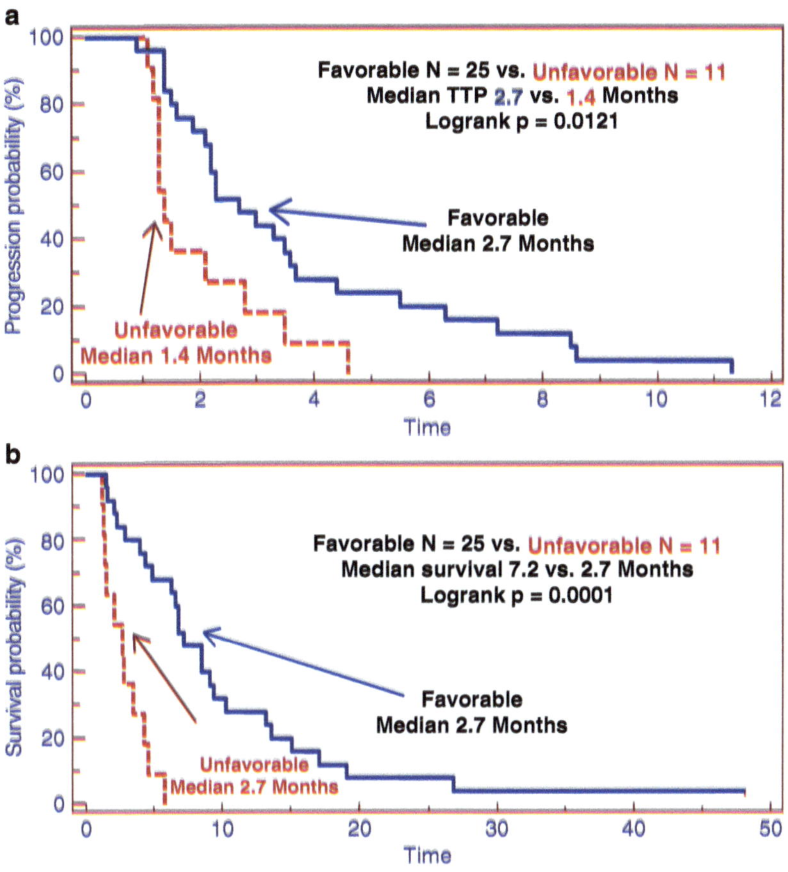

Fig. 11.8 Tumor p53 biomarker profiles identify statistically significant increases in time to progression (TTP) (Fig. 11.8a) and survival (Fig. 11.8b) following intra-tumoral *TP53* gene therapy in recurrent squamous cell carcinoma of the head and neck. In contrast, these biomarker profiles predictive of *TP53* gene therapy efficacy did not predict TTP or survival outcomes in the control group treated with methotrexate (Nemunaitis 2009)

component in the transformation of premalignant oral leukoplakia lesions into squamous cell carcinomas of the head and neck (SCCHN) (Lippman 1997). The important role of p53 dysfunction in the pathogenesis of SCCHN and the safety of adenoviral *TP53* gene therapy in previous clinical trials supported the evaluation of *TP53* gene transfer in the prevention of oral cancers. Li et al. (2009) evaluated the feasibility, safety, and biological activity of intraepithelial injections of recombinant adenovirus *TP53* in 22 patients with dysplastic oral leukoplakia. Sixteen patients showed clinical response to the treatment, and 14 patients showed clear histopathologic improvement. Similar findings with resolution of some leukoplakia lesions were also observed by Clayman (2011, personal communication). Li reported that posttreatment lesion biopsies demonstrated increased p53 and p21CIP/WAF protein expression while bcl-2 protein demonstrated low levels of expression. Hence, intraepithelial injections of recombinant human adenovirus *TP53* were safe, feasible, and biologically active in patients with dysplastic oral leukoplakia and may find future application in preventing the transformation of these lesions to malignant tumors.

6 Conclusion

The broad potential for *TP53* gene therapy in cancer treatment is based on the central role that p53 plays in cancer pathogenesis including its fundamental function in regulating cell proliferation and senescence, inhibiting angiogenesis, and inducing apoptotic cell death in response to DNA damaging radiation or chemotherapy. The observed clinical efficacy of *TP53* gene therapy in several tumor types is consistent with these known mechanisms of p53 action which have been documented in tumor specimens from treated patients. A growing body of clinical data supports the utility of *TP53* treatment in several tumor types, consistent with its ability to target one of the most common and fundamental molecular abnormalities responsible for cancer progression and treatment resistance. As radiation and chemotherapy exert their cytotoxic effects via p53 mediated DNA damage-induced apoptosis, *TP53* gene therapy has demonstrated enhanced therapeutic benefits in combination with standard radiation and chemotherapy regimens. Importantly, the exemplary safety profile of *TP53* gene therapy supports applications of *TP53* gene transfer for cancer prevention and enhances the efficacy of other therapeutic modalities with minimal added toxicity. In this regard, the clinical data generated to date indicate that *TP53* gene therapy improves outcomes in selected cancer patients receiving standard radiation and chemotherapy and provides new treatment options for evaluation in patients with unmet medical needs. In conclusion, *TP53* gene therapy has begun to realize its potential as an important cancer therapy targeting one of the most fundamental and common molecular defects in cancer. Clinical experience suggests that the future of *TP53* gene therapy may be improved by intra-arterial administration and by the use of tumor p53 biomarkers predictive of treatment efficacy. Further refinements in *TP53* gene therapy will be advanced by increased knowledge of normal and mutated p53 functions to guide future clinical evaluations and applications.

Conflict of Interest The following authors have or had employment/advisor (E/A) or investigator (I) relationships with the companies developing Advexin (p53 Inc/Introgen), Gendicine (SiBiono) and SCH58500 (Schering-Plough/Merck):

p53 Inc: Robert E. Sobol(E/A); Wei-Wei Zhang(E/A); Kerstin B. Menander(E/A); Sunil Chada(E/A); Daniel C. Maneval (E/A); Gary L. Clayman(I); Stephen G. Swisher(I); W. Jarrard Goodman(I); John Nemunaitis(I); Jack A. Roth(I/E/A)

SiBiono: Zhaohui Peng(E/A); Wei-Wei Zhang(E/A); Yong-Song Guan(I); Long-Jiang Li(I)

Schering-Plough/Merck: Daniel C. Maneval(E/A); Jane A. Horowitz(E/A); Robert Warren(I)

Acknowledgments This manuscript is dedicated with deep appreciation and gratitude to Eleanor McGlaughlin who exemplifies the finest qualities of an inspirational teacher—the source of all achievements.

References

Atencio IA, Grace M, Bordens R, Fritz M, Horowitz JA, Hutchins B, Indelicato S, Jacobs S, Kolz K, Maneval D, Musco ML, Shinoda J, Venook A, Wen S, Warren R (2006) Biological activities of a recombinant adenovirus p53 (SCH 58500) administered by hepatic arterial infusion in a Phase 1 colorectal cancer trial. Cancer Gene Ther 13(2):169–181

Bischoff JR, Kirn DH, Williams A, Heise C, Horn S, Muna M, Ng L, Nye JA et al (1996) An adenovirus mutant that replicates selectively in p53- deficient human tumor cells. Science 274(5286):373–376

Clayman GL, el-Naggar AK, Lippman SM, Henderson YC, Frederick M, Merritt JA, Zumstein LA, Timmons TM, Liu TJ, Ginsberg L, Roth JA, Hong WK, Bruso P, Goepfert H (1998) Adenovirus-mediated p53 gene transfer in patients with advanced recurrent head and neck squamous cell carcinoma. J Clin Oncol 16(6):2221–2232

Gjerset RA, Turla ST, Sobol RE, Scalise JJ, Mercola D, Collins H, Hopkins PJ (1995) Use of wild-type p53 to achieve complete treatment sensitization of tumor cells expressing endogenous mutant p53. Mol Carcinog 14(4):275–285

Gjerset RA, Sobol RE (1997) Treatment resistance, apoptosis and p53 tumor suppressor gene therapy. In: Bertino J (ed) Encyclopedia of Cancer. Academic Press, Waltham, Massachusetts pp 1785–1791

Guan YS, Liu Y, Zou Q, He Q, La Z, Yang L, Hu Y (2009) Adenovirus-mediated wild-type p53 gene transfer in combination with bronchial arterial infusion for treatment of advanced non-small-cell lung cancer, one year follow-up. J Zhejiang Univ Sci B 10(5):331–340

Guan YS, Liu Y, He Q, Li X, Yang L, Hu Y, La Z (2011) p53 gene therapy in combination with transcatheter arterial chemoembolization for HCC: One-year follow-up. World J Gastroenterol 17(16):2143–2149

Khuri F, Nemunaitis J, Ganly I, Arseneau J, Tannock I, Romel L, Gore M, Ironside J et al (2000) A controlled trial of intratumoral ONYX-015, a selectively-replicating adenovirus, in combination with cisplatin and 5-fluorouracil in patients with recurrent head and neck cancer. Nature Med 6(8):879–885

Lane D, Levine A (2010) p53 research: the past thirty years and the next thirty years. Cold Spring Harb Perspect Biol 2(12):a000893

Li LJ, Ding HY, Wang LJ, Ming WY, Han B (2006) A combination therapy of selective intraarterial adenoviral p53 infusion with chemotherapy for locally advanced head and neck carcinoma: a prospective clinical phase II study. Molecular Ther, Volume 13, Supplement 1, May

Li Y, Li LJ, Zhang ST, Wang LJ, Zhang Z, Gao N, Zhang YY, Chen QM (2009) In vitro and clinical studies of gene therapy with recombinant human adenovirus-p53 injection for oral leukoplakia. Clin Cancer Res 15(21):6724–6731

Lippman SM (1997) Head and neck chemoprevention: recent advances. Cancer Control 4(2):128–135

Malkin D, Li FP, Strong LC, Fraumeni JF Jr, Nelson CE, Kim DH, Kassel J, Gryka MA, Bischoff FZ, Tainsky MA et al (1990) Germ line p53 mutations in a familial syndrome of breast cancer, sarcomas, and other neoplasms. Science 250(4985):1233–1238

Nemunaitis J, Clayman G, Agarwala SS, Hrushesky W, Wells JR, Moore C, Hamm J, Yoo G, Baselga J, Murphy BA, Menander KA, Licato LL, Chada S, Gibbons RD, Olivier M, Hainaut P, Roth JA, Sobol RE, Goodwin WJ (2009) Biomarkers predict p53 gene therapy efficacy in recurrent squamous cell carcinoma of the head and neck. Clin Cancer Res 15(24): 7719–7725

Nielsen LL, Maneval DC (1998) P53 tumor suppressor gene therapy for cancer. Cancer Gene Ther 5:52–63

Nishizaki M, Fujiwara T et al (1999) Recombinant adenovirus expressing wild-type p53 is antiangiogenic: a proposed mechanism for bystander effect. Clin Cancer Res 5:1015–1023

Nishizaki M, Meyn RE, Levy LB, Atkinson EN, White RA, Roth JA, Ji L (2001) Synergistic inhibition of human lung cancer cell growth by adenovirus-mediated wild-type p53 gene transfer in combination with docetaxel and radiation therapeutics in vitro and in vivo. Clin Cancer Res 7(9):2887–2897

Peng Z (2005) Current status of Gendicine in China: recombinant human Ad-p53 agent for treatment of cancers. Hum Gene Ther 16(9):1016–1027

Reid T, Warren R, Kirn D (2002) Intravascular adenoviral agents in cancer patients: lessons from clinical trials. Cancer Gene Ther 9(12):979–986

Roth JA, Nguyen D, Lawrence DD, Kemp BL, Carrasco CH, Ferson DZ, Hong WK, Komaki R, Lee JJ, Nesbitt JC, Pisters KM, Putnam JB, Schea R, Shin DM, Walsh GL, Dolormente MM, Han CI, Martin FD, Yen N, Xu K, Stephens LC, McDonnell TJ, Mukhopadhyay T, Cai D (1996) Retrovirus-mediated wild-type p53 gene transfer to tumors of patients with lung cancer. Nat Med 2(9):985–991

Senzer N, Nemunaitis J, Nemunaitis M, Lamont J, Gore M, Gabra H, Eeles R, Sodha N, Lynch FJ, Zumstein LA, Menander KB, Sobol RE, Chada S (2007) p53 therapy in a patient with Li-Fraumeni syndrome. Mol Cancer Ther 6(5):1478–1482

Swisher SG, Roth JA, Komaki R, Gu J, Lee JJ, Hicks M, Ro JY, Hong WK, Merritt JA, Ahrar K, Atkinson NE, Correa AM, Dolormente M, Dreiling L, El-Naggar AK, Fossella F, Francisco R, Glisson B, Grammer S, Herbst R, Huaringa A, Kemp B, Khuri FR, Kurie JM, Liao Z, McDonnell TJ, Morice R, Morello F, Munden R, Papadimitrakopoulou V, Pisters KM, Putnam JB Jr, Sarabia AJ, Shelton T, Stevens C, Shin DM, Smythe WR, Vaporciyan AA, Walsh GL, Yin M (2003) Induction of p53-regulated genes and tumor regression in lung cancer patients after intratumoral delivery of adenoviral p53 (INGN 201) and radiation therapy. Clin Cancer Res 9(1):93–101

Tolcher AW, Hao D, de Bono J, Miller A, Patnaik A, Hammond LA, Smetzer L, Van Wart HJ, Merritt J, Rowinsky EK, Takimoto C, Von Hoff D, Eckhardt SG (2006) Phase I, pharmacokinetic, and pharmacodynamic study of intravenously administered Ad5CMV-p53, an adenoviral vector containing the wild-type p53 gene, in patients with advanced cancer. J Clin Oncol 24(13):2052–2058

Wills KN, Maneval DC, Menzel P, Harris MP, Sutjipto S, Vaillancourt MT et al (1994) Development and characterization of recombinant adenoviruses encoding human p53 for gene therapy of cancer. Hum Gene Ther 5(9):1079–1088

Xue W, Zender L et al (2007) Senescence and tumour clearance is triggered by p53 restoration in murine liver carcinomas. Nature 445(7128):656–660

Yu W, Fang H (2007) Clinical trials with oncolytic adenovirus in China. Curr Cancer Drug Targets 7(2):141–8

Zambetti G, Levine A (1993) A comparison of the biological activities of wild-type and mutant p53. FASEB J 7:855–865

Zhang WW, Fang X, Mazur W, French BA, Georges RN, Roth JA (1994) High-efficiency gene transfer and high-level expression of wild-type p53 in human lung cancer cells mediated by recombinant adenovirus. Cancer Gene Ther 1(1):5–13

Zhang WW, Roth JA (1994) Anti-oncogene and tumor suppressor gene therapy–examples from a lung cancer animal model. Vivo 8(5):755–69

Chapter 12
Upstream Targets in the p53 Pathway

Anna R. McCarthy and Sonia Lain

1 Introduction

p53 is the best studied tumour suppressor to date and its importance is highlighted by the diverse biological processes it is involved in. The prevention of tumour development by p53 is clear and in the majority of cancers its antitumour function is impaired either via mutation of the *TP53* gene or through altered expression of its regulators (for example, Mdm2 and p14^{ARF}), or the action of viral oncogenes (Lain and Lane 2003). Reactivation of p53 function by small molecules is accepted as a useful strategy in the development of therapeutics. Using compounds to target proteins that regulate the activation of wild-type p53 function is in contrast to developing molecules that target the protein itself or reactivate mutant forms and is the focus of this review. In this review we summarise small molecules that are known to activate wild-type p53 via a variety of upstream targets (see Fig. 12.1) and discuss their mode of action. The upstream pathways for which wild-type p53 activators have been reported include the p53 ubquitination cascade, protein deacetylation, nuclear transport, ribonucleotide synthesis and mitosis. There are also a class of activators that are apparent non-genotoxic DNA intercalators. Because the focus of this review is on reactivation of wild-type p53 via the modification of upstream targets, we only report small molecules that show the ability to activate p53 in cells and preclinical tumour models (where available) with emphasis on assessment of genotoxicity.

A.R. McCarthy • S. Lain (✉)
Department of Microbiology, Tumour and Cell Biology, Karolinska Institutet,
Nobels väg 16, 171 77 Stockholm, Sweden
e-mail: sonia.lain@ki.se

P. Hainaut et al. (eds.), *p53 in the Clinics*, DOI 10.1007/978-1-4614-3676-8_12,

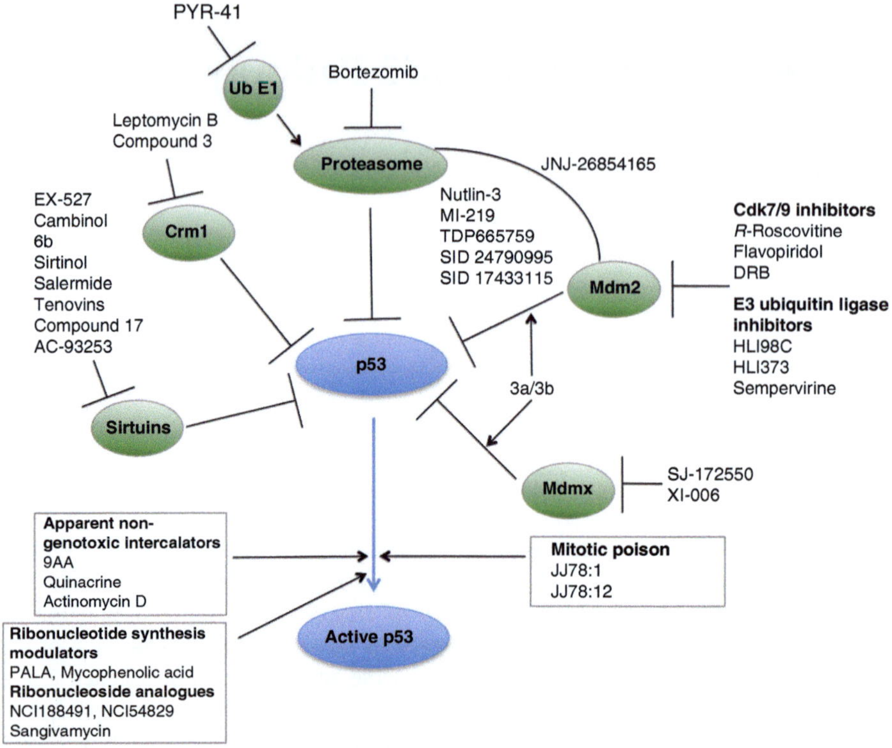

Fig. 12.1 Small molecule activators of wild-type p53

2 Inhibition of p53 Ubiquitination/Degradation

p53 stability and thus function is primarily regulated by the ubiquitin-proteasome system. Poly-ubiquitination targets p53 for degradation by the 26S proteasome imposing a short half-life on the protein in normal conditions. Stabilisation of p53 is increased upon cellular stress and in a majority of cases correlates to an increase in the transcriptional activity of p53. As a result, genes regulating a myriad of cellular processes are induced or suppressed, leading to a p53-mediated response. Therefore, proteins controlling the stability of p53 and its activity are of interest as targets for therapeutic intervention. Small molecules that activate p53 via modification of targets within the ubiquitination cascade have been developed. These targets include Mdm2, Mdmx, the 26S proteasome and other ubiquitin transfer-related enzymes. Compound modifiers of these targets that result in p53 activation will be discussed in more detail in the following section.

2.1 *Mdm2 Inhibition*

Mdm2 is a critical negative regulator of p53 and is overexpressed in many cancers, suppressing wild-type p53 function. Mdm2 regulates p53 at the level of stability, function and subcellular location (Dickens et al. 2010). The stability of p53 is predominantly controlled via the E3 ubiquitination ligase function of Mdm2 which targets p53 for degradation by the proteasome. Furthermore, Mdm2 blocks p53 transcriptional function by interacting with the transcription activation domain of p53. Nuclear export of p53 into the cytoplasm is also promoted by Mdm2. These regulatory functions of Mdm2 can be viewed as targets for therapeutic intervention towards activating p53 in a specific manner. Extensive validation of Mdm2 as a target for reactivation of p53 has been carried out and has been reviewed by Brown et al. (2009) and by Dickens et al. (2010).

2.1.1 Mdm2–p53 Interaction Inhibition by Small Molecules

The development of nutlins and other compounds as inhibitors of the Mdm2–p53 interaction confirms its therapeutic index in reactivation of p53. Inhibitors of this interaction have been reviewed extensively and the reader is directed to a series of recent reviews (Berg 2008; Dickens et al. 2010; Domling 2008; Hu and Hu 2008; Weber 2010). Here we focus on the three most important classes of Mdm2–p53 inhibitors which all mimic the N-terminal residues of p53 required for Mdm2 binding, thus competing with p53, as well as the most recently discovered inhibitors. The most important classes of Mdm2–p53 interaction inhibitors include the nutlins, spiro-oxindoles and benzodiazepinediones, which are only active in wild-type p53-expressing cells. The nutlins (see Fig. 12.2 for the structure of nutlin-3) were the first small molecule antagonists of Mdm2 to be reported (Vassilev et al. 2004) and are the best studied amongst the known small molecule inhibitors of the Mdm2–p53 interaction. Nutlins are able to displace p53 from Mdm2 *in vitro* at nanomolar concentration. These *cis*-imidazoline-based compounds bind to the p53 binding pocket of Mdm2 and mimic the molecular interaction between Mdm2 and three important residues in the N-terminus of p53 (phenylalanine 19, tryptophan 23 and leucine 26) as determined by the crystal structure of nutlin-2 bound to Mdm2. Wild-type p53 expressing cancer cells either undergo G1/G2 arrest or apoptosis in the presence of nutlin at micromolar concentration. In particular, nutlin-3 (of which the active enantiomer is 3a) induces apoptosis in hematologic malignant-derived cells such as acute myeloid leukaemia (AML), acute lymphoblastic leukaemia, myeloma and B-cell chronic lymphocytic leukaemia (Kojima et al. 2005; Secchiero et al. 2008). Nutlins are also effective in animal models with nutlin-3 capable of decreasing tumour size in mouse xenografts. The therapeutic relevance of activation of p53 by nutlins is supported by evidence that normal cells and tissues do not exhibit

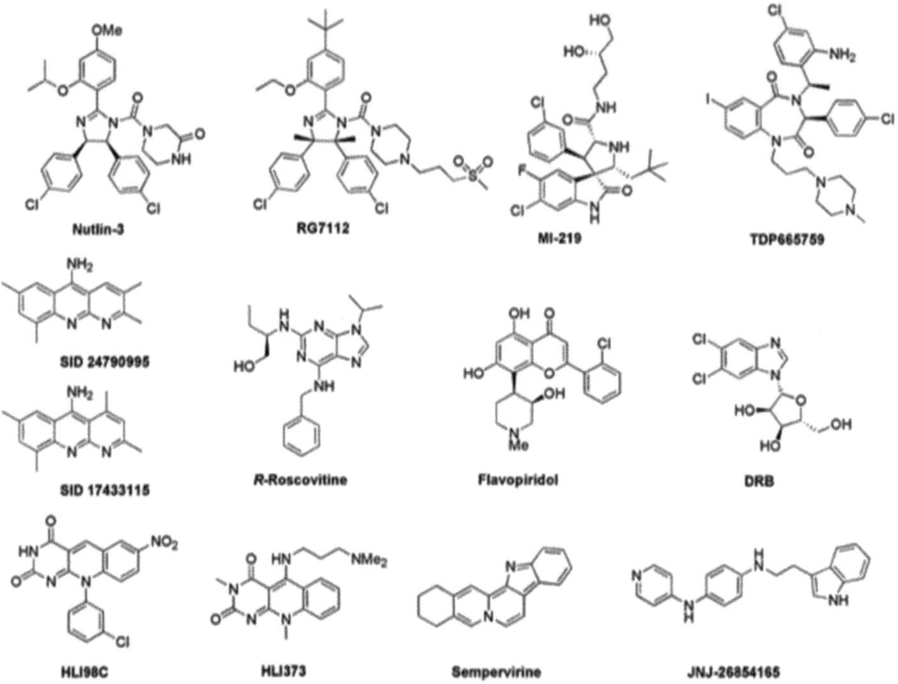

Fig. 12.2 Small molecule modulators of Mdm2 function

the toxicity seen upon treatment of tumour cells. Nutlin has been subjected to preclinical studies in the treatment of samples from patients with AML and neuroblastoma (Cheok et al. 2011). Treatment of xenografts derived from neuroblastoma and osteosarcoma resulted in regression of tumour growth from similar oral dosage regimes. With regards to the clinical development of nutlins, one compound from this family (RG7112, Hoffman-La Roche, Fig. 12.2) has entered phase I trials in patients with hematologic neoplasms (NCT00623870) and advanced solid tumours (NCT00559533). Encouragingly, upon oral dosing this compound is well tolerated and increases levels of the biomarker MIC-1, suggesting a potential mechanistic pharmacodynamic relationship (Beryozkina et al. 2011).

Whilst the nutlins were discovered via a biochemical screen for inhibition of the Mdm2–p53 interaction, another class of inhibitor, the spiro-oxindoles, was developed using a structure-based approach in combination with the Mdm2–p53 complex crystal structure (Ding et al. 2005, 2006). These spiro-oxindoles were designed to mimic the structure of tryptophan 23 found in the Mdm2 binding site of p53 and display potent activity in the nanomolar range *in vitro*. Within this class, a compound named MI-219 (Fig. 12.2) has been licensed by Ascenta Therapeutics and is active in decreasing tumour growth in prostate cancer and osteosarcoma-derived xenograft models in mice. Accumulation of p53 was observed as well as the upregulation of p53 target genes. MI-219 induces apoptosis in various cancer cell lines in a p53-dependent manner and inhibits growth of xenograft tumours as a single agent (Shangary et al. 2008a, b; Shangary and Wang 2008). Analogues of MI-219 have also shown anti-tumour activity, some in combination with other chemotherapeutics (Cheok et al. 2011).

Another important class of molecules that inhibits the Mdm2–p53 interaction are the benzodiazepinediones which have *in vitro* IC_{50} values in the range of 0.5–2 μM (biochemical assay) (Grasberger et al. 2005). In wild-type p53-containing cell lines these compounds are able to decrease cell growth. One compound from this class, TDP665759 (Fig. 12.2), is able to increase the expression of p21 after dosage in mice (Koblish et al. 2006).

Two other compounds that inhibit the Mdm2–p53 interaction are currently in preclinical trials (PXN727 and PXN822, Priaxon (Cheok et al. 2011)) and show efficacy in suppressing tumour growth (structures not available).

Current developments in the discovery of novel inhibitors of the Mdm2–p53 interaction include a screen reported recently (Dudgeon et al. 2010). Around 220,000 compounds were screened using a novel Mdm2–p53 protein–protein interaction biosensor assay. A series of three related methylbenzo-naphthyridin-5-amines showed reproducible activity and had a dose-dependent effect on disrupting the Mdm2–p53 interaction. Therefore, these three compounds were chosen for further characterisation. One compound (SID 24790995, Fig. 12.2) was found to stabilise not only p53 but also increased Mdm2 and p21 levels in HCT116 cells at 50 μM, albeit less than the effect observed for nutlin-3 treatment. The DNA-damaging ability of compound SID 17433115 (Fig. 12.2) at 50 μM was assessed. Not surprisingly, the levels of p53 phosphorylated at Ser15 were slightly increased although to a significantly lesser degree than the levels observed after camptothecin treatment at 100 nM. Similar effects were seen for the phosphorylation of Ser139 of H2AX. Inhibition of cell proliferation in isogenic HCT-116 $p53^{+/+}$ and $p53^{-/-}$ cell lines by SID 24790995 was found to be partially p53-dependent, as was the apoptotic effect of SID 17433115. However, induction of the G1 cell cycle arrest was found to be totally dependent on p53. The precise mechanism of action of these amines remains to be determined. In a different study a series of somewhat structurally related acridine orange compounds have been reported to stabilise p53 via inhibition of Hdm2-mediated ubquitination of p53 (Wang et al. 2005), but this effect was not assessed in this study.

2.1.2 Mdm2 Expression Inhibition

Downregulation of Mdm2 is known to stabilise p53 and its transcriptional activity, resulting in apoptosis (Juliano et al. 2008; Shen 2008). Thus, inhibition of Mdm2 expression is another strategy to reduce the destabilising effects of Mdm2 on p53. Inhibition of transcription leads to p53 activation in the presence of Cdk inhibitors *R*-roscovitine, flavopiridol and 5,6-dichloro-1-β-*D*-ribofuranosylbenzimidazole (DRB, see Fig. 12.2). These compounds inhibit Cdk 7 and 9 which are responsible for phosphorylation of RNA polymerase. They have also been shown to downregulate Mdm2 expression (Blattner et al. 1999; Demidenko and Blagosklonny 2004; Lu et al. 2001), although it is not yet clear whether this is necessary to observe effects on p53 (O'Hagan and Ljungman 2004).

2.1.3 Mdm2 E3 Ubiquitin Ligase Function Inhibition

Mdm2 is an E3 ubiquitin ligase which binds to the N-terminus of p53 blocking its transcriptional activity. Poly-ubiquitination of p53 by Mdm2 targets it for degradation. Inhibition of this E3 ligase activity is another strategy to stabilise wild-type p53 and increase its transcriptional activity. Indeed inhibitors of Mdm2 ligase activity have been developed with the ability to activate p53. A family of related 7-nitro-5-deazaflavin compounds named HLI98 were discovered via high-throughput screening for Mdm2 auto-ubiquitination (Davydov et al. 2004). One of these compounds, HLI98C (Fig. 12.2), was shown to reactivate p53 function and induce p53-dependent apoptosis in tumour cells. However, this family of compounds causes DNA damage via intercalation with DNA (Burgstaller et al. 1997), and the generation of a nitro anion radical due to the redox potential of the 5-deazaflavin scaffold (Kawamoto et al. 1995a, b). Further work has revealed that the nitro group of the HLI98 family is not required for p53 activation and less genotoxic analogues with retained ability to activate p53 have been developed, including a more potent and water-soluble analogue (HLI373, Fig. 12.2). HLI373 causes p53-dependent apoptosis in various cell lines containing wild-type p53 (Kitagaki et al. 2008).

The natural product sempervirine (Fig. 12.2) also inhibits the E3 ligase activity of Mdm2 and was discovered by high-throughput screening in an Mdm2 electrochemiluminescent assay (Sasiela et al. 2008). Treatment of wild-type p53 containing tumour cells with sempervirine resulted in accumulation of p53 and apoptosis.

Whilst targeting of the E3 ligase activity of Mdm2 represents an alternative approach to activating p53 and avoiding genotoxic effects of anticancer drugs, potential problems associated with this strategy as a therapeutic approach can be foreseen. For example, inhibition of Mdm2's E3 ligase function could lead to stabilisation of Mdm2 via the blockage of ubiquitination of Mdm2 itself, thus inhibiting p53 activation. Evidence suggests that Mdm2 auto-ubiquitination can be differentiated from Mdm2-dependent p53 ubiquitination (Allende-Vega and Saville 2010) and selective inhibitors for ubiquitination of p53 have been identified (Lai et al. 2002). A more recent screen was carried out to identify selective inhibitors of Mdm2-dependent p53 ubiquitination from a collection of over 600,000 compounds. Several selective inhibitors were identified although with weak potency and the structures were not disclosed (Murray et al. 2007). A better understanding of the catalytic mechanism of Mdm2 E3 ligase activity is required which could lead to further development of more potent and better characterised Mdm2 E3 ligase inhibitors.

JNJ-26854165 (Fig. 12.2) is another type of Mdm2 inhibitor that activates p53 which is currently being assessed in phase I clinical trials for treatment of solid tumours. This compound is not an inhibitor of the Mdm2–p53 interaction nor does it inhibit the E3 ligase function but blocks association of Mdm2 with the proteasome. It is not known how this compound increases p53 levels in tumour cell lines (Arts et al. 2008).

2.2 Proteasome Inhibition

The 26S proteasome recognises ubiquitinated proteins and carries out protein degradation. Inhibition of the proteasome's proteolytic activity leads to apoptosis and as such proteasome inhibitors have been developed. The selective proteasome inhibitor Bortezomib reduces tumour cell growth, induces apoptosis and is currently approved for use in the treatment of multiple myeloma (Orlowski and Kuhn 2008). However, the effects of proteasome inhibition on p53 activity are not fully understood. Some studies report that the effect of proteasome inhibition is p53 dependent whilst p53 independency is reported by others (Boccadoro et al. 2005; Concannon et al. 2007; Ding et al. 2007; Nair et al. 2006). Evidence of p53 activation upon treatment with proteasome inhibitors is also variable and different markers for p53 activation have been assessed. An increase in the amount of p53 associated with mitochondria has been reported in one study (Marchenko et al. 2007) whilst p53 was not observed at the mitochondria in another (Williams and McConkey 2003). Levels of p53 target gene proteins have also been monitored as evidence for p53 activation, as have mRNA levels from which it was concluded that p53 transcriptional activity was increased (Concannon et al. 2007; Williams and McConkey 2003). In another study the levels of some p53 responsive genes were increased whilst others were not (Nair et al. 2006). Evidence also exists for the stabilisation of p53 without increasing its transcriptional activity (Siliciano et al. 1997; Stommel and Wahl 2004; Zhu et al. 2007). Inhibition of Mdm2-mediated proteasomal degradation has been proposed as a mechanism contributing to the inability of proteasome inhibition to increase p53 transcriptional activity in some cases. One study proposes that proteasome inhibition does not lead to an increase of p53 transcriptional activity but that disruption of the Mdm2–p53 interaction causes an increase in the levels of p53 target gene mRNA levels (Stommel and Wahl 2004). Prevention of destabilisation of Mdm2 and Mdmx is another mechanism suggested for the blockage of p53 activation in response to proteasome inhibition (Stommel and Wahl 2004; Wang et al. 2007). As presented, evidence for the effects of proteasome inhibition on p53 activity is variable and further understanding is required in order to develop effective proteasome inhibitors as anti-tumour therapeutics with a defined mechanism of p53 activation.

2.3 Ubiquitin Activator Inhibition

The first step in the ubiquitination pathway is performed by ubiquitin-activating enzymes named ubiquitination E1. Because inhibition of the proteasome has shown therapeutic value, then inhibition of ubiquitination E1 activity also has the potential to be of use as a target for the development of anti-tumour agents. To date only one inhibitor of E1 with effects on p53 in cells has been reported in the literature (Yang et al. 2007). In an effort to find compounds with improved potency compared to the deazaflavin series described earlier, a larger library was screened and the pyrazone derivative PYR-41 (Fig. 12.3) was identified as a selective inhibitor of E1. PYR-41

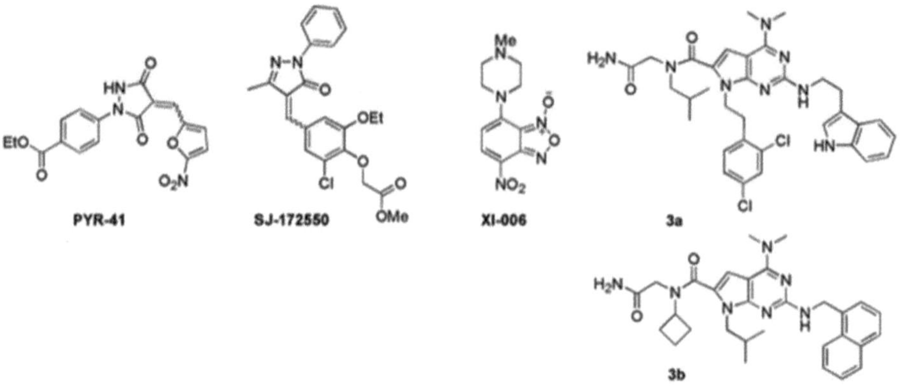

Fig. 12.3 Ubiquitin-activation inhibitor PYR-41 and novel Mdmx-targeting activators of p53

is the first cell permeable inhibitor of E1 and blocks p53 degradation and increases the level and activity of p53. Apoptosis in wild-type p53 containing cells was also induced by PYR-41. Interestingly, PYR-41 also inhibits NF-κB activation, a property which in combination with p53 activation makes it attractive as an anti-tumour agent. However, the precise mechanism of action of PYR-41 remains to be understood, the study of which will aid in understanding the therapeutic value of E1 inhibitors as anti-tumour agents.

2.4 Mdmx Inhibition

Whilst efforts towards the development of small molecules that activate p53 have focused on targeting Mdm2, its non-redundant homologue Mdmx is also a potential target. Mdmx also possesses a p53-binding domain and is another key regulator of p53. Whilst Mdm2 is involved in regulating p53 stability and localisation within the cell, Mdmx is primarily involved in regulating p53 transcriptional activity. Mdmx does not have the E3 ligase function of Mdm2 and its expression is not regulated by p53, thus making it an alternative target for p53 activation. Mdmx is amplified in some breast, colon and lung carcinomas, head and neck squamous carcinomas, retinoblastomas and some gliomas (Danovi et al. 2004; Laurie et al. 2006; Valentin-Vega et al. 2007). Peptides have been developed to antagonise the Mdmx–p53 interaction but as the focus of this review is on small molecules that activate p53 via upstream targets, these will not be discussed here (for recent examples see references Joseph et al. 2010; Li et al. 2009; Marine 2010; Phan et al. 2010). In terms of a therapeutic strategy, it may be necessary to inhibit both Mdm2 and Mdmx in tumour cells that express wild-type p53 in order to activate a p53 response (Gu et al. 2002; Hu et al. 2006; Toledo et al. 2006). Due to differences in the p53 binding pocket between Mdm2 and Mdmx (Böttger et al. 1999; Popowicz et al. 2007, 2008), it would seem probable that Mdm2 inhibitors do not bind as efficiently to Mdmx and this is indeed the case for nutlin-3a (Laurie et al. 2006). Non-peptide small molecule inhibitors of Mdmx have only been reported very recently.

Interestingly, the first crystal structure of a small molecule inhibitor bound to Mdmx has also been reported recently. The inhibitor in this case, named W298, is related to a 3-imidazoyl-indole family of compounds that bind to Mdm2 with nanomolar potency *in vitro* (Popowicz et al. 2010). However, although with less affinity (tenfold), W298 also binds to Mdmx. This compound is claimed to be the first antagonist of the Mdmx–p53 interaction but no cell activity data is currently available. The co-crystal structure data allows for comparisons to be made between the binding of small molecules to the p53 pocket of Mdm2 and Mdmx which will allow the rational development of compounds that inhibit Mdm2–p53 binding into more potent Mdmx–p53 antagonists.

The first reported small molecule inhibitor of Mdmx, SJ-172550 (Fig. 12.3), was discovered via a high-throughput screen of almost 300,000 compounds for their ability to disrupt binding of a p53 peptide to a GST-Mdmx (mouse) fusion protein (Reed et al. 2010). SJ-172550 was selected after prioritisation studies including measurement of the binding affinity of screen hits for Mdmx and Mdm2, cell-based assays using Mdmx amplified retinoblastoma cells (Weri1) and also p53 null cells (SJmRbl-8). Activity profiles for different scaffolds were generated and compounds chosen for further analysis. These scaffolds were also evaluated from a medicinal chemistry point of view for the purpose of eliminating those unsuitable for use as leads in drug development, an important process often neglected in the prioritisation of hit compounds generated from high-throughput screens. Indicative of Mdmx interaction, SJ-172550 was able to induce a shift in the melting temperature of GST-Mdmx (residues 1–185). Mass spectrometry studies indicate that it does not covalently modify the p53 binding domain of Mdmx. In retinoblastoma and leukaemia cell lines, SJ-172550 did not cause p53 induction but initiated apoptosis that was p53-dependent. The authors of this study suggest that the mechanism of action is not through p53 stabilisation. Whilst HCT-116 cells were sensitive to the drug, the corresponding p53 null cells were not. Co-immunoprecipitation was used to show that the Mdmx–p53 interaction was partially inhibited in cells. SJ-172550 also binds Mdm2, but less effectively. However, this study suggests that SJ-172550 warrants further study into its precise mechanism of action and could also be useful as a chemical tool to study Mdmx biology related to p53.

Another small molecule inhibitor of Mdmx has more recently been reported (Wang et al. 2011). Because Mdmx overexpression in cancer cells is predominantly caused by abnormal transcription (Gilkes et al. 2008), Wang and co-workers screened the NCI Diversity-Set to find small molecules that inhibit Mdmx transcription via a luciferase reporter activity. Of the initial 44 hit compounds, three were prioritised for further study by their ability to inhibit Mdmx expression in HT1080 cells. The predicted least toxic of these compounds, XI-006 (Fig. 12.3), was chosen for further investigation. Not only did XI-006 effect a decrease in the levels of Mdmx mRNA and protein in MCF-7 cells, but it also induced expression of p21 and Mdm2. p53 levels appeared to be increased also, as were the levels of pro-apoptotic gene expression (puma, bax and pig-3), indicating that a small molecule inhibitor of Mdmx was activating wild-type p53. Knockdown of Mdmx expression resulted in impaired p53 levels and p53-responsive genes, indicating that XI-006 could activate p53 via Mdmx inhibition in part. The induction of apoptosis by XI-006 was determined to be dependent on Mdmx inhibition and p53 activation via knockdown experiments looking at induction of PARP-mediated cleavage. XI-006 was found to work with nutlin-3a in an additive manner to activate p53 and decrease cell survival.

Of particular interest is the latest report to be published in the field of small molecule inhibition of Mdmx (Lee et al. 2010). Lee et al. have detailed a study into the rational design of small molecule α-helix mimetics for the disruption of α-helix-mediated protein–protein interactions, which includes the Mdm2/Mdmx–p53 interaction. A novel class of α-helix mimetics consisting of a pyrrolopyrimidine-based scaffold was designed from previously reported α-helix mimetics that are known to disrupt α-helix-mediated protein–protein interactions (Cummings and Hamilton 2010). A solid phase synthetic route was designed to access these novel pyrrolopyrimidine compounds that allowed for the synthesis of many analogues. To determine the biological use of these compounds as α-helix motif mimetics, their ability to disrupt the p53–Mdmx interaction was determined. A library of 900 pyrrolopyrimidines was synthesised to incorporate moieties to mimic the side chains of the key three amino acids of p53 known to be crucial for binding to Mdm2/Mdmx. Compounds were assayed for their ability to displace a rhodamine-labelled p53-based peptide from the Mdmx protein using fluorescence polarisation. Two "hits" (3a and 3b, Fig. 12.3) were selected for further characterisation. Most interestingly, these two compounds were able to inhibit binding of p53 to both Mdmx and Mdm2 with similar binding affinities measured for each interaction (high nanomolar K_i). Dual Mdmx/Mdm2 inhibitors are highly desirable for the reasons discussed earlier in this section. Both 3a and 3b were able to increase the levels of p53 and p21 in wild-type p53-expressing H460 cells, indicating that they are indeed cell permeable and capable of activating p53. Compound 3a also induced apoptosis in H460 cells but not p53-null H1299 cells, suggesting that the effect is p53-dependent. The exciting discovery of small molecules that show dual potent inhibition of Mdmx/Mdm2 binding to p53 with p53-activating ability in cells provides evidence that this can be achieved. However, as pointed out further mechanism of action studies need to be carried out, in particular to understand their mode of action as dual inhibitors.

3 Inhibiting p53 Deacetylation

Post-translational modification of p53 is involved in regulating its function as a transcription factor. Deacetylation of p53 by SirT1, a member of the sirtuin family of class III histone deacetylases, is thought to decrease p53 stability and DNA binding and thus its activity as a transcription factor (Langley et al. 2002; Luo et al. 2001, 2004; Vaziri et al. 2001). Several SirT1 inhibitors do indeed cause an increase

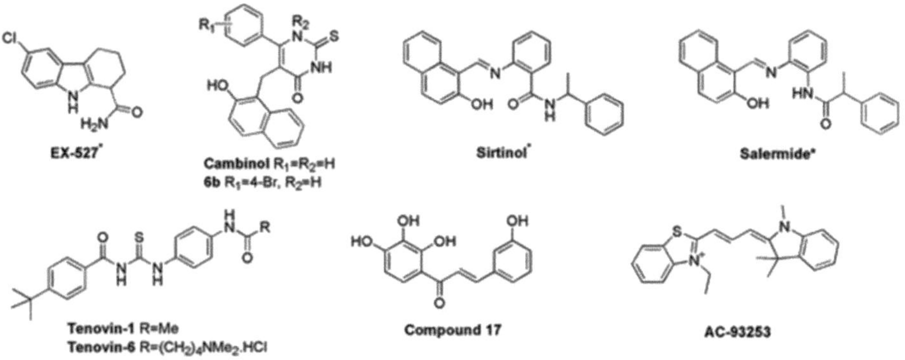

Fig. 12.4 Sirtuin inhibitors known to activate p53 in cells. *Asterisk* denotes stereochemistry not defined in the original report

in p53 levels in cells. EX-527 (Fig. 12.4) is a potent SirT1 inhibitor in vitro and causes accumulation of p53 and K382-acetylated p53 at a concentration of 1 μM, but only in the presence of the DNA-damaging agent etoposide (Solomon et al. 2006). Cambinol (Fig. 12.4) also inhibits SirT1 and increases levels of p53 and K382-acetylated p53 at high concentration, however, this is also only observed in the presence of etoposide (Heltweg et al. 2006). Interestingly, cambinol is tolerated as a single agent by epithelial cancer cells but is toxic to Burkitt lymphoma cells and decreases the growth of xenograft tumours derived from a Bcl-6 expressing Burkitt lymphoma cell line. A structure–activity relationship study carried out more recently by Medda et al. has resulted in an analogue of cambinol (6b, Fig. 12.4) that exhibits more potency in comparison (Medda et al. 2009). This novel analogue of cambinol also induces p53 and K382-acetylated p53 accumulation in cells, albeit in the presence of etoposide. However, some accumulation of p53 and K382-acetylated p53 is observed for SirT1 inhibitors sirtinol and its more active analogue salermide (at 50 μM, see Fig. 12.4 for structures) as single agents in a recent paper comparing the cellular effects of three sirtuin inhibitors (Peck et al. 2010). Sirtinol and salermide were also shown to increase the levels of a SirT2 target, K40-acetylated α-tubulin. However, the data for salermide's effect on K382-acetylated p53 is variable as in the original report it did not increase p53 levels (Lara et al. 2009). This effect seems to depend on the cell line utilised and perhaps the levels of acetylated p53 present for detection. Whilst salermide shows promising anti-tumour properties, its proapoptotic effect in cancer cells was deemed to be p53-independent.

Although unintentionally, sirtuin inhibitors possessing potent p53 activating ability were discovered as a result of a p53-based phenotypic screen (Lain et al. 2008). Tenovin-1 was identified as a hit compound and led to the development of a more water-soluble analogue, tenovin-6 (see Fig. 12.4 for structures). Tenovins rapidly increase the levels of p53 in cells at low micromolar concentrations and display effective activity against tumour growth in a mouse xenograft model of melanoma (derived from ARN8 cells). This promising antitumour activity prompted target identification studies and a yeast genetic screen identified the sirtuins as cellular targets of the tenovins. Inhibition of purified human SirT1 and SirT2 and the accumulation of their respective substrates in cells (acetylated p53 and acetylated tubulin) validated sirtuins as targets of tenovins. The discovery and characterisation of the tenovins highlights the fact that sirtuin inhibitors can be discovered as a result of screening for activation of wild-type p53 function, as a result of modifying protein targets upstream of p53.

In the last 2 years, many other examples of small molecules with sirtuin inhibition properties have been reported. Only studies that include evidence of p53 activation in cells are included here. Kahyo et al. have presented a novel chalcone polyphenol, compound 17 (3,2′,3′,4′-tetrahydroxychalcone, Fig. 12.4), which inhibits SirT1 activity in a biochemical assay ($IC_{50}=40\ \mu M$) and also in cells by inducing p53 K382-acetylation, and also decreases the growth of HEK293T cells (Kahyo et al. 2008). Because SirT1 overexpression is known to suppress p21 expression, the levels of p21 in HEK293T cells treated with compound 17 were analysed and indeed an accumulation of p21 levels was observed, also indicating induction of p53 transcriptional function. In addition, Zhang et al. reported AC-93253 (Fig. 12.4) as a selective SirT2 inhibitor (SirT2 $IC_{50}=6\ \mu M$, SirT1 $IC_{50}=45.3\ \mu M$ in vitro) (Zhang et al. 2009). Increased levels of K40-acetylated α-tubulin and K382-acetylated p53 were detected, showing that in cells AC-93253 inhibits SirT2 as well as SirT1. Interestingly the levels of total p53 were also slightly increased, as also seen for tenovin-6 (Lain et al. 2008). The ability of AC-93253 as a single agent to kill cells was assessed using a panel of tumour and normal cell lines. It was significantly more active against tumour cells, and primarily triggered apoptosis.

It is apparent that there are contradictions in the cellular activity of the sirtuin inhibitors discussed. Initial experimental evidence shows that sirtuin inhibitors can lead to the activation of p53 but in some instances (EX-527 and cambinol) concomitant induction of DNA damage by etoposide is required. This has led to controversy in this field. However, recent data on EX-527 shows that this compound can increase p53 acetylation as a single agent to a small level and that its effects are potentiated not only in combination with etoposide but also when combined with trichostatin A (an inhibitor of histone deacetylases also targeting p53) (Robers et al. 2011). On the other hand, Peck et al. suggest that sirtuin inhibitors need to target SirT1 in combination with SirT2 in order to induce p53 acetylation most efficiently (Peck et al. 2010). Apparent contradictions can also be found amongst non-sirtuin (NS-HDAC) inhibitors. Discrepancies in this case may be explained by the work of Peltonen and colleagues who reported that the NS-HDAC inhibitor trichostatin A can stabilise p53 but reduce its mRNA levels (Peltonen et al. 2005). It is also interesting to note that NS-HDACs have substrates in common with sirtuins, for example HDAC1 is also known to deacetylate p53, but are activated under different conditions. Therefore, it is possible that NS-HDAC activity may account for the discrepancies discussed above. It is clear that further studies are required in order to understand fully the mechanisms by which p53 is activated via sirtuin inhibition. A greater understanding of the interplay between sirtuins and NS-HDACs may help to explain the discrepancies also.

4 Apparent Non-genotoxic DNA Intercalators

A screen carried out by Gurova and colleagues identified compounds able to activate p53 function in renal cell carcinoma cells containing wild-type p53 (Gurova et al. 2005). Compounds were selected for their ability to induce p53-dependent transcription via β-galactosidase reporter activity. The most potent activator, 30d9, was chosen for preliminary structure-activity studies and it was found that its metabolite, 9AA (9-aminoacridine, Fig. 12.5), was in fact responsible for the observed effect in cells. Acridines have been widely used in the clinic for a range of diseases and they also show anti-tumour activity for which the mechanism of action remains unclear (Demeunynck et al. 2001; Denny 2002; Denny and Baguley 2003; Su 2002). 9AA was able to increase p53 levels, as well as induce p21 and Mdm2. The best induction of p53 by 9AA was observed in human fibrosarcoma cells, in which p53-dependent growth arrest or apoptosis occurred, depending on the dosage. This screen also identified a structurally related compound, quinacrine, as a p53 activator and it displayed similar biochemical and cellular activity to that of 9AA. Quinacrine (Fig. 12.5) also showed anti-tumour activity in comparison with 5-fluorouracil in mouse xenograft models of renal carcinoma (ACHN-derived). Inhibition of phosphorylation at serine-536 of the p65 subunit of NF-κB was proposed as a mechanism of action for

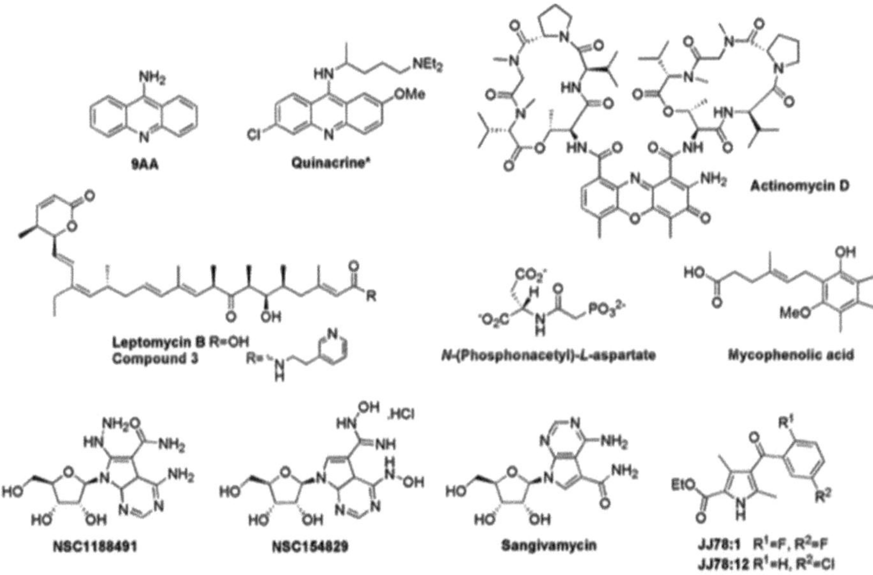

Fig. 12.5 Activators of p53 that are DNA intercalators (9AA, quinacrine and actinomycin D), inhibit nuclear transport (leptomycin B and analogue), modulators of ribonucleotide synthesis/ ribonucleoside analogues (*N*-(phosphonacetyl)-*L*-aspartate, mycophenolic acid, NSC1188491 and NSC154829, sangivamycin and the mitotic poison JJ78:1 and its optimised analogue JJ78:12. *Asterisk* denotes stereochemistry not defined in the original report

9AA, possibly through inhibition of the kinase responsible for this, IκB kinase 1. More recent studies have expanded on this proposed mechanism of action for 9AA in which it was found to inhibit the PI3K/AKT/mTOR signalling pathway. Quinacrine has similar effects to 9AA on these pathways and is currently in clinical trials for the treatment of hormone refractory prostate cancer (Guo et al. 2009), the results of which will give insight into the potential of 9AA as a therapeutic.

The planar nature and amino functional groups of acridines such as 9AA and quinacrine suggests they could activate p53 via DNA damage. However, the DNA-damaging ability of 9AA was assessed by analysing levels of p53 phosphorylated at Ser15 and Ser20 (no increase), H2AX phosphorylation (not increased) and ATM/ATR activation (no activation). Levels of p53 phosphorylated at Ser392 were increased, the phosphorylation of which is carried out by the kinase CK2 and not DNA damage inducible kinases. These results suggested that 9AA is not activating p53 via a DNA damage mechanism. Also 9AA does not inhibit topoisomerase II, as does the structurally related compound amsacrine (Gurova et al. 2005). The effect of these acridines is reminiscent of actinomycin D at low dosage (see below), a DNA intercalator that is specific for activating p53 at low dose, but is toxic at high dose (Choong et al. 2009). A study reported by Wang et al. confirmed that a series of acridines does not induce DNA damage (Wang et al. 2005). These compounds were in fact found to block p53 ubiquitination and induced apoptosis in a p53- and bax-dependent manner. A more recent study on the anti-tumour effect of quinacrine has been carried out using nuclear magnetic resonance spectroscopy to provide evidence of its binding with Bcl-xL in the groove used to interact with pro-apoptotic proteins (Orzaez et al. 2009).

Actinomycin D (Fig. 12.5) is a natural product and a potent activator of p53. At high dose, actinomycin D is toxic and induces markers of DNA damage, but at low dose is very specific for inducing p53 activity and is clinically approved. Efficacy has been identified for the treatment of Wilm's tumour (Green et al. 1998). It binds to adjacent GC pairs in DNA, thus at low doses it preferentially targets GC-rich regions in ribosomal DNA causing nucleolar disruption. As a result ribosome biogenesis is blocked, inducing the release of ribosomal proteins L11 and L5 that inactivate Mdm2.

5 Nuclear Transport Inhibitors

Leptomycin B (LMB, Fig. 12.5) was discovered as a potent p53 activator via the testing of small molecules known to target proteins that impinge on p53 activity. LMB inhibits nuclear export protein Crm1 and has been shown to induce apoptosis in cancer cells and reduce tumour growth in cancer models containing wild-type p53 (Chen et al. 2010). Although LMB is an extremely potent and specific inhibitor of Crm1 and has reached phase I clinical trials, its high toxicity and lack of efficacy resulted in the discontinuation of its clinical development. However, inhibitors of nuclear export are still of therapeutic interest and new improved Crm1 inhibitors based on LMB have been reported recently (Mutka et al. 2009). These derivatives synthesised via modification of LMB retain the potency of the parent compound but are better tolerated in mouse models. Compound 3 (Fig. 12.5), which showed more than 15-fold improvement in the maximum tolerated dose, was chosen for further characterisation. Apoptosis was induced by compound 3 in tumour cells (SiHa cells) but in normal human lung fibroblasts it caused reversible cell cycle arrest. The ability of compound 3 to inhibit nuclear export in different wild-type p53 cell lines correlated with the observation that p53 is upregulated in a sustained manner and localised to the nucleus. The apoptotic effect of compound 3 led to testing in mouse xenograft models (SiHa-derived tumours). Significant tumour regression was observed in comparison to LMB. Compound 3 also showed efficacy in other tumour models with evidence of better tolerance compared to LMB.

6 Modulators of Ribonucleotide Synthesis and Ribonucleoside Analogues

Stabilisation of p53 is known to occur upon treatment of cells with inhibitors of deoxyribonucleotide triphosphate (dNTP) synthesis. These inhibitors cause the accumulation of cells in S phase of the cell cycle, leading to DNA damage. Stabilisation of p53 under these conditions results in the expression of certain downstream effectors such as Pig3, but not p21 or Mdm2 (Gottifredi et al. 2001). On the other hand, ribonucleotide synthesis can be modulated resulting in p53 activation without DNA damage. *N*-phosphonacetyl-*L*-aspartate (PALA, Fig. 12.5) inhibits the ribonucleotide synthesis pathway, as does mycophenolic acid, and both induce the p53-dependent accumulation of cells in G1 or G0 of the cell cycle as well as an increase in p53 levels without inducing chromosomal abberations. DRB (Fig. 12.1), an mRNA synthesis inhibitor, also increases p53 levels and transcriptional activity (Ljungman et al. 1999; te Poele et al. 1999). This observation provides evidence that p53 may detect ribonucleotide depletion, and hence explain activation of p53 without DNA damage.

Adenosine analogues have been found to increase p53 levels and transcriptional activity in cells. Two such compounds, NCI188491 and NCI54829 (Fig. 12.5), were found via high-throughput screening for activators of wild-type p53 (Berkson et al. 2005). Assessment of the DNA-damaging ability of these compounds did not reveal any detectable damage. The structure of these NCI compounds is similar to that of the ribonucleoside analogue sangivamycin (Fig. 12.5); however, the mechanism by which all three compounds activate p53 is still unknown and there is no data on their effect in preclinical models.

7 Mitotic Poisons

Compounds able to disrupt mitosis are widely known to increase p53 levels. Examples of such mitotic poisons include taxol and the vinca alkaloids although how p53 is activated in response to these drugs is not yet completely understood. Not surprisingly, screens for p53 activators have led to the discovery of mitotic poisons.

One such compound, JJ78:1 (Fig. 12.5), was identified by this manner in the same screen from which the tenovins were discovered and was found to arrest cells via decreasing tubulin polymerisation (Staples et al. 2008). JJ78:12 (Fig. 12.5), an optimised analogue of JJ78:1, is able to induce p53 transcriptional function at low nanomolar concentrations and is more active in cells than the parent compound. A significant reduction in the growth of ARN8-derived tumours was also achieved by administration of JJ78:12.

8 Conclusion

The prevalence of mutant or inactive wild-type p53 in tumours renders it an important target for the development of new anti-cancer drugs. Small molecule activators of p53 are playing an increasingly important role in this regard and several have reached the clinical trial stage of testing. In this review we have specifically focussed on small molecules that activate wild-type p53 through modulating pathways that regulate p53 activity. As wild-type p53 is inactivated via a range of different mechanisms, this provides plentiful opportunity to develop anticancer therapies for tumours of differing status. The mechanisms by which such small molecules activate p53 have been discussed with consideration of the available data but it remains clear that further work is required in some cases to completely understand how targeting upstream regulatory factors leads to p53 activation and how they would be therapeutically relevant. Moreover, some parts of the p53 signalling pathway are still not fully understood and small molecule modulators of relevant targets upstream of p53 have additional use in that they may be useful to explore unknown mechanisms using chemical biology techniques.

References

Allende-Vega N, Saville MK (2010) Targeting the ubiquitin-proteasome system to activate wild-type p53 for cancer therapy. Semin Cancer Biol 20:29–39

Arts J, Page, M Valckx, A, Blattner C, Kulikov R, Floren W et al (2008) JNJ-26854165-a novel hdm2 antagonist in clinical development showing broad spectrum pre-clinical antitumor activity against solid malignancies. Proc Am Assoc Cancer Res 49: 1592

Berg T (2008) Small-molecule inhibitors of protein-protein interactions. Curr Opin Drug Discov Devel 11:666–674

Berkson RG, Hollick JJ, Westwood NJ, Woods JA, Lane DP, Lain S (2005) Pilot screening programme for small molecule activators of p53. Int J Cancer 115:701–710

Beryozkina A, Nichols GL, Reckner M, Vassilev LT, Rueger R, Jukofsky L et al (2011) Pharmacokinetics (PK) and pharmacodynamics (PD) of RG7112, an oral murine double minute 2 (MDM2) antagonist, in patients with leukemias and solid tumors. J Clin Oncol (Meeting Abstracts) 29:15_suppl 3039

Blattner C, Sparks A, Lane D (1999) Transcription factor E2F-1 is upregulated in response to DNA damage in a manner analogous to that of p53. Mol Cell Biol 19:3704–3713

Boccadoro M, Morgan G, Cavenagh J (2005) Preclinical evaluation of the proteasome inhibitor bortezomib in cancer therapy. Cancer Cell Int 5:18

Böttger V, Böttger A, Garcia-Echeverria C, Ramos Y, van der Eb AJ, Jochemsen AG, Lane DP (1999) Comparative study of the p53-mdm2 and p53-MDMX interfaces. Oncogene 18:189–199

Brown CJ, Lain S, Verma CS, Fersht AR, Lane DP (2009) Awakening guardian angels: drugging the p53 pathway. Nat Rev Cancer 9:862–873

Burgstaller P, Hermann T, Huber C, Westhof E, Famulok M (1997) Isoalloxazine derivatives promote photocleavage of natural RNAs at G.U base pairs embedded within helices. Nucleic Acids Res 25:4018–4027

Chen F, Wang W, El-Deiry WS (2010) Current strategies to target p53 in cancer. Biochem Pharmacol 80:724–730

Cheok CF, Verma CS, Baselga J, Lane DP (2011) Translating p53 into the clinic. Nat Rev Clin Oncol 8:25–37

Choong ML, Yang H, Lee MA, Lane DP (2009) Specific activation of the p53 pathway by low dose actinomycin D: a new route to p53 based cyclotherapy. Cell Cycle 8:2810–2818

Concannon CG, Koehler BF, Reimertz C, Murphy BM, Bonner C, Thurow N, Ward MW, Villunger A, Strasser A, Kogel D et al (2007) Apoptosis induced by proteasome inhibition in cancer cells: predominant role of the p53/PUMA pathway. Oncogene 26:1681–1692

Cummings CG, Hamilton AD (2010) Disrupting protein-protein interactions with non-peptidic, small molecule alpha-helix mimetics. Curr Opin Chem Biol 14:341–346

Danovi D, Meulmeester E, Pasini D, Migliorini D, Capra M, Frenk R, de Graaf P, Francoz S, Gasparini P, Gobbi A et al (2004) Amplification of Mdmx (or Mdm4) directly contributes to tumor formation by inhibiting p53 tumor suppressor activity. Mol Cell Biol 24: 5835–5843

Davydov IV, Woods D, Safiran YJ, Oberoi P, Fearnhead HO, Fang S, Jensen JP, Weissman AM, Kenten JH, Vousden KH (2004) Assay for ubiquitin ligase activity: high-throughput screen for inhibitors of HDM2. J Biomol Screen 9:695–703

Demeunynck M, Charmantray F, Martelli A (2001) Interest of acridine derivatives in the anticancer chemotherapy. Curr Pharm Des 7:1703–1724

Demidenko ZN, Blagosklonny MV (2004) Flavopiridol induces p53 via initial inhibition of Mdm2 and p21 and, independently of p53, sensitizes apoptosis-reluctant cells to tumor necrosis factor. Cancer Res 64:3653–3660

Denny WA (2002) Acridine derivatives as chemotherapeutic agents. Curr Med Chem 9:1655–1665

Denny WA, Baguley BC (2003) Dual topoisomerase I/II inhibitors in cancer therapy. Curr Top Med Chem 3:339–353

Dickens MP, Fitzgerald R, Fischer PM (2010) Small-molecule inhibitors of MDM2 as new anticancer therapeutics. Semin Cancer Biol 20:10–18

Ding K, Lu Y, Nikolovska-Coleska Z, Qiu S, Ding Y, Gao W, Stuckey J, Krajewski K, Roller PP, Tomita Y et al (2005) Structure-based design of potent non-peptide MDM2 inhibitors. J Am Chem Soc 127:10130–10131

Ding K, Lu Y, Nikolovska-Coleska Z, Wang G, Qiu S, Shangary S, Gao W, Qin D, Stuckey J, Krajewski K et al (2006) Structure-based design of spiro-oxindoles as potent, specific small-molecule inhibitors of the MDM2-p53 interaction. J Med Chem 49:3432–3435

Ding WX, Ni HM, Chen X, Yu J, Zhang L, Yin XM (2007) A coordinated action of Bax, PUMA, and p53 promotes MG132-induced mitochondria activation and apoptosis in colon cancer cells. Mol Cancer Ther 6:1062–1069

Domling A (2008) Small molecular weight protein-protein interaction antagonists: an insurmountable challenge? Curr Opin Chem Biol 12:281–291

Dudgeon DD, Shinde S, Hua Y, Shun TY, Lazo JS, Strock CJ, Giuliano KA, Taylor DL, Johnston PA (2010) Implementation of a 220,000-compound HCS campaign to identify disruptors of the interaction between p53 and hDM2 and characterization of the confirmed hits. J Biomol Screen 15:766–782

Gilkes DM, Pan Y, Coppola D, Yeatman T, Reuther GW, Chen J (2008) Regulation of MDMX expression by mitogenic signaling. Mol Cell Biol 28:1999–2010

Gottifredi V, Shieh S, Taya Y, Prives C (2001) p53 accumulates but is functionally impaired when DNA synthesis is blocked. Proc Natl Acad Sci USA 98:1036–1041

Grasberger BL, Lu T, Schubert C, Parks DJ, Carver TE, Koblish HK, Cummings MD, LaFrance LV, Milkiewicz KL, Calvo RR et al (2005) Discovery and cocrystal structure of benzodiazepinedione HDM2 antagonists that activate p53 in cells. J Med Chem 48:909–912

Green DM, Breslow NE, Beckwith JB, Finklestein JZ, Grundy PE, Thomas PR, Kim T, Shochat SJ, Haase GM, Ritchey ML et al (1998) Comparison between single-dose and divided-dose administration of dactinomycin and doxorubicin for patients with Wilms' tumor: a report from the National Wilms' Tumor Study Group. J Clin Oncol 16:237–245

Gu J, Kawai H, Nie L, Kitao H, Wiederschain D, Jochemsen AG, Parant J, Lozano G, Yuan ZM (2002) Mutual dependence of MDM2 and MDMX in their functional inactivation of p53. J Biol Chem 277:19251–19254

Guo C, Gasparian AV, Zhuang Z, Bosykh DA, Komar AA, Gudkov AV, Gurova KV (2009) 9-Aminoacridine-based anticancer drugs target the PI3K/AKT/mTOR, NF-kappaB and p53 pathways. Oncogene 28:1151–1161

Gurova KV, Hill JE, Guo C, Prokvolit A, Burdelya LG, Samoylova E, Khodyakova AV, Ganapathi R, Ganapathi M, Tararova ND et al (2005) Small molecules that reactivate p53 in renal cell carcinoma reveal a NF-kappaB-dependent mechanism of p53 suppression in tumors. Proc Natl Acad Sci USA 102:17448–17453

Heltweg B, Gatbonton T, Schuler AD, Posakony J, Li H, Goehle S, Kollipara R, DePinho RA, Gu Y, Simon JA et al (2006) Antitumor activity of a small-molecule inhibitor of human silent information regulator 2 enzymes. Cancer Res 66:4368–4377

Hu B, Gilkes DM, Farooqi B, Sebti SM, Chen J (2006) MDMX overexpression prevents p53 activation by the MDM2 inhibitor Nutlin. J Biol Chem 281:33030–33035

Hu CQ, Hu YZ (2008) Small molecule inhibitors of the p53-MDM2. Curr Med Chem 15:1720–1730

Joseph TL, Lane D, Verma CS (2010) Stapled peptides in the p53 pathway: computer simulations reveal novel interactions of the staples with the target protein. Cell Cycle 9:4560–4568

Juliano R, Alam MR, Dixit V, Kang H (2008) Mechanisms and strategies for effective delivery of antisense and siRNA oligonucleotides. Nucleic Acids Res 36:4158–4171

Kahyo T, Ichikawa S, Hatanaka T, Yamada MK, Setou M (2008) A novel chalcone polyphenol inhibits the deacetylase activity of SIRT1 and cell growth in HEK293T cells. J Pharmacol Sci 108:364–371

Kawamoto T, Ikeuchi Y, Hiraki J, Eikyu Y, Shimizu K, Tomishima M, Bessho K, Yoneda F, Mikata Y, Nishida M et al (1995a) Evaluation of differential hypoxic cytotoxicity and electrochemical studies of nitro 5-deazaflavins. Bioorg Med Chem Lett 5:2115–2118

Kawamoto T, Ikeuchi Y, Hiraki J, Eikyu Y, Shimizu K, Tomishima M, Bessho K, Yoneda F, Mikata Y, Nishida M et al (1995b) Synthesis and evaluation of nitro 5-deazaflavins as novel bioreductive antitumor agents. Bioorg Med Chem Lett 5:2109–2114

Kitagaki J, Agama KK, Pommier Y, Yang Y, Weissman AM (2008) Targeting tumor cells expressing p53 with a water-soluble inhibitor of Hdm2. Mol Cancer Ther 7:2445–2454

Koblish HK, Zhao S, Franks CF, Donatelli RR, Tominovich RM, LaFrance LV, Leonard KA, Gushue JM, Parks DJ, Calvo RR et al (2006) Benzodiazepinedione inhibitors of the Hdm2:p53 complex suppress human tumor cell proliferation in vitro and sensitize tumors to doxorubicin in vivo. Mol Cancer Ther 5:160–169

Kojima K, Konopleva M, Samudio IJ, Shikami M, Cabreira-Hansen M, McQueen T, Ruvolo V, Tsao T, Zeng Z, Vassilev LT et al (2005) MDM2 antagonists induce p53-dependent apoptosis in AML: implications for leukemia therapy. Blood 106:3150–3159

Lai Z, Yang T, Kim YB, Sielecki TM, Diamond MA, Strack P, Rolfe M, Caligiuri M, Benfield PA, Auger KR et al (2002) Differentiation of Hdm2-mediated p53 ubiquitination and Hdm2 autoubiquitination activity by small molecular weight inhibitors. Proc Natl Acad Sci USA 99:14734–14739

Lain S, Hollick JJ, Campbell J, Staples OD, Higgins M, Aoubala M, McCarthy A, Appleyard V, Murray KE, Baker L et al (2008) Discovery, in vivo activity, and mechanism of action of a small-molecule p53 activator. Cancer Cell 13:454–463

Lain S, Lane D (2003) Improving cancer therapy by non-genotoxic activation of p53. Eur J Cancer 39:1053–1060

Langley E, Pearson M, Faretta M, Bauer UM, Frye RA, Minucci S, Pelicci PG, Kouzarides T (2002) Human SIR2 deacetylates p53 and antagonizes PML/p53-induced cellular senescence. EMBO J 21:2383–2396

Lara E, Mai A, Calvanese V, Altucci L, Lopez-Nieva P, Martinez-Chantar ML, Varela-Rey M, Rotili D, Nebbioso A, Ropero S et al (2009) Salermide, a Sirtuin inhibitor with a strong cancer-specific proapoptotic effect. Oncogene 28:781–791

Laurie NA, Donovan SL, Shih CS, Zhang J, Mills N, Fuller C, Teunisse A, Lam S, Ramos Y, Mohan A et al (2006) Inactivation of the p53 pathway in retinoblastoma. Nature 444:61–66

Lee JH, Zhang Q, Jo S, Chai SC, Oh M, Im W, Lu H, Lim HS (2010) Novel pyrrolopyrimidine-based alpha-helix mimetics: cell-permeable inhibitors of protein-protein interactions. J Am Chem Soc 133:676–679

Li C, Pazgier M, Liu M, Lu WY, Lu W (2009) Apamin as a template for structure-based rational design of potent peptide activators of p53. Angew Chem Int Ed Engl 48:8712–8715

Ljungman M, Zhang F, Chen F, Rainbow AJ, McKay BC (1999) Inhibition of RNA polymerase II as a trigger for the p53 response. Oncogene 18:583–592

Lu W, Chen L, Peng Y, Chen J (2001) Activation of p53 by roscovitine-mediated suppression of MDM2 expression. Oncogene 20:3206–3216

Luo J, Li M, Tang Y, Laszkowska M, Roeder RG, Gu W (2004) Acetylation of p53 augments its site-specific DNA binding both in vitro and in vivo. Proc Natl Acad Sci USA 101:2259–2264

Luo J, Nikolaev AY, Imai S, Chen D, Su F, Shiloh A, Guarente L, Gu W (2001) Negative control of p53 by Sir2alpha promotes cell survival under stress. Cell 107:137–148

Marchenko ND, Wolff S, Erster S, Becker K, Moll UM (2007) Monoubiquitylation promotes mitochondrial p53 translocation. EMBO J 26:923–934

Marine JC (2010) Pharmacological rescue of p53 in cancer therapy: widening the sensitive tumor spectrum by targeting MDMX. Cancer Cell 18:399–400

Medda F, Russell RJ, Higgins M, McCarthy AR, Campbell J, Slawin AM, Lane DP, Lain S, Westwood NJ (2009) Novel cambinol analogs as sirtuin inhibitors: synthesis, biological evaluation, and rationalization of activity. J Med Chem 52:2673–2682

Murray MF, Jurewicz AJ, Martin JD, Ho TF, Zhang H, Johanson KO, Kirkpatrick RB, Ma J, Lor LA, Thrall SH et al (2007) A high-throughput screen measuring ubiquitination of p53 by human mdm2. J Biomol Screen 12:1050–1058

Mutka SC, Yang WQ, Dong SD, Ward SL, Craig DA, Timmermans PB, Murli S (2009) Identification of nuclear export inhibitors with potent anticancer activity in vivo. Cancer Res 69:510–517

Nair VD, McNaught KS, Gonzalez-Maeso J, Sealfon SC, Olanow CW (2006) p53 mediates non-transcriptional cell death in dopaminergic cells in response to proteasome inhibition. J Biol Chem 281:39550–39560

O'Hagan HM, Ljungman M (2004) Nuclear accumulation of p53 following inhibition of transcription is not due to diminished levels of MDM2. Oncogene 23:5505–5512

Orlowski RZ, Kuhn DJ (2008) Proteasome inhibitors in cancer therapy: lessons from the first decade. Clin Cancer Res 14:1649–1657

Orzaez M, Mondragon L, Garcia-Jareno A, Mosulen S, Pineda-Lucena A, Perez-Paya E (2009) Deciphering the antitumoral activity of quinacrine: binding to and inhibition of Bcl-xL. Bioorg Med Chem Lett 19:1592–1595

Peck B, Chen CY, Ho KK, Di Fruscia P, Myatt SS, Coombes RC, Fuchter MJ, Hsiao CD, Lam EW (2010) SIRT inhibitors induce cell death and p53 acetylation through targeting both SIRT1 and SIRT2. Mol Cancer Ther 9:844–855

Peltonen K, Kiviharju TM, Jarvinen PM, Ra R, Laiho M (2005) Melanoma cell lines are susceptible to histone deacetylase inhibitor TSA provoked cell cycle arrest and apoptosis. Pigment Cell Res 18:196–202

Phan J, Li Z, Kasprzak A, Li B, Sebti S, Guida W, Schonbrunn E, Chen J (2010) Structure-based design of high affinity peptides inhibiting the interaction of p53 with MDM2 and MDMX. J Biol Chem 285:2174–2183
Popowicz GM, Czarna A, Holak TA (2008) Structure of the human Mdmx protein bound to the p53 tumor suppressor transactivation domain. Cell Cycle 7:2441–2443
Popowicz GM, Czarna A, Rothweiler U, Szwagierczak A, Krajewski M, Weber L, Holak TA (2007) Molecular basis for the inhibition of p53 by Mdmx. Cell Cycle 6:2386–2392
Popowicz GM, Czarna A, Wolf S, Wang K, Wang W, Domling A, Holak TA (2010) Structures of low molecular weight inhibitors bound to MDMX and MDM2 reveal new approaches for p53-MDMX/MDM2 antagonist drug discovery. Cell Cycle 9:1104–1111
Reed D, Shen Y, Shelat AA, Arnold LA, Ferreira AM, Zhu F, Mills N, Smithson DC, Regni CA, Bashford D et al (2010) Identification and characterization of the first small molecule inhibitor of MDMX. J Biol Chem 285:10786–10796
Robers MB, Loh C, Carlson CB, Yang H, Frey EA, Hermanson SB, Bi K (2011) Measurement of the cellular deacetylase activity of SIRT1 on p53 via LanthaScreen(R) technology. Mol Biosyst 7:59–66
Sasiela CA, Stewart DH, Kitagaki J, Safiran YJ, Yang Y, Weissman AM, Oberoi P, Davydov IV, Goncharova E, Beutler JA et al (2008) Identification of inhibitors for MDM2 ubiquitin ligase activity from natural product extracts by a novel high-throughput electrochemiluminescent screen. J Biomol Screen 13:229–237
Secchiero P, di Iasio MG, Gonelli A, Zauli G (2008) The MDM2 inhibitor Nutlins as an innovative therapeutic tool for the treatment of haematological malignancies. Curr Pharm Des 14:2100–2110
Shangary S, Ding K, Qiu S, Nikolovska-Coleska Z, Bauer JA, Liu M, Wang G, Lu Y, McEachern D, Bernard D et al (2008a) Reactivation of p53 by a specific MDM2 antagonist (MI-43) leads to p21-mediated cell cycle arrest and selective cell death in colon cancer. Mol Cancer Ther 7:1533–1542
Shangary S, Qin D, McEachern D, Liu M, Miller RS, Qiu S, Nikolovska-Coleska Z, Ding K, Wang G, Chen J et al (2008b) Temporal activation of p53 by a specific MDM2 inhibitor is selectively toxic to tumors and leads to complete tumor growth inhibition. Proc Natl Acad Sci USA 105:3933–3938
Shangary S, Wang S (2008) Targeting the MDM2-p53 interaction for cancer therapy. Clin Cancer Res 14:5318–5324
Shen Y (2008) Advances in the development of siRNA-based therapeutics for cancer. IDrugs 11:572–578
Siliciano JD, Canman CE, Taya Y, Sakaguchi K, Appella E, Kastan MB (1997) DNA damage induces phosphorylation of the amino terminus of p53. Genes Dev 11:3471–3481
Solomon JM, Pasupuleti R, Xu L, McDonagh T, Curtis R, DiStefano PS, Huber LJ (2006) Inhibition of SIRT1 catalytic activity increases p53 acetylation but does not alter cell survival following DNA damage. Mol Cell Biol 26:28–38
Staples OD, Hollick JJ, Campbell J, Higgins M, McCarthy AR, Appleyard V, Murray KE, Baker L, Thompson A, Ronseaux S et al (2008) Characterization, chemical optimization and anti-tumour activity of a tubulin poison identified by a p53-based phenotypic screen. Cell Cycle 7:3417–3427
Stommel JM, Wahl GM (2004) Accelerated MDM2 auto-degradation induced by DNA-damage kinases is required for p53 activation. EMBO J 23:1547–1556
Su TL (2002) Development of DNA topoisomerase II-mediated anticancer agents, 3-(9-acridinylamino)-5-hydroxymethylanilines (AHMAs) and related compounds. Curr Med Chem 9:1677–1688
te Poele RH, Okorokov AL, Joel SP (1999) RNA synthesis block by 5, 6-dichloro-1-beta-D-ribofuranosylbenzimidazole (DRB) triggers p53-dependent apoptosis in human colon carcinoma cells. Oncogene 18:5765–5772
Toledo F, Krummel KA, Lee CJ, Liu CW, Rodewald LW, Tang M, Wahl GM (2006) A mouse p53 mutant lacking the proline-rich domain rescues Mdm4 deficiency and provides insight into the Mdm2-Mdm4-p53 regulatory network. Cancer Cell 9:273–285

Valentin-Vega YA, Barboza JA, Chau GP, El-Naggar AK, Lozano G (2007) High levels of the p53 inhibitor MDM4 in head and neck squamous carcinomas. Hum Pathol 38:1553–1562

Vassilev LT, Vu BT, Graves B, Carvajal D, Podlaski F, Filipovic Z, Kong N, Kammlott U, Lukacs C, Klein C et al (2004) In vivo activation of the p53 pathway by small-molecule antagonists of MDM2. Science 303:844–848

Vaziri H, Dessain SK, Ng Eaton E, Imai SI, Frye RA, Pandita TK, Guarente L, Weinberg RA (2001) hSIR2(SIRT1) functions as an NAD^+-dependent p53 deacetylase. Cell 107:149–159

Wang H, Ma X, Ren S, Buolamwini JK, Yan C (2011) A small-molecule inhibitor of MDMX activates p53 and induces apoptosis. Mol Cancer Ther 10:69–79

Wang W, Ho WC, Dicker DT, MacKinnon C, Winkler JD, Marmorstein R, El-Deiry WS (2005) Acridine derivatives activate p53 and induce tumor cell death through Bax. Cancer Biol Ther 4:893–898

Wang YV, Wade M, Wong E, Li YC, Rodewald LW, Wahl GM (2007) Quantitative analyses reveal the importance of regulated Hdmx degradation for p53 activation. Proc Natl Acad Sci USA 104:12365–12370

Weber L (2010) Patented inhibitors of p53-Mdm2 interaction (2006–2008). Expert Opin Ther Pat 20:179–191

Williams SA, McConkey DJ (2003) The proteasome inhibitor bortezomib stabilizes a novel active form of p53 in human LNCaP-Pro5 prostate cancer cells. Cancer Res 63:7338–7344

Yang Y, Kitagaki J, Dai RM, Tsai YC, Lorick KL, Ludwig RL, Pierre SA, Jensen JP, Davydov IV, Oberoi P et al (2007) Inhibitors of ubiquitin-activating enzyme (E1), a new class of potential cancer therapeutics. Cancer Res 67:9472–9481

Zhang Y, Au Q, Zhang M, Barber JR, Ng SC, Zhang B (2009) Identification of a small molecule SIRT2 inhibitor with selective tumor cytotoxicity. Biochem Biophys Res Commun 386:729–733

Zhu Q, Wani G, Yao J, Patnaik S, Wang QE, El-Mahdy MA, Praetorius-Ibba M, Wani AA (2007) The ubiquitin-proteasome system regulates p53-mediated transcription at p21waf1 promoter. Oncogene 26:4199–4208

Chapter 13
p53-Reactivating Molecules as Research Tools and Anticancer Drugs

Vera V. Grinkevich, Andreas Warnecke, and Galina Selivanova

1 Introduction

Diverse oncogenic events result in the activation of the p53 tumor suppressor and induction of p53-dependent programmed cell death or senescence. Among others, these include activation of oncogenic Ras, Myc, β-catenin, cdk4 or cyclins D/E, and/or inactivation of tumor suppressors p16, Rb, or APC (Vogelstein et al. 2000). Therefore, it appears that only those tumor cells that have acquired additional alterations blocking the p53 response will survive and continue to expand. Indeed, mutations that block apoptosis induction downstream of p53 are frequently found in tumors. However, p53 regulates a multitude of redundant genes suppressing tumor growth (Vogelstein et al. 2000). Therefore inactivation of p53 activity along with its widespread downstream network is a highly efficient way for cancer cells to disable the p53-induced response.

p53 can halt the propagation of cells carrying oncogenic lesions via a multitude of pathways, i.e., induction of growth arrest and senescence, induction of apoptosis, modulation of tumor stroma, angiogenesis and metabolism, as well as the block of invasion and metastasis (Vogelstein et al. 2000). This explains why loss of p53 function is selected for during tumor development, resulting in p53 inactivation in the majority of human tumors. Approximately half of all tumors carry point mutations in the *TP53* gene, which abolish p53 binding to its target promoters (Petitjean et al. 2007). In tumors without *TP53* mutations the cellular control of p53 activity is negatively deranged, leading to functional silencing of the tumor suppressor (Vogelstein et al. 2000). However, in contrast to other tumor suppressors—such as

V.V. Grinkevich • A. Warnecke • G. Selivanova (✉)
Department of Microbiology, Tumor and Cell Biology, Karolinska Institutet, 17177 Stockholm, Sweden
e-mail: Galina.Selivanova@ki.se

P. Hainaut et al. (eds.), *p53 in the Clinics*, DOI 10.1007/978-1-4614-3676-8_13,

Rb and p16—the p53 protein, although inactive, is still expressed in most tumors. This makes reinstatement of the p53 function by small molecules an attractive strategy to combat cancer.

Therapies aimed to restore the function of inactivated p53 in tumors are expected to reestablish the missing link between oncogene-induced pro-apoptotic signaling and cell death machinery. This should trigger massive apoptosis in presensitized tumor cells leaving normal cells largely unaffected (Selivanova 2010).

Although inhibition of oncogenic factors appears to be a more straightforward strategy for the development of target-specific therapies than reactivation of dormant tumor suppressors, the past few years have witnessed significant advances in restoring p53 function.

That the revival of p53 is a feasible approach to suppress tumor progression has been demonstrated by restoring p53 activity in established tumors using transgenic mouse models. Reinstatement of p53 in already established tumors of three different types, lymphoma, sarcoma, and hepatocellular carcinoma, induces tumor regression (Martins et al. 2006; Ventura et al. 2007; Xue et al. 2007). Notably, recent studies addressing the question how p53 reactivation affects developing nonsmall cell lung cancer (NSCLC) revealed that reinstatement of p53 efficiently eliminates advanced lesions whereas leaving early lesions unperturbed (Feldser et al. 2010; Junttila et al. 2010). This may be attributed to a more efficient activation of p53 upon its reinstatement in highly malignant tumor cells, as these assumingly display a higher degree of overall stress, compared to the low-grade tumors. These studies raise the question whether restoration of p53 by pharmacological means will suffice to kill all lesions or only eliminate the more advanced lesions from a heterogeneous tumor. This is of extreme clinical relevance, as killing of aggressive tumors will likely halt disease progression; however, sparing of early lesions might entail recurrence. On the upside, by the time of diagnosis, human tumors are usually more advanced resembling those whose regression was triggered by p53 reactivation, as described in earlier papers (Martins et al. 2006; Ventura et al. 2007; Xue et al. 2007). In addition, low-grade tumors respond well to conventional chemotherapy and thus could be efficiently eliminated. It is important to emphasize at this point that the tumors in the studies mentioned above were evolving in the absence of p53. It remains an important and relevant question how p53 reactivation will affect tumors that have undergone selection for p53 inactivation.

Introducing functional p53 into tumors using adenovirus vectors showed some effect in clinical trials (Edelman and Nemunaitis 2003) and *TP53* gene therapy has been approved in China in 2003 (Lane et al. 2010). In the recent years, the idea of pharmacological reactivation of p53 has been attracting increasing interest both in academia and in industry. The major strategies include restoration of proper folding to mutant p53 and prevention of p53 proteasomal degradation mediated by Mdm2. This chapter will focus on the strategies of pharmacological targeting of wild-type p53 in tumors and application of small molecules as research tools to study p53 biology.

2 Mechanisms Controlling p53 Activity Converge on Mdm2

Stress signals converge on the interaction of p53 with its negative regulators, mainly Mdm2 (also called Hdm2 in humans) and MdmX (also called HdmX and Mdm4 in mice). The overall outcome is liberation of active p53, which can manifest in different biological outcomes, depending on the type of tissue and initiating stress. The latest advances in understanding p53 regulation are extensively discussed in several excellent reviews (Levine and Oren 2009; Vousden and Prives 2009) and chapters in this issue. Here, we will briefly describe the major negative regulators of p53.

The E3-ligase Mdm2 is known as the most crucial inhibitor of p53. Mdm2 ubiquitinylates p53 and marks it for destruction by the proteasome, thus keeping p53 at bay in the absence of stress. In addition, Mdm2 can inhibit p53 transcriptional activity at the promoters (Momand et al. 1992). On the other hand, the Mdm2 paralog MdmX has no E3-ligase activity, but is involved in positively modulating Mdm2-mediated inhibition of p53 and can suppress the transactivation of p53 target genes independently of Mdm2 (Wang et al. 2007). The most recent models argue that the maximal activation of p53 requires both the disruption of Mdm2/p53 interaction and the degradation of MdmX (Wade et al. 2010).

The interaction between Mdm2 and p53 is modulated by a number of upstream regulators, such as p14ARF and several kinases (see Chap. 12 in this issue, for more details).

As displayed in Fig. 13.1, multiple genetic alterations which occur in cancer converge on the Mdm2/p53 interaction and enable Mdm2 to inhibit p53. This can occur, e.g., by amplification of Mdm2 itself, or via enhanced positive regulation of Mdm2 by activated oncogenes such as Akt. Alternatively, Mdm2 is deregulated

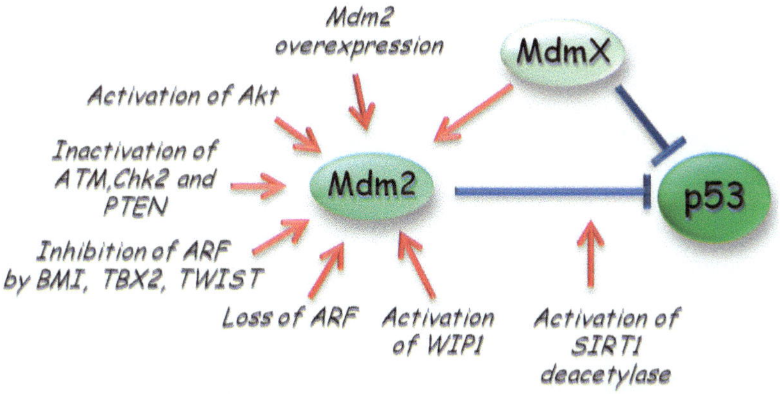

Fig. 13.1 Multiple genetic alterations which occur in cancer converge on the Mdm2/p53 interaction. These include overexpression of Mdm2 itself, activation of positive regulators of Mdm2 (e.g. Akt and MdmX), as well as inactivation of negative regulators due to the inhibition of p14ARF expression or mutations in ATM, Chk2, or PTEN. In addition, enhanced activity of SIRT1 deacetylase promotes Mdm2 interaction with p53. Since Mdm2/p53 interaction is the major switch for p53 inactivation, this complex is an attractive target for pharmacological modulation

due to abatement of negative upstream regulation by p14ARF or checkpoint kinases. As a result, the dependence of tumors on Mdm2-mediated inhibition of p53 is very common. Since the inhibition by Mdm2 appears to be the major regulatory step in p53 biology, targeting this axis is likely to elicit the most effective response. Thus, the Mdm2–p53 interaction is an excellent target for therapeutic intervention. Recent advances in the development of Mdm2 inhibitors provide strong evidence supporting the feasibility of this approach, as we will discuss below (see Chap. 12 in this issue).

Targeting other regulators of p53, such as the phosphatase Wip1, which removes phosphorylation marks preventing p53/Mdm2 interaction, is also conceivable (Le Guezennec and Bulavin 2010). Moreover, sirtuins, in particular SIRT1, have emerged as alternative targets as they modulate the Mdm2/p53 interaction by removing acetylation marks on p53 and, thereby, upholding the affinity of Mdm2 to p53 (Lain et al. 2008).

Another distinctive mechanism of p53 inhibition in tumors with wild-type p53 is the expression of the human papilloma virus (HPV)-encoded E6 protein, which, independently of Mdm2, holds p53 in check by conferring ubiquitinylation to p53 (Scheffner et al. 1990, 1993).

Below we will discuss different approaches and compounds that have been developed up to date to revive dormant wild-type p53 in tumors.

3 Wild-Type p53 as a Target in Cancer

A number of p53-reactivating molecules have been identified during the recent years, originating both from rational design or screening of chemical libraries in biochemical or phenotypic setups (Table 13.1). These small molecules either disrupt the interaction of p53 with its negative regulators, e.g. Mdm2, MdmX, E6, or inhibit negative regulation of p53 upstream of Mdm2/MdmX, e.g. sirtuins (reviewed in Lane et al. 2010). Molecules that reactivate wild-type p53 have been shown to elevate p53 protein level and to restore its transcriptional activity, faithfully reproducing classical tumor suppressor aspects of p53. Reactivation of p53 results in the induction of apoptosis or growth arrest in tumor cells. Moreover, most compounds do not display drastic effects in normal cells (in contrast to tumor cells) thus providing a sufficient therapeutic window.

3.1 Inhibition of Mdm2

Given the central role of Mdm2 in the control of p53 function, the potential for non-genotoxic cancer therapies based on the inhibition of Mdm2 is considerable, although the consequences of Mdm2 inhibition in vivo require more detailed investigation. As shown in elegant mouse model, in which p53 was restored in an Mdm2-null

Table 13.1 Examples of molecules restoring p53 function via targeting p53 or p53-binding proteins

Name	Effect on p53		Target	Mechanism of action	Effect in vitro/in vivo	Method of identification	References
	wt	mut					
Mdm2-binding molecules							
Nutlin	+	–	Mdm2, p53 binding domain	Prevents p53/Mdm2 interaction	Growth arrest or apoptosis/tumor suppression	In vitro screen for Mdm2 binding	Vassilev et al. (2004)
MI219	+	–	Mdm2, p53 binding domain	Prevents p53/Mdm2 interaction	Growth arrest or apoptosis/tumor suppression	Rational design	Shangary et al. (2008)
HLI98	+	ND	Mdm2	Inhibit s Mdm2 E3-ligase activity	Apoptosis/ND	In vitro screen for the inhibition of MDM2 autoubiquitination	Yang et al. (2005)
SAH-p53 peptide	+	ND	Mdm2, p53 binding domain	Prevents p53/Mdm2 interaction	Apoptosis/ND	Rational design	Bernal et al. (2007)
MdmX-binding molecules							
SJ172550	+	ND	MdmX	Prevents p53/Mdmx interaction	Apoptosis/ND	Biochemical and cell-based screen	Reed et al. (2010)
Stapled peptide SAH-p53-8	+	+	MdmX	Prevents p53/MdmX and p53/Mdm2 interaction	Apoptosis/tumor suppression	Rational design	Bernal et al. (2010)
p53-binding molecules							
RITA	+	+	p53, N-terminus	Prevents p53/Mdm2 interaction	Apoptosis/tumor suppression	Cell-based screen of chemical library	Issaeva et al. (2004)
SCH529074	+	+	p53, DNA-binding domain	Restores wt p53 conformation	Apoptosis/ND	In vitro screen for p53 DNA binding of quinozolines	Demma et al. (2010)
PRIMA-1^{MET}/ APR-246	$+^{*}$	+	p53	Restores wt p53 conformation	Apoptosis/tumor suppression	Cell-based screen of chemical library	Bykov et al. (2002)

(continued)

Table 13.1 (continued)

Name	Effect on p53		Target	Mechanism of action	Effect in vitro/in vivo	Method of identification	References
	wt	mut					
Peptide CDB3	+*	+	p53, DNA-binding domain	Restores wt p53 conformation	Weak apoptosis/ND	Rational design	Friedler et al. (2002), Issaeva et al. (2003)
CP31398	+*	+	p53, DNA-binding domain	Restores wt p53 conformation	Apoptosis/tumor suppression	In vitro screen	Foster et al. (1999)
Other							
Tenovin	+	+	Sirtuin 1 and 2	Induces p53 activating acetylation	Apoptosis/tumor suppression	Cell-based reporter screen of chemical library	Lain et al. (2008)

* effect is observed in certain cancer cell lines

background in adult mice, unrestrained p53 activation induces incredible havoc to all tissues (Ringshausen et al. 2006). Thus, it is imperative to maintain tumor selectivity when developing cancer treatments based on p53 reactivation. However, small molecules are unlikely to have the same effects as those produced by true null mutations. Therefore caution needs to be exercised when extrapolating effects caused by null mutations or effects of RNAi to predict the action of small molecules.

Different strategies of Mdm2 inhibition could be envisioned, including the inhibition of Mdm2 E3 ligase activity or blocking its interaction with p53, the reduction of Mdm2 mRNA or the inactivation of Mdm2 by means of negative cellular regulators such as the RPL (Ribosomal Protein, Large subunit) family. In addition, Mdm2-induced degradation of p53 can be blocked by proteasomal inhibitors or via prevention of p53 nuclear export by leptomycin B (Hainaut and Wiman 2005). Each strategy has both advantages and drawbacks; the success of a given strategy may depend on the mechanism by which the individual tumor achieved p53 inactivation.

Prevention of p53/Mdm2 interaction appears to be a promising approach to restore the p53 activity in tumors. The first evidence supporting this concept came from a study using a small p53-based peptide designed to bind the p53-binding pocket in Mdm2 with high affinity. The expression of this p53-derived peptide in cells results in p53 activation and induction of growth arrest (Bottger et al. 1997). Another synthetic peptide derived from p53 (AP peptide) inhibits the p53–Mdm2 interaction and induces p53 accumulation and activation followed by apoptosis in tumor cells with overexpressed Mdm2 (Chene et al. 2000). Although peptides are attractive candidates for stabilizing or disrupting protein–protein interactions, major obstacles to their efficacy in vivo are the loss of secondary structure, susceptibility to proteolytic degradation, and inefficient cellular uptake. Synthetic approaches such as hydrocarbon stapling, which reinforce native peptide sequences, provide an alternative way to manipulate protein–protein interactions via peptides. This is exemplified by SAH-p53 (*s*tabilized *a*lpha *h*elix of p53)—a peptide modeled after the transactivation domain of p53. SAH-p53 binds Mdm2 and reactivates the p53 pathway in Mdm2-overexpressing osteosarcoma cells (Bernal et al. 2007). This approach is promising, but remains quite challenging when applied in vivo.

Though historically protein–protein interactions that constitute the major key nodes in many signaling pathways have been regarded to be difficult targets for small molecules, the identification and development of small molecule inhibitors of the p53/Mdm2 interaction provide an inspiring example.

Several classes of small molecules that bind to Mdm2 and inhibit its interaction with p53 have been successfully identified. Nutlins are a group of cis-imidazoline analogs that mimic key residues in the p53 N-terminal alpha-helix critical for p53-binding to Mdm2 (Linke et al. 2008; Vassilev et al. 2004). Nutlin3a sterically blocks the hydrophobic p53-binding cleft in Mdm2, thus displacing p53 from Mdm2. This leads to p53 accumulation and restoration of its transcriptional activity (Vassilev et al. 2004). Nutlin3a promotes potent growth inhibition in a number of tumor cell lines of different origin, including colon, lung, breast and skin carcinomas, melanoma, lymphomas, and others. Furthermore, Nutlin3a suppresses tumor growth in vivo (Cheok et al. 2011).

Another small molecule, MI-219 (also known as AT-219; Ascent Therapeutics, Malvern, PA, USA), has been rationally designed based on the crystal structure of the p53/Mdm2 complex. MI-219 fits in the p53-binding pocket of Mdm2 in the same manner as Nutlin3a, mimicking the key p53 residues interacting with Mdm2. It activates the p53 pathway in wild-type p53-expressing cancer cells and leads to the induction of cell cycle arrest or apoptosis. Although MI-219 activates p53 in normal tissues, p53 accumulation is minimal and not harmful when applied in animals. It is a potent, highly selective, and orally active compound which can inhibit tumor growth in vivo (Shangary et al. 2008). Both Nutlin3a and MI-219 do not appear to display a substantial off-target activity.

Furthermore, several compounds targeting the p53/Mdm2 interaction are currently undergoing clinical trials: JnJ-26854165 (Johnson & Johnson, USA), PXn727 and PXn822 (Priaxon, Munich, Germany), and RG7112—a compound from the Nutlin3a series (F. Hoffmann–la Roche, Basel, Switzerland) (Cheok et al. 2011).

3.2 Inhibition of MdmX

The Mdm2 homologue MdmX inhibits p53-mediated transcription via direct interaction with p53 (Vousden and Lu 2002). Several studies have highlighted the importance of developing inhibitors toward MdmX. MdmX is frequently overexpressed in different types of tumors (Danovi et al. 2004). It has been shown that MdmX forms heterodimers with Mdm2, resulting in enhanced E3 ligase activity of Mdm2 toward p53. MdmX depletion leads to elevated p53 levels (Linares et al. 2003), suppresses colony outgrowth (Danovi et al. 2004) and reduces tumor xenograft growth (Gilkes et al. 2006).

Nutlin3a has a 30-fold lower efficiency in disrupting the p53/MdmX interaction compared to the p53/Mdm2 complex (Hu et al. 2006; Laurie et al. 2006; Patton et al. 2006; Wade et al. 2006). This also applies to MI-219, which is 10^4-fold more selective for Mdm2 over MdmX.

The efficacy of Mdm2 inhibition can be compromised by overexpression of MdmX, while downregulation of MdmX in combination with Nutlin3a significantly enhances p53 activity (Hu et al. 2006; Patton et al. 2006; Wade et al. 2006; Xia et al. 2008). These studies underscore the concept that full activation of p53 requires the restriction of activity of both Mdm2 and MdmX.

A recently identified chemical inhibitor of MdmX/p53 binding, SJ-172550, confers an additive growth suppression effect in cancer cells when combined with Mdm2 inhibitors (Reed et al. 2010). Further research will be necessary to clarify if chemical MdmX inhibitors as single agents or in combination will show clinical benefits.

There have also been reports about short peptides inhibiting the interaction between MdmX and p53 (Hu et al. 2007; Pazgier et al. 2009). For example, a stapled peptide SAH-p53-8 has been designed to preferentially target MdmX (Bernal et al. 2010). It blocks the formation of inhibitory p53–MdmX complexes, induces p53-dependent transcriptional activation, and thereby overcomes MdmX-mediated p53 suppression in vitro and in vivo. The successful development of stapled peptides targeting Mdm2 or MdmX is an encouraging option and may open the way for novel anticancer therapies.

3.3 Direct Targeting of p53

The most developed strategy for restoring p53 is the inhibition of Mdm2 and it is currently actively pursued both in academia and industry. In contrast, strategies aimed at direct targeting of p53 have received less attention. The major obstacle in this approach is the question whether a small molecule is capable of binding p53 and protecting it from degradation without affecting regular p53 function. This is especially challenging, as the p53 protein is conformationally flexible. In particular, the p53 N-terminus—where the Mdm2 binding site resides—is considered to be largely unstructured in solution (Dawson et al. 2003). Although the structures of several p53 domains have been solved, the full-length p53 protein eluded numerous attempts of crystallization, due to its intrinsic structural flexibility. The absence of the structural data or the full-length p53 protein limits the possibility of rational drug design (Selivanova 2010).

Via phenotypic screening of NCI chemical library we identified the thiophen-derived small molecule RITA, which activates p53 and induces apoptosis in cancer cell lines of a different origin. RITA was the first compound shown to bind the p53 N-terminus (residues 1–63), thereby preventing the binding of negative regulators such as Mdm2 and others (Issaeva et al. 2004).

Notably, several p53-binding molecules that rescue mutant p53 have been shown to activate the function of wild-type p53 as well. These include CDB3 (Issaeva et al. 2003), SCH529074 (Demma et al. 2010), CP-31398 (Luu et al. 2002), and PRIMA-1^{MET} (Bao et al. 2011) (Table 13.1). The mechanism of wild-type p53 reactivation by these molecules is not completely understood. At least some of them appear to inhibit the p53/Mdm2 interaction (see below for more details). Since all these molecules restore mutant p53 function by binding to p53 and affecting its conformation, it is tempting to speculate that the wild-type p53 could be reactivated by induction of certain conformational changes as well.

3.4 Blocking E6 to Reactivate p53 in Cervical Cancers

In over 90 % of cervical cancers, the p53 tumor suppressor pathway is inactivated by human papillomavirus-encoded oncogene HPV E6. HPV E6 promotes the degradation of wild-type p53 by recruiting the cellular E3 ligase E6-associated protein (E6AP), thus inhibiting p53 in a manner similar to Mdm2 (Scheffner et al. 1993). Mdm2 plays a minor role in p53 regulation in the presence of HPV E6 (Hengstermann et al. 2001). Prevention of p53 destruction by targeting HPV E6 appears to be an appealing strategy to combat cervical carcinoma, since the survival of HPV-infected cancer cells depends on p53 inactivation by HPV E6. Inhibition of HPV E6 can be achieved by a synthetic peptide (Dymalla et al. 2009), by depleting E6 mRNA, (Goodwin and DiMaio 2000; Jiang and Milner 2002; von Knebel Doeberitz et al. 1992), blocking of E6 expression by the combination of leptomycin B and actinomycin D or by small molecules ChemBridge 5254239 and 5228473 (Hietanen et al. 2000; Kochetkov et al. 2007). In addition, inhibiting E6AP by RNAi or a dominant-negative mutant has been shown to induce p53 (Kelley et al. 2005; Talis et al. 1998). Interestingly, we found that the p53-targeting molecule RITA prevents the interaction of p53 and E6-associated protein E6AP, which is required for HPV E6-mediated degradation. RITA induces the accumulation of p53 in cells containing high-risk HPV16 and HPV18 (Zhao et al. 2010b). In several cervical cancer cell lines RITA activates the transcription of pro-apoptotic p53 targets Noxa, PUMA, and Bax, while repressing the expression of pro-proliferative factors CyclinB1, CDC2, and CDC25C. This results in p53-dependent apoptosis and cell cycle arrest. Importantly, RITA suppresses the growth of cervical carcinoma xenografts in vivo, suggesting that p53-targeted compounds could be developed for the treatment of cervical carcinomas (Zhao et al. 2010b).

3.5 Inhibition of Sirtuins

Sirtuins are NAD^+-dependent deacetylases targeting a variety of proteins, including p53 (Haigis and Guarente 2006). Several inhibitors of sirtuin deacetylase activity have been described, including sirtinol, cambinol, and EX-527 (Lain et al. 2008). Tenovin 1 and its more soluble derivative Tenovin 6 are two novel sirtuin inhibitors identified by Lain et al. using a p53-dependent reporter as readout for the screening of a chemical library (Lain et al. 2008). Tenovin 1 inhibits Sirtuin 1, thus inducing acetylation of p53 at Lysine 382 and activating p53 (see Chap. 12 in this issue). It suppresses the growth of various tumor cell lines and inhibits the growth of highly aggressive melanoma xenograft tumors without causing DNA damage (Lain et al. 2008).

4 The Cellular Response to p53 Activation: What Determines the Outcome?

The cellular response upon p53 activation can range from transient growth arrest to cell death. Since the induction of cell cycle arrest might confer survival for some cancer cells, leading to tumor recurrence, we must understand the mechanism of this decision if we are to clinically exploit drugs activating p53. Although several compounds reactivating p53 are currently being tested in the clinical setting, there is still much to learn as to how we can manipulate p53 in a desirable way. Small molecules activating p53 could serve as excellent tools to address these burning questions of p53 biology.

Nutlin3a and RITA prevent the interaction between Mdm2 and p53, leading to p53 accumulation and restoration of its transcriptional activity and tumor suppressor function in cancer cells and tumor xenografts in vivo (Issaeva et al. 2004; Vassilev et al. 2004). Notably, despite targeting the same protein/protein complex, i.e., p53/Mdm2, these molecules induce different biological outcomes in tumor cells. Nutlin3a predominantly induces p53-dependent cell cycle arrest in cell lines in which RITA induces apoptosis and vice versa (Enge et al. 2009; Rinaldo et al. 2009). Elucidation of the mechanisms mediating these differences between RITA- and Nutlin3a-induced responses might help to predict the consequences of p53 reactivation in different genetic backgrounds. To this end, it will be necessary to decipher the detailed mechanisms of action and metabolism/pharmacokinetic (DMPK) properties of the individual compounds. This will also be a critical information when using the compounds either as drugs or research tools.

One of the reasons for the distinct outcomes might be the targeting of different partners of the Mdm2–p53 complex, namely Mdm2 by Nutlin3a and p53 by RITA. We have shown that RITA also affects p53's interaction with other proteins, such as the N-terminal binding p300 and iASPP or even the C-terminal binding Parc (Issaeva et al. 2004). These observations imply that RITA treatment may affect the global conformational folding of the p53 protein and thus alter its interaction with several partners. Likewise, the binding of Nutlin3a to Mdm2 promotes the displacement of several Mdm2 partner proteins such as E2F, p73, p21, and hnRNP K (Ambrosini et al. 2007; Enge et al. 2009; Rinaldo et al. 2009).

Thus, it appears that disruption of the p53/Mdm2 complex is not the sole effect of these compounds: binding of Nutlin3a changes the repertoire of Mdm2-binding partners, whereas RITA changes the p53 interactome. It is conceivable that these differences might be the cause of differential responses induced by p53 upon its release from Mdm2 by RITA or Nutlin3a.

Therefore, Nutlin3a and RITA may help to address the mechanisms governing the p53 decision process resulting in either growth arrest or apoptosis.

4.1 The Mdm2 Interactome Is Important for the Life/Death Decision

Following Nutlin3a or RITA treatment, a rapid release of Mdm2 is expected to form a pool of Mdm2 in cells. As depicted in Fig. 13.2, Mdm2 has a number of interacting proteins besides p53 (Bouska and Eischen 2009; Iwakuma and Lozano 2003; Manfredi 2010; Marine and Lozano 2010; Meek and Hupp 2010; Wade et al. 2010; Wang and Li 2010). Thus, it is possible that upon its displacement from p53, Nutlin3a-bound Mdm2 will affect their partners differently than free Mdm2 upon RITA treatment.

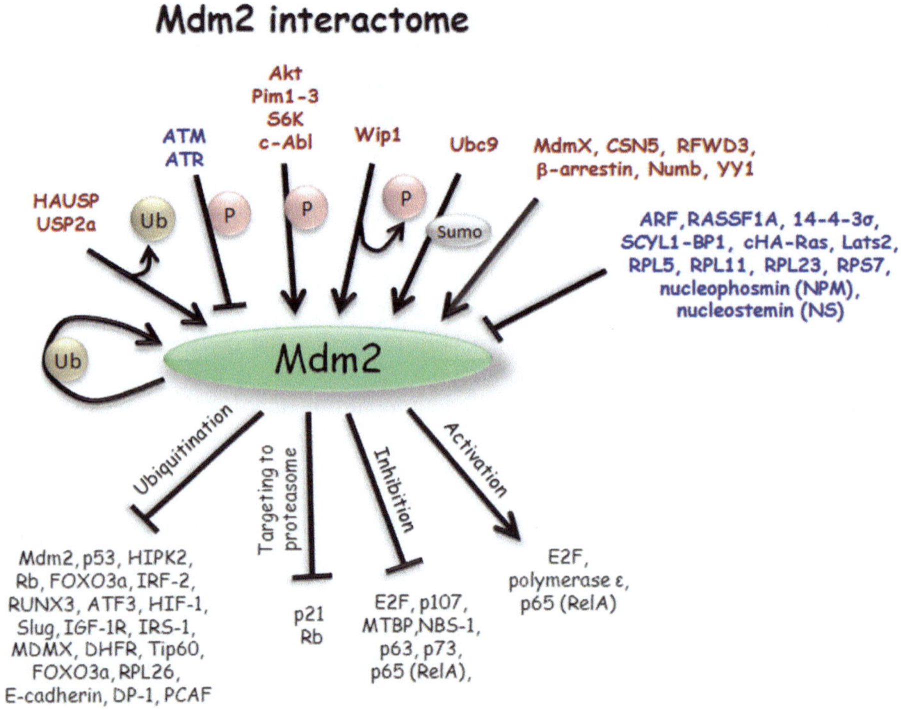

Fig. 13.2 Mdm2 interactome. A number of factors regulate Mdm2 and its interaction with p53, including enzymes such as kinases or phosphatases which modify Mdm2 thus enhancing its interaction with p53 (Akt, Pim1-3, S6K, Wip1) or inhibiting it (ATM, ATR, c-abl). De-ubiquitinating enzymes HAUSP and USP2a or SUMO ligase Ubc2 stabilize Mdm2, whereas a number of its interacting partners exert either positive or negative effects. *Red font* indicates positive regulators of Mdm2, *blue font* indicates negative ones. In turn, Mdm2 inhibits a number of proteins by promoting their degradation via direct ubiquitination or by targeting them to proteasome. Mdm2 can also regulate protein function in a positive or negative way via direct binding. Of note, E2F and p65 (Rel A) have been reported to be both activated and inhibited by Mdm2

We have tested this idea and found that in several cancer cell lines Mdm2 released from p53 upon RITA treatment promotes degradation of p21, the major p53 target responsible for cell cycle arrest (Enge et al. 2009). Moreover, Mdm2 also induces degradation of hnRNP K, a p53 cofactor required for the induction of p21 and other cell cycle arrest genes. As a result, the absence of robust p21 induction tilts the p53-triggered response toward apoptosis. In contrast, Mdm2 bound by Nutlin3a does not degrade hnRNP K or p21, leading to a strong p21 upregulation and induction of growth arrest (Enge et al. 2009), as depicted in Fig. 13.3. Transcription-independent induction of apoptosis by Nutlin3a in these cells, via the action of p53 at mitochondria (Marchenko and Moll 2007) might be reason why depletion of p21 does not enhance apoptosis induction by Nutlin3a in a number of cell lines (Xia et al. 2011). However, in AML cells impaired induction of p21 as a consequence of XIAP inhibition potentiates Nutlin3a-induced apoptosis, suggesting that the contribution of p21

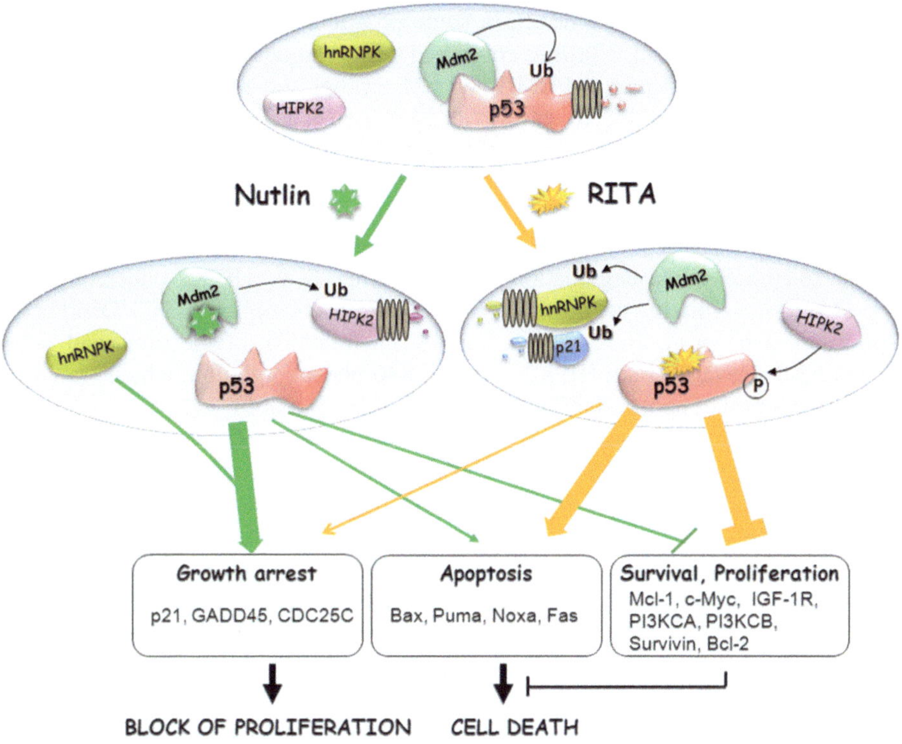

Fig. 13.3 Differential binding of Mdm2 to its substrates impinges on the biological response induced by p53. Upon treatment with Nutlin3a or RITA, Mdm2 differentially affects its substrate proteins which have impact on the biological response induced by these two compounds. *Right panel*: Mdm2 released from p53 upon RITA treatment promotes degradation of p21, the major p53 target responsible for cell cycle arrest. Mdm2 also induces degradation of hnRNP K, a p53 cofactor required for the induction of p21 and other cell cycle arrest genes. In the absence of robust induction of cell cycle arrest, the p53-induced response is shifted toward apoptosis. *Left panel*: In contrast, Mdm2 bound by Nutlin3a does not degrade hnRNP K or p21, leading to a strong p21 upregulation and induction of growth arrest. In addition, HIPK2 kinase (which phosphorylates p53 at Ser46—a modification required for the induction of pro-apoptotic genes by p53) is targeted for proteasomal degradation by Mdm2 released by Nutlin3a. In contrast, upon RITA treatment the level of HIPK2 and Ser46 phosphorylation of p53 is induced (*right panel*). The binding of Mdm2 to its partner proteins such as hnRNP K, p21, or HIPK2 can serve as a key switch regulating cell fate decision upon p53 reactivation

to the balance between growth arrest and apoptosis might be cell type-dependent (Carter et al. 2010).

Further, several lines of evidence were obtained demonstrating that inhibition of Mdm2 by small molecules affects its interaction with other substrates besides p21 and hnRNP K. Recent studies indicate that Nutlin3a disrupts the interaction between Mdm2 and its partner proteins E2F1 and p73 (Ambrosini et al. 2007; Kitagawa et al. 2008; Lau et al. 2008; Peirce and Findley 2009; Ray et al. 2011; Wang et al. 2011). Both factors interact with the N-terminal domain of Mdm2, where the Nutlin3a binding site is located. As a result, in some p53-defective tumor cell lines (both p53-null and p53-mutant) Nutlin3a combined with chemotherapeutic drugs induces the expression of p53 targets PUMA and p21 through the activation of E2F1 and p73 (Ambrosini et al. 2007; Kitagawa et al. 2008; Lau et al. 2008; Peirce and Findley 2009; Ray et al. 2011; Wang et al. 2011). Nutlin3a may therefore provide a therapeutic benefit in some tumors with mutant p53 when combined with chemotherapy (Ambrosini et al. 2007).

Another group discovered that Mdm2 released by Nutlin3a targets the HIPK2 kinase for proteasomal degradation (Rinaldo et al. 2009). HIPK2 phosphorylates p53 at Ser46, facilitating the induction of pro-apoptotic genes by p53 (D'Orazi et al. 2002; Hofmann et al. 2002). In contrast, RITA treatment of the same cell lines increased the level of both HIPK2 and p53-S46 phosphorylation, which lead to a potent induction of apoptosis (Fig. 13.3) (Rinaldo et al. 2009).

Interestingly, osteosarcoma SJSA and colon carcinoma RKO cell lines, which undergo apoptosis upon Nutlin3a treatment, are significantly less sensitive to RITA (Spinnler et al. 2011; Vassilev et al. 2004). The mechanism of the preferential induction of apoptosis by Nutlin3a in these cell lines awaits further investigation. It is plausible that the facilitation of the cytochrome C release from mitochondria by p53 in transcription-independent manner could influence the cellular outcome. As mentioned above, in some cell lines Nutlin3a-activated p53 induces apoptosis due to its interaction with Bcl-2 family proteins in the cytoplasm/mitochondria (Vaseva et al. 2009).

Further, MdmX plays an important role in the cellular response upon p53 reactivation. Recent studies indicate that the levels of MdmX determine the sensitivity of cells to Nutlin3a (Lam et al. 2010; Patton et al. 2006). More intriguing is the fact that the levels of both Mdm2 and MdmX eventually subside after RITA treatment (Spinnler et al. 2011; Yang et al. 2009). Although the mechanisms underlying the decreased levels of Mdm2 and MdmX are only partially understood, these studies suggest that certain branches of the negative feedback loop induced by p53 are relieved after RITA.

Further, it is possible that Mdm2, p53, and/or other cellular factors display different activities upon RITA and Nutlin3a due to posttranslational modifications induced by upstream pathways. We have found recently that the phosphatase Wip1, which removes activating marks from p53 and other proteins, is repressed upon RITA treatment (Spinnler et al. 2011). Further, p53-dependent DNA damage signaling is induced in cells by RITA (Ahmed et al. 2011; Yang et al. 2009). The mechanisms behind the p53-dependent DNA damage response triggered by RITA have not been fully elucidated and are currently under investigation. Prior to the discovery of p53-dependent effects of RITA, earlier studies suggested that cross-linking ability is a critical determinant for cancer cell sensitivity to this compound (Nieves-Neira et al. 1999; Rivera et al. 1999). However, the p53-dependence and nondetectable DSBs upon RITA treatment are intriguing and call for further investigation (Ahmed et al. 2011; Yang et al. 2009). Though little has been reported about the DMPK properties of either Nutlin3a or RITA (Phillip et al. 2002), it may prove vital to assess to which extent the differential uptake or metabolism may account for differences in the cellular responses.

In summary, a number of studies suggest that the Mdm2 interactome altered by small molecules disrupting p53/Mdm2 complex can contribute to the cell fate decisions upon p53 reactivation (Figs. 13.2 and 13.3). Proteomics and functional studies will be required in a future to elucidate all Mdm2 partners that can enhance the induction of apoptosis in cancer cells upon p53 reactivation.

4.2 Ablation of Survival Signaling Is an Important Component of p53-Induced Apoptosis

One of the major mechanisms which can render cells resistant to cell death is overexpression of factors blocking apoptosis downstream of p53, such as anti-apoptotic Mcl-1, Bcl-2, or other survival factors. It is believed that when survival signals prevail, p53 activation will more likely result in growth arrest (Lowe et al. 2004; Oren 2003).

Our studies reveal the ability of p53 to unleash transcriptional repression of major proliferative and survival factors as a potent aspect of the p53 response. p53 activation by RITA results in transcriptional repression of several proliferative and survival factors including key components of the IGFR–PI3K–Akt pathway. Rapid decline of crucial oncogenes, including c-Myc, Mcl-1, and survivin, due to their inhibition at several levels , along with blocking the Akt pathway, result in acute ablation of pro-proliferative and survival pathways (Grinkevich et al. 2009). This plays a critical role in p53-induced apoptosis by reducing the cancer cell's ability to buffer pro-apoptotic signals. The multitude of oncogenes, inhibited by RITA-reactivated p53 at multiple levels creates a robust p53 response. This redundancy might allow p53 to cope with the daunting challenge of anticancer therapy—multiple genetic abnormalities in individual cancers. As an example, we show the pleiotropic effect of RITA-reactivated p53 on c-Myc: repression of c-Myc transcription, block of its translation, and induction of proteasomal degradation. It is possible that such redundancy has evolved to secure downregulation of c-Myc irrespective of the particular type of c-Myc deregulation in a given cancer cell. Thus, if one of the mechanisms of c-Myc inhibition by p53 will be blocked, it will be compensated by other branches in the hierarchy (Grinkevich et al. 2009). Since tumors are often "addicted" to deregulated c-Myc expression (Felsher and Bishop 1999; Savino et al. 2011; Soucek et al. 2008), its inhibition might be an essential component of anticancer therapies targeting p53.

Although Nutlin3a treatment of the same cancer cell lines failed to show the same degree of repressing oncogenes, combination of Nutlin3a with inhibition of Mcl-1 and c-Myc acts synergistically to induce an apoptotic response.

Notably, it appears that the induction of pro-apoptotic genes and inhibition of oncogenes are differentially regulated by p53 (Grinkevich et al. 2009). These two branches of p53-mediated apoptosis may require distinct cofactors and/or posttranslational modifications and create an additional level regulating the choice between life and death upon p53 activation.

Thus, the ability of reactivated p53 to repress transcription, induce degradation, and block translation of several important oncogenes adds a new dimension to the mechanism of tumor suppression upon p53 activation by small molecules.

5 Small Molecules Simultaneously Targeting Wild-Type and Mutant p53

Mdm2-binding compounds such as Nutlin3a and MI-219 have been shown to be efficient mainly in tumors with wild-type p53, but not with mutant p53. A recent study of a large cohort of primary CLL samples found a direct correlation between wild-type *TP53* status and Nutlin3a or MI-219-induced cytotoxicity (Saddler et al. 2008). This correlation was, however, not as clear in an analysis of 60 NCI cell lines (Liu et al. 2009). As discussed above, despite the inability of Nutlin3a to restore function to mutant p53, it can induce growth suppression in some p53-null or mutant p53-carrying tumors by activating E2F and p73 in combination with cytotoxic drugs (Ambrosini et al. 2007; Kitagawa et al. 2008; Lau et al. 2008; Peirce and Findley 2009; Ray et al. 2011; Wang et al. 2011).

The evidence suggesting that RITA treatment impinges on p53 conformation inspired us to test whether RITA is capable of restoring mutant p53 activity. We found that RITA suppresses the growth and induces apoptosis in a variety of mutant p53 carrying cell lines of different origin in a p53-dependent manner (Zhao et al. 2010a). This is accompanied by the induction of the known p53 target genes, p21, BAX, PUMA, and GADD45, suggesting that mutant p53 restored by RITA is functional (Zhao et al. 2010a). These findings are supported by studies showing that RITA is a potent inducer of apoptosis in CLL and AML patient samples that expressed mutant p53 (Nahi et al. 2008), as well as tests in head and neck squamous cell carcinoma cell lines (Roh et al. 2011). Altogether, these studies promote the idea of developing compounds capable of simultaneously targeting wild-type and mutant p53 in a broad set of tumors.

In line with this notion, several compounds which have been shown to restore the function of mutant p53 by interaction with its DNA-binding domain (DBD) also reactivate wild-type p53 by interfering with Mdm2-mediated degradation. These compounds include the synthetic peptide CDB3 and small molecules CP-31398 and SCH529074 (Demma et al. 2010). It could be hypothesized that the binding of these molecules to the DBD induces a conformational block that prevents the docking of the central domain of Mdm2. Since the efficient ubiquitination of p53 by Mdm2 requires the interaction of Mdm2 with both the N-terminus and DBD of p53 (Wallace et al. 2006), preventing DBD/Mdm2 interaction may be sufficient to impede the inhibitory effects of Mdm2 (Demma et al. 2010). It is also possible that these compounds might act via blocking degradation-independent mechanisms of Mdm2-mediated inhibition of p53. Mdm2 can prevent p53/DNA interaction by inducing a partial unfolding of the DBD, leading to mutant-like conformation (Cross et al. 2011). Thus, restoration of conformation to Mdm2-bound p53 might allow p53 to bind DNA and regulate transcription. Further studies will be required to test these hypotheses.

The mechanisms of activation of wild-type p53 by two other mutant p53-reactivating molecules, PRIMA-1^{Met} (APR-246) (Bykov et al. 2002) and p53R3 (Weinmann et al. 2008), are currently less clear. p53R3 enhances the recruitment of both wild-type and mutant p53 to target promoters and triggers the expression of a number of p53 target genes (Weinmann et al. 2008), but its effects on wild-type p53 stability and/or binding to Mdm2 have not been tested so far.

Intriguingly, we have recently found that PRIMA-1^{MET}/APR-246 activates p53 transcriptional activity and induces p53-dependent apoptosis in wild-type p53 carrying melanoma cells in vitro and in vivo without apparent impact on p53 level or stability (Bao et al. 2011). Previously, we reported that PRIMA-1^{MET}/APR-246 can induce apoptosis in AML patient samples with wild-type p53 (Nahi et al. 2008). However, PRIMA-1^{MET}/APR-246 does not appear to affect Mdm2 or E6-mediated ubiquitination of p53 (Demma et al. 2010). It has been put forward that alkylation of p53 by PRIMA-1^{MET}/APR-246 (Lambert et al. 2009) may play a causative role, but the exact mechanism awaits further investigation. This is further discussed by Shen et al. in Chap. 14.

Notably, a novel Mdm2-independent mechanism of wild-type p53 inactivation has been recently proposed. It occurs via partial unfolding of the DBD upon depletion of HIPK2 or phosphorylation of Ser269 (Fraser et al. 2010a, b; Puca et al. 2008). It is therefore conceivable that the rescue of wild-type p53 by mutant p53-reactivating drugs might occur via prevention of p53 unfolding caused by other factors than Mdm2.

In conclusion, compounds which directly target p53 mentioned above might be applicable for the treatment of a wider range of tumors, irrespective of mutant or wild-type p53 status. In most cases, however, the mechanism remains elusive and awaits further investigation. A detailed elucidation of the structural features of the full-length p53 may be helpful in order to improve the understanding of the mechanism of action of p53-reactivating compounds. Due to its high conformational flexibility, p53 has proven to be a challenge for structural biologists to date. However, the first important milestones have been achieved in this direction (Okorokov et al. 2006; Tidow et al. 2007).

6 Drug Resistance and p53-Reactivating Compounds

One of the main causes of failure of anticancer treatments is the development of de novo drug resistance by cancer cells. This is a very serious clinical problem that entails disease recurrence which may lead to patients' deaths. Tumors often acquire resistance not only to the initially applied drugs, but simultaneously to others they have not been exposed to before. Multiple mechanisms contribute to drug resistance, including among others: mutation in the drug target itself or its upstream regulators/downstream effectors, loss of cell surface receptors or transporters, and changes in drug metabolism (Gottesman et al. 2002).

Treatment with Mdm2-targeting compounds may create a selective pressure for p53 mutations as a mechanism of de novo resistance. In addition, a recent study of AML samples from over 100 patients indicated a preexisting, primary resistance to Mdm2 inhibitors, despite the presence of wild-type *TP53* (Long et al. 2010). Absence of a response seems to be due to the insufficient p53 induction or compromised p53-dependent apoptosis. These observations complicate the translation of Mdm2 inhibitors into the clinic.

Clearly, translating the use of Nutlin3a and other p53-reactivating compounds to the clinic will be very much promoted by identifying molecular pathways that confer sensitivity or resistance to treatment. shRNA screens could be a very valuable approach, allowing to expose nodes whose inactivation can lead to the loss or gain of sensitivity to the drug. Brummelkamp et al. performed a genome-wide shRNA screen to gain insight into the mechanism of Nutlin3a action (Brummelkamp et al. 2006). Inhibition of 14 genes resulted in the resistance of MCF7 cells to Nutlin3a, including primarily p53 itself and its regulators 53BP1 and hnRNP K, several p53 transcription targets such as the Bcl-2 antagonist BAD. Other factors previously not implicated in the regulation of the p53 pathway such as TIEG (TGFβ pathway) and USP28 (ubiquitin-specific protease) were also revealed (Brummelkamp et al. 2006). These findings are very intriguing, but further shRNA screens will need to be performed in several cancer cell lines of different origin using various molecules activating p53 in order to boost the understanding of how to use p53-reactivating compounds with the highest efficiency and to avoid resistance.

7 Combinational Treatment

To minimize drug resistance, cocktails of drugs with different mechanisms of action and entry into cells are applied. The use of drug combinations is a well-established pharmaceutical principle, based on the knowledge that the probability of a given cell being simultaneously resistant to a combination of drugs is much lower than to each of the individual components. The choice of drugs to combine might be based on the knowledge of molecular pathways and their interactions and/or empirical testing. For example, systematically studying combinations of active agents for additive or synergistic effects may identify the most effective combinations.

Consequently, it will be a future challenge to find most potent combinational treatments that synergize with p53 activation in tumors without putting normal cells at risk. Recent studies explored this possibility using combinations of Nutlin3a and chemotherapeutic drugs. Nutlin3a has been shown to synergize with a wide variety of cytotoxic agents including those targeting tubulin (vinblastine), DNA-damaging agents (doxorubicin), S-phase agents (fludarabine and gemcitabine), and radiation (reviewed in Cheok et al. 2011). The pitfall of combining specific p53 activation with conventional chemotherapy is, however, the narrowing therapeutic window, since such combinations might stimulate p53 activity to a greater extent in normal tissues as well, leading to increased side effects. In fact, the involvement of p53 in tissue toxicity upon chemotherapy prompted researchers to develop a somewhat paradoxical approach: to minimize side effects in normal tissues by small molecules inhibiting p53 (Gudkov and Komarova 2010). Data obtained in mouse models support this idea; the benefit of this approach in a clinical setting awaits further investigation.

It seems worthwhile to explore the combinations of p53-reactivating compounds with those treatments that specifically target other cancer-related pathways, but could display synergistic or synthetic lethal effects in tumors upon pharmacological activation of p53. Synergistic effects of combinations of Nutlin3a with several target-specific drugs have been recently demonstrated. These include cyclin-dependent kinase inhibitors (roscovitine), aurora kinase inhibitors (VX-6807), and tumor necrosis factor α-related apoptosis-inducing ligand (TRAIL) in AML (reviewed in Cheok et al. 2011). Moreover, another concept of applying drug combinations involving p53-reactivating compounds is cyclotherapy (Lane et al. 2010). It is possible that the induction of reversible growth arrest by p53-activating drugs in normal cells will protect them from antimitotic drugs, thus selectively exposing tumor cells with mutant p53 to the effects of S-phase agents or antimitotic drugs. Again, this concept awaits detailed investigation.

A parallel strategy to kill resistance in its cradle is the application of "dirty drugs" that affect multiple oncogenic targets at once, but still display selectivity toward tumors. In fact, it is very likely that many small molecules will have more than one target, which can produce undesirable off-target effects. On the other hand, off-target effect might be beneficial, as it was for example shown for Gleevec (imatinib), which in addition to Bcr-abl also targets c-Kit, allowing its application for treatment of GIST tumors (Heinrich et al. 2000). Interestingly, we found that RITA inhibits TrxR1 activity in vitro and in cells (Hedstrom et al. 2009), thereby challenging the cellular redox systems. Although as already mentioned above, RITA has been suggested to produce cross-links (Nieves-Neira et al. 1999), RITA incites the DNA damage response only in p53-positive HCT116 *TP53*$^{+/+}$, but not in p53-null HCT116 *TP53*$^{-/-}$ cells (Yang et al. 2009). Notably, inhibition of TrxR and induction of DNA damage signaling are not observed in nontransformed cells, nor is there a significant increase of γH2AX foci in nontransformed cells (Hedstrom et al. 2009; Yao Shi, Fedor Nikulenkov, and Galina Selivanova, manuscript in preparation), suggesting that RITA-induced γH2AX depends on the presence of p53 and activated oncogenes. The underlying mechanism calls for detailed investigation, which possibly will inspire the design of novel compounds with similar properties.

In a clinical setting, it may prove pivotal to stratify patient samples by screening their sensitivity to all available anticancer drugs and their combinations. This would allow to determine the optimal combination of drugs and may be a first step toward customized cancer treatment that aims to achieve the maximum response.

8 Concluding Remarks

Our growing understanding of p53 biology has led to the discovery of a number of p53-activating drugs, including those targeting p53 directly. However, we still have much to learn, as the exact molecular mechanisms of action of many compounds are still unclear. In particular, it is very intriguing how p53-binding compounds can restore the function of both mutant and wild-type p53.

Further, it is difficult to predict the cellular response to p53 activation by different compounds, which can range from cell cycle arrest to apoptosis. Deciphering the underlying mechanisms may facilitate the development of novel drugs and/or combinations of existing drugs to achieve synergy and avoid resistance. In particular, existing compounds can serve as valuable research tools to study p53 biology.

The idea of p53-targeting therapy is very attractive and has met important milestones on its way to a successful translation to the clinic. The development of tumor-specific drugs is the art of cancer therapy and we believe that reinstatement of p53 function in tumors has the potential to constitute the core of future cancer therapy.

References

Ahmed A, Yang J, Maya-Mendoza A, Jackson DA, Ashcroft M (2011) Pharmacological activation of a novel p53-dependent S-phase checkpoint involving CHK-1. Cell Death Dis 2:e160

Ambrosini G, Sambol EB, Carvajal D, Vassilev LT, Singer S, Schwartz GK (2007) Mouse double minute antagonist Nutlin-3a enhances chemotherapy-induced apoptosis in cancer cells with mutant p53 by activating E2F1. Oncogene 26:3473–3481

Bao W, Chen M, Zhao X, Kumar R, Spinnler C, Thullberg M, Issaeva N, Selivanova G, Stromblad S (2011) PRIMA-1 (Met)/APR-246 induces wild-type p53-dependent suppression of malignant melanoma tumor growth in 3D culture and in vivo. Cell Cycle 10:301–307

Bernal F, Tyler AF, Korsmeyer SJ, Walensky LD, Verdine GL (2007) Reactivation of the p53 tumor suppressor pathway by a stapled p53 peptide. J Am Chem Soc 129:2456–2457

Bernal F, Wade M, Godes M, Davis TN, Whitehead DG, Kung AL, Wahl GM, Walensky LD (2010) A stapled p53 helix overcomes HDMX-mediated suppression of p53. Cancer Cell 18:411–422

Bottger A, Bottger V, Sparks A, Liu WL, Howard SF, Lane DP (1997) Design of a synthetic Mdm2-binding mini protein that activates the p53 response in vivo. Curr Biol 7:860–869

Bouska A, Eischen CM (2009) Murine double minute 2: p53-independent roads lead to genome instability or death. Trends Biochem Sci 34:279–286

Brummelkamp TR, Fabius AW, Mullenders J, Madiredjo M, Velds A, Kerkhoven RM, Bernards R, Beijersbergen RL (2006) An shRNA barcode screen provides insight into cancer cell vulnerability to MDM2 inhibitors. Nat Chem Biol 2:202–206

Bykov VJ, Issaeva N, Shilov A, Hultcrantz M, Pugacheva E, Chumakov P, Bergman J, Wiman KG, Selivanova G (2002) Restoration of the tumor suppressor function to mutant p53 by a low-molecular-weight compound. Nat Med 8:282–288

Carter BZ, Mak DH, Schober WD, Koller E, Pinilla C, Vassilev LT, Reed JC, Andreeff M (2010) Simultaneous activation of p53 and inhibition of XIAP enhance the activation of apoptosis signaling pathways in AML. Blood 115:306–314

Chene P, Fuchs J, Bohn J, Garcia-Echeverria C, Furet P, Fabbro D (2000) A small synthetic peptide, which inhibits the p53-hdm2 interaction, stimulates the p53 pathway in tumour cell lines. J Mol Biol 299:245–253

Cheok CF, Verma CS, Baselga J, Lane DP (2011) Translating p53 into the clinic. Nat Rev Clin Oncol 8:25–37

Cross B, Chen L, Cheng Q, Li B, Yuan ZM, Chen J (2011) Inhibition of p53 DNA binding function by the MDM2 protein acidic domain. J Biol Chem 286:16018–16029

D'Orazi G, Cecchinelli B, Bruno T, Manni I, Higashimoto Y, Saito S, Gostissa M, Coen S, Marchetti A, Del Sal G et al (2002) Homeodomain-interacting protein kinase-2 phosphorylates p53 at Ser 46 and mediates apoptosis. Nat Cell Biol 4:11–19

Danovi D, Meulmeester E, Pasini D, Migliorini D, Capra M, Frenk R, de Graaf P, Francoz S, Gasparini P, Gobbi A et al (2004) Amplification of Mdmx (or Mdm4) directly contributes to tumor formation by inhibiting p53 tumor suppressor activity. Mol Cell Biol 24:5835–5843

Dawson R, Muller L, Dehner A, Klein C, Kessler H, Buchner J (2003) The N-terminal domain of p53 is natively unfolded. J Mol Biol 332:1131–1141

Demma M, Maxwell E, Ramos R, Liang L, Li C, Hesk D, Rossman R, Mallams A, Doll R, Liu M et al (2010) SCH529074, a small molecule activator of mutant p53, which binds p53 DNA binding domain (DBD), restores growth-suppressive function to mutant p53 and interrupts HDM2-mediated ubiquitination of wild type p53. J Biol Chem 285:10198–10212

Dymalla S, Scheffner M, Weber E, Sehr P, Lohrey C, Hoppe-Seyler F, Hoppe-Seyler K (2009) A novel peptide motif binding to and blocking the intracellular activity of the human papillomavirus E6 oncoprotein. J Mol Med 87:321–331

Edelman J, Nemunaitis J (2003) Adenoviral p53 gene therapy in squamous cell cancer of the head and neck region. Curr Opin Mol Ther 5:611–617

Enge M, Bao W, Hedstrom E, Jackson SP, Moumen A, Selivanova G (2009) MDM2-dependent downregulation of p21 and hnRNP K provides a switch between apoptosis and growth arrest induced by pharmacologically activated p53. Cancer Cell 15:171–183

Feldser DM, Kostova KK, Winslow MM, Taylor SE, Cashman C, Whittaker CA, Sanchez-Rivera FJ, Resnick R, Bronson R, Hemann MT, Jacks T (2010) Stage-specific sensitivity to p53 restoration during lung cancer progression. Nature 468:572–575

Felsher DW, Bishop JM (1999) Reversible tumorigenesis by MYC in hematopoietic lineages. Mol Cell 4:199–207

Foster BA, Coffey HA, Morin MJ, Rastinejad F (1999) Pharmacological rescue of mutant p53 conformation and function. Science 286: 2507–2510

Fraser JA, Madhumalar A, Blackburn E, Bramham J, Walkinshaw MD, Verma C, Hupp TR (2010a) A novel p53 phosphorylation site within the MDM2 ubiquitination signal: II. a model in which phosphorylation at SER269 induces a mutant conformation to p53. J Biol Chem 285:37773–37786

Fraser JA, Vojtesek B, Hupp TR (2010b) A novel p53 phosphorylation site within the MDM2 ubiquitination signal: I. phosphorylation at SER269 in vivo is linked to inactivation of p53 function. J Biol Chem 285:37762–37772

Friedler A, Hansson LO, Veprintsev DB, Freund SM, Rippin TM, Nikolova PV, Proctor MR, Rudiger S, Fersht AR (2002) A peptide that binds and stabilizes p53 core domain: chaperone strategy for rescue of oncogenic mutants. Proc Natl Acad Sci USA 99:937–942

Gilkes DM, Chen L, Chen J (2006) MDMX regulation of p53 response to ribosomal stress. EMBO J 25:5614–5625

Goodwin EC, DiMaio D (2000) Repression of human papillomavirus oncogenes in HeLa cervical carcinoma cells causes the orderly reactivation of dormant tumor suppressor pathways. Proc Natl Acad Sci USA 97:12513–12518

Gottesman MM, Fojo T, Bates SE (2002) Multidrug resistance in cancer: role of ATP-dependent transporters. Nat Rev Cancer 2:48–58

Grinkevich VV, Nikulenkov F, Shi Y, Enge M, Bao W, Maljukova A, Gluch A, Kel A, Sangfelt O, Selivanova G (2009) Ablation of key oncogenic pathways by RITA-reactivated p53 is required for efficient apoptosis. Cancer Cell 15:441–453

Gudkov AV, Komarova EA (2010) Radioprotection: smart games with death. J Clin Invest 120:2270–2273

Haigis MC, Guarente LP (2006) Mammalian sirtuins–emerging roles in physiology, aging, and calorie restriction. Genes Dev 20:2913–2921

Hainaut P, Wiman K (2005) 25 years of p53 research. Springer, Dordrecht

Hedstrom E, Eriksson S, Zawacka-Pankau J, Arner ES, Selivanova G (2009) p53-dependent inhibition of TrxR1 contributes to the tumor-specific induction of apoptosis by RITA. Cell Cycle 8:3576–3583

Heinrich MC, Griffith DJ, Druker BJ, Wait CL, Ott KA, Zigler AJ (2000) Inhibition of c-kit receptor tyrosine kinase activity by STI 571, a selective tyrosine kinase inhibitor. Blood 96:925–932

Hengstermann A, Linares LK, Ciechanover A, Whitaker NJ, Scheffner M (2001) Complete switch from Mdm2 to human papillomavirus E6-mediated degradation of p53 in cervical cancer cells. Proc Natl Acad Sci USA 98:1218–1223

Hietanen S, Lain S, Krausz E, Blattner C, Lane DP (2000) Activation of p53 in cervical carcinoma cells by small molecules. Proc Natl Acad Sci USA 97:8501–8506

Hofmann TG, Moller A, Sirma H, Zentgraf H, Taya Y, Droge W, Will H, Schmitz ML (2002) Regulation of p53 activity by its interaction with homeodomain-interacting protein kinase-2. Nat Cell Biol 4:1–10

Hu B, Gilkes DM, Chen J (2007) Efficient p53 activation and apoptosis by simultaneous disruption of binding to MDM2 and MDMX. Cancer Res 67:8810–8817

Hu B, Gilkes DM, Farooqi B, Sebti SM, Chen J (2006) MDMX overexpression prevents p53 activation by the MDM2 inhibitor Nutlin. J Biol Chem 281:33030–33035

Issaeva N, Bozko P, Enge M, Protopopova M, Verhoef LG, Masucci M, Pramanik A, Selivanova G (2004) Small molecule RITA binds to p53, blocks p53-HDM-2 interaction and activates p53 function in tumors. Nat Med 10:1321–1328

Issaeva N, Friedler A, Bozko P, Wiman KG, Fersht AR, Selivanova G (2003) Rescue of mutants of the tumor suppressor p53 in cancer cells by a designed peptide. Proc Natl Acad Sci USA 100:13303–13307

Iwakuma T, Lozano G (2003) MDM2, an introduction. Mol Cancer Res 1:993–1000

Jiang M, Milner J (2002) Selective silencing of viral gene expression in HPV-positive human cervical carcinoma cells treated with siRNA, a primer of RNA interference. Oncogene 21:6041–6048

Junttila MR, Karnezis AN, Garcia D, Madriles F, Kortlever RM, Rostker F, Brown Swigart L, Pham DM, Seo Y, Evan GI, Martins CP (2010) Selective activation of p53-mediated tumour suppression in high-grade tumours. Nature 468:567–571

Kelley ML, Keiger KE, Lee CJ, Huibregtse JM (2005) The global transcriptional effects of the human papillomavirus E6 protein in cervical carcinoma cell lines are mediated by the E6AP ubiquitin ligase. J Virol 79:3737–3747

Kitagawa M, Aonuma M, Lee SH, Fukutake S, McCormick F (2008) E2F-1 transcriptional activity is a critical determinant of Mdm2 antagonist-induced apoptosis in human tumor cell lines. Oncogene 27:5303–5314

Kochetkov DV, Il'inskaia GV, Komarov PG, Strom E, Agapova LS, Ivanov AV, Budanov AV, Frolova EI, Chumakov PM (2007) Transcriptional inhibition of human papilloma virus in cervical carcinoma cells reactivates functions of the tumor suppressor p53. Mol Biol (Mosk) 41:515–523

Lain S, Hollick JJ, Campbell J, Staples OD, Higgins M, Aoubala M, McCarthy A, Appleyard V, Murray KE, Baker L et al (2008) Discovery, in vivo activity, and mechanism of action of a small-molecule p53 activator. Cancer Cell 13:454–463

Lam S, Lodder K, Teunisse AF, Rabelink MJ, Schutte M, Jochemsen AG (2010) Role of Mdm4 in drug sensitivity of breast cancer cells. Oncogene 29:2415–2426

Lambert JM, Gorzov P, Veprintsev DB, Soderqvist M, Segerback D, Bergman J, Fersht AR, Hainaut P, Wiman KG, Bykov VJ (2009) PRIMA-1 reactivates mutant p53 by covalent binding to the core domain. Cancer Cell 15:376–388

Lane DP, Cheok CF, Lain S (2010) p53-based cancer therapy. Cold Spring Harb Perspect Biol 2:a001222

Lau LM, Nugent JK, Zhao X, Irwin MS (2008) HDM2 antagonist Nutlin-3 disrupts p73-HDM2 binding and enhances p73 function. Oncogene 27:997–1003

Laurie NA, Donovan SL, Shih CS, Zhang J, Mills N, Fuller C, Teunisse A, Lam S, Ramos Y, Mohan A et al (2006) Inactivation of the p53 pathway in retinoblastoma. Nature 444:61–66

Le Guezennec X, Bulavin DV (2010) WIP1 phosphatase at the crossroads of cancer and aging. Trends Biochem Sci 35:109–114

Levine AJ, Oren M (2009) The first 30 years of p53: growing ever more complex. Nat Rev Cancer 9:749–758

Linares LK, Hengstermann A, Ciechanover A, Muller S, Scheffner M (2003) HdmX stimulates Hdm2-mediated ubiquitination and degradation of p53. Proc Natl Acad Sci USA 100: 12009–12014

Linke K, Mace PD, Smith CA, Vaux DL, Silke J, Day CL (2008) Structure of the MDM2/MDMX RING domain heterodimer reveals dimerization is required for their ubiquitylation in trans. Cell Death Differ 15:841–848

Liu W, He L, Ramirez J, Ratain MJ (2009) Interactions between MDM2 and TP53 genetic alterations, and their impact on response to MDM2 inhibitors and other chemotherapeutic drugs in cancer cells. Clin Cancer Res 15:7602–7607

Long J, Parkin B, Ouillette P, Bixby D, Shedden K, Erba H, Wang S, Malek SN (2010) Multiple distinct molecular mechanisms influence sensitivity and resistance to MDM2 inhibitors in adult acute myelogenous leukemia. Blood 116:71–80

Lowe SW, Cepero E, Evan G (2004) Intrinsic tumour suppression. Nature 432:307–315

Luu Y, Bush J, Cheung KJ Jr, Li G (2002) The p53 stabilizing compound CP-31398 induces apoptosis by activating the intrinsic Bax/mitochondrial/caspase-9 pathway. Exp Cell Res 276:214–222

Manfredi JJ (2010) The Mdm2-p53 relationship evolves: Mdm2 swings both ways as an oncogene and a tumor suppressor. Genes Dev 24:1580–1589

Marchenko ND, Moll UM (2007) The role of ubiquitination in the direct mitochondrial death program of p53. Cell Cycle 6:1718–1723

Marine JC, Lozano G (2010) Mdm2-mediated ubiquitylation: p53 and beyond. Cell Death Differ 17:93–102

Martins CP, Brown-Swigart L, Evan GI (2006) Modeling the therapeutic efficacy of p53 restoration in tumors. Cell 127:1323–1334

Meek DW, Hupp TR (2010) The regulation of MDM2 by multisite phosphorylation—opportunities for molecular-based intervention to target tumours? Semin Cancer Biol 20:19–28

Momand J, Zambetti GP, Olson DC, George D, Levine AJ (1992) The mdm-2 oncogene product forms a complex with the p53 protein and inhibits p53-mediated transactivation. Cell 69:1237–1245

Nahi H, Selivanova G, Lehmann S, Mollgard L, Bengtzen S, Concha H, Svensson A, Wiman KG, Merup M, Paul C (2008) Mutated and non-mutated TP53 as targets in the treatment of leukaemia. Br J Haematol 141:445–453

Nieves-Neira W, Rivera MI, Kohlhagen G, Hursey ML, Pourquier P, Sausville EA, Pommier Y (1999) DNA protein cross-links produced by NSC 652287, a novel thiophene derivative active against human renal cancer cells. Mol Pharmacol 56:478–484

Okorokov AL, Sherman MB, Plisson C, Grinkevich V, Sigmundsson K, Selivanova G, Milner J, Orlova EV (2006) The structure of p53 tumour suppressor protein reveals the basis for its functional plasticity. EMBO J 25:5191–5200

Oren M (2003) Decision making by p53: life, death and cancer. Cell Death Differ 10:431–442

Patton JT, Mayo LD, Singhi AD, Gudkov AV, Stark GR, Jackson MW (2006) Levels of HdmX expression dictate the sensitivity of normal and transformed cells to Nutlin-3. Cancer Res 66:3169–3176

Pazgier M, Liu M, Zou G, Yuan W, Li C, Li C, Li J, Monbo J, Zella D, Tarasov SG, Lu W (2009) Structural basis for high-affinity peptide inhibition of p53 interactions with MDM2 and MDMX. Proc Natl Acad Sci USA 106:4665–4670

Peirce SK, Findley HW (2009) The MDM2 antagonist nutlin-3 sensitizes p53-null neuroblastoma cells to doxorubicin via E2F1 and TAp73. Int J Oncol 34:1395–1402

Petitjean A, Achatz MI, Borresen-Dale AL, Hainaut P, Olivier M (2007) TP53 mutations in human cancers: functional selection and impact on cancer prognosis and outcomes. Oncogene 26:2157–2165

Phillip LR, Jorden JL, Rivera MI, Wolfe TL, Upadhyayb K, Stinson SF (2002) Identification of the major metabolite of 2,5-bis(5-hydroxymethyl-2-thienyl)furan (NSC 652287), an antitumor agent, in the S9 subcellular fraction of dog liver cells. J Chromatogr B Analyt Technol Biomed Life Sci 767:27–33

Puca R, Nardinocchi L, Gal H, Rechavi G, Amariglio N, Domany E, Notterman DA, Scarsella M, Leonetti C, Sacchi A et al (2008) Reversible dysfunction of wild-type p53 following homeodomain-interacting protein kinase-2 knockdown. Cancer Res 68:3707–3714

Ray RM, Bhattacharya S, Johnson LR (2011) Mdm2 inhibition induces apoptosis in p53 deficient human colon cancer cells by activating p73- and E2F1-mediated expression of PUMA and Siva-1. Apoptosis 16:35–44

Reed D, Shen Y, Shelat AA, Arnold LA, Ferreira AM, Zhu F, Mills N, Smithson DC, Regni CA, Bashford D et al (2010) Identification and characterization of the first small molecule inhibitor of MDMX. J Biol Chem 285:10786–10796

Rinaldo C, Prodosmo A, Siepi F, Moncada A, Sacchi A, Selivanova G, Soddu S (2009) HIPK2 regulation by MDM2 determines tumor cell response to the p53-reactivating drugs nutlin-3 and RITA. Cancer Res 69:6241–6248

Ringshausen I, O'Shea CC, Finch AJ, Swigart LB, Evan GI (2006) Mdm2 is critically and continuously required to suppress lethal p53 activity in vivo. Cancer Cell 10:501–514

Rivera MI, Stinson SF, Vistica DT, Jorden JL, Kenney S, Sausville EA (1999) Selective toxicity of the tricyclic thiophene NSC 652287 in renal carcinoma cell lines: differential accumulation and metabolism. Biochem Pharmacol 57:1283–1295

Roh JL, Kang SK, Minn I, Califano JA, Sidransky D, Koch WM (2011) p53-Reactivating small molecules induce apoptosis and enhance chemotherapeutic cytotoxicity in head and neck squamous cell carcinoma. Oral Oncol 47:8–15

Saddler C, Ouillette P, Kujawski L, Shangary S, Talpaz M, Kaminski M, Erba H, Shedden K, Wang S, Malek SN (2008) Comprehensive biomarker and genomic analysis identifies p53 status as the major determinant of response to MDM2 inhibitors in chronic lymphocytic leukemia. Blood 111:1584–1593

Savino M, Annibali D, Carucci N, Favuzzi E, Cole MD, Evan GI, Soucek L, Nasi S (2011) The action mechanism of the myc inhibitor termed omomyc may give clues on how to target myc for cancer therapy. PLoS One 6:e22284

Scheffner M, Huibregtse JM, Vierstra RD, Howley PM (1993) The HPV-16 E6 and E6-AP complex functions as a ubiquitin-protein ligase in the ubiquitination of p53. Cell 75:495–505

Scheffner M, Werness BA, Huibregtse JM, Levine AJ, Howley PM (1990) The E6 oncoprotein encoded by human papillomavirus types 16 and 18 promotes the degradation of p53. Cell 63:1129–1136

Selivanova G (2010) Therapeutic targeting of p53 by small molecules. Semin Cancer Biol 20:46–56

Shangary S, Qin D, McEachern D, Liu M, Miller RS, Qiu S, Nikolovska-Coleska Z, Ding K, Wang G, Chen J et al (2008) Temporal activation of p53 by a specific MDM2 inhibitor is selectively toxic to tumors and leads to complete tumor growth inhibition. Proc Natl Acad Sci USA 105:3933–3938

Soucek L, Whitfield J, Martins CP, Finch AJ, Murphy DJ, Sodir NM, Karnezis AN, Swigart LB, Nasi S, Evan GI (2008) Modelling Myc inhibition as a cancer therapy. Nature 455:679–683

Spinnler C, Hedstrom E, Li H, de Lange J, Nikulenkov F, Teunisse AF, Verlaan-de Vries M, Grinkevich V, Jochemsen AG, Selivanova G (2011) Abrogation of Wip1 expression by RITA-activated p53 potentiates apoptosis induction via activation of ATM and inhibition of HdmX. Cell Death Differ 18(11):1736–1745

Talis AL, Huibregtse JM, Howley PM (1998) The role of E6AP in the regulation of p53 protein levels in human papillomavirus (HPV)-positive and HPV-negative cells. J Biol Chem 273: 6439–6445

Tidow H, Melero R, Mylonas E, Freund SM, Grossmann JG, Carazo JM, Svergun DI, Valle M, Fersht AR (2007) Quaternary structures of tumor suppressor p53 and a specific p53 DNA complex. Proc Natl Acad Sci USA 104:12324–12329

Wade M, Wang YV, Wahl GM (2010) The p53 orchestra: Mdm2 and Mdmx set the tone. Trends Cell Biol 20:299–309

Wade M, Wong ET, Tang M, Stommel JM, Wahl GM (2006) Hdmx modulates the outcome of p53 activation in human tumor cells. J Biol Chem 281:33036–33044

Wallace M, Worrall E, Pettersson S, Hupp TR, Ball KL (2006) Dual-site regulation of MDM2 E3-ubiquitin ligase activity. Mol Cell 23:251–263

Wang J, Zheng T, Chen X, Song X, Meng X, Bhatta N, Pan S, Jiang H, Liu L (2011) MDM2 antagonist can inhibit tumor growth in hepatocellular carcinoma with different types of p53 in vitro. J Gastroenterol Hepatol 26:371–377

Wang YV, Wade M, Wong E, Li YC, Rodewald LW, Wahl GM (2007) Quantitative analyses reveal the importance of regulated Hdmx degradation for p53 activation. Proc Natl Acad Sci USA 104:12365–12370

Wang Z, Li B (2010) Mdm2 links genotoxic stress and metabolism to p53. Protein Cell 1:1063–1072

Vaseva AV, Marchenko ND, Moll UM (2009) The transcription-independent mitochondrial p53 program is a major contributor to nutlin-induced apoptosis in tumor cells. Cell Cycle 8:1711–1719

Vassilev LT, Vu BT, Graves B, Carvajal D, Podlaski F, Filipovic Z, Kong N, Kammlott U, Lukacs C, Klein C et al (2004) In vivo activation of the p53 pathway by small-molecule antagonists of MDM2. Science 303:844–848

Weinmann L, Wischhusen J, Demma MJ, Naumann U, Roth P, Dasmahapatra B, Weller M (2008) A novel p53 rescue compound induces p53-dependent growth arrest and sensitises glioma cells to Apo2L/TRAIL-induced apoptosis. Cell Death Differ 15:718–729

Ventura A, Kirsch DG, McLaughlin ME, Tuveson DA, Grimm J, Lintault L, Newman J, Reczek EE, Weissleder R, Jacks T (2007) Restoration of p53 function leads to tumour regression in vivo. Nature 445:661–665

Vogelstein B, Lane D, Levine AJ (2000) Surfing the p53 network. Nature 408:307–310

von Knebel DM, Rittmuller C, zur Hausen H, Durst M (1992) Inhibition of tumorigenicity of cervical cancer cells in nude mice by HPV E6-E7 anti-sense RNA. Int J Cancer 51:831–834

Vousden KH, Lu X (2002) Live or let die: the cell's response to p53. Nat Rev Cancer 2:594–604

Vousden KH, Prives C (2009) Blinded by the light: the growing complexity of p53. Cell 137:413–431

Xia M, Knezevic D, Tovar C, Huang B, Heimbrook DC, Vassilev LT (2008) Elevated MDM2 boosts the apoptotic activity of p53-MDM2 binding inhibitors by facilitating MDMX degradation. Cell Cycle 7:1604–1612

Xia M, Knezevic D, Vassilev LT (2011) p21 does not protect cancer cells from apoptosis induced by nongenotoxic p53 activation. Oncogene 30:346–355

Xue W, Zender L, Miething C, Dickins RA, Hernando E, Krizhanovsky V, Cordon-Cardo C, Lowe SW (2007) Senescence and tumour clearance is triggered by p53 restoration in murine liver carcinomas. Nature 445:656–660

Yang J, Ahmed A, Ashcroft M (2009) Activation of a unique p53-dependent DNA damage response. Cell Cycle 8:1630–1632

Yang Y, Ludwig RL, Jensen JP, Pierre SA, Medaglia M V, Davydov IV, Safiran YJ, Oberoi P, Kenten JH, Phillips AC et al (2005) Small molecule inhibitors of HDM2 ubiquitin ligase activity stabilize and activate p53 in cells. Cancer Cell 7:547–559

Zhao CY, Grinkevich VV, Nikulenkov F, Bao W, Selivanova G (2010a) Rescue of the apoptotic-inducing function of mutant p53 by small molecule RITA. Cell Cycle 9:1847–1855

Zhao CY, Szekely L, Bao W, Selivanova G (2010b) Rescue of p53 function by small-molecule RITA in cervical carcinoma by blocking E6-mediated degradation. Cancer Res 70:3372–3381

Chapter 14
Targeting Mutant p53 for Improved Cancer Therapy

Jinfeng Shen, Vladimir J.N. Bykov, and Klas G. Wiman

1 Introduction

Over the last decades, major efforts have been directed towards the design of targeted cancer therapy. This is based on the improved understanding of the hallmarks of cancer and genetic alterations that drive tumor growth (Hanahan and Weinberg 2011). Inhibition of the bcr-abl kinase with imatinib (Gleevec) has dramatically improved therapy and prognosis in chronic myeloid leukemia (Druker et al. 1996; Gambacorti-Passerini 2008). Other examples of targeted therapy include gefitinib (Iressa) that inhibits the EGF receptor in certain types of lung cancer (Sordella et al. 2004). More recently, the BRAF kinase inhibitor vemurafenib (PLX4032) was shown to improve survival of patients with metastatic melanoma carrying the BRAF V600E mutation (Chapman et al. 2011). A number of other kinase inhibitors are currently being tested in clinical trials or are on their way to the clinic.

Tumors carry multiple genetic defects and potential targets for novel therapy. Among these, the p53 tumor suppressor has a unique position as the most frequently mutated gene in cancer (Soussi and Wiman 2007). At least 50 % of all human malignancies carry mutant p53. Wild-type p53 triggers cell cycle arrest, senescence, and/or cell death by apoptosis in response to cellular stress, e.g., DNA damage, hypoxia, and oncogene activation (Vousden and Prives 2009). This serves to eliminate incipient tumor cells. p53 induces these biological responses through transcriptional

J. Shen
Department of Oncology-Pathology, Cancer Center Karolinska (CCK), Karolinska Institutet, 171 76, Stockholm, Sweden

Department of Hematology, Institute of Stem Cell Biology and Regenerative Medicine, Stanford University, Stanford, CA, USA

V.J.N. Bykov • K.G. Wiman (✉)
Department of Oncology-Pathology, Cancer Center Karolinska (CCK), Karolinska Institutet, 171 76, Stockholm, Sweden
e-mail: Klas.Wiman@ki.se

P. Hainaut et al. (eds.), *p53 in the Clinics*, DOI 10.1007/978-1-4614-3676-8_14,

transactivation of target genes, including p21 and 14-3-3sigma (cell cycle arrest), and Bax, Noxa, and Puma (apoptosis). p53 can also promote apoptosis through transcription-independent activities (Chipuk et al. 2004; Mihara et al. 2003). Moreover, it is becoming increasingly clear that p53 has roles in a wide range of biological processes, including metabolism and aging (Jones et al. 2005; Matheu et al. 2007), and recent studies have shown that p53 inhibits the induction of pluripotent stem cells (Hong et al. 2009; Kawamura et al. 2009; Marion et al. 2009).

p53 belongs to a gene family with two other structurally and functionally related members, p63 and p73 (Bourdon 2007). Each p53 family protein is expressed as at least six isoforms, due to alternative promoters in the N terminus (TA and dN) and alternative splicing that generates different C-termini (alpha, beta, and gamma). The full-length forms are transcriptionally active, while the dN forms lack the N-terminal transactivation domain and therefore inhibit the transcriptional transactivation activity of the full-length isoforms by forming inactive hetero-tetramers and by competing for DNA binding (Dotsch et al. 2010; Ratovitski et al. 2006).

The frequent mutation of p53 in human tumors indicates a strong selective pressure to inactivate p53-mediated cell cycle arrest, senescence, and apoptosis during tumor development (Soussi and Wiman 2007); see also http://www-p53.iarc.fr; http://p53.free.fr. Mutant p53 is often expressed at elevated levels in tumors, partially due to the failure of mutant p53 to activate transcription of its own negative regulator Mdm2. Importantly, p53 mutations may not only inactivate wild-type p53 function but also confer gain-of-function effects, such as illegitimate regulation of transcription of genes that promote cell proliferation or inhibit apoptosis, or activation of the pro-inflammatory NF-kB pathway (Brosh and Rotter 2009). Furthermore, several mutant p53 proteins can physically interact with TAp63 and TAp73, thus inhibiting their transcriptional transactivation activities (Di Como et al. 1999; Gaiddon et al. 2001). Mutant p53 complexing with p63 stimulates tumor cell migration by affecting integrin and epidermal growth factor receptor (EGFR) trafficking (Muller et al. 2009). Therefore, many mutant p53 alleles can be regarded as oncogenes whose activity is essential for sustained neoplastic growth.

In contrast to the frequent mutation of p53, mutations in p63 or p73 rarely occur in human tumors (Moll and Slade 2004). p63 is mainly involved in the development of epithelial tissues whereas p73 is important for neuronal development and the immune system (Flores 2011; Mills et al. 1999; Yang et al. 1999, 2000, 2002). However, accumulating evidence suggests that both p63 and p73 may have roles in promotion or inhibition of tumor development. Overexpression of dNp63alpha was observed in primary tumors, in particular squamous cell carcinoma (Flores et al. 2005). Other studies show that dNp63 is downregulated during tumor progression and metastasis (Comperat et al. 2006; Fukushima et al. 2009; Urist et al. 2002). TAp63 regulates a series of microRNAs and inhibits metastasis (Su et al. 2010). The ratio between TAp73 and dNp73 expression levels is reduced in many tumors. TAp73 is silenced by methylation in lymphoblastic leukemias and Burkitt lymphomas (Corn et al. 1999; Kawano et al. 1999). Consistent with these observations, TAp73 null mice develop spontaneous tumors and show increased sensitivity to carcinogens (Tomasini et al. 2008).

2 p53 As Therapeutic Target

Loss of p53 function is a common and perhaps universal event during the formation of tumors. As discussed above, many tumors carry mutant p53, often expressed at high levels. In a sizeable fraction of human tumors, the p53 gene itself is intact but the p53 pathway is nonetheless inactivated by other alterations, for example overexpression of Mdm2 (Manfredi 2010), or infection by human papilloma viruses (HPV) whose E6 protein binds p53 and induces p53 degradation (Scheffner et al. 1990). Regardless of how p53 function has been disrupted in a tumor, p53 reactivation is clearly an attractive therapeutic strategy. Restoration of wild-type p53 function should trigger massive apoptosis and tumor regression. This idea has received strong experimental support from mouse studies showing that reactivation of wild-type p53 expression in vivo leads to efficient elimination of the tumor (Martins et al. 2006; Ventura et al. 2007; Xue et al. 2007). Thus, these findings argue convincingly that p53 reactivation is sufficient for induction of tumor cell apoptosis and/or senescence even in the presence of other genetic alterations and in the context of tumor evolution in vivo.

For tumors that carry wild-type p53, the p53–Mdm2 complex is an appealing therapeutic target. A group of small molecules called Nutlins were designed based on detailed structural analysis of the p53-binding cavity in Mdmd2 (Vassilev et al. 2004). They bind with high affinity to Mdm2 and disrupt p53–Mdm2 complexes, leading to non-genotoxic activation of p53. One of these compounds has recently been tested in clinical trials. Therapeutic reactivation of wild-type p53 is discussed in further detail in Chaps. 12 and 13 by S. Lain and G. Selivanova.

Abundantly expressed point mutant p53 is a promising target for novel therapy in mutant p53-carrying tumors. However, pharmacological reactivation of a mutant transcription factor such as p53 is a daunting mission. Functional reactivation will require structural restoration to allow specific DNA binding and transcriptional transactivation of target genes. This is complicated by the structural heterogeneity of tumor-associated mutant p53 proteins, which raises the question whether a small molecule can restore more than a limited set of mutant p53 proteins. An alternative and more modest therapeutic goal that might nonetheless produce significant clinical benefit is to block mutant p53 gain-of-function activities without full restoration of wild-type p53 function. Despite the major challenges associated with functional restoration of p53 as compared to for instance inhibition of an activated kinase, several mutant p53-targeting molecules have been identified using various approaches (Wiman 2010). Here we shall highlight some of the recent progress in the development of mutant p53-targeting compounds and discuss their potential for improved cancer therapy.

3 Identification of Mutant p53-Reactivating Small Molecules

Already over a decade ago, work by Hupp and colleagues demonstrated that short peptides derived from the C terminus of p53 can enhance wild-type p53 DNA binding (Hupp et al. 1995). Interestingly, similar peptides as well as antibodies directed against the p53 C terminus (e.g., PAb421) were shown to reactivate DNA binding of mutant p53. Peptide 46, corresponding to residues 361–382 in human p53, activates p53 DNA binding and triggers apoptosis upon transfer into mutant p53-carrying tumor cells (Selivanova et al. 1997). CDB3, a 9-mer peptide derived from the p53-binding protein 53BP2 or ASPP, also restores wild-type conformation and enhances p53 DNA binding but through a different mechanism that involves a chaperone effect (Friedler et al. 2002a; Issaeva et al. 2003). Although not suitable for further drug development, these peptides have demonstrated convincingly that mutant p53 is amenable to reactivation by small ligands, and stimulated efforts to identify novel small mutant p53-reactivating molecules for drug development.

Approaches to identify small molecules that reactivate mutant p53 can be divided into those that are based on rational design and those that rely on screening of chemical libraries. Furthermore, random screening can be carried out using either recombinant mutant p53 in vitro or cell-based assays, for instance human tumor cells carrying mutant p53. Chemical libraries have been screened using both approaches. This has led to the identification of compounds such as CP-31398 (Foster et al. 1999), PRIMA-1 (Bykov et al. 2002), and P53R3 (Weinmann et al. 2008).

The styrylquinazoline molecule CP-31398 prevents wild-type and mutant p53 unfolding in vitro, and inhibits xenograft tumor growth in mice (Foster et al. 1999). It was also shown to block UVB-induced skin carcinogenesis by induction of apoptosis, presumably by targeting mutant p53 (Tang et al. 2007). CP-31398 stabilizes p53 by a mechanism that involves inhibition of ubiquitination (Wang et al. 2003), and induces expression of classical p53 target genes like p21 and Bax, but also induces p53-independent cell death (Wischhusen et al. 2003). NMR analysis indicated that CP-31398 does not bind directly to the p53 core domain (Rippin et al. 2002). Thus, the exact mechanism of action for this compound is still unclear.

The compound WR1065 is an active metabolite of amifostine, an agent used in the clinic to counteract side effects of chemotherapy and radiotherapy. The ability of amifostine (WR2721) to reactivate mutant p53 was first shown in a yeast transcription assay (Maurici et al. 2001). Further studies demonstrated that WR1065 promotes wild-type conformation of Met-272 temperature-sensitive mutant p53 and restores its transcription transactivation capacity in esophageal carcinoma cells (North et al. 2002). This appears to require direct binding of WR1065 to p53, possibly through the formation of disulfide bridges between cysteines in p53 and the thiol in WR1065 (Shen et al. 2001). WR1065 induced wild-type p53-dependent changes in gene expression colon carcinoma cells, including upregulation of the p53 target gene p21, and triggered p53-dependent G1 cell cycle arrest (Mann and Hainaut 2005).

PRIMA-1 and its structural analog PRIMA-1^{MET}, also denoted APR-246, restore wild-type conformation and DNA binding to mutant p53, induce mutant p53-dependent apoptosis, and inhibit xenograft tumor growth in mice (Bykov et al. 2002; Liang et al. 2011; Shen et al. 2008; Zache et al. 2008a) (Fig. 14.1). PRIMA-1 was shown to also induce apoptosis in a transcription-independent manner (Chipuk et al. 2003). APR-246/PRIMA-1^{MET} acts synergistically with chemotherapeutic drugs, both in vitro and in vivo (Bykov et al. 2005a). Such synergy is presumably due to

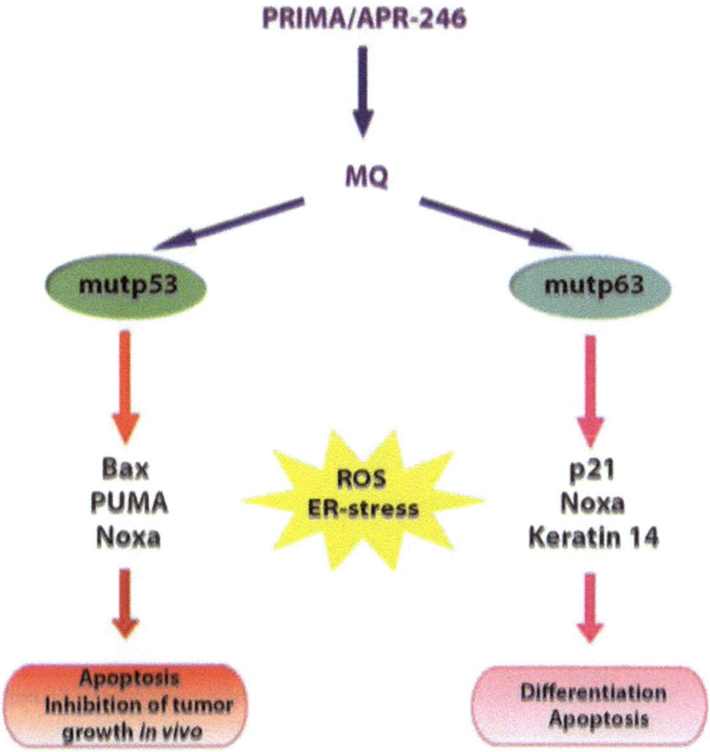

Fig. 14.1 Reactivation of mutant p53 or mutant p63 by PRIMA-1/APR-246. The compounds PRIMA-1 and APR-246/PRIMA-1^{MET} are converted to the biologically active Michael acceptor MQ that binds covalently to cysteines in mutant p53. This leads to upregulation of proapoptotic target genes, induction of apoptosis, and inhibition of tumor growth in vivo. Mutant p53 reactivation by APR-246/PRIMA-1^{MET} also induces ROS and ER stress. In cells that express mutant TAp63alpha or gamma, APR-246/PRIMA-1^{MET} induces expression of the target genes p21 and Noxa, as well as the differentiation marker Keratin 14, followed by apoptosis and/or differentiation

restoration of wild-type p53 function and the DNA damage response in the tumor cells. It is also conceivable that enhanced mutant p53 expression levels upon treatment with DNA-damaging chemotherapeutic drugs makes the tumor cells more sensitive to APR-246/PRIMA-1^{MET}.

Both PRIMA-1 and APR-246/PRIMA-1^{MET} are converted to the compound MQ, a Michael acceptor that binds covalently to p53 core domain cysteines (Lambert et al. 2009) (Fig. 14.2). Protein transfer experiments demonstrated that covalent modification by MQ per se is sufficient to reactivate mutant p53 and trigger apoptosis in human tumor cells. Unfolded wild-type p53 can also be covalently modified by MQ and functionally restored (Lambert et al. 2009). This provides an explanation for the observed effect of PRIMA-1 and APR-246/PRIMA-1^{MET} on wild-type p53-carrying tumor cells. APR-246/PRIMA-1^{MET} inhibits growth of wild-type p53-carrying human melanoma cells in 3D culture and in vivo in mice (Bao et al. 2011).

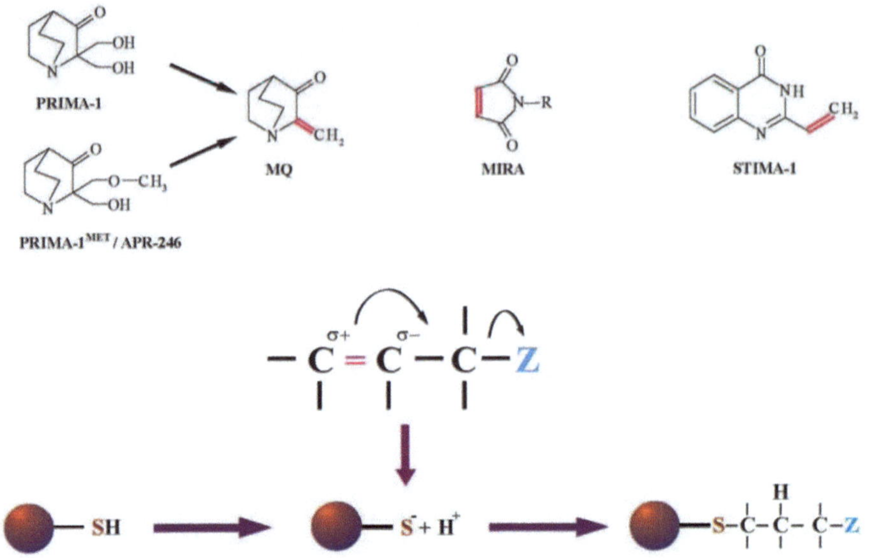

Fig. 14.2 Michael acceptor activity of mutant p53-targeting compounds. *Upper panel*: PRIMA-1 and APR-246/PRIMA-1^{MET} are converted to the active compound and Michael acceptor MQ (Lambert et al. 2009). The mutant p53-targeting compounds MIRA, including MIRA-1 (Bykov et al. 2005b) and STIMA-1 (Zache et al. 2008b), are also Michael acceptors. *Lower panel*: the carbon–carbon double bond is destabilized by the electron-withdrawing group Z, resulting in covalent binding to SH groups

APR-246/PRIMA-1^{MET} triggers multiple proapoptotic pathways and ER stress in a mutant p53-dependent manner (Lambert et al. 2010) (Fig. 14.1). Treatment with APR-246/PRIMA-1^{MET} leads to a redistribution of mutant p53 from the nucleoplasm to nucleoli, along with the proteins PML, CBP, and Hsp70 (Rokaeus et al. 2007). PRIMA-1-induced nucleolar relocalization appears to be associated with enhanced p53 degradation via ubiquitylation in breast cancer cells (Russo et al. 2010). Mutant p53 reactivation by PRIMA-1 and APR-246/PRIMA-1^{MET} results in upregulation of the proapoptotic p53 target genes Bax, Puma, and Noxa (Lambert et al. 2009; Shen et al. 2008) (Fig. 14.1). PRIMA-1 may also trigger apoptosis by affecting microRNA expression. Treatment with PRIMA-1 was shown to induce the p53 target miRNA-34a in mutant p53-expressing human lung cancer cells (Duan et al. 2010). The safety of APR-246/PRIMA-1^{MET} has recently been tested in a phase I clinical trial.

The compound P53R3 was identified by screening of a chemical library using an in vitro DNA binding assay (Weinmann et al. 2008). It inhibits growth of glioma cells expressing exogenous mutant p53, enhances binding of wild-type and mutant p53 to p53 target promoters, and induces expression of p53 targets such as p21, Puma, and BAI1 in human T98G glioma cells carrying Ile237 mutant p53, but also in wild-type p53-expressing LNT-229 glioma cells.

Using the same assay, Demma and colleagues subsequently identified SCH529074 from a library of quinazolines (Demma et al. 2010). This compound binds to the core domain of p53 and restores wild-type conformation and function to mutant p53 as demonstrated by the wild-type conformation-specific antibody PAb1620 and chromatin immunoprecipitation to assess binding to p53 target gene promoters. SCH529074 was also shown to inhibit xenograft tumor growth in vivo. The molecular mechanism of SCH529074 appears to involve a chaperone effect similar to that of the previously described peptide CDB3 (Friedler et al. 2002b). Interestingly, P53R3 and SCH529074 share striking structural similarities. They both contain piperasin and quinazoline radicals, and both have some resemblance to the previously identified compound CP-31398.

The small molecule RETRA was identified by screening of a chemical library using a cellular p53 reporter assay. RETRA selectively inhibits growth of mutant p53-carrying tumor cells in vitro and human tumor xenografts in vivo in nude mice (Kravchenko et al. 2008). This effect is mediated by disruption of complexes between mutant p53 and p73, resulting in enhanced levels of p73 protein and p73-dependent induction of apoptosis.

The compound Phi-Kan-083 was designed based on detailed structural analysis of a specific mutant p53 protein that carries a Tyr to Cys substitution at position 220 (Boeckler et al. 2008). PhiKan083 stabilizes the Cys220 mutant in a concentration-dependent manner, raising the melting temperature and increasing the thermodynamic stability of Cys220 mutant p53. Detailed structural information about interactions between the compound and the protein is available, and specific conformational changes upon binding have been identified. There are as yet no data on the effects of PhiKan083 on cancer cells in vitro and in vivo.

4 Role of Cysteine Modification for Mutant p53 Reactivation

The exact molecular mechanism behind mutant p53 reactivation has remained unknown for most of the identified compounds. This is a critical question, since insights into the mechanism may facilitate the design of novel compounds with higher potency, selectivity, and/or improved pharmacokinetic properties. In general, there is little information about the stability of the mutant p53-targeting compounds and therefore, it remains unclear whether they are active themselves or if an active product is derived from the original compound by decomposition or metabolic activation. However, recent results have indicated that cysteine modification by the so-called Michael addition may play a key role. This type of nucleophilic addition was described by Arthur Michael at the end of the nineteenth century (Michael 1887), referring to compounds with a functional group featuring a carbon–carbon double bond coupled to an electron-withdrawing group (Fig. 14.2). Such functional groups allow compounds to act as electrophiles and form covalent bonds with nucleophilic molecules. In living cells, the prime target for such modification is thiol groups in cysteine-containing proteins.

The prostaglandin and Michael acceptor 15-deoxy-delta (12,14)-prostaglandin was shown to stabilize wild-type p53 through covalent binding to cysteines in the core domain, although this modification inactivates rather than activates p53 (Kim et al. 2010). Fersht and colleagues used an NMR-based screening for compounds that bind to Y220C mutant p53 and identified Michael acceptor compounds, i.e., 3-benzoylacrylic acid and its derivative (E)-4-(4-fluorophenyl)-4-oxobut-2-enoic acid, that stabilize mutant p53 via alkylation of core domain cysteines (Kaar et al. 2010). The p53 core domain has ten cysteine residues. The reactivity of a thiol group is a function of its accessibility for a modifying electrophile and nucleophilicity (pKa). Also, since protein structure is flexible, modification of one cystein may affect modification of other cysteines, either positively or negatively. Cysteines C124 and C141 in p53 were the first cysteines to be modified by 3-benzoylacrylic acid. Residues that are more deeply buried in the p53 core, such as C135, were modified only at higher concentrations. C176, C238, and C242 were most effectively shielded by the hydrophobic core and by coordination with Zn (Kaar et al. 2010). Although the observed cysteine modification increased p53 protein stability, it did not rescue DNA binding. Thus, the identified molecules do not seem to restore p53 function as a transcription factor. However, it remains to be determined whether stabilization of p53 conformation could rescue transcription-independent proapoptotic activity of p53.

The demonstration that both PRIMA-1 and APR-246/PRIMA-1^{MET} are converted to the Michael acceptor MQ (methylene quinuclidinone) that binds covalently to cysteines in the p53 core domain indicates that cysteine modification can in fact restore both wild-type conformation and function to mutant p53. That MQ is the active compound is consistent with the observations that adduct formation with *N*-acetyl cysteine (NAC) completely blocks the biological effect of PRIMA-1 and APR-246/PRIMA-1^{MET}, and that the structural analog PRIMA-Dead that cannot be converted to MQ is biologically inactive (Lambert et al. 2009). This notion is also further supported by the observation that the compound MIRA-1, a maleimide and Michael acceptor that modifies cysteines (Fig. 14.2), reactivates mutant p53, and induces apoptosis in mutant p53-expressing cells, although with faster kinetics than PRIMA-1 (Bykov et al. 2005b). Importantly, MIRA-1 analogs that lack a reactive double bond fail to trigger a mutant p53-dependent biological response, demonstrating that the Michael acceptor activity is critical for biological activity of MIRA compounds. STIMA-1, another mutant p53-targeting compound (Zache et al. 2008b), has Michael acceptor activity as well (Fig. 14.2), and as mentioned above, the compound WR1065 binds to p53 via thiol groups.

The identification of MQ as the compound responsible for the biological effects of PRIMA-1 and APR-246/PRIMA-1^{MET} raises questions regarding mutant p53 selectivity. At first, the electrophilic properties of MQ suggest that PRIMA-1 and APR-246/PRIMA-1^{MET} will potentially modify thiol groups in multiple proteins. However, it is likely that docking effects may result in a more restricted range of protein targets. As illustrated in Fig. 14.3, the structural context at a specific cysteine residue may allow or prevent covalent binding of a Michael acceptor such as MQ. Moreover, cysteine modification by MQ might have minor functional consequences in many

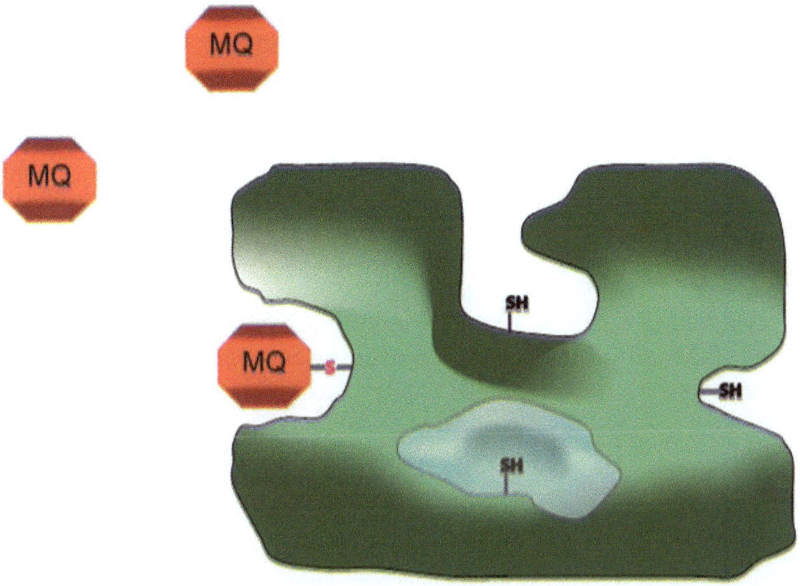

Fig. 14.3 Model for the effect of structural context on covalent binding of a Michael acceptor such as MQ to thiols of specific cysteines. In a protein with several cysteine residues, accessibility of different cysteines to a thiol-binding compound is expected to vary, due to docking effects depending on the structural environment at each cysteine. Cysteines that are not exposed on the outer surface will not be modified

proteins. Modification of albumin by MQ did not result in any cytotoxicity upon transfer of the modified protein into living cells (Lambert et al. 2009). This does not exclude the possibility that other cellular proteins can be modified by MQ and that such modifications can contribute to the biological effects of PRIMA-1 and APR-246/PRIMA-1^{MET}, but suggests that the number of such critical targets is limited.

5 Michael Acceptors with Antiproliferative and Anticancer Activity

Leptomycin B, a compound that induces p53 expression and inhibits cell proliferation, has Michael acceptor activity. This compound was originally discovered in a screen for novel antifungal antibiotics (Hamamoto et al. 1983). Later studies showed that Leptomycin B also inhibits proliferation of mammalian cells (Yoshida et al. 1990), and CRM1/exportin 1 was identified as its cellular target (Nishi et al. 1994). The main function of CRM1/exportin 1 is nuclear export of proteins carrying a nuclear export signal (NES). Leptomycin B inactivates CRM1/exportin 1 by covalent binding via Michael addition to cysteine 529 (Kudo et al. 1999). The potent Michael acceptor *N*-ethylmaleimide has been shown to inactivate CRM1/

exportin 1 by a similar mechanism (Holaska and Paschal 1998). The high specificity of Leptomycin B for its target CRM1 is in agreement with the proposed docking effects in Fig. 14.3.

A major mechanism of drug resistance against EGFR inhibitors such as gefitinib (Iressa) and erlotinib (Tarceva) is an acquisition of mutations that affect drug affinity for the ATP-binding pocket of the tyrosine kinase (Sharma et al. 2007). One way to enhance the potency of these compounds is to make them irreversible inhibitors by introducing Michael acceptor functionality. This not only improves potency (Fry et al. 1998), but is also sufficient to overcome resistance due to acquired mutations (Li et al. 2008; Wu et al. 2010). One example is HKI-272 that most likely modifies cysteine 773 in the EFGR (Kwak et al. 2005). Conceivably, covalent binding can compensate for the reduced affinity to the binding pocket in mutated EGFR proteins.

It is interesting to note that Michael acceptors are widely represented in our environment and in our diet, and that some have antitumor activity (Talalay et al. 1988). Cinnamaldehyde, a Michael acceptor that gives flavor to cinnamon oil and is widely used in cooking, has anticancer properties in mouse models, probably by inducing oxidative stress and inhibiting NFkB signaling (Cabello et al. 2009). Its cytotoxic properties are at least in part mediated by inhibition of thioredoxin reductase (TrxR) (Chew et al. 2010). Another Michael acceptor, curcumin, gives flavor and color to curries. Epidemiological studies have revealed a reduced risk of several common types of cancer in populations with high amounts of curry-spiced food in their diet (Lopez-Lazaro 2008). Curcumin effectively targets different tumor cells as compared to normal untransformed cells. Clinical trials with curcumin did not show any significant effect, but this was possibly due to poor bioavailability of curcumin upon oral administration (Lopez-Lazaro 2008). Again, inhibition of TrxR appears to be an important mechanism for the cytotoxic activity (Fang et al. 2005).

6 Targeting Mutant p53 Family Members

The high degree of structural similarity among p53 family proteins, particularly in their DNA-binding core domains, prompted us to test if APR-246/PRIMA-1^{MET} is able to target mutant forms of p63 and p73. Using human tumor cells carrying exogenous temperature-sensitive mutant TAp73alpha, TAp73beta, or TAp63gamma, we showed that APR-246/PRIMA-1^{MET} suppresses cell proliferation and induces apoptosis in tumor cells in a p73beta- and p63gamma-dependent manner (Rokaeus et al. 2010) (Fig. 14.1). Furthermore, APR-246/PRIMA-1^{MET} inhibits growth of human tumor cells carrying Tet-inducible TAp63gamma with mutations in the core domain (R204W, R304W) that are associated with the ectrodactyly–ectodermal dysplasia–cleft (EEC) syndrome and correspond to classical tumor-associated p53 core domain mutations (R175H, R273H) (Rokaeus et al. 2010). Treatment with APR-246/PRIMA-1^{MET} induced the p53 target genes p21 and Noxa as well as the differentiation marker Keratin 14 (Fig. 14.1), and triggered nucleolar redistribution of mutant

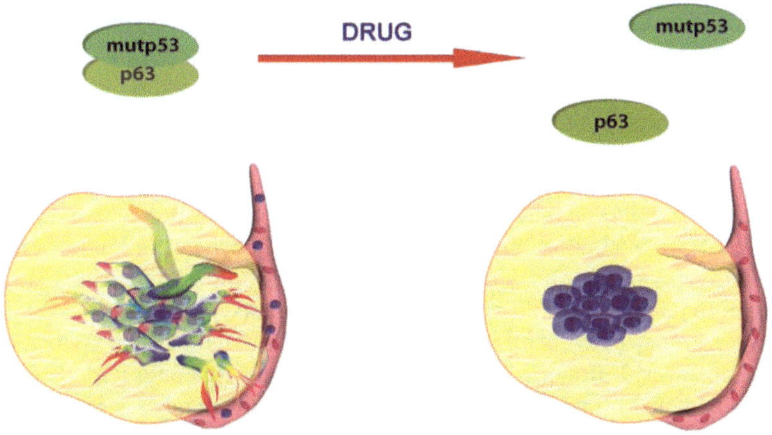

Fig. 14.4 Therapeutic strategy based on pharmacological disruption of mutant p53 binding to TAp63 and restoration of TAp63 tumor suppressor activity. Disruption of mutant p53–TAp63 complexes will release TAp63 and allow TAp63-mediated tumor suppression by various mechanisms. For instance, release of TAp63 might restore regulation of tumor cell migration and invasion

TAp63gamma similar to the redistribution of mutant p53 in APR-246/PRIMA-1^{MET}-treated cells (Rokaeus et al. 2007, 2010). These observations suggest that APR-246/PRIMA-1^{MET}-mediated reactivation of mutant p53 and mutant p53 family members p63 and p73 involve similar structural elements with their core domains.

As discussed above, the antitumor activity of full-length p63 and p73 may be inhibited due to complex formation with mutant p53. Therefore, such complexes are potential targets for novel therapy (Fig. 14.4). Peptides or low molecular weight compounds have been shown to rescue wild-type p73 by disrupting its complex formation with mutant p53 (Di Agostino et al. 2008; Kravchenko et al. 2008). Similarly, small molecules that disrupt mutant p53 complexing with TAp63 should restore TAp63 function and could for instance block mutant p53-mediated tumor cell migration and invasion (Fig. 14.4).

7 Conclusions and Future Perspectives

Significant progress has been made towards more efficient cancer therapy by targeting mutant p53. Numerous studies have demonstrated that at least a fraction of mutant p53 proteins can be rescued for DNA binding and transcriptional transactivation through various low molecular weight compounds. Improved understanding of the structural consequences of different p53 mutations has clearly been critical for the rational design of mutant p53-interacting compounds. So far, however, few of the rationally designed compounds have been thoroughly tested in cultured cells or in suitable mouse models, and therefore their antitumor efficacy in vivo is still unknown.

In parallel, screening of chemical libraries using cellular assays has allowed the identification of small molecules with a desired mutant p53-dependent biological effect, i.e., cell death by apoptosis, but with as yet poorly understood molecular mechanisms. Thus, this approach may produce candidate drugs that efficiently eliminate tumor cells and that can be taken to preclinical development but is limited by the lack of a solid understanding of the molecular mechanism. Future screens of improved diversified chemical libraries should rely on multiple filtering assays that examine both p53 protein folding and cellular endpoints such as apoptosis.

A primary goal is to identify and develop compounds that can restore a wide range of mutant p53 proteins. Nonetheless, compounds that only target a subset of specific mutant p53 proteins are worthwhile pursuing; the number of cancer patients worldwide with tumors that carry such particular p53 mutations may still be substantial. Furthermore, it should be emphasized that although complete restoration of wild-type conformation and function is the ultimate aim, inhibition of mutant p53 gain-of-function activities may have considerable therapeutic effect. This could be achieved by approaches that inhibit mutant p53 expression, presumably an easier task than restoration of wild-type function. Another example is disruption of complexes between mutant p53 and p63 or p73. Thus, major progress could be made by targeting different mutant p53 gain-of-function activities.

The safety of one mutant p53-targeting compound, APR-246/PRIMA-1^{MET}, has been tested in the clinic. Further clinical studies of mutant p53-targeting compounds will hopefully be performed during the next few years. An area where significant therapeutic advances could probably be made is a combination therapy with conventional chemotherapeutic drugs that show selectivity for wild-type p53-carrying tumors. Mutant p53 reactivation may greatly increase the sensitivity of chemoresistant tumors to such drugs. In vitro studies have already shown that mutant p53 reactivation by APR-246/PRIMA-1^{MET} can synergize with cisplatin and other chemotherapeutic drugs. Moreover, combination treatment with mutant p53-targeting compounds and novel therapeutic agents, for example, kinase inhibitors, proteasome inhibitors, or HDAC inhibitors, may reveal unexpected synergies. Once such synergies have been established in cultured tumor cells and appropriate mouse models, they should be further explored in clinical studies. Finally, a great challenge will be the implementation of novel p53-targeting drugs in the clinic. This will require robust and affordable routine methods for molecular characterization of p53 itself as well as the p53 pathway, in order to select the most optimal therapy. Thus, there are many remaining difficulties on the road ahead, but if we are successful the benefits for public health may be immense.

Acknowledgments We thank the Swedish Cancer Society (Cancerfonden), the Swedish Medical Research Council (VR), the Cancer Society of Stockholm, the Stockholm County Council (ALF), the EU FP6 framework program, and Karolinska Institutet for generous support. V.J.N.B. and K.G.W. are cofounders and shareholders of the company Aprea AB that develops p53-based cancer therapy including APR-246. K.G.W. is a member of its board.

References

Bao W, Chen M, Zhao X, Kumar R, Spinnler C, Thullberg M, Issaeva N, Selivanova G, Stromblad S (2011) PRIMA-1^{Met}/APR-246 induces wild-type p53-dependent suppression of malignant melanoma tumor growth in 3D culture and in vivo. Cell Cycle 10:301–307

Boeckler FM, Joerger AC, Jaggi G, Rutherford TJ, Veprintsev DB, Fersht AR (2008) Targeted rescue of a destabilized mutant of p53 by an in silico screened drug. Proc Natl Acad Sci USA 105:10360–10365

Bourdon JC (2007) p53 Family isoforms. Curr Pharm Biotechnol 8:332–336

Brosh R, Rotter V (2009) When mutants gain new powers: news from the mutant p53 field. Nat Rev Cancer 9:701–713

Bykov VJ, Issaeva N, Shilov A, Hultcrantz M, Pugacheva E, Chumakov P, Bergman J, Wiman KG, Selivanova G (2002) Restoration of the tumor suppressor function to mutant p53 by a low-molecular-weight compound. Nat Med 8:282–288

Bykov VJ, Zache N, Stridh H, Westman J, Bergman J, Selivanova G, Wiman KG (2005a) PRIMA-1(MET) synergizes with cisplatin to induce tumor cell apoptosis. Oncogene 24:3484–3491

Bykov VJ, Issaeva N, Zache N, Shilov A, Hultcrantz M, Bergman J, Selivanova G, Wiman KG (2005b) Reactivation of mutant p53 and induction of apoptosis in human tumor cells by maleimide analogs. J Biol Chem 280:30384–30391

Cabello CM, Bair WB III, Lamore SD, Ley S, Bause AS, Azimian S, Wondrak GT (2009) The cinnamon-derived Michael acceptor cinnamic aldehyde impairs melanoma cell proliferation, invasiveness, and tumor growth. Free Radic Biol Med 46:220–231

Chapman PB, Hauschild A, Robert C, Haanen JB, Ascierto P, Larkin J, Dummer R, Garbe C, Testori A, Maio M et al (2011) Improved survival with vemurafenib in melanoma with BRAF V600E mutation. N Engl J Med 364:2507–2516

Chew EH, Nagle AA, Zhang Y, Scarmagnani S, Palaniappan P, Bradshaw TD, Holmgren A, Westwell AD (2010) Cinnamaldehydes inhibit thioredoxin reductase and induce Nrf2: potential candidates for cancer therapy and chemoprevention. Free Radic Biol Med 48:98–111

Chipuk JE, Maurer U, Green DR, Schuler M (2003) Pharmacologic activation of p53 elicits Bax-dependent apoptosis in the absence of transcription. Cancer Cell 4:371–381

Chipuk JE, Kuwana T, Bouchier-Hayes L, Droin NM, Newmeyer DD, Schuler M, Green DR (2004) Direct activation of Bax by p53 mediates mitochondrial membrane permeabilization and apoptosis. Science 303:1010–1014

Comperat E, Camparo P, Haus R, Chartier-Kastler E, Bart S, Delcourt A, Houlgatte A, Francois R, Capron F, Vieillefond A (2006) Immunohistochemical expression of p63, p53 and MIB-1 in urinary bladder carcinoma. A tissue microarray study of 158 cases. Virchows Arch 448:319–324

Corn PG, Kuerbitz SJ, van Noesel MM, Esteller M, Compitello N, Baylin SB, Herman JG (1999) Transcriptional silencing of the p73 gene in acute lymphoblastic leukemia and Burkitt's lymphoma is associated with 5' CpG island methylation. Cancer Res 59:3352–3356

Demma M, Maxwell E, Ramos R, Liang L, Li C, Hesk D, Rossman R, Mallams A, Doll R, Liu M et al (2010) SCH529074, a small molecule activator of mutant p53, which binds p53 DNA binding domain (DBD), restores growth-suppressive function to mutant p53 and interrupts HDM2-mediated ubiquitination of wild type p53. J Biol Chem 285:10198–10212

Di Agostino S, Cortese G, Monti O, Dell'Orso S, Sacchi A, Eisenstein M, Citro G, Strano S, Blandino G (2008) The disruption of the protein complex mutantp53/p73 increases selectively the response of tumor cells to anticancer drugs. Cell Cycle 7:3440–3447

Di Como CJ, Gaiddon C, Prives C (1999) p73 function is inhibited by tumor-derived p53 mutants in mammalian cells. Mol Cell Biol 19:1438–1449

Dotsch V, Bernassola F, Coutandin D, Candi E, Melino G (2010) p63 and p73, the ancestors of p53. Cold Spring Harb Perspect Biol 2:a004887

Druker BJ, Tamura S, Buchdunger E, Ohno S, Segal GM, Fanning S, Zimmermann J, Lydon NB (1996) Effects of a selective inhibitor of the Abl tyrosine kinase on the growth of Bcr-Abl positive cells. Nat Med 2:561–566

Duan W, Gao L, Wu X, Wang L, Nana-Sinkam SP, Otterson GA, Villalona-Calero MA (2010) MicroRNA-34a is an important component of PRIMA-1-induced apoptotic network in human lung cancer cells. Int J Cancer 127:313–320

Fang J, Lu J, Holmgren A (2005) Thioredoxin reductase is irreversibly modified by curcumin: a novel molecular mechanism for its anticancer activity. J Biol Chem 280:25284–25290

Flores ER (2011) p73 is critical for the persistence of memory. Cell Death Differ 18:381–382

Flores ER, Sengupta S, Miller JB, Newman JJ, Bronson R, Crowley D, Yang A, McKeon F, Jacks T (2005) Tumor predisposition in mice mutant for p63 and p73: evidence for broader tumor suppressor functions for the p53 family. Cancer Cell 7:363–373

Foster BA, Coffey HA, Morin MJ, Rastinejad F (1999) Pharmacological rescue of mutant p53 conformation and function. Science 286:2507–2510

Friedler A, Hansson LO, Veprintsev DB, Freund SM, Rippin TM, Nikolova PV, Proctor MR, Rudiger S, Fersht AR (2002) A peptide that binds and stabilizes p53 core domain: chaperone strategy for rescue of oncogenic mutants. Proc Natl Acad Sci USA 99:937–942

Fry DW, Bridges AJ, Denny WA, Doherty A, Greis KD, Hicks JL, Hook KE, Keller PR, Leopold WR, Loo JA et al (1998) Specific, irreversible inactivation of the epidermal growth factor receptor and erbB2, by a new class of tyrosine kinase inhibitor. Proc Natl Acad Sci USA 95:12022–12027

Fukushima H, Koga F, Kawakami S, Fujii Y, Yoshida S, Ratovitski E, Trink B, Kihara K (2009) Loss of DeltaNp63alpha promotes invasion of urothelial carcinomas via N-cadherin/Src homology and collagen/extracellular signal-regulated kinase pathway. Cancer Res 69:9263–9270

Gaiddon C, Lokshin M, Ahn J, Zhang T, Prives C (2001) A subset of tumor-derived mutant forms of p53 down-regulate p63 and p73 through a direct interaction with the p53 core domain. Mol Cell Biol 21:1874–1887

Gambacorti-Passerini C (2008) Part I: milestones in personalised medicine – imatinib. Lancet Oncol 9:600

Hamamoto T, Gunji S, Tsuji H, Beppu T (1983) Leptomycins A and B, new antifungal antibiotics. I. Taxonomy of the producing strain and their fermentation, purification and characterization. J Antibiot (Tokyo) 36:639–645

Hanahan D, Weinberg RA (2011) Hallmarks of cancer: the next generation. Cell 144:646–674

Holaska JM, Paschal BM (1998) A cytosolic activity distinct from crm1 mediates nuclear export of protein kinase inhibitor in permeabilized cells. Proc Natl Acad Sci USA 95:14739–14744

Hong H, Takahashi K, Ichisaka T, Aoi T, Kanagawa O, Nakagawa M, Okita K, Yamanaka S (2009) Suppression of induced pluripotent stem cell generation by the p53–p21 pathway. Nature 460:1132–1135

Hupp TR, Sparks A, Lane DP (1995) Small peptides activate the latent sequence-specific DNA binding function of p53. Cell 83:237–245

Issaeva N, Friedler A, Bozko P, Wiman KG, Fersht AR, Selivanova G (2003) Rescue of mutants of the tumor suppressor p53 in cancer cells by a designed peptide. Proc Natl Acad Sci USA 100:13303–13307

Jones RG, Plas DR, Kubek S, Buzzai M, Mu J, Xu Y, Birnbaum MJ, Thompson CB (2005) AMP-activated protein kinase induces a p53-dependent metabolic checkpoint. Mol Cell 18:283–293

Kaar JL, Basse N, Joerger AC, Stephens E, Rutherford TJ, Fersht AR (2010) Stabilization of mutant p53 via alkylation of cysteines and effects on DNA binding. Protein Sci 19:2267–2278

Kawamura T, Suzuki J, Wang YV, Menendez S, Morera LB, Raya A, Wahl GM, Belmonte JCI (2009) Linking the p53 tumour suppressor pathway to somatic cell reprogramming. Nature 460:1140–1144

Kawano S, Miller CW, Gombart AF, Bartram CR, Matsuo Y, Asou H, Sakashita A, Said J, Tatsumi E, Koeffler HP (1999) Loss of p73 gene expression in leukemias/lymphomas due to hypermethylation. Blood 94:1113–1120

Kim DH, Kim EH, Na HK, Sun Y, Surh YJ (2010) 15-Deoxy-Delta(12,14)-prostaglandin J(2) stabilizes, but functionally inactivates p53 by binding to the cysteine 277 residue. Oncogene 29:2560–2576

Kravchenko JE, Ilyinskaya GV, Komarov PG, Agapova LS, Kochetkov DV, Strom E, Frolova EI, Kovriga I, Gudkov AV, Feinstein E, Chumakov PM (2008) Small-molecule RETRA suppresses mutant p53-bearing cancer cells through a p73-dependent salvage pathway. Proc Natl Acad Sci USA 105:6302–6307

Kudo N, Matsumori N, Taoka H, Fujiwara D, Schreiner EP, Wolff B, Yoshida M, Horinouchi S (1999) Leptomycin B inactivates CRM1/exportin 1 by covalent modification at a cysteine residue in the central conserved region. Proc Natl Acad Sci USA 96:9112–9117

Kwak EL, Sordella R, Bell DW, Godin-Heymann N, Okimoto RA, Brannigan BW, Harris PL, Driscoll DR, Fidias P, Lynch TJ et al (2005) Irreversible inhibitors of the EGF receptor may circumvent acquired resistance to gefitinib. Proc Natl Acad Sci USA 102:7665–7670

Lambert JM, Gorzov P, Veprintsev DB, Soderqvist M, Segerback D, Bergman J, Fersht AR, Hainaut P, Wiman KG, Bykov VJ (2009) PRIMA-1 reactivates mutant p53 by covalent binding to the core domain. Cancer Cell 15:376–388

Lambert JM, Moshfegh A, Hainaut P, Wiman KG, Bykov VJ (2010) Mutant p53 reactivation by PRIMA-1(MET) induces multiple signaling pathways converging on apoptosis. Oncogene 29:1329–1338

Li D, Ambrogio L, Shimamura T, Kubo S, Takahashi M, Chirieac LR, Padera RF, Shapiro GI, Baum A, Himmelsbach F et al (2008) BIBW2992, an irreversible EGFR/HER2 inhibitor highly effective in preclinical lung cancer models. Oncogene 27:4702–4711

Liang Y, Besch-Williford C, Benakanakere I, Thorpe PE, Hyder SM (2011) Targeting mutant p53 protein and the tumor vasculature: an effective combination therapy for advanced breast tumors. Breast Cancer Res Treat 125:407–420

Lopez-Lazaro M (2008) Anticancer and carcinogenic properties of curcumin: considerations for its clinical development as a cancer chemopreventive and chemotherapeutic agent. Mol Nutr Food Res 52(Suppl 1):S103–S127

Manfredi JJ (2010) The Mdm2-p53 relationship evolves: Mdm2 swings both ways as an oncogene and a tumor suppressor. Genes Dev 24:1580–1589

Mann K, Hainaut P (2005) Aminothiol WR1065 induces differential gene expression in the presence of wild-type p53. Oncogene 24:3964–3975

Marion RM, Strati K, Li H, Murga M, Blanco R, Ortega S, Fernandez-Capetillo O, Serrano M, Blasco MA (2009) A p53-mediated DNA damage response limits reprogramming to ensure iPS cell genomic integrity. Nature 460:1149–1153

Martins CP, Brown-Swigart L, Evan GI (2006) Modeling the therapeutic efficacy of p53 restoration in tumors. Cell 127:1323–1334

Matheu A, Maraver A, Klatt P, Flores I, Garcia-Cao I, Borras C, Flores JM, Vina J, Blasco MA, Serrano M (2007) Delayed ageing through damage protection by the Arf/p53 pathway. Nature 448:375–379

Maurici D, Monti P, Campomenosi P, North S, Frebourg T, Fronza G, Hainaut P (2001) Amifostine (WR2721) restores transcriptional activity of specific p53 mutant proteins in a yeast functional assay. Oncogene 20:3533–3540

Michael A (1887) Ueber die Addition von Natriumacetessig- und natriummalonsäureäthern zu den aethern ungesättigter säuren. J Prakt Chem 35:349–356

Mihara M, Erster S, Zaika A, Petrenko O, Chittenden T, Pancoska P, Moll UM (2003) p53 has a direct apoptogenic role at the mitochondria. Mol Cell 11:577–590

Mills AA, Zheng B, Wang XJ, Vogel H, Roop DR, Bradley A (1999) p63 is a p53 homologue required for limb and epidermal morphogenesis. Nature 398:708–713

Moll UM, Slade N (2004) p63 and p73: roles in development and tumor formation. Mol Cancer Res 2:371–386

Muller PA, Caswell PT, Doyle B, Iwanicki MP, Tan EH, Karim S, Lukashchuk N, Gillespie DA, Ludwig RL, Gosselin P et al (2009) Mutant p53 drives invasion by promoting integrin recycling. Cell 139:1327–1341

Nishi K, Yoshida M, Fujiwara D, Nishikawa M, Horinouchi S, Beppu T (1994) Leptomycin B targets a regulatory cascade of crm1, a fission yeast nuclear protein, involved in control of higher order chromosome structure and gene expression. J Biol Chem 269:6320–6324

North S, Pluquet O, Maurici D, El-Ghissassi F, Hainaut P (2002) Restoration of wild-type conformation and activity of a temperature-sensitive mutant of p53 (p53(V272M)) by the cytoprotective aminothiol WR1065 in the esophageal cancer cell line TE-1. Mol Carcinog 33:181–188

Ratovitski E, Trink B, Sidransky D (2006) p63 and p73: teammates or adversaries? Cancer Cell 9:1–2

Rippin TM, Bykov VJ, Freund SM, Selivanova G, Wiman KG, Fersht AR (2002) Characterization of the p53-rescue drug CP-31398 in vitro and in living cells. Oncogene 21:2119–2129

Rokaeus N, Klein G, Wiman KG, Szekely L, Mattsson K (2007) PRIMA-1(MET) induces nucleolar accumulation of mutant p53 and PML nuclear body-associated proteins. Oncogene 26:982–992

Rokaeus N, Shen J, Eckhardt I, Bykov J, Wiman KG, Wilhelm MT (2010) PRIMA-1^{MET}/APR-246 targets mutant forms of p53 family members p63 and p73. Oncogene 29(49):6442–6451

Russo D, Ottaggio L, Penna I, Foggetti G, Fronza G, Inga A, Menichini P (2010) PRIMA-1 cytotoxicity correlates with nucleolar localization and degradation of mutant p53 in breast cancer cells. Biochem Biophys Res Commun 402:345–350

Scheffner M, Werness BA, Huibregtse JM, Levine AJ, Howley PM (1990) The E6 oncoprotein encoded by human papillomavirus types 16 and 18 promotes the degradation of p53. Cell 63:1129–1136

Selivanova G, Iotsova V, Okan I, Fritsche M, Strom M, Groner B, Grafstrom RC, Wiman KG (1997) Restoration of the growth suppression function of mutant p53 by a synthetic peptide derived from the p53 C-terminal domain. Nat Med 3:632–638

Sharma SV, Bell DW, Settleman J, Haber DA (2007) Epidermal growth factor receptor mutations in lung cancer. Nat Rev Cancer 7:169–181

Shen H, Chen ZJ, Zilfou JT, Hopper E, Murphy M, Tew KD (2001) Binding of the aminothiol WR-1065 to transcription factors influences cellular response to anticancer drugs. J Pharmacol Exp Ther 297:1067–1073

Shen J, Vakifahmetoglu H, Stridh H, Zhivotovsky B, Wiman KG (2008) PRIMA-1^{MET} induces mitochondrial apoptosis through activation of caspase-2. Oncogene 27:6571–6580

Sordella R, Bell DW, Haber DA, Settleman J (2004) Gefitinib-sensitizing EGFR mutations in lung cancer activate anti-apoptotic pathways. Science 305:1163–1167

Soussi T, Wiman KG (2007) Shaping genetic alterations in human cancer: the p53 mutation paradigm. Cancer cell 12:303–312

Su X, Chakravarti D, Cho MS, Liu L, Gi YJ, Lin YL, Leung ML, El-Naggar A, Creighton CJ, Suraokar MB et al (2010) TAp63 suppresses metastasis through coordinate regulation of Dicer and miRNAs. Nature 467:986–990

Talalay P, De Long MJ, Prochaska HJ (1988) Identification of a common chemical signal regulating the induction of enzymes that protect against chemical carcinogenesis. Proc Natl Acad Sci USA 85:8261–8265

Tang X, Zhu Y, Han L, Kim AL, Kopelovich L, Bickers DR, Athar M (2007) CP-31398 restores mutant p53 tumor suppressor function and inhibits UVB-induced skin carcinogenesis in mice. J Clin Invest 117:3753–3764

Tomasini R, Tsuchihara K, Wilhelm M, Fujitani M, Rufini A, Cheung CC, Khan F, Itie-Youten A, Wakeham A, Tsao MS et al (2008) TAp73 knockout shows genomic instability with infertility and tumor suppressor functions. Genes Dev 22:2677–2691

Urist MJ, Di Como CJ, Lu ML, Charytonowicz E, Verbel D, Crum CP, Ince TA, McKeon FD, Cordon-Cardo C (2002) Loss of p63 expression is associated with tumor progression in bladder cancer. Am J Pathol 161:1199–1206

Vassilev LT, Vu BT, Graves B, Carvajal D, Podlaski F, Filipovic Z, Kong N, Kammlott U, Lukacs C, Klein C et al (2004) In vivo activation of the p53 pathway by small-molecule antagonists of MDM2. Science 303:844–848

Ventura A, Kirsch DG, McLaughlin ME, Tuveson DA, Grimm J, Lintault L, Newman J, Reczek EE, Weissleder R, Jacks T (2007) Restoration of p53 function leads to tumour regression in vivo. Nature 445:661–665

Vousden KH, Prives C (2009) Blinded by the light: the growing complexity of p53. Cell 137: 413–431

Wang W, Takimoto R, Rastinejad F, El-Deiry WS (2003) Stabilization of p53 by CP-31398 inhibits ubiquitination without altering phosphorylation at serine 15 or 20 or MDM2 binding. Mol Cell Biol 23:2171–2181

Weinmann L, Wischhusen J, Demma MJ, Naumann U, Roth P, Dasmahapatra B, Weller M (2008) A novel p53 rescue compound induces p53-dependent growth arrest and sensitises glioma cells to Apo2L/TRAIL-induced apoptosis. Cell Death Differ 15:718–729

Wiman KG (2010) Pharmacological reactivation of mutant p53: from protein structure to the cancer patient. Oncogene 29:4245–4252

Wischhusen J, Naumann U, Ohgaki H, Rastinejad F, Weller M (2003) CP-31398, a novel p53-stabilizing agent, induces p53-dependent and p53-independent glioma cell death. Oncogene 22:8233–8245

Wu CH, Coumar MS, Chu CY, Lin WH, Chen YR, Chen CT, Shiao HY, Rafi S, Wang SY, Hsu H et al (2010) Design and synthesis of tetrahydropyridothieno[2,3-d]pyrimidine scaffold based epidermal growth factor receptor (EGFR) kinase inhibitors: the role of side chain chirality and Michael acceptor group for maximal potency. J Med Chem 53:7316–7326

Xue W, Zender L, Miething C, Dickins RA, Hernando E, Krizhanovsky V, Cordon-Cardo C, Lowe SW (2007) Senescence and tumour clearance is triggered by p53 restoration in murine liver carcinomas. Nature 445:656–660

Yang A, Schweitzer R, Sun D, Kaghad M, Walker N, Bronson RT, Tabin C, Sharpe A, Caput D, Crum C, McKeon F (1999) p63 is essential for regenerative proliferation in limb, craniofacial and epithelial development. Nature 398:714–718

Yang A, Walker N, Bronson R, Kaghad M, Oosterwegel M, Bonnin J, Vagner C, Bonnet H, Dikkes P, Sharpe A et al (2000) p73-deficient mice have neurological, pheromonal and inflammatory defects but lack spontaneous tumours. Nature 404:99–103

Yang A, Kaghad M, Caput D, McKeon F (2002) On the shoulders of giants: p63, p73 and the rise of p53. Trends Genet 18:90–95

Yoshida M, Nishikawa M, Nishi K, Abe K, Horinouchi S, Beppu T (1990) Effects of leptomycin B on the cell cycle of fibroblasts and fission yeast cells. Exp Cell Res 187:150–156

Zache N, Lambert JM, Wiman KG, Bykov VJ (2008a) PRIMA-1^{MET} inhibits growth of mouse tumors carrying mutant p53. Cell Oncol 30:411–418

Zache N, Lambert JM, Rokaeus N, Shen J, Hainaut P, Bergman J, Wiman KG, Bykov VJ (2008b) Mutant p53 targeting by the low molecular weight compound STIMA-1. Mol Oncol 2:70–80

Chapter 15
p53 Immunotherapy of Cancer

Hakim Echchannaoui and Matthias Theobald

1 p53: Guardian of the Genome

Cancer is usually associated with deregulated cell cycle progression and impaired apoptosis induction due to the activation of proto-oncogenes and/or inactivation of tumor suppressor genes (Hanahan and Weinberg 2000). Mutation of the *TP53* tumor suppressor gene, the "guardian of the genome," is found in ~50 % of human cancers. The p53 molecule plays a critical role in tumor suppression mainly by triggering growth arrest, apoptosis, and senescence. Mutation abrogates the regulatory function of p53 on cell cycle control and in many cases leads to increased half-life of the otherwise very unstable p53 molecule and thereby to accumulation of p53 protein (Hollstein et al. 1991). The persistent overexpression of mutated p53 in tumors makes this protein a potential target antigen for immunotherapy of cancer. Furthermore, in wt p53 tumors, p53 can be functionally inhibited, for instance as a result of human papillomavirus (HPV) E6 or murine double minute-2 (MDM2) oncoprotein expression mediating p53 degradation (Haupt et al. 1997; Kubbutat et al. 1997; Scheffner et al. 1993). Such abnormalities can also arise from alterations in the expression of other genes closely related to *TP53*, such as cyclin-dependent kinase inhibitor 2A (*CDKN2A*) encoding for $p14^{ARF}$ protein, which acts as a tumor suppressor, binds to MDM2, and blocks MDM2-induced degradation of p53 (Ashcroft and Vousden 1999). Altogether, perturbation of the p53 regulatory pathway occurs in most human cancers, which makes p53 widely attractive target for the immunotherapy of cancer.

H. Echchannaoui • M. Theobald (✉)
Department of Hematology, Oncology and Pneumology, Johannes Gutenberg-University Medical Center, Langenbeckstrasse 1, 55131 Mainz, Germany
e-mail: direktor-3med@unimedizin-mainz.de

P. Hainaut et al. (eds.), *p53 in the Clinics*, DOI 10.1007/978-1-4614-3676-8_15,

2 p53 As a Target for Immunotherapy

Because p53 is an intracellular protein, endogenously p53-derived peptides are processed by the major histocompatibility complex (MHC) class I pathway. Thus, targeting tumors through the p53 antigen will rely on the recognition of p53-processed epitopes by T lymphocytes in the context of MHC molecules at the surface of malignant cells. As mutation in the *TP53* gene is a frequent event in cancer, two types of p53 epitopes can be generated: mutated and wt peptides. Mutant p53 peptides would then represent true tumor-antigens (non-self). However, mutations in p53 are highly diverse (Hollstein et al. 1991). Vaccination approaches targeting p53 mutated epitopes are applicable to only a limited number of patients that harbor these mutations. Consequently, immunotherapy based on the mutated part of p53 would therefore require prior identification of the p53 mutation in patient tumors, followed by generating "custom-tailored" vaccines for each patient (Carbone et al. 2005). Furthermore, there remains the question of the immunogenicity of such mutated processed peptides. To date, only few examples of mutant p53-derived T-cell epitopes have been reported in mice (Noguchi et al. 1994; Yanuck et al. 1993) or humans (Ito et al. 2007). Therefore, research has been focused on targeting wt p53 epitopes as they can be presented by many tumor cells, including those that exhibit abnormalities in p53 function due to other causes than mutations in p53. An important advantage of using wt p53-based T-cell antigens is their broad applicability.

3 MHC-Restricted p53-Specific Cytotoxic T Lymphocytes

Because the majority of solid tumors lack expression of MHC class II, p53-specific T-cell response has been mainly focused on MHC class I-restricted cytotoxic T lymphocytes (CTL). Several groups, including ourselves, have found evidence that wt p53-derived peptides can be presented in the context of murine and human MHC class I molecules. Accordingly, we demonstrated that tumor cells presenting such epitopes can be recognized by wt p53-specific CTLs in vitro (Kuball et al. 2002; Kuckelkorn et al. 2002; Theobald et al. 1995, 1997, 1998) and that such CTLs are able to efficiently eradicate tumors in vivo (Vierboom et al. 1997, 2000). These preclinical mouse studies revealed that p53-specific CTLs, generated in p53-deficient mice, were capable of eradicating established tumors in p53-proficient mice without any detectable damage to normal tissue (Vierboom et al. 1997).

Studies were also conducted in humans to analyze p53-specific T-cell responses against tumors. We demonstrated that proteasomal processing of the natural HLA-A2.1(A2.1)-restricted CTL epitope p53(264–272) and of a putative precursor peptide is profoundly affected by a mutational p53 hotspot (R to H) at the COOH-terminal flanking residue 273. As a consequence, target cells that overexpress p53 harboring the 273 R to H mutation are not susceptible to in vitro and in vivo lysis by A2.1-restricted CTLs specific for the flanking p53(264–272) epitope (Theobald et al. 1998).

A2.1-restricted wt p53(264–272) CTLs generated from circulating precursor T cells of A2.1(+) healthy donors and patients with squamous cell carcinoma of the head and neck (SCCHN) (Hoffmann et al. 2000) revealed that patients with SCCHN had a significantly higher frequency of p53-specific CD8+ T cells yet it did not correlate with p53 accumulation in the tumor (Hoffmann et al. 2002). These observations are consistent with the argument that high steady-state levels of p53 are not a prerequisite for tumor regression by p53-specific CTL (Vierboom et al. 2000). Instead, p53 turnover appeared to be an important factor affecting the susceptibility of malignant cells to CTL (Vierboom et al. 2000). These results suggest that in vivo, wt p53(264–272)-specific CTL might play a role in the eradication of tumor cells expressing this epitope and ignore tumor cells which lost the epitope. Altogether, tumors expressing wt p53 may also serve as targets for killing by p53-specific CTL. Thus, CTL-mediated immunotherapy targeting p53 can be extended to a wider range of tumors.

4 Self-Tolerance to p53

An effective p53 CTL-mediated immunotherapy needs to circumvent self-tolerance to this ubiquitously expressed molecule (Rogel et al. 1985). Studies involving adoptive transfer of p53-specific CTL into tumor-bearing mice report CTL-mediated antitumor responses in vivo. In some studies, the CTL were capable of delaying but not preventing tumor outgrowth (Hilburger and Abrams 2001; Peralta et al. 1999), while other reports showed that such CTL can entirely eradicate tumors (Vierboom et al. 1997, 2000). It should be emphasized that CTL isolated from p53-proficient mice were used in the two former studies, whereas the latter reports took advantage of CTL generated in p53-deficient animals. In fact, we have unambiguously demonstrated that a considerable degree of p53-specific self-tolerance exists in p53-proficient as opposed to p53-deficient mice at the MHC class I-restricted T lymphocyte level (Kuball et al. 2002; Theobald et al. 1997). This p53-specific self-tolerance results in a peripheral p53-reactive T-cell repertoire that is devoid of high-avidity MHC class I-restricted CTL with only low-avidity, if any, T lymphocytes left (Kuball et al. 2002; Theobald et al. 1997; Vierboom et al. 1997; Hernandez et al. 2000, 2001). In another study, tolerance to human p53 was addressed in a transgenic mouse model expressing the core domain of the human p53 gene; this mouse is referred to as *hu*man *p*53 *k*nock-*in* (Hupki) (Luo et al. 2001). In contrast to normal Balb/c, Hupki mice immunized with modified vaccinia virus Ankara (MVA) expressing human p53 did not develop p53-specific CTL responses (Song et al. 2007). Repetitive immunizations followed by several rounds of in vitro stimulation were required to generate p53-specific CTL responses in Hupki mice.

The wt p53 protein is expressed in the thymus (Rogel et al. 1985). Although steady-state level of p53 in thymic antigen presenting cells (APC) is low, due to the high sensitivity of wt p53 to ubiquitin/proteasome-mediated degradation, it is also likely to result in processing of p53-derived epitopes into MHC class I molecules of thymic APC. It is therefore conceivable that high-avidity immature T cells encounter wt p53 epitopes in the context of MHC class I in the thymus and thus are deleted (negative selection) from the repertoire. Regardless of whether or not MHC class I-restricted high-avidity p53-specific CTL are affected primarily by negative selection in the thymus (central tolerance) or by peripheral mechanisms of tolerance induction (Heath et al. 1998; Walker and Abbas 2002), it is obvious that this p53-reactive CTL immune repertoire is almost only limited to low-avidity CTL.

5 Circumventing p53 Self-Tolerance by T-Cell Antigen Receptor Gene Transfer

Although a wide range of vaccines targeting wt p53 epitopes implemented in preclinical murine models have shown evidence for their in vivo efficacy (Song et al. 2007; Nikitina et al. 2002), in human clinical trials peptide-based vaccine could effectively generate a quantifiable T-cell-specific immune response in the PBMC of cancer patients; however, such a response was often associated with rare clinical responses (Antonia et al. 2006) or did not correlate with a clinically evident regression of tumors (Lee et al. 1999; Leffers et al. 2009; Speetjens et al. 2009).

The adoptive transfer of tumor-reactive cells is a promising approach in the treatment of human malignancies, but the challenge of isolating T cells with high avidity for tumor antigens in each patient has limited application of this approach. The transfer into T cells of T-cell antigen receptor (TCR) genes encoding high-affinity TCRs recognizing defined tumor-associated antigens, using retroviral gene transfer technology (Fig. 15.1), can potentially circumvent this obstacle (Schumacher 2002). We demonstrated that TCR gene transfer can be used to circumvent self-tolerance of autologous T cells to tumor-associated antigens by transferring a humanized MDM2-specific TCR selected from A2.1 transgenic mice (Stanislawski et al. 2001). Thus, it is potentially possible to generate within a short period of time a large numbers of T cells with antigen-specific receptors such as tumor-reactive T cells. In preclinical mouse models including immunodeficient NOD/SCID mice, it has been demonstrated that redirected human T cells prevented engraftment of a leukemia cell line as well as autologous primary leukemic cells (Xue et al. 2005).

In the first clinical trial using TCR gene transfer, T cells engineered to express a A2.1-restricted TCR specific for glycoprotein 100 (gp100) were adoptively transferred into lymphodepleted patients with progressive metastatic melanoma, resulting in cancer regression yet, the persistence of engineered T cells was observed only in some of the patients (Morgan et al. 2006). There are several factors that might affect the function of adoptively transferred T cells. It is clear now that increasing TCR expression has a profound effect on in vivo antitumor activity (de Witte et al. 2008).

We used A2.1 transgenic mice, in which the mouse CD8 molecule cannot efficiently interact with the α3 domain of A2.1 (Sherman et al. 1992), to generate a

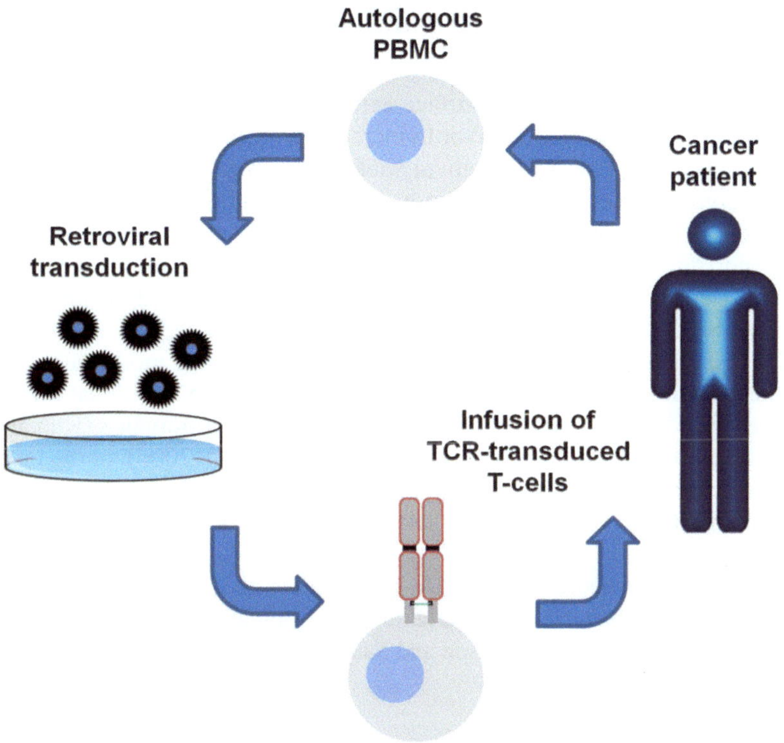

Fig. 15.1 For the generation of retroviruses, packaging cell lines (Phoenix) are transfected with retroviral constructs encoding for antigen-specific TCRs. The retroviral supernatant is used to transduce autologous T lymphocytes isolated from cancer patients. TCR-gene modified T cells are then infused back into lymphodepleted patients to induce tumor regression

high-affinity, CD8-independent TCR specific for the tumor-associated human p53(264–272)-antigen. Retroviral expression of this CD8-independent, p53-specific TCR into human T cells imparted the CD8+ T lymphocytes with broad tumor-specific CTL activity and turned CD4+ T cells into potent tumor-reactive, p53A2.1-specific Th cells (Kuball et al. 2005). Using a similar CD8-independent, A2.1-restricted murine anti-p53 TCR, Cohen and coworkers could demonstrate the recognition and killing of primary tumors from melanoma patients by CD8+ and CD4+ TCR-transduced lymphocytes (Cohen et al. 2005). Because wt p53 epitopes are expressed in most tissues, adoptive transfer of T cells directed against p53 (264–272)-peptide can be associated with "on-target" toxicity.

However, such p53 autoimmunity is unlikely to occur with p53TCR-transduced T cells as we and others did not observe reactivity against normal A2.1+ cells, including hepatocytes, fibroblasts, or PBMC (Kuball et al. 2005; Cohen et al. 2005).

6 Optimizing TCR for p53 Immunotherapy

Strategies to enhance TCR cell surface expression (Fig. 15.2) using codon optimized TCR (Scholten et al. 2006) or to increase the affinity and functional avidity of TCR-transduced T cells by removing defined *N*-glycosylation sites in the TCR constant domain have been shown to be effective with mouse and human TCRs specific for different antigens (Kuball et al. 2009). However, a safety concern of TCR gene transfer is the potential pairing of endogenous and exogenous TCR chains that could result in the formation of new TCRs directed against self-antigens (off-target autoimmunity), as shown in a mouse model of adoptive T-cell transfer (Bendle et al. 2010). We have shown that sequence modifications of the introduced TCR constant chains like the introduction of an additional inter-chain disulfide bond between the TCR α and β chain constant domains (Kuball et al. 2007), the use of human-murine hybrid TCR (Cohen et al. 2006), or the inversion of amino acid residues in the constant region of the TCR α and β chains that form the TCR interface (Voss et al. 2008) can reduce the formation of TCR mixed heterodimers.

We further generated a murine CD8-independent p53(264–272)-specific single chain (sc)TCR by connecting the variable TCRα domain to the TCRβ chain via a short peptide linker (SL7) (Robinson and Sauer 1998). However, despite detectable

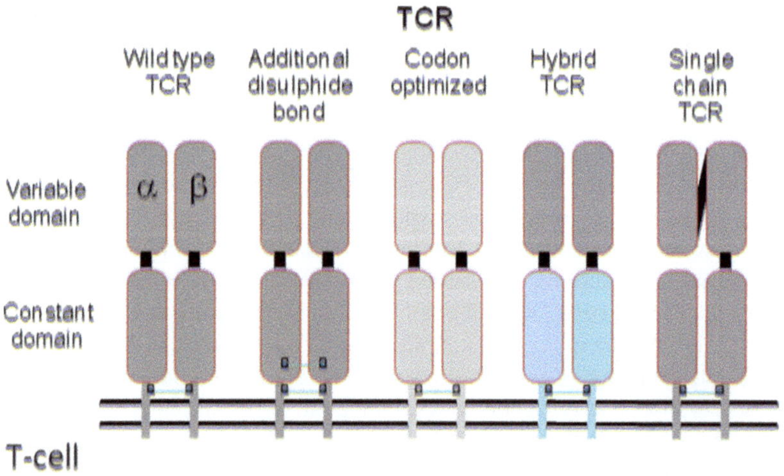

Fig. 15.2 Strategies to improve the expression of the introduced TCR include additional cysteine residues into the α and β constant chains to induce an additional disulfide bond between the modified TCR chains, codon modification of TCR to avoid mRNA instability motifs, human/murine hybrid TCR and single chain TCR co-expressed with the constant α chain. These molecular modifications aim to reduce the potential pairing of endogenous and exogenous TCR chains that could result in the formation of new TCRs with unknown specificity

surface expression of p53(264–272) scTCR in human T cells, specific cytolytic activity was not observed. Therefore, we increased the stability of the scTCR molecule by coexpression of a truncated constant TCR α (Cα) domain. This strategy resulted in p53(264–272) scTCR constructs that mediated efficient antitumor reactivity in human T lymphocytes (Voss et al. 2010).

7 T Cell Subsets for Adoptive T Cell Transfer

Several groups have reported CD4+ Th responses against p53 in mice and humans (Fedoseyeva et al. 2000; van der Burg et al. 2003; Zwaveling et al. 2002). A p53-specific memory T-cell response was even reported in tumor infiltrating T cells and in tumor draining lymph nodes in ovarian cancer patients (Lambeck et al. 2007), suggesting that p53-specific CD4+ Th cells are less affected by self-tolerance (Lauwen et al. 2008). Several immunogenic HLA class II-restricted wt p53 epitopes have been identified (Chikamatsu et al. 2003; Fujita et al. 1998; Ito et al. 2006).

Adoptive transfer of p53-specific CD4+ Th cells can induce efficient tumor eradication in mice bearing p53-overexpressing tumors (Zwaveling et al. 2002; Lauwen et al. 2008). Tumor-specific CD4+ T cells have been shown to induce recruitment of CD8+ T cells within the tumor milieu and to enhance tumor lysis (Wong et al. 2008). CD4+ T cells increase the number and cytolytic function of tumor infiltrating CD8+ T cells by IFNg and IL-2 secretion which enhance granzyme B expression (Bos and Sherman 2010). In a recent study, adoptive transfer of naive tumor-specific CD4+ T cells into lymphopenic animals bearing large established melanoma caused tumor regression. Moreover, this antitumor effect was independent of vaccination, cytokine treatment, and CD8+, B, NK, and NKT cells (Xie et al. 2010).

In several other mouse models, similar studies found that tumor-specific CD4+ Th cells critically contribute to the development and efficacy of antitumor immune responses (Hung et al. 1998). Th lymphocytes were also shown to exert their tumor-reactive effect independently of CTL via recruitment of innate immune effector cells, such as tumoricidal macrophages and eosinophils (Hung et al. 1998). In our laboratory, we could demonstrate that p53TCR-engineered CD4+ Th cells and CD8+ CTL were cooperative and interacted synergistically with dendritic cell intermediates and tumor targets (Kuball et al. 2005). By triggering professional APC through CD40–CD40L interaction (Bennett et al. 1998; Ridge et al. 1998; Schoenberger et al. 1998) and potentially by IFNg release, Th delivers an essential activation signal to APC. As a result, the APC will display its entire array of antigen presenting and co-stimulatory functions, the latter of which are considered key for inducing full activation of CTL. Whereas the delivery of this type of T-cell help for CTL requires the APC to present both the MHC class I- and class II-restricted peptides, these antigens do not have to be derived necessarily from the very same protein. These studies highlight the important role of CD4+ Th cells in antitumor responses mediated by CD8+ CTL and innate immune cells.

The role of regulatory T cells (Tregs) in p53 immunotherapy needs to be addressed. Several studies described the presence of Treg at the tumor sites of head and neck cancer (HNC) patients (Albers et al. 2005) and in the periphery of patients with breast cancer (Svane et al. 2008). These cells may hamper effective antitumor specific responses and thus contribute to tumoral immune escape and disease progression.

An additional challenge of adoptive therapy is to establish a persistent memory T-cell response to prevent disease recurrence. The persistence of transferred T cells is crucial for an effective adoptive therapy, as it correlates with cancer regression in patients receiving cell transfer therapy (Robbins et al. 2004). CD8+ T cells exist as naïve (T_N), central memory (T_{CM}), and effector memory (T_{EM}) populations, with distinct phenotype and function (Sallusto et al. 1999). Studies with primates demonstrated that adoptive transfer only of T_{CM} cells leads to long-term reconstitution of T_{EM} in these animals (Berger et al. 2008). The long persistence of T_{CM} has been confirmed in a mouse model of adoptive T-cell transfer (Hinrichs et al. 2009); however, effector cells derived from T_N were shown to possess superior proliferation capacity, cytokine release, and tumor treatment efficacy after transfer.

Likewise, strategies, such as in vitro depletion of Tregs and selection of "effective" T cells (T_N and/or T_{CM}) prior adoptive transfer, may confer a superior efficacy for the treatment of malignant diseases.

8 Adoptive p53 Immunotherapy in Cancer Patients

Based on preclinical studies, therapeutic trials of adoptive transfer of p53TCR-specific T lymphocytes were conducted in cancer patients. The group of Steven A. Rosenberg (National Cancer Institute, National Institutes of Health) has recently completed two clinical trials in cancer patients treated with anti-p53 TCR-engineered peripheral blood lymphocytes (http://clinicaltrials.gov/).

- Phase II Study of Metastatic Cancer That Overexpresses p53 Using Lymphodepleting Conditioning Followed by Infusion of Anti-p53 TCR-Gene Engineered Lymphocytes (NCT00393029). In this study, patients with a variety of epithelial cancers were treated with nonmyeloablative, lymphodepleting chemotherapy followed by adoptive transfer of lymphocytes genetically engineered to express murine T-cell receptors specific for p53(264–272) and infusion of high-dose IL-2. Partial results of this clinical trial have been reported recently (Davis et al. 2010). Among ten patients enrolled in this study, one patient experienced objective, partial tumor regression. Importantly, TCR-gene modified cells were detectable in the peripheral blood of all patients at 1 month after adoptive transfer.
- Phase II Study of Metastatic Cancer That Overexpresses p53 Using Lymphodepleting Conditioning Followed by Infusion of Anti-p53 TCR-Gene Engineered Lymphocytes and Dendritic Cell Vaccination (NCT00704938). The results of this study have not been published so far.

9 Conclusion

It is evident that wt p53-specific CTLs are a highly desirable tool for targeted immunotherapy of a wide range of cancers. Although self-tolerance appears to hamper the generation of high-avidity anti-p53 CTL under various conditions, alternative strategies, such as TCR gene transfer, can circumvent this obstacle. Furthermore, transfer of high-affinity CD8-independent TCR recognizing wt p53-derived CTL epitopes by both CD4+ and CD8+ T cells represents a promising approach in adoptive T-cell therapy for cancer patients as efficient immune response against malignant disease requires a full cooperation between both T-cell subsets.

Acknowledgments This work was supported in part by a grant from the Deutsche Forschungsgemeinschaft (DFG) (KFO183 TP1 to H.E. and M.T.).

References

Albers AE, Ferris RL, Kim GG, Chikamatsu K, DeLeo AB, Whiteside TL (2005) Immune responses to p53 in patients with cancer: enrichment in tetramer+p53 peptide-specific T cells and regulatory T cells at tumor sites. Cancer Immunol Immunother 54:1072–1081

Antonia SJ, Mirza N, Fricke I, Chiappori A, Thompson P, Williams N, Bepler G, Simon G, Janssen W, Lee JH, Menander K, Chada S, Gabrilovich DI (2006) Combination of p53 cancer vaccine with chemotherapy in patients with extensive stage small cell lung cancer. Clin Cancer Res 12:878–887

Ashcroft M, Vousden KH (1999) Regulation of p53 stability. Oncogene 18:7637–7643

Bendle GM, Linnemann C, Hooijkaas AI, Bies L, de Witte MA, Jorritsma A, Kaiser AD, Pouw N, Debets R, Kieback E, Uckert W, Song JY, Haanen JB, Schumacher TN (2010) Lethal graft-versus-host disease in mouse models of T cell receptor gene therapy. Nat Med 16(565–70):1p

Bennett SR, Carbone FR, Karamalis F, Flavell RA, Miller JF, Heath WR (1998) Help for cytotoxic-T-cell responses is mediated by CD40 signalling. Nature 393:478–480

Berger C, Jensen MC, Lansdorp PM, Gough M, Elliott C, Riddell SR (2008) Adoptive transfer of effector CD8+ T cells derived from central memory cells establishes persistent T cell memory in primates. J Clin Invest 118:294–305

Bos R, Sherman LA (2010) CD4+ T-cell help in the tumor milieu is required for recruitment and cytolytic function of CD8+ T lymphocytes. Cancer Res 70:8368–8377

Carbone DP, Ciernik IF, Kelley MJ, Smith MC, Nadaf S, Kavanaugh D, Maher VE, Stipanov M, Contois D, Johnson BE, Pendleton CD, Seifert B, Carter C, Read EJ, Greenblatt J, Top LE, Kelsey MI, Minna JD, Berzofsky JA (2005) Immunization with mutant p53- and K-ras-derived peptides in cancer patients: immune response and clinical outcome. J Clin Oncol 23:5099–5107

Chikamatsu K, Albers A, Stanson J, Kwok WW, Appella E, Whiteside TL, DeLeo AB (2003) P53(110-124)-specific human CD4+ T-helper cells enhance in vitro generation and antitumor function of tumor-reactive CD8+ T cells. Cancer Res 63:3675–3681

Cohen CJ, Zheng Z, Bray R, Zhao Y, Sherman LA, Rosenberg SA, Morgan RA (2005) Recognition of fresh human tumor by human peripheral blood lymphocytes transduced with a bicistronic retroviral vector encoding a murine anti-p53 TCR. J Immunol 175:5799–5808

Cohen CJ, Zhao Y, Zheng Z, Rosenberg SA, Morgan RA (2006) Enhanced antitumor activity of murine-human hybrid T-cell receptor (TCR) in human lymphocytes is associated with improved pairing and TCR/CD3 stability. Cancer Res 66:8878–8886

Davis JL, Theoret MR, Zheng Z, Lamers CH, Rosenberg SA, Morgan RA (2010) Development of human anti-murine T-cell receptor antibodies in both responding and nonresponding patients enrolled in TCR gene therapy trials. Clin Cancer Res 16:5852–5861

de Witte MA, Jorritsma A, Kaiser A, van den Boom MD, Dokter M, Bendle GM, Haanen JB, Schumacher TN (2008) Requirements for effective antitumor responses of TCR transduced T cells. J Immunol 181:5128–5136

Fedoseyeva EV, Boisgerault F, Anosova NG, Wollish WS, Arlotta P, Jensen PE, Ono SJ, Benichou G (2000) CD4+ T cell responses to self- and mutated p53 determinants during tumorigenesis in mice. J Immunol 164:5641–5651

Fujita H, Senju S, Yokomizo H, Saya H, Ogawa M, Matsushita S, Nishimura Y (1998) Evidence that HLA class II-restricted human CD4+ T cells specific to p53 self peptides respond to p53 proteins of both wild and mutant forms. Eur J Immunol 28:305–316

Hanahan D, Weinberg RA (2000) The hallmarks of cancer. Cell 100:57–70

Haupt Y, Maya R, Kazaz A, Oren M (1997) Mdm2 promotes the rapid degradation of p53. Nature 387:296–299

Heath WR, Kurts C, Miller JF, Carbone FR (1998) Cross-tolerance: a pathway for inducing tolerance to peripheral tissue antigens. J Exp Med 187:1549–1553

Hernandez J, Lee PP, Davis MM, Sherman LA (2000) The use of HLA A2.1/p53 peptide tetramers to visualize the impact of self tolerance on the TCR repertoire. J Immunol 164:596–602

Hernandez J, Ko A, Sherman LA (2001) CTLA-4 blockade enhances the CTL responses to the p53 self-tumor antigen. J Immunol 166:3908–3914

Hilburger RM, Abrams SI (2001) Characterization of CD8+ cytotoxic T lymphocyte/tumor cell interactions reflecting recognition of an endogenously expressed murine wild-type p53 determinant. Cancer Immunol Immunother 49:603–612

Hinrichs CS, Borman ZA, Cassard L, Gattinoni L, Spolski R, Yu Z, Sanchez-Perez L, Muranski P, Kern SJ, Logun C, Palmer DC, Ji Y, Reger RN, Leonard WJ, Danner RL, Rosenberg SA, Restifo NP (2009) Adoptively transferred effector cells derived from naive rather than central memory CD8+ T cells mediate superior antitumor immunity. Proc Natl Acad Sci USA 106:17469–17474

Hoffmann TK, Nakano K, Elder EM, Dworacki G, Finkelstein SD, Appella E, Whiteside TL, DeLeo AB (2000) Generation of T cells specific for the wild-type sequence p53(264-272) peptide in cancer patients: implications for immunoselection of epitope loss variants. J Immunol 165:5938–5944

Hoffmann TK, Donnenberg AD, Finkelstein SD, Donnenberg VS, Friebe-Hoffmann U, Myers EN, Appella E, DeLeo AB, Whiteside TL (2002) Frequencies of tetramer+T cells specific for the wild-type sequence p53(264-272) peptide in the circulation of patients with head and neck cancer. Cancer Res 62:3521–3529

Hollstein M, Sidransky D, Vogelstein B, Harris CC (1991) p53 mutations in human cancers. Science 253:49–53

Hung K, Hayashi R, Lafond-Walker A, Lowenstein C, Pardoll D, Levitsky H (1998) The central role of CD4(+) T cells in the antitumor immune response. J Exp Med 188:2357–2368

Ito D, Albers A, Zhao YX, Visus C, Appella E, Whiteside TL, DeLeo AB (2006) The wild-type sequence (wt) p53(25-35) peptide induces HLA-DR7 and HLA-DR11-restricted CD4+ Th cells capable of enhancing the ex vivo expansion and function of anti-wt p53(264-272) peptide CD8+ T cells. J Immunol 177:6795–6803

Ito D, Visus C, Hoffmann TK, Balz V, Bier H, Appella E, Whiteside TL, Ferris RL, DeLeo AB (2007) Immunological characterization of missense mutations occurring within cytotoxic T cell-defined p53 epitopes in HLA-A*0201+ squamous cell carcinomas of the head and neck. Int J Cancer 120:2618–2624

Kuball J, Schuler M, Antunes FE, Herr W, Neumann M, Obenauer-Kutner L, Westreich L, Huber C, Wolfel T, Theobald M (2002) Generating p53-specific cytotoxic T lymphocytes by recombinant adenoviral vector-based vaccination in mice, but not man. Gene Ther 9:833–843

Kuball J, Schmitz FW, Voss RH, Ferreira EA, Engel R, Guillaume P, Strand S, Romero P, Huber C, Sherman LA, Theobald M (2005) Cooperation of human tumor-reactive CD4+ and CD8+ T cells after redirection of their specificity by a high-affinity p53A2.1-specific TCR. Immunity 22:117–129

Kuball J, Dossett ML, Wolfl M, Ho WY, Voss RH, Fowler C, Greenberg PD (2007) Facilitating matched pairing and expression of TCR chains introduced into human T cells. Blood 109:2331–2338

Kuball J, Hauptrock B, Malina V, Antunes E, Voss RH, Wolfl M, Strong R, Theobald M, Greenberg PD (2009) Increasing functional avidity of TCR-redirected T cells by removing defined N-glycosylation sites in the TCR constant domain. J Exp Med 206:463–475

Kubbutat MH, Jones SN, Vousden KH (1997) Regulation of p53 stability by Mdm2. Nature 387:299–303

Kuckelkorn U, Ferreira EA, Drung I, Liewer U, Kloetzel PM, Theobald M (2002) The effect of the interferon-gamma-inducible processing machinery on the generation of a naturally tumor-associated human cytotoxic T lymphocyte epitope within a wild-type and mutant p53 sequence context. Eur J Immunol 32:1368–1375

Lambeck A, Leffers N, Hoogeboom BN, Sluiter W, Hamming I, Klip H, ten Hoor K, Esajas M, van Oven M, Drijfhout JW, Platteel I, Offringa R, Hollema H, Melief K, van der Burg S, van der Zee A, Daemen T, Nijman H (2007) P53-specific T cell responses in patients with malignant and benign ovarian tumors: implications for p53 based immunotherapy. Int J Cancer 121:606–614

Lauwen MM, Zwaveling S, de Quartel L, Ferreira Mota SC, Grashorn JA, Melief CJ, van der Burg SH, Offringa R (2008) Self-tolerance does not restrict the CD4+ T-helper response against the p53 tumor antigen. Cancer Res 68:893–900

Lee KH, Wang E, Nielsen MB, Wunderlich J, Migueles S, Connors M, Steinberg SM, Rosenberg SA, Marincola FM (1999) Increased vaccine-specific T cell frequency after peptide-based vaccination correlates with increased susceptibility to in vitro stimulation but does not lead to tumor regression. J Immunol 163:6292–6300

Leffers N, Lambeck AJ, Gooden MJ, Hoogeboom BN, Wolf R, Hamming IE, Hepkema BG, Willemse PH, Molmans BH, Hollema H, Drijfhout JW, Sluiter WJ, Valentijn AR, Fathers LM, Oostendorp J, van der Zee AG, Melief CJ, van der Burg SH, Daemen T, Nijman HW (2009) Immunization with a P53 synthetic long peptide vaccine induces P53-specific immune responses in ovarian cancer patients, a phase II trial. Int J Cancer 125:2104–2113

Luo JL, Yang Q, Tong WM, Hergenhahn M, Wang ZQ, Hollstein M (2001) Knock-in mice with a chimeric human/murine p53 gene develop normally and show wild-type p53 responses to DNA damaging agents: a new biomedical research tool. Oncogene 20:320–328

Morgan RA, Dudley ME, Wunderlich JR, Hughes MS, Yang JC, Sherry RM, Royal RE, Topalian SL, Kammula US, Restifo NP, Zheng Z, Nahvi A, de Vries CR, Rogers-Freezer LJ, Mavroukakis SA, Rosenberg SA (2006) Cancer regression in patients after transfer of genetically engineered lymphocytes. Science 314:126–129

Nikitina EY, Chada S, Muro-Cacho C, Fang B, Zhang R, Roth JA, Gabrilovich DI (2002) An effective immunization and cancer treatment with activated dendritic cells transduced with full-length wild-type p53. Gene Ther 9:345–352

Noguchi Y, Chen YT, Old LJ (1994) A mouse mutant p53 product recognized by CD4+ and CD8+ T cells. Proc Natl Acad Sci USA 91:3171–3175

Peralta EA, Liu X, McCarthy TM, Wilson TG, Diamond DJ, Ellenhorn JD (1999) Immunotherapy of bladder cancer targeting P53. J Urol 162:1806–1811

Ridge JP, Di Rosa F, Matzinger P (1998) A conditioned dendritic cell can be a temporal bridge between a CD4+ T-helper and a T-killer cell. Nature 393:474–478

Robbins PF, Dudley ME, Wunderlich J, El Gamil M, Li YF, Zhou J, Huang J, Powell DJ Jr, Rosenberg SA (2004) Cutting edge: persistence of transferred lymphocyte clonotypes correlates with cancer regression in patients receiving cell transfer therapy. J Immunol 173: 7125–7130

Robinson CR, Sauer RT (1998) Optimizing the stability of single-chain proteins by linker length and composition mutagenesis. Proc Natl Acad Sci USA 95:5929–5934

Rogel A, Popliker M, Webb CG, Oren M (1985) p53 cellular tumor antigen: analysis of mRNA levels in normal adult tissues, embryos, and tumors. Mol Cell Biol 5:2851–2855

Sallusto F, Lenig D, Forster R, Lipp M, Lanzavecchia A (1999) Two subsets of memory T lymphocytes with distinct homing potentials and effector functions. Nature 401:708–712
Scheffner M, Huibregtse JM, Vierstra RD, Howley PM (1993) The HPV-16 E6 and E6-AP complex functions as a ubiquitin-protein ligase in the ubiquitination of p53. Cell 75:495–505
Schoenberger SP, Toes RE, van der Voort EI, Offringa R, Melief CJ (1998) T-cell help for cytotoxic T lymphocytes is mediated by CD40-CD40L interactions. Nature 393:480–483
Scholten KB, Kramer D, Kueter EW, Graf M, Schoedl T, Meijer CJ, Schreurs MW, Hooijberg E (2006) Codon modification of T cell receptors allows enhanced functional expression in transgenic human T cells. Clin Immunol 119:135–145
Schumacher TN (2002) T-cell-receptor gene therapy. Nat Rev Immunol 2:512–519
Sherman LA, Hesse SV, Irwin MJ, La Face D, Peterson P (1992) Selecting T cell receptors with high affinity for self-MHC by decreasing the contribution of CD8. Science 258:815–818
Song GY, Gibson G, Haq W, Huang EC, Srivasta T, Hollstein M, Daftarian P, Wang Z, Diamond D, Ellenhorn JD (2007) An MVA vaccine overcomes tolerance to human p53 in mice and humans. Cancer Immunol Immunother 56:1193–1205
Speetjens FM, Kuppen PJ, Welters MJ, Essahsah F, Voet van den Brink AM, Lantrua MG, Valentijn AR, Oostendorp J, Fathers LM, Nijman HW, Drijfhout JW, van de Velde CJ, Melief CJ, van der Burg SH (2009) Induction of p53-specific immunity by a p53 synthetic long peptide vaccine in patients treated for metastatic colorectal cancer. Clin Cancer Res 15:1086–1095
Stanislawski T, Voss RH, Lotz C, Sadovnikova E, Willemsen RA, Kuball J, Ruppert T, Bolhuis RL, Melief CJ, Huber C, Stauss HJ, Theobald M (2001) Circumventing tolerance to a human MDM2-derived tumor antigen by TCR gene transfer. Nat Immunol 2:962–970
Svane IM, Pedersen AE, Nikolajsen K, Zocca MB (2008) Alterations in p53-specific T cells and other lymphocyte subsets in breast cancer patients during vaccination with p53-peptide loaded dendritic cells and low-dose interleukin-2. Vaccine 26:4716–4724
Theobald M, Biggs J, Dittmer D, Levine AJ, Sherman LA (1995) Targeting p53 as a general tumor antigen. Proc Natl Acad Sci USA 92:11993–11997
Theobald M, Biggs J, Hernandez J, Lustgarten J, Labadie C, Sherman LA (1997) Tolerance to p53 by A2.1-restricted cytotoxic T lymphocytes. J Exp Med 185:833–841
Theobald M, Ruppert T, Kuckelkorn U, Hernandez J, Haussler A, Ferreira EA, Liewer U, Biggs J, Levine AJ, Huber C, Koszinowski UH, Kloetzel PM, Sherman LA (1998) The sequence alteration associated with a mutational hotspot in p53 protects cells from lysis by cytotoxic T lymphocytes specific for a flanking peptide epitope. J Exp Med 188:1017–1028
van der Burg SH, Menon AG, Redeker A, Franken KL, Drijfhout JW, Tollenaar RA, Hartgrink HH, van de Velde CJ, Kuppen PJ, Melief CJ, Offringa R (2003) Magnitude and polarization of P53-specific T-helper immunity in connection to leukocyte infiltration of colorectal tumors. Int J Cancer 107:425–433
Vierboom MP, Nijman HW, Offringa R, van der Voort EI, van Hall T, van den Broek L, Fleuren GJ, Kenemans P, Kast WM, Melief CJ (1997) Tumor eradication by wild-type p53-specific cytotoxic T lymphocytes. J Exp Med 186:695–704
Vierboom MP, Zwaveling S, Bos GMJ, Ooms M, Krietemeijer GM, Melief CJ, Offringa R (2000) High steady-state levels of p53 are not a prerequisite for tumor eradication by wild-type p53-specific cytotoxic T lymphocytes. Cancer Res 60:5508–5513
Voss RH, Willemsen RA, Kuball J, Grabowski M, Engel R, Intan RS, Guillaume P, Romero P, Huber C, Theobald M (2008) Molecular design of the Calphabeta interface favors specific pairing of introduced TCRalphabeta in human T cells. J Immunol 180:391–401
Voss RH, Thomas S, Pfirschke C, Hauptrock B, Klobuch S, Kuball J, Grabowski M, Engel R, Guillaume P, Romero P, Huber C, Beckhove P, Theobald M (2010) Coexpression of the T-cell receptor constant alpha domain triggers tumor reactivity of single-chain TCR-transduced human T cells. Blood 115:5154–5163
Walker LS, Abbas AK (2002) The enemy within: keeping self-reactive T cells at bay in the periphery. Nat Rev Immunol 2:11–19
Wong SB, Bos R, Sherman LA (2008) Tumor-specific CD4+ T cells render the tumor environment permissive for infiltration by low-avidity CD8+ T cells. J Immunol 180:3122–3131

Xie Y, Akpinarli A, Maris C, Hipkiss EL, Lane M, Kwon EK, Muranski P, Restifo NP, Antony PA (2010) Naive tumor-specific CD4(+) T cells differentiated in vivo eradicate established melanoma. J Exp Med 207:651–667

Xue SA, Gao L, Hart D, Gillmore R, Qasim W, Thrasher A, Apperley J, Engels B, Uckert W, Morris E, Stauss H (2005) Elimination of human leukemia cells in NOD/SCID mice by WT1-TCR gene-transduced human T cells. Blood 106:3062–3067

Yanuck M, Carbone DP, Pendleton CD, Tsukui T, Winter SF, Minna JD, Berzofsky JA (1993) A mutant p53 tumor suppressor protein is a target for peptide-induced CD8+ cytotoxic T-cells. Cancer Res 53:3257–3261

Zwaveling S, Vierboom MP, Ferreira Mota SC, Hendriks JA, Ooms ME, Sutmuller RP, Franken KL, Nijman HW, Ossendorp F, van der Burg SH, Offringa R, Melief CJ (2002) Antitumor efficacy of wild-type p53-specific CD4(+) T-helper cells. Cancer Res 62:6187–6193

Chapter 16
Designing p53 Trials: A Surgical Oncologists View

Daniela Kandioler, Sonja Kappel, and Brigitte Wolf

1 Designing p53 Trials: A Surgical Oncologist's Point of View

The moderate efficacy of cancer therapy, being responsible for over- and undertreatment, is the major challenge in current surgical oncology. Reports from exploratory studies suggest a number of markers that may be helpful in providing more individualised and optimised cancer therapy. However, just a few markers are currently used in clinical practice. The absence of a disciplined approach has been repeatedly cited as the main reason for the slow and hesitant introduction of markers for clinical use (Sargent et al. 2005). A number of initiatives were at least successful in recommending guidelines how to correctly report on prognostic tumour marker studies: e.g. the CONSORT initiative for randomised clinical trials, the STARD statement for studies of diagnostic test accuracy and the REMARK guidelines for reporting tumour marker prognostic studies (Moher et al. 2001; Bossuyt et al. 2003; McShane et al. 2005). However, continuousness in marker evaluation has not been achieved and a correct strategic approach how to translate a predictive marker to clinical use is to be discussed and finally defined.

In the treatment of early cancers, complete surgical resection remains the primary goal because it is the only chance of cure. In locally advanced and/or lymph-node-positive cancers, preoperative systemic therapy may be successful in downstaging tumours in some patients and permitting subsequent resection. A significant pathological response to preoperative systemic therapy appears to be the most important factor of long-term survival in patients with advanced cancers. However, currently <15% of preoperatively treated patients experience a major

D. Kandioler (✉) • S. Kappel • B. Wolf
Department of Surgery, Medical University of Vienna,
Waehringer Guertel 18-20, 1090 Wien, Austria
e-mail: daniela.kandioler@meduniwien.ac.at

P. Hainaut et al. (eds.), *p53 in the Clinics*, DOI 10.1007/978-1-4614-3676-8_16,

Table 16.1 Differences in the aim of clinical trials evaluating new treatments versus new markers

	New treatment		New marker	
Clinical phase	Aim of study		Aim of study	
Phase I study	Pharmacokinetics	Safe dosage range	Robustness	Specificity prevalence hypothesis
Phase II study	Proof of concept	Efficacy, safety	Reproducibility	Confirm test
Phase III study	Pivotal study	Confirm efficacy, compare with standard drugs	Relevance	Confirm magnitude of effect, compare with standard treatment

pathological response. The moderate improvement in response rates evoked by sophisticated new drugs has failed to keep pace with the cost of these treatments. Individualised and tailored cancer therapy has been suggested as a possible exit strategy. However, translating this concept into clinical practice requires two things: first the availability of reliable markers to predict the benefit of treatment, and second a revolutionary approach on the part of researchers, clinicians, the pharmaceutical industry, official authorities and health care providers.

1.1 How Does One Develop a Marker?

Gene expression technology and the availability of human tissue and clinical data banks can detect a plethora of potential markers being somehow related to carcinogenesis, bad prognosis or treatment failure. Once a potential marker has been identified with one of these exploratory approaches, a number of questions must be answered before the marker can be subjected to clinical testing.

There is widespread agreement among experts as to how anticancer drugs should be evaluated before they can be used in standard therapy. The sequence of clinical trials as well as the level of evidence provided by each type of trial has been exactly defined. However, a clear algorithm for clinical evaluation of a potential marker is lacking, although the requirements might be quite similar to those for evaluation of a new drug (Table 16.1).

In the following, a generally applicable algorithm for clinical evaluation of a marker is described, and the algorithm is applied to the marker p53.

1.2 Phase I Marker Studies

The aim of phase I marker studies is to demonstrate the robustness of a potential new marker. The robustness of a marker is defined by its specificity, prevalence and hypothesis. These aspects can be evaluated in observational studies.

1.2.1 Specificity of a Marker

The specificity of a marker with reference to a specific type of cancer can be evaluated by the use of high-throughput technologies. Tissue banks may serve as sources of material. The cancer of interest is studied in relation to normal tissue, other diseases and other cancers.

For the p53 gene, Vogelstein et al. reported the first missense mutations in colorectal cancers in 1989. Since this discovery, a myriad of publications have described tumour-specific inactivation of p53 by gene mutations (Soussi 2007; Vousden and Lane 2007).

1.2.2 Prevalence of Markers

The next information required for the development of markers is an exact definition of the target population (cancer type) and the frequency of the aberrant marker status within the target population. The above-mentioned technologies can provide these data. The prevalence of a marker is also crucial for generating a hypothesis as well as designing a phase III trial.

Since the first listing of p53 mutations in 1994 (Hollstein et al. 1996), at least three p53 mutation databases have been established (De Vries et al. 1996; Hainaut et al. 1997; Beroud and Soussi 2003). Currently the IARC *TP53* database (http://www-p53.iarc.fr/p53aim.html) appears to be the most comprehensive one (Petitjean et al. 2007). The data collected in these databases strengthen the p53 marker hypothesis and provide actual prevalence data for most cancer types. The fact that the prevalence figures in the databases are still rising shows that prevalence depends on the sensitivity of analysis methods, among other factors (Olivier et al. 2010).

1.2.3 The Marker Hypothesis

Undirected genetic screening technologies combined with statistical analyses of clinical data have yielded hundreds of potential markers. However, in the absence of a hypothesis concerning the type of interaction with cancer, these markers have not been subjected to clinical evaluation.

The hypothesis is the most crucial aspect of a clinical trial because it delineates the central question to be tested. The type of question, in turn, determines the design of a trial (Nottage and Siu 2002; Mandrekar and Sargent 2009). Therefore, in the absence of a hypothesis no marker can be tested in a clinical trial.

For entirely unknown markers the investigator will have to move a step back into basic research and develop a hypothesis. This could be done, for instance, by identifying gene products, interactions of proteins, or protein binding sites. When this cannot be done, a marker is not valid for testing.

The definition of a hypothesis for p53 poses no difficulty. As a transcription factor, p53 controls the majority of the pathways involved in the life and death of cells

(Bates and Vousden 1999). p53 is directly involved in apoptosis, cell cycle regulation, angiogenesis, antioxidant defence, metabolic homeostasis and senescence (Levine and Oren 2009) providing background for a number of hypotheses. However, this does not make things easier. Too many hypotheses might be as helpful as no hypothesis. The question of rating the different pathways and their probable breakdown—not only in the context of p53 itself but also in the context of the vast network of pathways interfering with p53—is yet to be answered. But the more we learn about these complex interactions, the farther we move away from a hypothesis that can be reliably tested in a clinical trial.

Clinical application requires strong simplification of complex issues. A large number of interactions and relations cannot be considered in the formulation of the marker hypothesis because they are simply unknown, too complex or too many. However, if the power of these interactions is insignificant, they will exert no major impact on marker testing. In order to proceed we need to simplify the questions and focus on the most powerful interactions.

1.3 Phase II Marker Studies

Phase II marker studies focus on the marker test. "Once a marker meets the criterion of 'promising', additional data must be gathered before initiating confirmatory studies to test its clinical utility (Sargent et al. 2005)".

Confidence in the marker test is a crucial requirement for developing a clinically useful marker. Therefore, the sensitivity and specificity of the marker test are the major endpoints of phase II marker studies. A phase II marker study mainly addresses problems relating to technology. A balanced distribution of marker-positive and -negative phenotypes is not required in order to answer phase II questions. Therefore, phase II marker studies may be performed retrospectively.

Some of the questions addressed by phase II marker studies are the following:

- Does the method measure the right thing?
- How does the marker separate the population?
- Is the test reproducible?

1.3.1 Does the Method Measure the Right Thing?

The method for determining marker status should closely relate to the hypothesis; "cheap and easy" might prove to be the wrong goal.

The cancer-specific aspect of the *TP53* gene is mutation. Therefore, gene mutation is the focal aspect of the hypothesis. Detecting the presence of the protein product instead of the gene mutation might be a cheap and easy method, but might turn out to be the wrong approach. The p53 protein is normally undetectable due to rapid degradation, but accumulation makes it visible via immunohistochemistry (IHC).

However, immunohistochemical staining does not differentiate between pathological accumulation due to gene mutation and physiological accumulation due to cellular stress. Consequently, physiological stabilisation will produce false positive IHC results. Besides, some mutations may sabotage IHC and produce false negative results (mutations located at the binding sites targeted by IHC antibodies or mutations forming premature STOP codons). Several studies confirmed this discrepancy between IHC and mutation analysis by DNA sequencing (Chakravarty et al. 1996; Smith et al. 1996; Kandioler-Eckersberger et al. 1999, 2000; Kandioler et al. 2002; de Jong et al. 2005; Singer et al. 2005; Hanazono et al. 2006; Karim and Ali 2009) (Table 16.5).

Therefore, the results of phase I and phase II IHC studies are likely to be biased. The wide acceptance of a cheap and easy but insensitive test almost ruined p53 as a potentially powerful marker.

Today DNA sequence analysis is accepted as the gold standard to detect *TP53* mutations (Soussi 2007; Robles and Harris 2010). However, even a gold standard needs to be standardised. Various protocols and technologies exist for sequence analysis. Their sensitivity and specificity in detecting all types of mutation must be proven before they can be safely tested in clinical trials.

1.3.2 How Does the Marker Split the Population?

Markers which express more than two conditions (more than normal and aberrant), are clinically not useful because they separate the population into more than two groups. Clinical decisions based on a marker would consist of whether treatment should be administered or not, and whether one should apply a specific treatment or not. A marker that demonstrates more than two phenotypes would not be useful in making such decisions.

For example, when marker conditions are determined by scoring systems, the cut-off points are initially defined arbitrarily, subsequently based on patient outcome data, and finally confirmed by validation. Visual assays such as IHC may lead to different interpretations caused by inter- and intra-observer variability (Henry and Hayes 2006).

In this context, p53 sequencing delivers a perfect result because it provides evidence of two clearly defined conditions, i.e. whether a mutation is present or not.

1.3.3 Is the Test Reproducible?

Finally, a useful marker should be such that the measurement can be reliably performed on several target specimens. Especially for testing tumour tissue in clinical trials, the availability of source material is crucial. The need for additional invasive tumour harvesting to gain fresh or deep-frozen tissue is likely to be a major handicap for a clinical trial. Fine-needle biopsies or surgical tumour biopsies are most

Table 16.2 Who to treat

Prognostic marker			
Predicts	Determines	Trial endpoint	Clinical value
Relapse	Need for adjuvant therapy	Freedom from relapse	Avoid undertreatment
No relapse	No need for adjuvant therapy		Avoid overtreatment

Table 16.3 How to treat

Predictive marker			
Predicts	Determines	Trial endpoint	Clinical value
Response	Benefit of a certain therapy	Response to treatment	Provide effective treatment
No response	No benefit of a certain therapy		Avoid ineffective treatment

commonly used to diagnose cancer. Therefore, paraffin-embedded diagnostic tumour biopsy specimens are the most reliable source material for marker analysis, as they are routinely available and permit the investigator to obtain microscopic evidence of tumour cell content.

TP53 sequencing based on DNA harvested from paraffin-embedded biopsies has proven to be sensitive, reproducible and feasible in a number of trials (Kandioler-Eckersberger et al. 2000; Kandioler et al. 2002, 2008; Hall and McCluggage 2006; Hutchison 2007; Bennett et al. 2009).

1.4 Phase III Marker Studies

The purpose of phase III marker trials is to test the clinical utility of a marker and confirm its clinical relevance, thus providing evidence of the marker being useful in optimising or individualising cancer therapy. Phase III marker trials are supposed to provide the highest level of evidence (level I) and must be prospective, randomised controlled investigations specifically designed to test the marker of interest.

The likelihood of a marker being able to guide a therapeutic decision determines its clinical relevance. A marker can aid two therapeutic decisions: “Who to treat” and “How to treat” (Tables 16.2 and 16.3).

In order to answer the question “Who to treat?” a marker should be able to identify patients who are at risk for relapse. A marker predicting risk for relapse is defined as a PROGNOSTIC MARKER.

In order to answer the question “How to treat?” a marker should be able to identify patients who will respond to a certain therapy. A marker predicting response to a certain therapy is defined as a PREDICTIVE MARKER.

Both marker types predict entirely different things, which implies that the projected trial endpoints will also differ (Tables 16.2 and 16.3).

Therefore, the first step in planning a phase III marker trial is to determine whether the marker of interest is prognostic or predictive (Fig. 16.1).

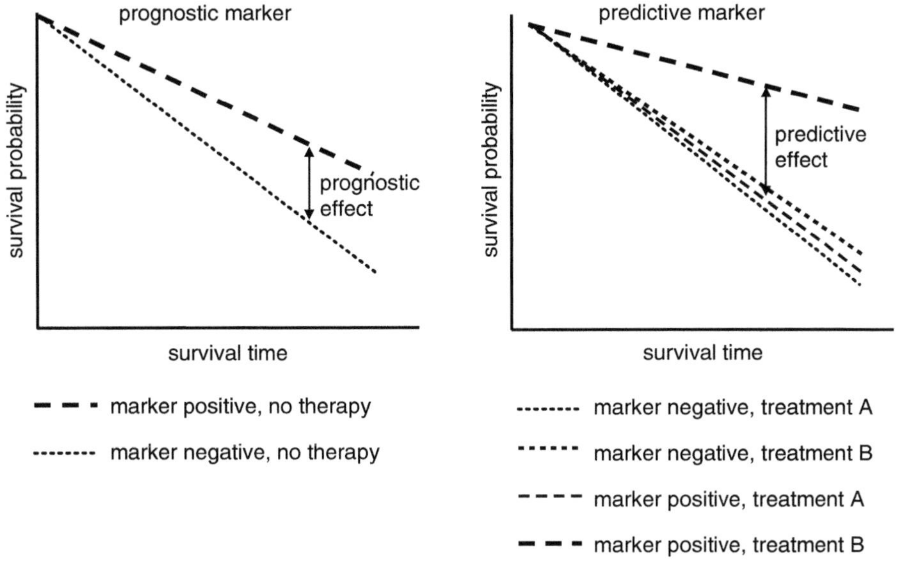

Fig. 16.1 Prognostic OR predictive biomarker. An effect (*arrow*) that is obviously related to marker positivity (*dashed line*) does not establish the prognostic value of a marker when there is any treatment involved. The positive effect on survival may be either due to the mere prognostic effect of the marker or due to interaction of the marker with therapy

It should be remembered that traditional prognostic markers (such as lymph node status) are diagnostic ones while predictive markers are therapeutic ones. Theoretically, a marker can be prognostic and predictive. However, this is rarely the case. A number of so-called prognostic markers might erroneously predict a poor prognosis due to a yet undetected interaction between the marker and the standard treatment.

1.4.1 Phase III PROGNOSTIC Marker Trials

A study comprising an "untreated" cohort is definitely qualified to assess whether a marker possesses prognostic power (Fig. 16.2). It will be crucial to exclude any possibility of interaction between a potential marker and treatment (confounding factor). However, today "untreated" cancer patients are rare.

On the other hand, we may well encounter cancer patients who have merely undergone surgical resection and received no neoadjuvant or adjuvant treatment (Fig. 16.2a). In this scenario, the prognostic marker will predict tumour relapse. Consequently, the trial endpoint would be relapse-free survival. A study arm of patients receiving any treatment in addition to surgery will fail to provide valid information concerning the prognostic power of the marker.

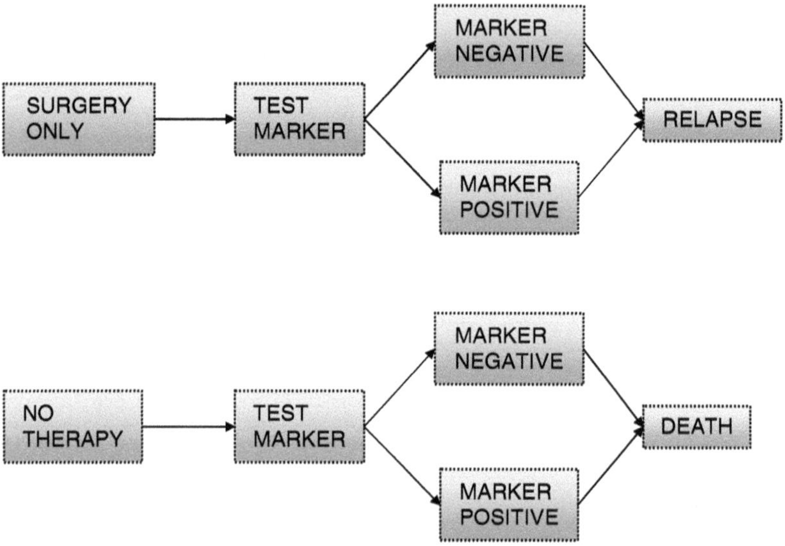

Fig. 16.2 Scenarios to test a prognostic marker. Completely untreated cohorts are rare (Fig. 16.2b) and permit only death as the endpoint. Death may be confounded by many factors

Relapse-free survival is frequently used as an endpoint in clinical phase III trials. Therefore, it has been suggested to test potentially prognostic markers retrospectively in patient cohorts derived from clinical phase III trials, to avoid prolonged recruitment and follow-up periods. However, the majority of clinical phase III trials are designed to prove the superiority of a certain therapy and do not include a surgery-only arm or an untreated control arm. Additionally, retrospective marker test results had to be included in post hoc multivariate analyses or subgroup analyses of such cohorts. These retrospective analyses are likely to be associated with problems concerning statistical power because the original study design did not include the marker test.

It is crucial to assess the prognostic value of a marker in a "surgery only" patient population (Fig. 16.2a). However, "surgery only" is a treatment option for early stage cancer patients, usually with negative lymph nodes; all other cancer patients are offered at least adjuvant therapy today. Relapses are rare in early stages of cancer. Therefore, evaluation of a prognostic marker will require investigation of large numbers of patients for several years. Even so, this approach will only detect powerful prognostic markers.

Entirely untreated cancer patients (even without surgical resection, Fig. 16.2b) are rare and not perfectly suitable to answer questions relating to prognostic markers. This approach would exclude "relapse" as an endpoint and permit only "death", which would be influenced by several other factors such as co-morbidities or supportive care.

Regarding the prognostic value of p53, most reports are based on observational studies or amendments to clinical trials designed for other purposes. Many of these studies do not even specify the chemotherapy or radiation therapy regimen used in the original study, although it is evident that there must have been one (Malats et al. 2005; Munro et al. 2005; Lan et al. 2007; Goebell et al. 2010). Only little information on the radiation and chemotherapy regimen has been provided by some authors (Yamashita et al. 2004; de Graeff et al. 2006; Popat et al. 2006; Dookeran et al. 2010; Jung et al. 2011). Although these reports include hazard ratios for disease progression or death, they are not qualified to yield reliable information on the prognostic power of p53 because they ignore potential interactions between the various treatments and the p53 status.

Just a few studies include "surgery only" cohorts, which qualifies them for the prognostic marker question. Nevertheless, the information yielded by the majority of these studies is meagre because of other factors such as the method of marker testing. Kuhadetti et al. described p53-positive IHC as a significant predictor for reduced disease-free and overall survival in prostate cancer (Cuzick et al. 2006; Kudahetti et al. 2009).

In summary, the question as to whether p53 possesses prognostic power has not been adequately addressed yet.

1.4.2 Phase III PREDICTIVE Marker Trials

Any attempt to prove an interaction between a marker and an anti-neoplastic drug in a clinical trial is associated with a number of problems mainly relating to current standards of cancer therapy.

Response as an Endpoint

The endpoint of a predictive marker trial—response to treatment—is difficult to assess because response is not a quantitative result and is prone to considerable subjectivity.

There is great controversy as to whether tumour shrinkage can be used as reliable endpoint for the efficacy of cytostatic agents (Stewart 2010), and whether RECIST criteria constitute an adequate tool to measure response in view of the limited sensitivity of currently available imaging techniques (Therasse et al. 2000, 2006). Besides, response might be highly tumour specific. The role of lymph node response is very unclear because imaging techniques are currently not sensitive enough to assess lymph node status with sufficient sensitivity and specificity. Other parameters such as tumour marker levels or standard uptake values of contrast agents may be quantitative. However, they are neither specific nor representative of response (Kinahan et al. 2009).

One approach to the problem of sensitive response assessment can be histological evaluation of the tumour after treatment. This requires the use of a neoadjuvant setting for a phase III predictive marker study. Preoperative therapy and subsequent surgical resection will permit histological evaluation. However, a number of

Table 16.4 Crucial aspects of designing a phase III predictive marker trial

Phase III predictive marker trial	
Prospective evaluation	The marker test is a crucial part of a phase III marker trial. Only inclusion of the marker test as a stratification criterion in the trial design will guarantee a balanced distribution of marker-positive and marker-negative patients in the study arms. Consequently, a phase III marker trial has to be a prospective trial
Sensitive marker test	Only marker tests whose sensitivity has been proven in phase II marker trials should be used in a phase III marker trial
Endpoint	A predictive marker predicts response to treatment. A reliable response endpoint should be assessable with high sensitivity and should be related to survival. For instance, complete pathological remission (CPR) may be recommended as an endpoint
Neoadjuvant setting	A neoadjuvant study setting permits pathohistological response assessment
Primary resectability	Pathological assessment of the tumour requires surgical resection after neoadjuvant therapy. Consequently, primary resectability should be considered as an inclusion criterion
Monotherapy	Treatments evaluated for potential interaction with the marker should preferentially be used as monotherapy in order to reduce complexity
Statistical planning and evaluation	A statistical test of interaction should be used to prove interaction between a marker and a drug

histological features may indicate response (including tumour viability, degree of apoptosis, T stage, N stage, vascular or lymph node invasion, etc.).

In any case, response to cancer therapy has to translate into significantly better survival for the patients; otherwise, the risk of the treatment would be unacceptable. There is just one feature which has been consistently reported as being related to significantly improved survival rates after cytostatic therapy, that is complete pathological remission (CPR) (Yeo et al. 2010; Bedikian et al. 2011; Kim et al. 2011; Topova et al. 2011). CPR can be established beyond any doubt in the surgical specimen. From that point of view, CPR qualifies as a sensitive, measurable and reliable parameter of response.

Recruitment

Recruitment of an adequate patient cohort is rendered difficult today. In order to exclude any (currently unknown) interaction between the marker and concomitantly used agents, it would make sense to evaluate cohorts being preoperatively treated with a single substance. This is difficult in view of the standards of combination chemotherapy. Furthermore, also new agents are used to be tested as an additive to standard therapy. A summary of the essential aspects to be considered when planning of a phase III predictive marker trial is given in Table 16.4.

Published data concerning the value of p53 as a predictive marker are subject to significant bias. Most studies are laboratory companion studies to clinical trials for new therapy regimens (Graziano et al. 2001; Brown et al. 2008; Sanchez-Munoz et al.

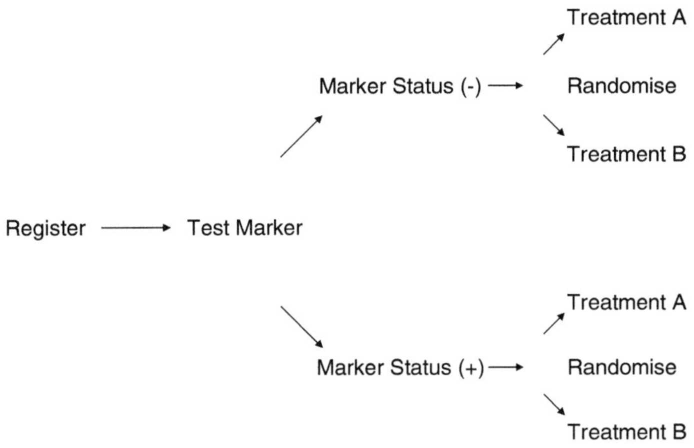

Fig. 16.3 Marker by treatment interaction design proposed by Sargent et al. (2005)

2010; Bonnefoi et al. 2011). In none of the studies the p53 status was used for stratification in order to provide a balanced distribution of normal and mutated tumours in the research arms. Furthermore, the application of drug combinations (rather than monotherapy) does not permit the investigator to detect the source of interaction. In summary, endpoints selected and trial designs used in published studies are not adequate to identify potential interaction between p53 status and tumour therapy.

To our knowledge there is just one clinical phase III marker trial aiming to evaluate the predictive power of p53, which has been adequately designed. The PANCHO trial (p53 adapted neoadjuvant chemotherapy for operable oesophageal cancer; ClinicalTrials.gov identifier NCT00525200), initiated by the p53research group, is a prospective randomised, phase III predictive marker trial (Kandioler 2008). For the first time the "marker by treatment interaction design", which has been proposed by Sargant to answer a predictive marker question (Sargent et al. 2005), is applied in a clinical trial (Fig. 16.3). The goal of the trial is to uncover the interaction between neoadjuvant chemotherapy and the *TP53* genotype in oesophageal cancer patients. The recruitment phase of the PANCHO trial was finished in 2011 (Table 16.5).

1.5 Summary

Both prognostic and predictive markers can enhance benefit of cancer therapy.

As long as effective treatments are not available, the fact that a patient is probably subject to a high risk of relapse is of minor clinical relevance. When predictive markers render treatments more effective, the value of prognostic information will be enhanced severalfold. In the absence of predictive markers, prognostic markers are of limited value.

Table 16.5 Clinical studies of the p53research group

Clinical cohort	Marker trial type	Cancer/study synonym	Type of treatment	Study status
Prospective randomised phase III	Phase III predictive marker trial	pANCHO/p53 adjusted neoadjuvant chemotherapy for potentially resectable oesophageal cancer (NCT00525200)	Neoadjuvant ct + surgery	Finished recruitment
Prospective phase II	Phase III predictive marker trial	PART1/p53 adapted preoperative radiotherapy for T2 and T3 rectal cancer	Radiation	Recruiting
Prospective phase II	Phase III predictive marker trial	p53 adapted induction chemotherapy in NSCLC	Neoadjuvant ct + surgery	Finished recruitment
Prospective phase II	Phase II marker trial	Oesophageal cancer	Neoadjuvant ct + surgery	Kandioler et al. (2007)
Prospective phase II	Phase II marker trial	Rectal cancer	Neoadjuvant ct + surgery	Kandioler et al. (2002)
Prospective randomised cohort	Phase III prognostic/predictive marker trial	Colorectal liver metastases	Neoadjuvant ct + surgery/surgery only	Manuscript in preparation
Prospective randomised cohort	Phase II marker trial	Lung cancer	Neoadjuvant ct + surgery	Kandioler et al. (2008)
Prospective randomised cohort	Phase II marker trial	Studie 90/colon cancer	Adjuvant ct	Manuscript in preparation
Observational	Phase I marker trial	Lung cancer	Neoadjuvant ct + surgery	Kandioler-Eckersberger et al. (1999)
Observational	Phase I marker trial	Oral cancer	Neoadjuvant ct + surgery	Manuscript in preparation
Observational	Phase I marker trial	Ovarian cancer	neoadjuvant ct + surgery	Manuscript in preparation
Observational	Phase I marker trial	Breast cancer	Neoadjuvant ct + surgery	Kandioler-Eckersberger et al. (2000)

ct chemotherapy

2 Conclusion/Wrap-Up Point on p53

To our opinion, p53 has great potential as a unique, broadly applicable predictive marker. The prognostic impact of p53 might be overestimated.

Studies clearly addressing the prognostic and predictive potential of p53 are on the way. Results from properly designed clinical studies assessing a potential interaction between *TP53* status and chemotherapy are also awaited. The demonstration of an interaction of the most frequently mutated gene in human cancer with commonly used chemotherapies could have deep impact on the standards of cancer therapy and the design of upcoming clinical trials. Additionally the clinical application of the "predictive marker *TP53*" will raise complex requirements on the sensitivity and specificity of the *TP53* marker test.

References

Bates S, Vousden KH (1999) Mechanisms of p53-mediated apoptosis. Cell Mol Life Sci 55:28–37

Bedikian AY, Johnson MM, Warneke CL et al (2011) Does complete response to systemic therapy in patients with stage IV melanoma translate into long-term survival? Melanoma Res 21:84–90

Bennett R, Schneider H, Estrella E et al (2009) Automated DNA mutation detection using universal conditions direct sequencing: application to ten muscular dystrophy genes. BMC Genet 10:66

Beroud C, Soussi T (2003) The UMD-p53 database: new mutations and analysis tools. Hum Mutat 21:176–181

Bonnefoi H, Piccart M, Bogaerts J et al (2011) TP53 status for prediction of sensitivity to taxane versus non-taxane neoadjuvant chemotherapy in breast cancer (EORTC 10994/BIG 1–00): a randomised phase 3 trial. Lancet Oncol 12:527–539

Bossuyt PM, Reitsma JB, Bruns DE et al (2003) The STARD statement for reporting studies of diagnostic accuracy: explanation and elaboration. Clin Chem 49:7–18

Brown PD, Krishnan S, Sarkaria JN et al (2008) Phase I/II trial of erlotinib and temozolomide with radiation therapy in the treatment of newly diagnosed glioblastoma multiforme: North Central Cancer Treatment Group Study N0177. J Clin Oncol 26:5603–5609

Chakravarty G, Redkar A, Mittra I (1996) A comparative study of detection of p53 mutations in human breast cancer by flow cytometry, single-strand conformation polymorphism and genomic sequencing. Br J Cancer 74:1181–1187

Cuzick J, Fisher G, Kattan MW et al (2006) Long-term outcome among men with conservatively treated localised prostate cancer. Br J Cancer 95:1186–1194

de Graeff P, Hall J, Crijns AP et al (2006) Factors influencing p53 expression in ovarian cancer as a biomarker of clinical outcome in multicentre studies. Br J Cancer 95:627–633

de Jong KP, Gouw AS, Peeters PM et al (2005) P53 mutation analysis of colorectal liver metastases: relation to actual survival, angiogenic status, and p53 overexpression. Clin Cancer Res 11:4067–4073

De Vries EM, Ricke DO, De Vries TN et al (1996) Database of mutations in the p53 and APC tumor suppressor genes designed to facilitate molecular epidemiological analyses. Hum Mutat 7:202–213

Dookeran KA, Dignam JJ, Ferrer K et al (2010) p53 as a marker of prognosis in African-American women with breast cancer. Ann Surg Oncol 17:1398–1405

Goebell PJ, Groshen SG, Schmitz-Drager BJ (2010) p53 immunohistochemistry in bladder cancer – a new approach to an old question. Urol Oncol 28:377–388

Graziano SL, Tatum A, Herndon JE II et al (2001) Use of neuroendocrine markers, p53, and HER2 to predict response to chemotherapy in patients with stage III non-small cell lung cancer: a Cancer and Leukemia Group B study. Lung Cancer 33:115–123

Hainaut P, Soussi T, Shomer B et al (1997) Database of p53 gene somatic mutations in human tumors and cell lines: updated compilation and future prospects. Nucleic Acids Res 25:151–157

Hall PA, McCluggage WG (2006) Assessing p53 in clinical contexts: unlearned lessons and new perspectives. J Pathol 208:1–6

Hanazono K, Natsugoe S, Stein HJ et al (2006) Distribution of p53 mutations in esophageal and gastric carcinomas and the relationship with p53 expression. Oncol Rep 15:821–824

Henry NL, Hayes DF (2006) Uses and abuses of tumor markers in the diagnosis, monitoring, and treatment of primary and metastatic breast cancer. Oncologist 11:541–552

Hollstein M, Shomer B, Greenblatt M et al (1996) Somatic point mutations in the p53 gene of human tumors and cell lines: updated compilation. Nucleic Acids Res 24:141–146

Hutchison CA (2007) DNA sequencing: bench to bedside and beyond. Nucleic Acids Res 35:6227–6237

Jung SY, Jeong J, Shin SH et al (2011) Accumulation of p53 determined by immunohistochemistry as a prognostic marker in node negative breast cancer; analysis according to st gallen consensus and intrinsic subtypes. J Surg Oncol 103:207–211

Kandioler D (2008) Personalized medicine – p53 gene analysis for prediction of response to neo-adjuvant therapy in esophageal cancer. MEMO 1:137–142

Kandioler D, Zwrtek R, Ludwig C et al (2002) TP53 genotype but not p53 immunohistochemical result predicts response to preoperative short-term radiotherapy in rectal cancer. Ann Surg 235:493–498

Kandioler D, Hejna M, Zwrtek R et al (2007) p53 Research Group Study. p53 adapted neoadjuvant therapy for esophageal cancer: pilot study. Annual Meeting 2007, American Society of Clinical Oncology, Abstract 4535

Kandioler D, Stamatis G, Eberhardt W et al (2008) Growing clinical evidence for the interaction of the p53 genotype and response to induction chemotherapy in advanced non-small cell lung cancer. J Thorac Cardiovasc Surg 135:1036–1041

Kandioler-Eckersberger D, Kappel S, Mittlbock M et al (1999) The TP53 genotype but not immunohistochemical result is predictive of response to cisplatin-based neoadjuvant therapy in stage III non-small cell lung cancer. J Thorac Cardiovasc Surg 117:744–750

Kandioler-Eckersberger D, Ludwig C, Rudas M et al (2000) TP53 mutation and p53 overexpression for prediction of response to neoadjuvant treatment in breast cancer patients. Clin Cancer Res 6:50–56

Karim S, Ali A (2009) Correlation of p53 over-expression and alteration in p53 gene detected by polymerase chain reaction-single strand conformation polymorphism in adenocarcinoma of gastric cancer patients from India. World J Gastroenterol 15:1381–1387

Kim T, Grobmyer SR, Smith R et al (2011) Esophageal cancer – the five year survivors. J Surg Oncol 103:179–183

Kinahan PE, Doot RK, Wanner-Roybal M et al (2009) PET/CT assessment of response to therapy: tumor change measurement, truth data, and error. Transl Oncol 2:223–230

Kudahetti S, Fisher G, Ambroisine L et al (2009) p53 immunochemistry is an independent prognostic marker for outcome in conservatively treated prostate cancer. BJU Int 104:20–24

Lan YT, Chang SC, Li AF et al (2007) p53 protein accumulation as a prognostic marker in sporadic colorectal cancer. Int J Colorectal Dis 22:499–506

Levine AJ, Oren M (2009) The first 30 years of p53: growing ever more complex. Nat Rev Cancer 9:749–758

Malats N, Bustos A, Nascimento CM et al (2005) P53 as a prognostic marker for bladder cancer: a meta-analysis and review. Lancet Oncol 6:678–686

Mandrekar SJ, Sargent DJ (2009) Clinical trial designs for predictive biomarker validation: theoretical considerations and practical challenges. J Clin Oncol 27:4027–4034

McShane LM, Altman DG, Sauerbrei W et al (2005) Reporting recommendations for tumor marker prognostic studies. J Clin Oncol 23:9067–9072

Moher D, Schulz KF, Altman D (2001) The CONSORT statement: revised recommendations for improving the quality of reports of parallel-group randomized trials. JAMA 285:1987–1991

Munro AJ, Lain S, Lane DP (2005) P53 abnormalities and outcomes in colorectal cancer: a systematic review. Br J Cancer 92:434–444

Nottage M, Siu LL (2002) Principles of clinical trial design. J Clin Oncol 20:42S–46S

Olivier M, Hollstein M, Hainaut P (2010) TP53 mutations in human cancers: origins, consequences, and clinical use. Cold Spring Harb Perspect Biol 2:a001008

Petitjean A, Mathe E, Kato S et al (2007) Impact of mutant p53 functional properties on TP53 mutation patterns and tumor phenotype: lessons from recent developments in the IARC TP53 database. Hum Mutat 28:622–629

Popat S, Chen Z, Zhao D et al (2006) A prospective, blinded analysis of thymidylate synthase and p53 expression as prognostic markers in the adjuvant treatment of colorectal cancer. Ann Oncol 17:1810–1817

Robles AI, Harris CC (2010) Clinical outcomes and correlates of TP53 mutations and cancer. Cold Spring Harb Perspect Biol 2:a001016

Sanchez-Munoz A, Duenas-Garcia R, Jaen-Morago A et al (2010) Is it possible to increase pCR in the neoadjuvant treatment with a dose-dense/sequential combination?: results from a phase II Trial combining epirubicin and cyclophosphamide followed by paclitaxel and gemcitabine +/– trastuzumab in stage II and III breast cancer patients. Am J Clin Oncol 33:432–437

Sargent DJ, Conley BA, Allegra C et al (2005) Clinical trial designs for predictive marker validation in cancer treatment trials. J Clin Oncol 23:2020–2027

Singer G, Stohr R, Cope L et al (2005) Patterns of p53 mutations separate ovarian serous borderline tumors and low- and high-grade carcinomas and provide support for a new model of ovarian carcinogenesis: a mutational analysis with immunohistochemical correlation. Am J Surg Pathol 29:218–224

Smith DR, Ji CY, Goh HS (1996) Prognostic significance of p53 overexpression and mutation in colorectal adenocarcinomas. Br J Cancer 74:216–223

Soussi T (2007) p53 alterations in human cancer: more questions than answers. Oncogene 26:2145–2156

Stewart DJ (2010) Randomized phase II trials: misleading and unreliable. J Clin Oncol 28:e649–e650, author reply e651–e653

Therasse P, Arbuck SG, Eisenhauer EA et al (2000) New guidelines to evaluate the response to treatment in solid tumors. European Organization for Research and Treatment of Cancer, National Cancer Institute of the United States, National Cancer Institute of Canada. J Natl Cancer Inst 92:205–216

Therasse P, Eisenhauer EA, Verweij J (2006) RECIST revisited: a review of validation studies on tumour assessment. Eur J Cancer 42:1031–1039

Topova L, Hellmich G, Puffer E et al (2011) Prognostic value of tumor response to neoadjuvant therapy in rectal carcinoma. Dis Colon Rectum 54:401–411

Vousden KH, Lane DP (2007) p53 in health and disease. Nat Rev Mol Cell Biol 8:275–283

Yamashita H, Nishio M, Toyama T et al (2004) Coexistence of HER2 over-expression and p53 protein accumulation is a strong prognostic molecular marker in breast cancer. Breast Cancer Res 6:R24–R30

Yeo SG, Kim DY, Kim TH et al (2010) Pathologic complete response of primary tumor following preoperative chemoradiotherapy for locally advanced rectal cancer: long-term outcomes and prognostic significance of pathologic nodal status (KROG 09-01). Ann Surg 252:998–1004

Chapter 17
p53 in the Clinic: A Pathologist's View

Philippe Bertheau, Jean-François Fléjou, Yves Allory, Pascale Varlet, Sylvie Lantuejoul, Louis-François Plassa, Hugues de Thé, Anne Janin, and Jacqueline Lehmann-Che

P. Bertheau (✉) • A. Janin
Department of Pathology, AP-HP, Hospital Saint-Louis, 1 av C. Vellefaux, Paris 75010, France

INSERM UMR_S728, Paris 75010, France

University Paris Diderot – Paris 7, Paris 75010, France
e-mail: philippe.bertheau@sls.aphp.fr; anne_janin@yahoo.com

J.-F. Fléjou
Department of Pathology, AP-HP, Hospital Saint-Antoine, Paris, France

Faculté de Médecine Pierre et Marie Curie, INSERM UMR_S 938,
Centre de Recherche Saint-Antoine, Paris, France
e-mail: jean-francois.flejou@sat.aphp.fr

Y. Allory
Department of Pathology, AP-HP, Hospital Henri Mondor, Créteil, France

Faculté de Médecine, INSERM U955, Institut Mondor de Recherche Biomédicale,
Université Paris Est Créteil, Créteil, France
e-mail: yves.allory@hmn.aphp.fr

P. Varlet
Department of Pathology, Hospital Sainte Anne, Paris, France
e-mail: p.varlet@ch-sainte-anne.fr

S. Lantuejoul
Department of Pathology, CHU A Michallon, Grenoble, France

INSERM U 823 Institut A Bonniot, Université J Fourier, Grenoble, France
e-mail: slantuejoul@chu-grenoble.fr

L.-F. Plassa
Laboratory of Biochemistry, AP-HP, Hospital Saint-Louis, Paris 75010, France
e-mail: lfplassa@yahoo.fr

H. de Thé • J. Lehmann-Che
University Paris Diderot – Paris 7, Paris 75010, France

Laboratory of Biochemistry, AP-HP, Hospital Saint-Louis, Paris 75010, France

CNRS UMR7212, INSERM U944, Paris 75010, France
e-mail: dethe@univ-paris-diderot.fr; jacqueline.lehmann-che@sls.aphp.fr

P. Hainaut et al. (eds.), *p53 in the Clinics*, DOI 10.1007/978-1-4614-3676-8_17,

1 Introduction

The *TP53* gene, described 30 years ago, has a prominent role in cancer and in human biology. p53 protein is not only pivotal in tumour suppression but is also crucial in many other cell functions (Levine and Oren 2009). p53 regulation, p53 targets, and all aspects of p53 biology are fully reviewed in this book and we will focus here, through several examples, on how pathologists use p53 status determination, and p53 family proteins determination, in their daily practice.

Based on gross examination and microscopical examination of tissue or cell samples, a good pathological diagnosis now often includes also a molecular analysis, done either by the pathologist himself (molecular pathologist) or in collaboration with molecular biologists. Indeed, molecular components of cells and tissues can be studied with morphological tests such as immunohistochemistry (IHC) and in situ hybridization or with in vitro tests performed on carefully preserved samples. How to analyse samples have been the subject of multiple guidelines and standards (Amin 2010), as well as how to preserve biological samples in biobanks (Bevilacqua et al. 2010) in order to perform subsequent tests on frozen samples. This workflow of tissue and cell samples has become particularly critical these last few years with the huge increase of targeted therapies that require the analysis of predictive biomarkers (Keller et al. 2010; Osamura 2009).

With its central role in the control of cell death and proliferation, and with many available techniques for its determination, p53 is a useful tool in the hands of the pathologist for helping in diagnosing and treating neoplasia.

To have an overall idea of how p53 is used by pathologists, we contacted 15 colleagues in several hospitals in France, each being specialized in one specific area of pathology. We asked them how often they use this marker, with which method, and for which purpose. Some of these colleagues had no use or just very little use of this marker while other colleagues had a significant use of p53 determination and accepted to be co-authors of this chapter (JFF, YA, PV, SL, AJ). However, it was not possible to cover all possible uses of p53 status in pathology in a single chapter and some fields could not been detailed in this review.

2 The Pathologist's Tools to Analyse p53 and p63

2.1 p53 IHC and Immunocytochemistry

p53 IHC and immunocytochemistry (ICC) in daily practice are now most often performed with the mouse monoclonal DO-7 antibody (Garg et al. 2010) or sometimes by the mouse monoclonal Pab1801 antibody (Dookeran et al. 2010). DO-7 and Pab1801 have been successfully evaluated more than 15 years ago (Baas et al. 1994; Horne et al. 1996; Jacquemier et al. 1994; Lambkin et al. 1994; Würl et al. 1997).

The epitopes of DO-7 and Pab1801 are mapping, respectively, between amino acids 1–45 and amino acids 32–79 of human p53 and there is no cross-reactivity with mouse or rabbit p53. They are used for the detection of wild-type and mutant p53 by IHC, ICC, and immunofluorescence. Antigen retrieval is used in formalin-fixed paraffin-embedded tissues. In tumours with missense *TP53* mutations, the nuclear staining is usually strong and well-defined, with rare and weak cytoplasmic staining.

Many other monoclonal and polyclonal anti-p53 antibodies have been tested in the literature in the 1990s (Baas et al. 1994; Bártek et al. 1993; Jacquemier et al. 1994; Lambkin et al. 1994; Pezzella et al. 1994; Renninson et al. 1994; Vojtěsek et al. 1993; Würl et al. 1997), but none of them was really better than DO-7 or Pab1801 for the correlation with mutation status assessed by in vitro molecular techniques, although this correlation is not ideal even with these two antibodies. More recently, new monoclonal rabbit anti-human p53 has been developed but still needs to be evaluated. A recent review lists the different commercially available antibodies and their specificity for several p53 isoforms (Khoury and Bourdon 2010).

2.2 *TP53 Molecular Analyses*

Since p53 IHC can miss some *TP53* gene mutations or give false-positive results, it is often necessary to perform in vitro molecular analyses if a precise assessment of *TP53* gene status is needed, and if enough tissue is available. These analyses, such as functional test in yeast, direct sequencing, or array profiling of target genes, are detailed in other parts of this book and will not be described here. These analyses are not part of the daily practice of most pathologists and are usually performed by molecular biologists or molecular pathologists. These molecular analyses require high-quality tissue samples, especially when using mRNA. Checking tissue quality and tumour cell quantity with microscopical examination is now an important mission of pathologists. These crucial so-called "preanalytical steps" also require that pathologists work in professional networks allowing an efficient workflow of information and tissue samples among oncologists, pathologists, and biologists.

2.3 *Correlation Between p53 IHC and Molecular Analyses*

In 1994, a comparison between six different antibodies reported a good overall concordance between IHC and gene mutational status but already mentioned some unexpected discordances, with IHC-positive cases without mutation and IHC-negative cases with mutation (Baas et al. 1994). Indeed, some mutations create stop codons so that the complete protein cannot be expressed and IHC is negative

(Alsner et al. 2008). In colorectal cancer, 50 % of tumours that display negative p53 IHC have in fact *TP53* mutation when using another method (Colomer et al. 2003). These discrepancies have been observed in many other tumour types (Coggi et al. 1997; Soong et al. 1996). However, most point missense mutations give rise to tumours that stain strongly with p53 IHC. The mutation may itself directly stabilizes the protein, or may inhibit the synthesis of MDM2 (p53 natural inhibitor) and therefore inhibit p53 protein degradation. Conversely, wild-type p53 can lead to positive IHC if activated by cellular stress but is usually of much weaker intensity as compared to that observed when *TP53* is mutated. Wild-type p53 proteins are usually not detectable in normal conditions because of permanent degradation by the proteasome after E3-ligase MDM2-dependent ubiquitination. The situation may be also complicated by the fact that p53 has several different isoforms, some of them being not recognized by transactivation domain-specific antibodies (Khoury and Bourdon 2010).

Although usually homogeneous, p53 immunostaining can be sometimes heterogeneous across tumour sections in a way that seems not to reflect simple tissue-processing artefacts.

The relationship between p53 function and p53 immunostaining remains complex and is a reason for not considering p53 IHC as a fully reliable technique to assess p53 status. However, even if null mutations escape IHC detection, a strong IHC staining remains a very good predictor of *TP53* missense mutation and is therefore used in many pathological situations as we will see below, sometimes using panels of antibodies (Nenutil et al. 2005). A real advantage of IHC remains its low cost and feasibility in all pathology laboratories.

2.4 p63 IHC and ICC

p53 belongs to a family of proteins that includes two other members, p63 and p73, implicated in normal development (Yang et al. 2000), and also in human tumourigenesis and tumour response to therapy (Stiewe 2007). p63 and p73 can heterodimerize with p53 and have to be taken into account in the "p53 world", especially p63 that is used in the daily practice of pathologists. p63 is most often used as a differentiation marker in several pathological areas such as breast tumours and prostate tumours and these applications will be discussed.

In normal tissues, p63 expression is restricted to the nucleus in epithelial cells of stratified epithelia, such as skin, oesophagus, exocervix, tonsil, and bladder, and to certain subpopulations of basal cells in glandular structures of prostate and breast (myoepithelial cells), as well as in bronchi (Di Como et al. 2002).

Several antibodies are available for this immunostaining (Barbareschi et al. 2001) and a strong and well-defined nuclear staining is usually obtained, allowing easy and reproducible conclusions.

3 p53 in the Daily Practice of Pathologists

3.1 *Differential Diagnosis of Reactive/Dysplastic/Benign/ Malignant Lesions*

In his or her daily practice, the pathologist often has to deal with borderline lesions. Regular H&E staining sometimes does not allow a reliable distinction between regenerative atypia (e.g. cellular atypia due to nearby inflammation) and atypia due to dysplasia or neoplasia. Even experienced pathologists may have trouble with these lesions, and it is often recommended that these diagnostics should be confirmed by a second pathologist. Biomarkers may be of great help for these lesions. However, most often only IHC can be performed on these microscopical lesions since it is hard to get enough tissue material for molecular analysis and most often impossible to get frozen tissue material. These microscopical lesions remain an important reason for keeping p53 IHC to assess p53 status.

3.1.1 Dysplasia and Digestive Tract

Gastrointestinal cancers are among the most frequent cancers in humans. For most of these cancers, a morphological sequence of events precedes invasive tumours, with increasing grades of malignant precursor lesions, called dysplasia or intraepithelial neoplasia (Odze et al. 2010).

Dysplasia can be observed either as mass-forming lesions called adenomas, or as flat areas, only detected by systematic sampling with biopsies. Dysplasia identification in mucosal biopsy specimen is the basis of surveillance programmes in patients with Barrett's oesophagus (BO) and inflammatory bowel disease (IBD), two chronic inflammatory conditions with an increased risk of cancer.

Pathologically, dysplasia is classified as negative, indefinite, or positive (low or high grade), both in squamous epithelium (native oesophageal mucosa) and cylindrical epithelium (metaplastic esophageal, gastric, intestinal) (Riddell et al. 1983). The natural history of dysplasia and cancer development has been well investigated in BO and IBD. However, several limitations exist for the use of this marker as a perfect indicator of an increased risk of cancer. They include the difficulty of establishing a precise estimation of the cancer risk at an individual level, the risk of sampling error, and the inter-observer variability of pathological diagnosis, due to uncertainties in distinguishing regeneration (non-neoplastic) from dysplasia (neoplastic), high-grade dysplasia from invasive (intramucosal) carcinoma, and low-grade from high-grade dysplasia. This diagnostic variability appears intrinsic to dysplasia, as this lesion develops on a continuous linear scale but not in discrete categories (Odze 2006).

Because of these difficulties, a number of alternate markers have been studied to complement or even replace dysplasia as a tool to identify patients at high risk of progression to cancer (Jankowski and Odze 2009). These biomarkers have to be expressed in accessible tissue, to be assessable with reproducible criteria, validated against acknowledged disease endpoints, and to have a predictive value confirmed in prospective studies. p53 is one of the very few biomarkers that fulfil most of these criteria, especially in patients with BO.

Barrett's Oesophagus

Among precancerous conditions of the digestive tract, BO can be considered as a model for transferring biomarkers into clinical practice, due to its frequency, and its relatively easy endoscopic examination, allowing repeating biopsies during the surveillance (Fléjou and Svrcek 2007). The most widely studied biomarkers in BO include DNA content abnormality, increased proliferation, epigenetic alterations, and alteration of tumour suppressor genes, especially *TP53* (Prasad et al. 2010).

TP53 gene mutations can be detected in the metaplastic oesophagus with molecular techniques. It has been demonstrated that *TP53* mutations are frequent in Barrett's adenocarcinoma, occurring in 70–80 % of cases (Hamelin et al. 1994; Keswani et al. 2006). Despite important limitations already discussed, p53 protein IHC has been established as a biomarker for malignant potential in numerous series of BO. It has been demonstrated that p53 protein overexpression was frequent in dysplastic epithelium adjacent to p53-positive adenocarcinomas, a result which suggested that *TP53* mutation was an important step in the progression towards cancer. Then it has been shown that p53 overexpression increased parallel to the degree of dysplasia (0–10 % in non-dysplastic mucosa, 10–20 % in low-grade dysplasia, 60–80 % in high-grade dysplasia and adenocarcinoma) [review in Keswani et al. (2006)]. More interestingly, it has been shown that p53 overexpression was a strong indicator for an increased risk of progression to high-grade dysplasia or cancer in patients with biopsies indefinite for dysplasia or with low-grade dysplasia (Bani-Hani et al. 2000; Sikkema et al. 2009).

However, even if all published series suggest the utility of p53 as a biomarker for evaluating the risk of cancer in BO, two major barriers persist that limit its use in general clinical practice, one being the limited sensitivity of IHC, and the other being the lack of standardization of IHC. Indeed, the criteria of positivity used in the literature are highly variable, with various cut-offs for positive cases (from 5 to 50 % of positive cells). Only cases with an intense positive staining in several adjacent epithelial cells should be considered as overexpressing p53 protein, with mild positivity of non-neoplastic epithelial structures present on the same slide serving as a positive control. Very interestingly, a recent report described a novel staining pattern of p53 in Barrett's dysplasia—the absent pattern—as another predictor of neoplastic progression (Kaye et al. 2010). This pattern may result from truncating mutation in *TP53* which cannot be recognized by the commonly used DO7 antibody.

Another potential use of p53 protein expression in BO is to reduce uncertainties in the histological diagnosis of dysplasia. Even if the diagnosis of dysplasia still relies on H&E stained sections, the presence of p53 overexpression reduces inter-observer variability and predicts progression to more severe lesions (Kaye et al. 2009).

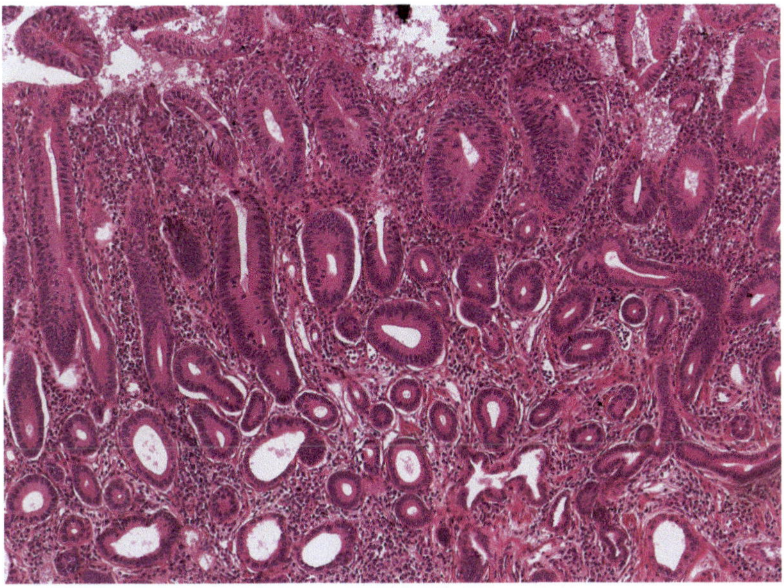

Fig. 17.1 Colonic crypts with low-grade and high-grade dysplasia in Crohn's disease (H&E, original magnification ×200)

Other Preneoplastic Lesions of the GI Tract

Among the other preneoplastic epithelial lesions of the gastrointestinal tract, dysplasia occurring in patients with IBD (Fig. 17.1) shares many common characteristics with dysplasia in BO (Harpaz and Polydorides 2010). The frequency of *TP53* gene mutation and resulting p53 protein overexpression is high in this context, but still not constant, both in ulcerative colitis and Crohn's disease, and *TP53* mutation is considered as an early event in the carcinogenic process (Harpaz and Polydorides 2010). Although the same limitations of IHC study of p53 protein exist as in BO, this marker is also considered as a useful adjunct to standard histology to detect high-risk patients (Marx et al. 2009) (Figs. 17.1 and 17.2).

In other dysplastic lesions of glandular mucosa of the GI tract (gastric flat of polypoid dysplasia, ampulla of Vater), the frequency of *TP53* gene mutations is lower, and therefore p53 protein overexpression has a lower sensitivity to detect high-risk patients. Nevertheless, a frank p53 overexpression represents a strong argument for the presence of true neoplasia in case of doubtful diagnosis of intraepithelial neoplasia. p53 is probably of lower interest in the commonest form of dysplasia of the GI tract, i.e. adenomas of the colon, which by definition represent dysplastic lesions. The major diagnostic challenge in colon adenomas is to diagnose malignant transformation, and p53 protein expression has no interest for this purpose.

Squamous intraepithelial neoplasia can be observed in the oesophagus and in the anus. In the oesophagus, p53 gene mutation is frequent, and p53 protein overexpression is a good argument for a high-grade lesion. Although p53 overexpression is

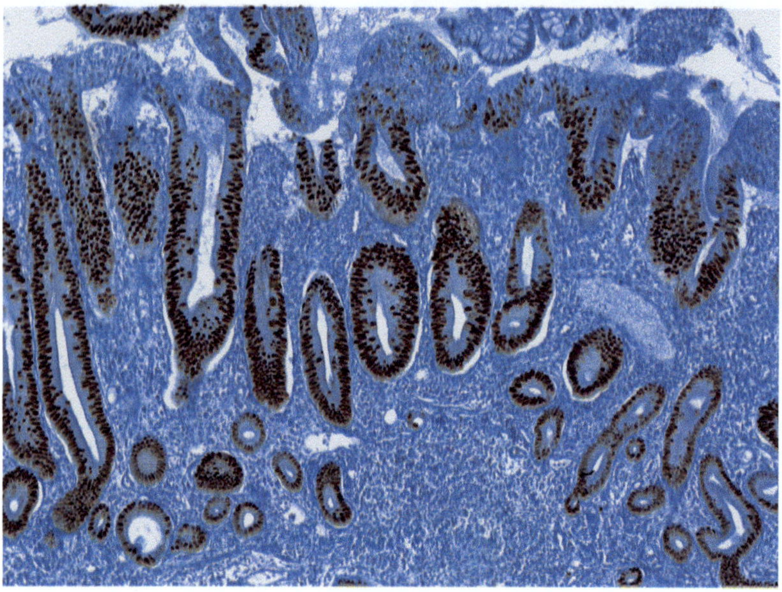

Fig. 17.2 Same area as in Fig. 17.1, immunostained for p53. The dysplastic area shows a strong nuclear staining (original magnification ×200)

relatively frequent in anal intraepithelial neoplasia (AIN), the best marker that can be used in complement to histology is p16 expression, as in other squamous neoplastic lesions due to HPV infections.

3.1.2 Dysplasia and Tumours of the Bladder

The p53 pathway plays a major role in urinary bladder carcinoma progression, and its exploration has been proposed for diagnostic and prognostic purposes in clinical practice. Urothelial carcinoma comprises two main pathways of tumour progression, the papillary tumour pathway and the carcinoma in situ (CIS) pathway, and p53 alterations should be considered separately for each pathway.

The papillary carcinoma, either respecting the basement membrane (stage Ta) or invading into the lamina propria (stage T1), has a 10–15 % overall risk to progress to muscle invasive disease (stage ≥ T2) (Sylvester et al. 2006). *TP53* mutations are rare in Ta low-grade carcinoma, and are frequently detected in tumour with high grade and/or invasive stage, suggesting a role in tumour progression (Berggren et al. 2001). This is also true at the protein level: nuclear p53 accumulation, associated with *TP53* gene mutations in 60–75 % of urinary bladder cancer cases, is more frequently observed in muscle invasive than in non-muscle invasive tumours (Bernardini et al. 1999). However, both *TP53* mutation analysis and p53 IHC are not recommended in clinical practice. Indeed, their potential prognostic values to predict the progression to muscle invasive disease or the survival, respectively, have been investigated in multiple studies with conflicting results. A meta-analysis based on the review of 117 studies showed that p53 changes were only weakly predictive of recurrence, progression, and mortality in bladder cancer, and concluded that evidence was not sufficient to conclude whether changes in p53 could act as outcome markers in patients with bladder cancer (Malats et al. 2005).

CIS of the urinary bladder is a flat lesion (stage Tis) considered as the most common precursor of muscle invasive disease, with a risk of progression to muscle invasion estimated around 50 %. *TP53* mutations are frequently detected in CIS, suggesting that the alterations occur earlier in CIS than in papillary pathway (Spruck et al. 1994). An intense nuclear p53 accumulation is also observed in CIS (Wagner et al. 1995). In clinical practice, the distinction of urothelial CIS from reactive atypia on the basis of standard H&E staining can be difficult, whereas this distinction is crucial in terms of prognosis and treatment. p53 IHC has become part of the immunohistochemical panel commonly used in this context, including also Ki-67, cytokeratin 20 (CK20), and CD44 (Sun et al. 2002; Wagner et al. 1995). In normal urothelium, there is no or only weak and patchy nuclear p53 staining, and reactive urothelium displays no p53 staining with occasional weak positivity in the basal and parabasal cells. In contrast, in up to 75 % cases, CIS shows intense nuclear p53 in the majority of tumour cells. It is commonly accepted that the staining should be intense and present in at least 20–50 % of cells to be significant and support the diagnosis of CIS. Urothelial dysplasia is considered as a preneoplastic lesion, representing an early morphological manifestation of progressive alterations between normal urothelium and CIS, but there is no real consensus on the morphological diagnosis of dysplasia (Hodges et al. 2010). Accordingly, a few studies have suggested that nuclear p53 is observed in up to 25 % cases of dysplasia and could support this diagnosis together with CK20 expression (Hodges et al. 2010; Sun et al. 2002; Wagner et al. 1995).

3.1.3 Bronchial Dysplasia

Several studies have shown that dysplastic bronchial epithelium contains *TP53* mutations (Kohno et al. 1999; Sozzi et al. 1992). *TP53* mutations seem to occur early in lung carcinogenesis and accumulation of abnormal p53 protein is proportional to the degree of dysplasia. Mild or moderate dysplasia accumulates p53 in 17–28 % while severe dysplasia shows p53 accumulation in 59–100 % of cases (Brambilla et al. 1998).

Moreover, dysplastic areas close to invasive carcinoma areas often display p53 accumulation while similar dysplastic lesions not associated with invasive cancer do not accumulate p53. This suggests that p53 mutation and/or accumulation in dysplastic areas is predictive of synchronous or metachronous carcinoma (Brambilla et al. 1998; Ponticiello et al. 2000).

3.1.4 Oral Lichen Planus-Like Lesions of Chronic Graft Versus Host Disease

Graft versus host disease (GVHD) is the main complication of hematopoietic stem cell transplantation (HSCT) (Choi et al. 2010) and lichen planus-like eruption is the hallmark of the chronic phase of GVHD (Janin-Mercier et al. 1981). The oral localization is one of the most constant features of patients with chronic GVHD (Saurat and Gluckman 1977). At the tissue level, apoptosis of basal keratinocytes, the target cells of the alloimmune reaction, is associated with large, actively synthesizing keratinocytes at the upper levels of the epithelium (Janin-Mercier et al. 1981). This association of cell damage and tissue repair within the epithelium is characteristic of a lichenoid reaction (Eady and Cowen 1977). The ulcerated form of idiopathic oral lichen planus is associated with a high risk of malignant transformation.

Squamous cell carcinomas (SCC) are the most frequent secondary malignancies occurring in long-term survivors after HSCT (Socié et al. 1993). Tumour cells in SCC of chronic GVHD patients often overexpress p53 (Janin et al. 2009). It is also observed in the basal layer of lichen planus-like lesions surrounding SCC. To further assess if this overexpression of p53 is found in lichen planus-like lesions of chronic GVHD, a retrospective study was done on successive biopsies performed in a series of nine patients with oral SCC after HSCT, followed-up over a period of 10 years (unpublished data). p53 expression was seen in more than 50 % of the basal cell nuclei in patients with lichen planus-like eruption with severe inflammatory reaction and in 20 % of patients with ulcerated idiopathic oral lichen planus, the form most at risk for SCC among idiopathic lichen planus. Therefore, a systematic detection of keratinocytes expressing p53 is now performed on sequential biopsies of patients with chronic GVHD and oral lichen planus for early detection of patients at risk of SCC. Since the prognosis of these SCC is poor, p53 expression could allow an earlier diagnosis and treatment of these lesions.

3.2 p53 Can Help to Better Characterize Malignant Tumours

3.2.1 Central Nervous System Tumours

Mutations of the *TP53* tumour suppressor gene were first implicated in gliomagenesis almost 20 years ago in Li–Fraumeni syndrome. Subsequently, *TP53* was identified as an early main driver in low-grade gliomas and secondary glioblastomas (Louis 1994). However, even if p53 accumulation is frequently evaluated in routine neurosurgical pathology in order to help discriminating between reactive/benign/malignant glial lesions, many results remain controversial particularly in gliomas. It is important to note that recommended cut-off values and interpretation guidelines have not been firmly established to permit reproducible and comparable analysis. This is particularly crucial in infiltrative gliomas in which the establishment of the real percentage of tumoural cells represents a challenge even for experienced pathologists.

This leads to further difficulties for the semi-quantitative interpretation of p53 immunostaining and the use of a tumour-adapted p53 positivity threshold.

3.2.2 Astrocytic Tumours

The high frequency of *TP53* mutations/overexpression in infiltrating low-grade gliomas (>50 %) is compatible with the view that this genetic alteration represents a common early genetic event in gliomagenesis. Knowing the rarity of p53 overexpression in pilocytic astrocytoma (PA, astrocytoma grade I), p53 immunostatus can be used as a valuable routine marker in the diagnostic evaluation of infiltrative gliomas versus PA. However, this criterion should not be used in an isolated fashion but integrated with other criteria such as imaging characteristics.

p53 is still a controversial prognostic factor in patients with grade II, III, or IV astrocytomas (glioblastoma, GBM) with a trend for worse outcome in *TP53*-mutated tumours (Okamoto et al. 2004; Peraud et al. 2002; Ständer et al. 2004; Weller et al. 2009). A recent study about anaplastic features in PA has shown that p53 is correlated with anaplasia and a worse prognosis (Rodriguez et al. 2010).

Oligodendroglioma

The classical genetic hallmark of adult oligodendroglioma is 1p/19q co-deletion, which is mutually exclusive with *TP53* mutation. Recently, integrated gene expression and epigenetic studies have led to propose three molecular subgroups in oligodendroglioma with prognostic implications. The intermediate subgroup is characterized by *TP53* mutation or LOH of 17p, IDH1 mutations, and GSTP1 promoter methylation (Ferrer-Luna et al. 2010).

Medulloblastoma

Recent contradictory results about the prognostic value of *TP53* mutation in sporadic medulloblastoma (MB) have led to identify two MB subgroups with an over-representation of *TP53* mutations. A subgroup with more favourable prognostic is associated with CTNNB1 mutation (encoding beta-catenin) and a subgroup with a worse prognostic is associated with MYCN amplification (Pfaff et al. 2010; Tabori et al. 2010a). Therefore, the prognostic impact of p53 status does not seem to be independent in MB.

Choroid Plexus Tumours

Choroid plexus tumours are rare, accounting for 0.3–0.6 % of all brain tumours, so prognostic factors are difficult to establish.

p53 dysfunction has been found to play a major role in choroid plexus carcinogenesis in a recent multi-institutional study (Tabori et al. 2010b). This study showed that p53 allowed patient stratification in terms of therapy response and prognosis. Based on their findings, the authors proposed the following recommendations: (1) accurate family history and genetic counselling is recommended for all patients with a choroid plexus carcinoma (CPC) as 50 % of them may present a germline *TP53* mutation; (2) If p53 overexpression is observed, a somatic and germline mutation should be searched for; (3) if *TP53* is not mutated, a therapeutic protocol comprising complete surgery and chemotherapy without radiotherapy should be provided and, if a *TP53* mutation is found, alternative therapies must be tested (Tabori et al. 2010b).

3.2.3 Endometrial Neoplasia

Endometrial carcinoma subtypes can be difficult to diagnose and positive p53 IHC has been shown to correlate with the serous subtype and to be helpful for prognostic assessment and therapeutic decision (Fayallah et al. 2010; Garg et al. 2010; Lee et al. 2010). Staining in serous adenocarcinomas is intense, diffuse, and homogeneous (Quddus et al. 1999). It has also been shown that *TP53* mutation is probably one of the most important factors to initiate the endometrial serous carcinogenesis (Jia et al. 2008).

3.2.4 Smooth Muscle Tumour of Uncertain Malignant Potential

The current World Health Organization classification indicates that a uterine smooth muscle tumour that cannot be histologically diagnosed as unequivocally benign or malignant should be termed "smooth muscle tumour of uncertain malignant potential" (STUMP). In a recent study (Ip et al. 2009), p53 and p16 IHC were performed in a group of STUMP and it was shown that these markers may be helpful in the prediction of the behaviour of STUMPs.

3.3 p53 Can Help to Predict Response to Treatment in Cancer

Given the key role of p53 in response to stress and its central role in apoptosis, senescence, and cell cycle arrest, the *TP53* status of a tumour should have deep implication for response to various genotoxic chemotherapeutic agents. The predictive value of *TP53* status has been extensively analysed in clinical samples, but no real agreement could be found. The conflicting conclusions could be explained by

(1) the heterogeneity of the p53 determination methods, (2) the heterogeneity of tumour types studied, (3) the types of *TP53* mutations considered, and (4) the large range of the drug associations and of the drug doses. As this aspect is covered in detail in other chapters of this book, we will here describe our experience with breast cancer. In breast cancer, chemotherapy has become an essential part of treatment in the adjuvant or neoadjuvant settings and response to chemotherapy is one of the most important prognostic factors, especially when pathological complete response is achieved (Bertheau et al. 2005; Guarneri et al. 2006).

Using the functional FASAY assay in yeast to detect *TP53* mutations, it has been showed that mutated *TP53* is required for obtaining pathological complete response in non-inflammatory locally advanced breast cancers (LABC) treated by dose-dense epirubicin–cyclophosphamide regimen (Bertheau et al. 2002, 2007).

To evaluate the impact of different treatment on homogeneous patient populations, three teams collected retrospectively three series of patients treated by epirubicin alone, FAC, or dose-dense EC protocol (ddEC) in the neoadjuvant setting and the *TP53* status was analysed by the functional FASAY assay (Lehmann-Che et al. 2010). It was found that absence of functional p53 is predictive of resistance to anthracyclines alone, but is highly predictive for response to dose-dense alkylating agents containing regimen. The predictive value of *TP53* mutation for ddEC protocol is particularly important in the oestrogen-negative subtype of breast cancer analysed. Thus, addition of dose-dense alkylating agent to epirubicin may reverse resistance of *TP53*-mutated tumours to anthracyclines alone. The *TP53* status determined by direct sequencing or functional assay could therefore be a predictive factor for response to particular chemotherapeutic agents and guide the therapeutic decision in breast cancer.

Many other tumour types have been investigated, most often without clear-cut conclusions (see other chapters in this book). All these examples illustrate the complexity of p53 effects in the treatment of solid tumours that depend largely on the types and the drug doses.

The contribution of pathologists in this setting is probably not so much the IHC analysis that does not seem to be a clear predictor but rather the tissue preservation and qualification before molecular analysis, if enough tumour tissue material is available.

3.4 Non-Tumour Pathology

p53 status determination is almost always used in neoplastic settings, but another possible application has been reported recently in the settings of infectious diseases and kidney transplantation (Wiesend et al. 2010). As p53 is upregulated during both early and late phases of BK virus infection, p53 IHC and SV40 staining were used to detect this infection that is a significant threat to renal transplant outcome.

4 p63 in the Daily Practice of Pathologists

4.1 p63 in Breast Pathology

Normal breast ducts and almost all benign epithelial breast lesions display an internal epithelial layer and an external discontinuous myoepithelial layer. Conversely, all invasive breast carcinomas lack a myoepithelial layer around the tumour structures. In situ breast carcinomas often, but not always, lack this myoepithelial layer. Myoepithelial markers are therefore a very important tool in breast pathology for the differential diagnosis between benign and malignant lesions.

p63 has been recognized as a myoepithelial marker in breast tissues in the early 2000 (Barbareschi et al. 2001; Wang et al. 2002). Previous myoepithelial markers such as smooth muscle actin were already used to distinguish invasive and non-invasive lesions. However, smooth muscle actin is also present in vessel walls and has therefore specificity limits. p63 staining is more specific of myoepithelial cells and was quickly used as a second myoepithelial marker in breast pathology. The p63 usefulness for distinguishing invasive carcinoma from its histological mimics in breast pathology has been largely confirmed in subsequent reports (Werling et al. 2003). p63 IHC, along with smooth muscle actin or CD10 IHC, is now performed routinely in order to show the presence of a myoepithelial layer in non-invasive lesions or to show its disappearance in invasive lesions, especially tubular carcinomas that mimic benign lesions (de Moraes Schenka et al. 2006).

Another important use of p63 staining in breast pathology area is the differential diagnosis of papillary lesions of the breast, helping for the distinction between papillomas that still have myoepithelial cells and papillary carcinomas that have lost myoepithelial cells (Tse et al. 2009).

A potential pitfall in using p63 staining to exclude malignancy is the secretory carcinoma that often displays cytoplasmic p63 staining (Bratthauer et al. 2005).

Molecular classifications of breast cancers have found that breast cancers can be divided into three main categories: the luminal, the HER2, and the basal one (Perou et al. 2000). It has been suggested that p63 staining could be a possible marker of basal differentiation (Ribeiro-Silva et al. 2005). However, this remains controversial since the basal entity itself has not a totally fixed definition (Badve et al. 2010).

Although originally considered not so useful in cytological samples (Reis-Filho et al. 2002), more recent works reported possible uses of p63 staining in breast cytology (Harton et al. 2007) especially for the diagnosis of fibro-epithelial lesions (Al-Maghraby et al. 2010).

4.2 p63 in Prostate Pathology

Distinguishing benign prostate glands from malignant ones, based purely on morphology, on prostatic core needle biopsy may prove difficult, in particular if the equivocal focus is small. The diagnosis of malignancy is often based on the absence of basal cells. High molecular weight cytokeratin (HMWCK) was historically the first marker used to detect the presence of prostate basal cells, but p63 has appeared as an alternate marker with specific advantages. p63 is expressed in the basal epithelial cells of normal prostatic glands. In contrast, p63 protein is not detected in most human prostate carcinomatous cells, and p63 immunostaining is a recognized and valuable tool in the differential diagnosis of benign versus malignant prostatic lesions (Signoretti et al. 2000). Occasional non-specific staining of prostate adenocarcinoma with HMWCK can lead to false-negative diagnoses, and p63 seems to be more specific for basal cells than HMWCK, with less labelling of tumour cells (Ali and Epstein 2008). Also, p63 expression in prostate basal cells appears to be more sensitive than HMWCK (Shah et al. 2002). Furthermore, the combined use of the two basal cell markers p63 and HMWCK increases the sensitivity of basal cell detection, which is a critical issue for identifying benign prostatic lesions that may have inconspicuous or even lack basal cell lining focally (Zhou et al. 2003).

Alpha-methylacyl-CoA racemase (AMACR) is another marker commonly used for the diagnosis of prostate cancer. It is expressed in the cytoplasm of malignant prostate cells and absent or weakly expressed in non-tumoural prostate cells. Due to their distinct subcellular expression, AMACR and p63 can be detected with a single immunostaining procedure using a cocktail of two antibodies. This combined technique spares the material and avoids the disappearance of ambiguous lesions in cases of small foci (Molinié et al. 2004). On the basis of these two markers, prostatic intraepithelial neoplasia is characterized by the presence of transformed luminal cells expressing AMACR and the maintenance of an outer basal cell layer with p63 positivity. Specifically, high-grade prostatic intraepithelial neoplasia and the ductal adenocarcinoma variant of prostate cancer can be distinguished with difficulty on morphological basis, and the absence of p63 and/or HMWCK positive basal cell layer supports in this context the diagnosis of ductal adenocarcinoma.

Some limitations for the contribution of p63 to prostate cancer diagnosis have to be mentioned. Partial atrophy is a benign mimicker of prostate cancer on needle biopsy both at morphological and IHC levels, with the p63 and/or HMWCK immunostainings failing to detect basal cells in 20–30 % of cases whereas AMACR can be expressed (Wang et al. 2008). The very rare variant of basal cell carcinoma of the prostate is another pitfall as p63 is detected in this tumour type (Ali and Epstein 2008). This tumour type should be recognized mainly on morphological features.

4.3 p63 in Lung Pathology

As in breast and prostate tumours, p63 is a useful tool to explore the differentiation of lung carcinomas, and particularly help in distinguishing SCC and adenocarcinomas. p63 is expressed in 97 % SCC, with a strong nuclear immunostaining, while it is present in <30 % adenocarcinomas with heterogeneous and weak staining. In order to increase the reliability of the SCC/adenocarcinoma differential diagnosis, p63 marker is now used in pathology laboratories with a cocktail of two other antibodies, anti-TTF1 (almost always negative in SCC and positive in 85 % adenocarcinomas) and anti-cytokeratin 5/6 (positive in almost all SCC) (Loo et al. 2010; Nicholson et al. 2010).

5 Conclusion

With so many crucial biological effects and with such a high prevalence of mutations in human cancer, p53 had everything to be a promising biomarker and therapeutic target. However, from the pathologist's point of view, it appears that the assessment of p53 status in tumours is not so frequent except in some clinicopathological settings we have described in this review. The better characterization of atypical and dysplastic lesions in Barret's oesophagus, in IBDs, and in bladder is one of the most frequent use of p53 IHC since these lesions are often only present at the microscopical level and are poorly accessible to in vitro molecular assays. Better classifications of central nervous system tumours and help for predicting response to treatment in cancer especially in breast cancer are also performed regularly in some pathology laboratories. Also, p63 IHC is frequently used in tumour diagnoses for the characterization of tumour differentiation. However, there are limits in using p53 in pathology laboratories, especially regarding p53 IHC, due to the underlying molecular mechanisms of protein overexpression, and the lack of standardized techniques and interpretation of IHC results. When sufficient amount of high-quality tissue samples can be obtained in pathology laboratories, in vitro molecular tests can be performed. These tests are more powerful and more reliable than IHC and allow a more accurate *TP53* status assessment.

Acknowledgements We thank our colleagues from several laboratories of pathology who kindly answered our survey: J. Brière, F. Capron, J.M. Coindre, A. Couvelard, J.F. Emile, K. Mokhtari, V. Paradis, M. Peuchmaur, M.C. Vacher-Lavenue, and L. Xerri.

References

Ali TZ, Epstein JI (2008) False positive labeling of prostate cancer with high molecular weight cytokeratin: p63 a more specific immunomarker for basal cells. Am J Surg Pathol 32:1890–1895

Al-Maghraby H, Ghorab Z, Khalbuss W, Wong J, Silverman JF, Saad RS (2010) The diagnostic utility of CK5/6 and p63 in fine-needle aspiration of the breast lesions diagnosed as proliferative fibrocystic lesion. Diagn Cytopathol 40(2):141–147

Alsner J, Jensen V, Kyndi M, Offersen BV, Vu P, Børresen-Dale AL, Overgaard J (2008) A comparison between p53 accumulation determined by immunohistochemistry and *TP53* mutations as prognostic variables in tumours from breast cancer patients. Acta Oncol 47:600–607

Amin MB (2010) The 2009 version of the cancer protocols of the college of American pathologists. Arch Pathol Lab Med 134:326–330

Baas IO, Mulder JW, Offerhaus GJ, Vogelstein B, Hamilton SR (1994) An evaluation of six antibodies for immunohistochemistry of mutant *TP53* gene product in archival colorectal neoplasms. J Pathol 172:5–12

Badve S, Dabbs DJ, Schnitt SJ, Baehner FL, Decker T, Eusebi V, Fox SB, Ichihara S, Jacquemier J, Lakhani SR, Palacios J, Rakha EA, Richardson AL, Schmitt FC, Tan PH, Tse GM, Weigelt B, Ellis IO, Reis-Filho JS (2010) Basal-like and triple-negative breast cancers: a critical review with an emphasis on the implications for pathologists and oncologists. Mod Pathol 24(2):157–167

Bani-Hani K, Martin IG, Hardie LJ, Mapstone N, Briggs JA, Forman D, Wild CP (2000) Prospective study of cyclin D1 overexpression in Barrett's esophagus: association with increased risk of adenocarcinoma. J Natl Cancer Inst 92:1316–1321

Barbareschi M, Pecciarini L, Cangi MG, Macrì E, Rizzo A, Viale G, Doglioni C (2001) p63, a p53 homologue, is a selective nuclear marker of myoepithelial cells of the human breast. Am J Surg Pathol 25:1054–1060

Bártek J, Bártková J, Lukás J, Stasková Z, Vojtěsek B, Lane DP (1993) Immunohistochemical analysis of the p53 oncoprotein on paraffin sections using a series of novel monoclonal antibodies. J Pathol 169:27–34

Berggren P, Steineck G, Adolfsson J, Hansson J, Jansson O, Larsson P, Sandstedt B, Wijkström H, Hemminki K (2001) p53 mutations in urinary bladder cancer. Br J Cancer 84:1505–1511

Bernardini S, Adessi GL, Billerey C, Chezy E, Carbillet JP, Bittard H (1999) Immunohistochemical detection of p53 protein overexpression versus gene sequencing in urinary bladder carcinomas. J Urol 162:1496–1501

Bertheau P, Plassa F, Espié M, Turpin E, de Roquancourt A, Marty M, Lerebours F, Beuzard Y, Janin A, de Thé H (2002) Effect of mutated *TP53* on response of advanced breast cancers to high-dose chemotherapy. Lancet 360:852–854

Bertheau P, Lerebours F, Mounier N, de Roquancourt A, Espié M, Clot P, Servant JM, Misset JL, Marty M, Janin A (2005) Prognostic significance of a combined clinicopathologic score for response to primary systemic therapy in locally advanced breast cancer. Oncol Rep 14:513–520

Bertheau P, Turpin E, Rickman DS, Espié M, de Reyniès A, Feugeas JP, Plassa LF, Soliman H, Varna M, de Roquancourt A, Lehmann-Che J, Beuzard Y, Marty M, Misset JL, Janin A, de Thé H (2007) Exquisite sensitivity of *TP53* mutant and basal breast cancers to a dose-dense epirubicin-cyclophosphamide regimen. PLoS Med 4:e90

Bevilacqua G, Bosman F, Dassesse T, Höfler H, Janin A, Langer R, Larsimont D, Morente MM, Riegman P, Schirmacher P, Stanta G, Zatloukal K, Caboux E, Hainaut P (2010) The role of the pathologist in tissue banking: European Consensus Expert Group Report. Virchows Arch 456:449–454

Brambilla E, Gazzeri S, Lantuejoul S, Coll JL, Moro D, Negoescu A, Brambilla C (1998) p53 mutant immunophenotype and deregulation of p53 transcription pathway (Bcl2, Bax, and Waf1) in precursor bronchial lesions of lung cancer. Clin Cancer Res 4:1609–1618

Bratthauer GL, Saenger JS, Strauss BL (2005) Antibodies targeting p63 react specifically in the cytoplasm of breast epithelial cells exhibiting secretory differentiation. Histopathology 47:611–616

Choi SW, Levine JE, Ferrara JL (2010) Pathogenesis and management of graft-versus-host disease. Immunol Allergy Clin North Am 30:75–101

Coggi G, Bosari S, Roncalli M, Graziani D, Bossi P, Viale G, Buffa R, Ferrero S, Piazza M, Blandamura S, Segalin A, Bonavina L, Peracchia A (1997) p53 protein accumulation and *TP53* gene mutation in esophageal carcinoma. A molecular and immunohistochemical study with clinicopathologic correlations. Cancer 79:425–432

Colomer A, Erill N, Verdú M, Roman R, Vidal A, Cordon-Cardo C, Puig X (2003) Lack of p53 nuclear immunostaining is not indicative of absence of *TP53* gene mutations in colorectal adenocarcinomas. Appl Immunohistochem Mol Morphol 11:130–137

de Moraes Schenka NG, Schenka AA, de Souza QL, de Almeida MM, Alvarenga M, Vassallo J (2006) p63 and CD10: reliable markers in discriminating benign sclerosing lesions from tubular carcinoma of the breast? Appl Immunohistochem Mol Morphol 14:71–77

Di Como CJ, Urist MJ, Babayan I, Drobnjak M, Hedvat CV, Teruya-Feldstein J, Pohar K, Hoos A, Cordon-Cardo C (2002) p63 expression profiles in human normal and tumor tissues. Clin Cancer Res 8:494–501

Dookeran KA, Dignam JJ, Ferrer K, Sekosan M, McCaskill-Stevens W, Gehlert S (2010) p53 as a marker of prognosis in African-American women with breast cancer. Ann Surg Oncol 17:1398–1405

Eady RA, Cowen T (1977) Epidermal repair in lichen planus: a light and electron microscopical study. Clin Exp Dermatol 2:323–334

Fayallah EA, Hemida RA, Gamal AM, Abd Elhady E, Anwar KI, Nada NA, Sherif LS, Sayed-Ahmed MT (2010) Pretreatment study of P53 overexpression for selection of candidates for pelvic lymphadenectomy in clinical stage I endometrial carcinoma: a randomized-controlled study. Arch Gynecol Obstet 283(3):617–622

Ferrer-Luna R, Núñez L, Piquer J, Arias E, Dasí F, Cervio A, Arakaki N, Sevlever G, Celda B, Martinetto H (2010) Whole-genomic survey of oligodendroglial tumors: correlation between allelic imbalances and gene expression profiles. J Neurooncol 103(1):71–85

Fléjou JF, Svrcek M (2007) Barrett's oesophagus – a pathologist's view. Histopathology 50:3–14

Garg K, Leitao MM Jr, Wynveen CA, Sica GL, Shia J, Shi W, Soslow RA (2010) p53 overexpression in morphologically ambiguous endometrial carcinomas correlates with adverse clinical outcomes. Mod Pathol 23:80–92

Guarneri V, Broglio K, Kau SW, Cristofanilli M, Buzdar AU, Valero V, Buchholz T, Meric F, Middleton L, Hortobagyi GN, Gonzalez-Angulo AM (2006) Prognostic value of pathologic complete response after primary chemotherapy in relation to hormone receptor status and other factors. J Clin Oncol 24:1037–1044

Hamelin R, Fléjou JF, Muzeau F, Potet F, Laurent-Puig P, Fékété F, Thomas G (1994) *TP53* gene mutations and p53 protein immunoreactivity in malignant and premalignant Barrett's esophagus. Gastroenterology 107:1012–1018

Harpaz N, Polydorides AD (2010) Colorectal dysplasia in chronic inflammatory bowel disease: pathology, clinical implications, and pathogenesis. Arch Pathol Lab Med 134:876–895

Harton AM, Wang HH, Schnitt SJ, Jacobs TW (2007) p63 immunocytochemistry improves accuracy of diagnosis with fine-needle aspiration of the breast. Am J Clin Pathol 128:80–85

Hodges KB, Lopez-Beltran A, Davidson DD, Montironi R, Cheng L (2010) Urothelial dysplasia and other flat lesions of the urinary bladder: clinicopathologic and molecular features. Hum Pathol 41:155–162

Horne GM, Anderson JJ, Tiniakos DG, McIntosh GG, Thomas MD, Angus B, Henry JA, Lennard TW, Horne CH (1996) p53 protein as a prognostic indicator in breast carcinoma: a comparison of four antibodies for immunohistochemistry. Br J Cancer 73:29–35

Ip PP, Cheung AN, Clement PB (2009) Uterine smooth muscle tumors of uncertain malignant potential (STUMP): a clinicopathologic analysis of 16 cases. Am J Surg Pathol 33:992–1005

Jacquemier J, Molès JP, Penault-Llorca F, Adélaide J, Torrente M, Viens P, Birnbaum D, Theillet C (1994) p53 immunohistochemical analysis in breast cancer with four monoclonal antibodies: comparison of staining and PCR-SSCP results. Br J Cancer 69:846–852

Janin A, Murata H, Leboeuf C, Cayuela JM, Gluckman E, Legrès L, Desveaux A, Varna M, Ratajczak P, Soulier J, de Thé H, Bertheau P, Socié G (2009) Donor-derived oral squamous cell carcinoma after allogeneic bone marrow transplantation. Blood 113:1834–1840

Janin-Mercier A, Saurat JH, Bourges M, Sohier J, Jean LD, Gluckman E (1981) The lichen planus like and sclerotic phases of the graft versus host disease in man: an ultrastructural study of six cases. Acta Derm Venereol 61:187–193

Jankowski JA, Odze RD (2009) Biomarkers in gastroenterology: between hope and hype comes histopathology. Am J Gastroenterol 104:1093–1096

Jia L, Liu Y, Yi X, Miron A, Crum CP, Kong B, Zheng W (2008) Endometrial glandular dysplasia with frequent *TP53* gene mutation: a genetic evidence supporting its precancer nature for endometrial serous carcinoma. Clin Cancer Res 14:2263–2269

Kaye PV, Haider SA, Ilyas M, James PD, Soomro I, Faisal W, Catton J, Parsons SL, Ragunath K (2009) Barrett's dysplasia and the Vienna classification: reproducibility, prediction of progression and impact of consensus reporting and p53 immunohistochemistry. Histopathology 54:699–712

Kaye PV, Haider SA, James PD, Soomro I, Catton J, Parsons SL, Ragunath K, Ilyas M (2010) Novel staining pattern of p53 in Barrett's dysplasia–the absent pattern. Histopathology 57:933–935

Keller U, von Bubnoff N, Peschel C, Duyster J (2010) Oncologist's/haematologist's view on the roles of pathologists for molecular targeted cancer therapy. J Cell Mol Med 14:805–817

Keswani RN, Noffsinger A, Waxman I, Bissonnette M (2006) Clinical use of p53 in Barrett's esophagus. Cancer Epidemiol Biomarkers Prev 15:1243–1249

Khoury MP, Bourdon JC (2010) The isoforms of the p53 protein. Cold Spring Harb Perspect Biol 2:a000927

Kohno H, Hiroshima K, Toyozaki T, Fujisawa T, Ohwada H (1999) p53 mutation and allelic loss of chromosome 3p, 9p of preneoplastic lesions in patients with nonsmall cell lung carcinoma. Cancer 85:341–347

Lambkin HA, Mothersill CM, Kelehan P (1994) Variations in immunohistochemical detection of p53 protein overexpression in cervical carcinomas with different antibodies and methods of detection. J Pathol 172:13–18

Lee EJ, Kim TJ, Kim DS, Choi CH, Lee JW, Lee JH, Bae DS, Kim BG (2010) p53 alteration independently predicts poor outcomes in patients with endometrial cancer: a clinicopathologic study of 131 cases and literature review. Gynecol Oncol 116:533–538

Lehmann-Che J, André F, Desmedt C, Mazouni C, Giacchetti S, Turpin E, Espié M, Plassa LF, Marty M, Bertheau P, Sotiriou C, Piccart M, Symmans WF, Pusztai L, de Thé H (2010) Cyclophosphamide dose intensification may circumvent anthracycline resistance of p53 mutant breast cancers. Oncologist 15:246–252

Levine AJ, Oren M (2009) The first 30 years of p53: growing ever more complex. Nat Rev Cancer 9:749–758

Loo PS, Thomas SC, Nicolson MC, Fyfe MN, Kerr KM (2010) Subtyping of undifferentiated non-small cell carcinomas in bronchial biopsy specimens. J Thorac Oncol 5:442–447

Louis DN (1994) The *TP53* gene and protein in human brain tumors. J Neuropathol Exp Neurol 53:11–21

Malats N, Bustos A, Nascimento CM, Fernandez F, Rivas M, Puente D, Kogevinas M, Real FX (2005) P53 as a prognostic marker for bladder cancer: a meta-analysis and review. Lancet Oncol 6:678–686

Marx A, Wandrey T, Simon P, Wewer A, Grob T, Reichelt U, Minner S, Simon R, Spehlmann M, Tigges W, Soehendra N, Seitz U, Seewald S, Izbicki JR, Yekebas E, Kaifi JT, Mirlacher M, Terracciano L, Fleischmann A, Raedler A, Sauter G (2009) Combined alpha-methylacyl coenzyme A racemase/p53 analysis to identify dysplasia in inflammatory bowel disease. Hum Pathol 40:166–173

Molinié V, Fromont G, Sibony M, Vieillefond A, Vassiliu V, Cochand-Priollet B, Hervé JM, Lebret T, Baglin AC (2004) Diagnostic utility of a p63/alpha-methyl-CoA-racemase (p504s) cocktail in atypical foci in the prostate. Mod Pathol 17:1180–1190

Nenutil R, Smardova J, Pavlova S, Hanzelkova Z, Muller P, Fabian P, Hrstka R, Janotova P, Radina M, Lane DP, Coates PJ, Vojtesek B (2005) Discriminating functional and non-functional p53 in human tumours by p53 and MDM2 immunohistochemistry. J Pathol 207:251–259

Nicholson AG, Gonzalez D, Shah P, Pynegar MJ, Deshmukh M, Rice A, Popat S (2010) Refining the diagnosis and EGFR status of non-small cell lung carcinoma in biopsy and cytologic material, using a panel of mucin staining, TTF-1, cytokeratin 5/6, and P63, and EGFR mutation analysis. J Thorac Oncol 5:436–441

Odze RD (2006) Diagnosis and grading of dysplasia in Barrett's oesophagus. J Clin Pathol 59:1029–1038

Odze RD, Riddell RH, Bosman FT, Carneiro F, Fléjou J-F, Geboes K et al (2010) Premalignant lesions of the digestive system. In: Bosman FT, Carneiro F, Hruban RH, Theise ND (eds) WHO classification of tumours of digestive system. IARC, Lyon

Okamoto Y, Di Patre PL, Burkhard C, Horstmann S, Jourde B, Fahey M, Schüler D, Probst-Hensch NM, Yasargil MG, Yonekawa Y, Lütolf UM, Kleihues P, Ohgaki H (2004) Population-based study on incidence, survival rates, and genetic alterations of low-grade diffuse astrocytomas and oligodendrogliomas. Acta Neuropathol 108:49–56

Osamura RY (2009) Roles of pathologists in molecular targeted cancer therapy. J Cell Mol Med 13:4286–4290

Peraud A, Kreth FW, Wiestler OD, Kleihues P, Reulen HJ (2002) Prognostic impact of *TP53* mutations and P53 protein overexpression in supratentorial WHO grade II astrocytomas and oligoastrocytomas. Clin Cancer Res 8:1117–1124

Perou CM, Sørlie T, Eisen MB, van de Rijn M, Jeffrey SS, Rees CA, Pollack JR, Ross DT, Johnsen H, Akslen LA, Fluge O, Pergamenschikov A, Williams C, Zhu SX, Lønning PE, Børresen-Dale AL, Brown PO, Botstein D (2000) Molecular portraits of human breast tumours. Nature 406:747–752

Pezzella F, Micklem K, Turley H, Pulford K, Jones M, Kocialkowski S, Delia D, Aiello A, Bicknell R, Smith K (1994) Antibody for detecting p53 protein by immunohistochemistry in normal tissues. J Clin Pathol 47:592–596

Pfaff E, Remke M, Sturm D, Benner A, Witt H, Milde T, von Bueren AO, Wittmann A, Schöttler A, Jorch N, Graf N, Kulozik AE, Witt O, Scheurlen W, von Deimling A, Rutkowski S, Taylor MD, Tabori U, Lichter P, Korshunov A, Pfister SM (2010) *TP53* mutation is frequently associated with CTNNB1 mutation or MYCN amplification and is compatible with long-term survival in medulloblastoma. J Clin Oncol 28:5188–5196

Ponticiello A, Barra E, Giani U, Bocchino M, Sanduzzi A (2000) P53 immunohistochemistry can identify bronchial dysplastic lesions proceeding to lung cancer: a prospective study. Eur Respir J 15:547–552

Prasad GA, Bansal A, Sharma P, Wang KK (2010) Predictors of progression in Barrett's esophagus: current knowledge and future directions. Am J Gastroenterol 105:1490–1502

Quddus MR, Sung CJ, Zheng W, Lauchlan SC (1999) p53 immunoreactivity in endometrial metaplasia with dysfunctional uterine bleeding. Histopathology 35:44–49

Reis-Filho JS, Milanezi F, Amendoeira I, Albergaria A, Schmitt FC (2002) p63 Staining of myoepithelial cells in breast fine needle aspirates: a study of its role in differentiating in situ from invasive ductal carcinomas of the breast. J Clin Pathol 55:936–939

Renninson J, Baker BW, McGown AT, Murphy D, Norton JD, Fox BW, Crowther D (1994) Immunohistochemical detection of mutant p53 protein in epithelial ovarian cancer using polyclonal antibody CMI: correlation with histopathology and clinical features. Br J Cancer 69:609–612

Ribeiro-Silva A, Ramalho LN, Garcia SB, Brandão DF, Chahud F, Zucoloto S (2005) p63 correlates with both BRCA1 and cytokeratin 5 in invasive breast carcinomas: further evidence for the pathogenesis of the basal phenotype of breast cancer. Histopathology 47:458–466

Riddell RH, Goldman H, Ransohoff DF, Appelman HD, Fenoglio CM, Haggitt RC, Ahren C, Correa P, Hamilton SR, Morson BC (1983) Dysplasia in inflammatory bowel disease: standardized classification with provisional clinical applications. Hum Pathol 14:931–968

Rodriguez FJ, Scheithauer BW, Burger PC, Jenkins S, Giannini C (2010) Anaplasia in pilocytic astrocytoma predicts aggressive behavior. Am J Surg Pathol 34:147–160

Saurat JH, Gluckman E (1977) Lichen-planus-like eruption following bone marrow transplantation: a manifestation of the graft-versus-host disease. Clin Exp Dermatol 2:335–344

Shah RB, Zhou M, LeBlanc M, Snyder M, Rubin MA (2002) Comparison of the basal cell-specific markers, 34betaE12 and p63, in the diagnosis of prostate cancer. Am J Surg Pathol 26:1161–1168

Signoretti S, Waltregny D, Dilks J, Isaac B, Lin D, Garraway L, Yang A, Montironi R, McKeon F, Loda M (2000) p63 is a prostate basal cell marker and is required for prostate development. Am J Pathol 157:1769–1775

Sikkema M, Kerkhof M, Steyerberg EW, Kusters JG, van Strien PM, Looman CW, van Dekken H, Siersema PD, Kuipers EJ (2009) Aneuploidy and overexpression of Ki67 and p53 as markers for neoplastic progression in Barrett's esophagus: a case–control study. Am J Gastroenterol 104:2673–2680

Socié G, Henry-Amar M, Bacigalupo A, Hows J, Tichelli A, Ljungman P, McCann SR, Frickhofen N, Van't Veer-Korthof E, Gluckman E (1993) Malignant tumors occurring after treatment of aplastic anemia. European Bone Marrow Transplantation-Severe Aplastic Anaemia Working Party. N Engl J Med 329:1152–1157

Soong R, Robbins PD, Dix BR, Grieu F, Lim B, Knowles S, Williams KE, Turbett GR, House AK, Iacopetta BJ (1996) Concordance between p53 protein overexpression and gene mutation in a large series of common human carcinomas. Hum Pathol 27:1050–1055

Sozzi G, Miozzo M, Donghi R, Pilotti S, Cariani CT, Pastorino U, Della Porta G, Pierotti MA (1992) Deletions of 17p and *TP53* mutations in preneoplastic lesions of the lung. Cancer Res 52:6079–6082

Spruck CH, Ohneseit PF, Gonzalez-Zulueta M, Esrig D, Miyao N, Tsai YC, Lerner SP, Schmütte C, Yang AS, Cote R (1994) Two molecular pathways to transitional cell carcinoma of the bladder. Cancer Res 54:784–788

Ständer M, Peraud A, Leroch B, Kreth FW (2004) Prognostic impact of *TP53* mutation status for adult patients with supratentorial World Health Organization Grade II astrocytoma or oligoastrocytoma: a long-term analysis. Cancer 101:1028–1035

Stiewe T (2007) The p53 family in differentiation and tumorigenesis. Nat Rev Cancer 7:165–168

Sun W, Zhang PL, Herrera GA (2002) p53 protein and Ki-67 overexpression in urothelial dysplasia of bladder. Appl Immunohistochem Mol Morphol 10:327–331

Sylvester RJ, van der Meijden AP, Oosterlinck W, Witjes JA, Bouffioux C, Denis L, Newling DW, Kurth K (2006) Predicting recurrence and progression in individual patients with stage Ta T1 bladder cancer using EORTC risk tables: a combined analysis of 2596 patients from seven EORTC trials. Eur Urol 49:466–475, discussion 475–477

Tabori U, Baskin B, Shago M, Alon N, Taylor MD, Ray PN, Bouffet E, Malkin D, Hawkins C (2010a) Universal poor survival in children with medulloblastoma harboring somatic *TP53* mutations. J Clin Oncol 28:1345–1350

Tabori U, Shlien A, Baskin B, Levitt S, Ray P, Alon N, Hawkins C, Bouffet E, Pienkowska M, Lafay-Cousin L, Gozali A, Zhukova N, Shane L, Gonzalez I, Finlay J, Malkin D (2010b) *TP53* alterations determine clinical subgroups and survival of patients with choroid plexus tumors. J Clin Oncol 28:1995–2001

Tse GM, Tan PH, Moriya T (2009) The role of immunohistochemistry in the differential diagnosis of papillary lesions of the breast. J Clin Pathol 62:407–413

Vojtěsek B, Fisher CJ, Barnes DM, Lane DP (1993) Comparison between p53 staining in tissue sections and p53 proteins levels measured by an ELISA technique. Br J Cancer 67:1254–1258

Wagner U, Sauter G, Moch H, Novotna H, Epper R, Mihatsch MJ, Waldman FM (1995) Patterns of p53, erbB-2, and EGF-r expression in premalignant lesions of the urinary bladder. Hum Pathol 26:970–978

Wang X, Mori I, Tang W, Nakamura M, Nakamura Y, Sato M, Sakurai T, Kakudo K (2002) p63 expression in normal, hyperplastic and malignant breast tissues. Breast Cancer 9:216–219

Wang W, Sun X, Epstein JI (2008) Partial atrophy on prostate needle biopsy cores: a morphologic and immunohistochemical study. Am J Surg Pathol 32:851–857

Weller M, Felsberg J, Hartmann C, Berger H, Steinbach JP, Schramm J, Westphal M, Schackert G, Simon M, Tonn JC, Heese O, Krex D, Nikkhah G, Pietsch T, Wiestler O, Reifenberger G, von Deimling A, Loeffler M (2009) Molecular predictors of progression-free and overall survival in patients with newly diagnosed glioblastoma: a prospective translational study of the German Glioma Network. J Clin Oncol 27:5743–5750

Werling RW, Hwang H, Yaziji H, Gown AM (2003) Immunohistochemical distinction of invasive from noninvasive breast lesions: a comparative study of p63 versus calponin and smooth muscle myosin heavy chain. Am J Surg Pathol 27:82–90

Wiesend WN, Parasuraman R, Li W, Farinola MA, Rooney MT, Hick SK, Samarapungavan D, Cohn SR, Reddy GH, Rocher LL, Dumler F, Lin F, Zhang PL (2010) Adjuvant role of p53 immunostaining in detecting BK viral infection in renal allograft biopsies. Ann Clin Lab Sci 40:324–329

Würl P, Taubert H, Meye A, Berger D, Lautenschläger C, Holzhausen HJ, Schmidt H, Kalthoff H, Rath FW, Dralle H (1997) Prognostic value of immunohistochemistry for p53 in primary soft-tissue sarcomas: a multivariate analysis of five antibodies. J Cancer Res Clin Oncol 123:502–508

Yang A, Walker N, Bronson R, Kaghad M, Oosterwegel M, Bonnin J, Vagner C, Bonnet H, Dikkes P, Sharpe A, McKeon F, Caput D (2000) p73-deficient mice have neurological, pheromonal and inflammatory defects but lack spontaneous tumours. Nature 404:99–103

Zhou M, Shah R, Shen R, Rubin MA (2003) Basal cell cocktail (34betaE12+p63) improves the detection of prostate basal cells. Am J Surg Pathol 27:365–371

Chapter 18
Genetic Counseling for *TP53* Germline Mutations

Maria Isabel Waddington Achatz and Patricia Ashton-Prolla

1 Introduction and Brief Review of LFS/LFL

Most genetic counseling situations that involve *TP53* germline mutations are related to Li–Fraumeni syndrome (LFS; OMIM # 151623), an autosomal dominant disorder associated with multiple early-onset cancers. Classic LFS and its variants, collectively named Li–Fraumeni-like (LFL) syndrome, are two recognized forms of the syndrome that share many but not all of the clinical features, including an increased predisposition to soft tissue sarcoma, osteosarcoma, premenopausal breast cancer, brain tumors, adrenocortical carcinoma (ACC), and a variety of other tumors.

The LFS phenotype was initially described in 1969 by Li and Fraumeni in families from children diagnosed with rhabdomyosarcoma. In 1988, the same investigators proposed the first definition for LFS as a cancer predisposition syndrome, which was subsequently confirmed by a series of epidemiological studies that detected a similar tumor pattern among family members (Li et al. 1988). The molecular basis of LFS remained unknown until its connection to the *TP53* tumor suppressor gene in 1990 (Malkin et al. 1990). Subsequent studies have found germline mutations in the *TP53* coding region of many, but not all, Li–Fraumeni families (Srivastava et al. 1990). It has been suggested that in these families, the cancer predisposition phenotype could

M.I.W. Achatz, M.D., Ph.D. (✉)
Department of Oncogenetics, Oncogenomics Program, Hospital A.C. Camargo – Fundação Antônio Prudente, Rua Professor Antonio Prudente, 211 – 1oSS, CEP 01509-010, São Paulo, SP, Brazil
e-mail: miachatz@cipe.accamargo.org.br

P. Ashton-Prolla, M.D., Ph.D.
Departamento de Genética, UFRGS e SGM-HCPA, Rua Ramiro Barcelos 2350, Porto Alegre, RS 90035-903, Brazil

Centro de Pesquisa Experimental, Hospital de Clínicas de Porto Alegre, Rua Ramiro Barcelos 2350, Porto Alegre, RS 90035-903, Brazil
e-mail: pprolla@hcpa.ufrgs.br

P. Hainaut et al. (eds.), *p53 in the Clinics*, DOI 10.1007/978-1-4614-3676-8_18,

be due to mutations outside the *TP53* coding regions (intronic, promoter region), gene rearrangements that escape standard detection methods which are based on the sequencing of individual exons, and mutations in other genes closely linked to the *TP53* signaling pathway. In addition, one could postulate that the LFS/LFL phenotype is genetically heterogeneous, with *TP53* mutations accounting for only a fraction of the families. Data from the IARC *TP53* database (Petitjean et al. 2007) show that in ~81% of classic LFS cases, and 67% of LFL cases, affected family members carry a germline mutation of one allele of *TP53*. Earlier reports on the possible role of *CHK2* germline mutations in these families (Bell et al. 1999; Vahteristo et al. 2001) have not been substantiated in further studies (Evans et al. 2008; Ruijs et al. 2009). Although a few families with LFS have been shown to carry mutations in *PTEN* (Tachibana et al. 2000) or *CDKN2A* genes (Brown et al. 2000), no other cancer predisposition gene has been systematically linked to the syndrome.

LFS is a highly penetrant cancer syndrome. A segregation analysis conducted on families with LFS revealed that carriers of a germline*TP53* mutation have, on average, a 50% risk to develop cancer before the age of 40 years, compared to 1% in the general population. It also demonstrated that the lifetime risk of cancer in an LFS patient that carries a germline *TP53* mutation is 90% by age 60 years (Lustbader et al. 1992). In addition, individuals who survive a first primary tumor are prone to develop other primaries, particularly if exposed to radiation therapy. In LFS, the most common childhood cancers are adrenocortical carcinoma (ACC) (infancy), choroid plexus tumors, and soft-tissue sarcomas in the first 5 years of life. Osteosarcomas often occur in adolescence, and acute leukemia and brain tumors occur throughout childhood and young adulthood. In young adults, premenopausal breast cancer is, by far, the most common neoplasm, but other common tumors identified in the LFS/LFL syndromes are melanoma, gastric, colorectal, lung, pancreatic, ovarian, prostate, thyroid, endometrial, cervical cancer, and lymphoma (Nichols et al. 2001; Wong et al. 2006). Additional tumors that have been described in LFS families and considered as part of the LFS tumor spectrum by some authors include germ cell tumors, choroid plexus papilloma, and Wilms' tumor. However, the exact tumor spectrum of the syndrome is still a matter of debate. Clearly defined personal and family history criteria characterize both LFS and LFL and these are summarized in Table 18.1. The frequency of germline mutations in *TP53* among families with the LFS/LFL phenotype varies according to the diagnostic criteria that are used to indicate testing, with higher mutation frequencies being observed in those families fulfilling the more strict clinical criteria. However, since only a few *TP53* mutation-positive families fulfill the classical LFS criteria, other testing criteria must be considered in the clinic. The modified Chompret criteria (Tinat et al. 2009) have been associated with the identification of a germline *TP53* mutation in 20% of the probands tested. Several authors have suggested that germline *TP53* mutations may be more common than previously thought; for instance, they could be responsible for a significant proportion of very early-onset breast cancer cases. Considering that the full tumor spectrum and mutation penetrance of the LF/LFL syndromes have not yet been determined, additional indications for *TP53* mutation testing may be considered in the future.

Table 18.1 Clinical criteria for classic Li–Fraumeni syndrome (LFS), LFS-like criteria (LFL), and Chompret criteria

Classic LFS criteria (Li et al. 1988)
Proband diagnosed with a sarcoma before age 45; AND
A first-degree[a] relative with cancer diagnosed before age 45; AND
Another first- or second-degree[b] relative on the same side of the family with cancer diagnosed before age 45 or a sarcoma at any age
LFS-like (LFL) criteria Birch (1994)
Proband with any childhood cancer or sarcoma, brain tumor, or adrenocortical carcinoma diagnosed before age 45; AND
First- or second-degree relative with a typical LFS cancer (sarcoma, breast cancer, brain tumor, leukemia, or adrenocortical carcinoma) diagnosed at any age; AND
A first- or second-degree relative on the same side of the family with any cancer diagnosed under age 60
Eeles definition of LFL (Eeles 1995)
Two first- or second-degree relatives with LFS-related malignancies (sarcoma, breast cancer, brain tumor, leukemia, adrenocortical tumor, melanoma, prostate cancer, pancreatic cancer) at any age
Chompret criteria for TP53 germline mutation screening (Tinat 2009)
Proband diagnosed with a tumor belonging to the LFS tumor spectrum (sarcoma, brain tumor, premenopausal breast cancer, adrenocortical carcinoma, leukemia, lung bronchoalveolar cancer) before age 46, and at least one first- or second-degree relative affected with a LFS tumor (other than breast cancer if the proband is affected by breast cancer) before age 56, or a relative with multiple primary tumors at any age; OR
A proband with multiple primary tumors (except multiple breast tumors), two of which belong to the LFS tumor spectrum and the first of which occurred before age 46, regardless of family history; OR
A proband with adrenocortical carcinoma or choroid plexus tumor, regardless of family history

[a]First-degree relative is defined as a parent, sibling, or child

[b]Second-degree relative is defined as a grandparent, aunt, uncle, niece, nephew, or grandchild

2 Formal Clinical Assessment and Genetic Counseling of Individuals and Families with Suspected LFS/LFL

2.1 Assessing a Suspected Patient and/or Family

A comprehensive family history must be obtained including a pedigree of at least three generations. All cases of suspected or proven cancer diagnosis in the family must be recorded. Information obtained should include number of affected generations in the family, number of deaths by cancer, age at tumor diagnosis, tumor type, tumor localization, and relationship to the proband. Whenever possible, tumor diagnoses should be confirmed by pathology reports, death certificates, or hospital/medical records. In all families with multiple cancer diagnoses, an attempt should be made to classify the pedigree according to the known criteria for LF/LFL syndrome and for other cancer predisposition syndromes.

2.2 Differential Diagnosis of LFS/LFL with Other Cancer Predisposition Syndromes and Overlapping Phenotypes

LFS/LFL syndromes may have a clinical presentation that overlaps other hereditary breast cancer predisposition syndromes such as the hereditary breast and ovarian cancer syndrome (HBOC) caused by germline mutations in the *BRCA1* and *BRCA2* genes as well as the recently identified hereditary breast and colon cancer (HBCC) syndrome caused by germline mutations in the *CHEK2* gene. Some LFS/LFL families may also resemble kindreds with familial brain tumors. Frequently, a family that fulfills criteria for LFL syndrome (either Birch or Eeles) also fulfills criteria for the clinical diagnosis of a HBOC or HBCC syndrome (Lindor et al. 2008).

2.3 Molecular Diagnosis

Currently, *TP53* is the only gene known to be associated with LFS. Molecular genetic testing for germline mutations in the *TP53* gene has been done in LFS/LFL families for the purposes of confirmation of the clinical suspicion, predisposition testing and prenatal or preimplantation genetic diagnoses, as well as for prognostic reasons, since a few genotype/phenotype correlations have been suggested (Birch 2004; Olivier 2003). Considering the significant molecular heterogeneity of LFS/LFL, testing should include mutation screening of the entire coding exons 2–11 and flanking intronic regions, as well as gene rearrangement screening. For interpreting sequence analysis results, the recommendations adopted by the ACMG working group (Richards et al. 2008) can be applied to *TP53*. When a mutation is detected, information on its pathogenicity can be retrieved from the IARC *TP53* database. The mutation may already have been described in LFS/LFL families, in which case the associated tumor spectrum can be reviewed (http://www-p53.iarc.fr/TumorCriteria.ASP). If the mutation has not previously been described, data on its frequency in sporadic cancers and biological impact (assessed in functional assays or predicted by sequence conservation models) can be obtained (http://www-p53.iarc.fr/MutationValidationCriteria.asp) to estimate its functional severity. For gene rearrangement testing, additional mutation screening methods such as quantitative multiplex PCR of short fluorescent fragments (QMPSF), comparative genomic hybridization (CGH), or multiplex ligation-dependent probe amplification (MLPA) should be employed. If no mutation is identified, the family history should be carefully checked for the possibility of other association with other cancer predisposition syndromes.

2.4 Genetic Counseling

LFS/LFL syndromes are inherited in an autosomal dominant manner. Although de novo mutations have been reported (see below), most of the mutation carriers identified to date have a significant family history of cancer. Genetic testing in a family with suspected LFS/LFL should be carried out as in other cancer predisposition syndromes. Whenever possible, a cancer-affected individual of the family (preferentially with an early-onset tumor of the LFS/LFL spectrum) should be tested initially. Once a mutation has been identified, mutation-specific testing of other at-risk relatives is indicated. In families with extensive cancer family histories, especially when tumors of the LFS spectrum are present in more than one branch, whole gene and not only site-specific mutation testing may be considered after one *TP53* germline mutation has been identified in the family.

Risk of occurrence. The incidence of the LFS/LFL syndromes has not been determined. It has been considered a rare cancer predisposition syndrome and germline *TP53* mutations have been estimated at a rate of 1:5,000 in the general population (Lalloo et al. 2003). However, since the full tumor spectrum and mutation penetrance of affected individuals remain to be determined, it is very likely an underdiagnosed condition. Germline de novo *TP53* mutations have been described in particular among children with ACC and choroid plexus carcinomas (Varley et al. 1999; Libé and Bertherat 2005). Recently, Gonzalez and collaborators (2009) have described de novo *TP53* mutations in 7% of a series of patients with early-onset tumors of the LFS spectrum.

Risk for recurrence in an affected family. For individuals diagnosed with a germline mutation in the *TP53* gene, the risk for siblings depends on the genetic status of the proband's parents (OMIM # 151623). If one of them is found to have the same disease-causing mutation, then each of the proband's siblings is at 50% risk of also being a carrier. If neither parent is affected by the mutation, the proband is assumed to have a de novo mutation and the risk for his/her siblings is probably low. However, the risk may be slightly increased over the risk for the general population due to parental germline mosaicism. In addition, possible nonmedical explanations must also be considered, such as alternate paternity or undisclosed adoption. If the family meets LFS/LFL criteria but no *TP53* germline mutation can be identified by the currently available mutation testing procedures, the proband's siblings and children are assumed to have a 50% recurrence risk and clinical management should be proposed accordingly. Each child of an individual with a disease-causing mutation in *TP53* is at 50% risk of inheriting the same mutation.

2.5 Clinical Management of LFS/LFL Syndrome

Clinical management of individuals with LFS/LFL is a major challenge due to the wide tumor spectrum. Screening for cancer should be focused in the most common tumors that occur in the syndrome and that are eligible for early detection and based to some extent on the specific pattern of cancer occurring in the family. All patients should be aware about the limitations of screening. Recently monitoring strategies have shown to reduce mortality in carriers of *TP53* germline mutations (Villani et al. 2010).

According to the National Comprehensive Cancer Network Guidelines (NCCN 2011, USA), all carriers of a *TP53* germline mutation should undergo annual comprehensive physical examination, including careful skin and neurological examination, blood counts, imaging studies, endoscopies, and/or biopsies. Clinicians should be aware of the high risk of rare, early-onset cancers as well as the high risk of additional primary malignancies in cancer survivors. All patients should receive education regarding signs and symptoms of cancer, particularly headaches, bone pain, or abdominal discomfort.

Breast cancer is the most common tumor that occurs in LFS/LFL adult women; thus screening premenopausal women is highly recommended in all LFS women. Breast cancer monitoring includes monthly self-breast examination and biannual clinical breast examination beginning at age 18 years; annual mammogram and annual breast MRI should begin at age 20–25 years or individualized based on the earliest age of onset in the family (Lindor et al. 2008, NCCN 2010). Females with a germline *TP53* mutation should be counseled about the option of prophylactic mastectomy to reduce the risk of breast cancer (Thull and Vogel 2004).

Targeted cancer surveillance based on individual family history should be addressed. Periodic screening for colorectal cancer with colonoscopy every 2–5 years is suggested, starting no later than age 25 years (NCCN 2010).

Intensive screening strategies in children include rapid whole-body MRI, brain MRI, abdominal ultrasound examination, and biochemical markers of adrenal cortical function (Villani et al. 2010). In adults with LFS, a pilot trial of screening FDG-PET/CT scans has been reported, but there are concerns about the amount of radiation exposure with the test (Masciari et al. 2008). Results may guide future screening recommendations.

3 Specific Genetic Counseling Situations Related to TP53

3.1 Psychological Aspects of Genetic Testing and Counseling in LFS/LFL

Genetic testing for LFS/LFL has been available since the identification of germline *TP53* mutations in LFS/LFL families (Malkin et al. 1990) and has been offered both as a diagnostic (i.e., to confirm a clinical diagnosis or suspicion of the syndrome) and a predictive genetic test (i.e., to identify mutation carriers among asymptomatic at-risk family members of an affected individual). *TP53* mutation testing in the clinic, especially in the predictive setting, has been considered a very challenging aspect of the management and genetic counseling process in LFS/LFL families. It is assumed that most *TP53* mutation carriers have a lifetime risk of developing cancer of close to 90% and many of them are diagnosed with multiple primary tumors, either synchronous or metachronous (Lindor et al. 2008; Gonzalez et al. 2009). In addition, regular cancer surveillance is difficult in these individuals since tumor spectrum is very broad and age at tumor onset may be variable, even within members of a family that carry the same germline mutation. Apart from recommendations for breast and colorectal cancer screening, and breast cancer prevention by risk-reducing mastectomy, there is currently no consensus regarding other screening or preventive interventions in mutation carriers (NCCN 2010).

A wealth of literature has been published on the psychological aspects involved in genetic testing and counseling and long-term management in the more common cancer predisposition syndromes such as HBOC for instance (Broadstock et al. 2000; Esplen et al. 2001; Vadaparampil et al. 2006; O'Neill et al. 2010; Morrison et al. 2010). Less information is available for LFS/LFL. Regarding genetic testing uptake, most studies show that in general, predictive testing is more frequently accepted among individuals at risk for a disease that can be easily prevented, and classical studies are those that have looked at testing uptake in Huntington disease families (15%) as opposed to those in common cancer predisposition syndromes such as HBOC and Lynch syndrome (~60%) (Craufurd et al. 1989; Evans et al. 1997; Ropka et al. 2006; Wagner et al. 2002). Considering that in LFS preventive and treatment options are still limited, one would expect less interest of at-risk individuals in testing. Two studies from the late 1990s in the UK and USA have shown that predictive testing uptake in LFS families was indeed low (<40%) (Evans et al. 1997; Patenaude et al. 1996). However, a more recent study from the Netherlands showed that 55% of asymptomatic at-risk LFS family members opted to know their genetic status, a figure that is closer to the one observed for other cancer predisposition syndromes with clear surveillance and/or preventive guidelines. Interestingly the authors demonstrate that although ~25% of the whole cohort reported significant distress warranting professional psychological help, there was no difference in distress levels between carriers and noncarriers (Lammens et al. 2010a). Regarding the effect of genetic testing, LFS/LFL patients seem to follow the same pattern observed in other cancer predisposition syndromes: noncarriers experience psychological benefits from genetic testing, while carriers do not demonstrate significant adverse events (Broadstock et al. 2000; Braithwaite et al. 2004).

Psychological well-being of close relatives of a mutation carrier, especially if cancer affected, is usually quite compromised. Spouses are usually particularly distressed, not only with the concern with their partners, but also with the potential risk for their children. In partners of women who carry BRCA mutations, this has been clearly demonstrated and routine psychological support has been considered very important (Metcalfe et al. 2002; Mireskandari et al. 2006; Sherman et al. 2010). In LFS/LFL families, Lammens et al. (2010b) have shown that in a series of 50 partners of *TP53* mutation carriers, approximately one-third reported clinically relevant psychological distress directly related to the diagnosis and most of them reported a desire and/or need for routine psychosocial support.

Finally, psychological well-being has been implied as an important factor in compliance with medical management recommendations in different disorders. In most cancer predisposition syndromes, periodic surveillance and interventions that result from such screening strategies are usually the mainstream of management (Jarvinen et al. 1995; Narod and Offitt 2005). Thus, compliance to these recommendations may be considered one of the most important aspects to define outcome (Palmquist et al. 2010; Robson et al. 2010; Stoffel et al. 2010). Since no consensus is currently available on the management and efficacy of screening recommendations for all tumors observed in LFS/LFL families, there is an obvious concern regarding adherence to such recommendations. In an interesting study that included 55 high-risk LFS patients in the Netherlands, Lammens et al. (2010c) report that most patients adhere to screening recommendations, even if clinical benefits of such interventions remain undetermined. In addition, the large majority of patients in this study believed in the value of cancer screening and reported that undergoing such surveillance provided "a sense of control and security."

3.2 Prenatal and Preimplantation Diagnosis in LFS/LFL

Prenatal diagnosis for pregnancies at risk for LF/LFL syndromes (i.e., pregnancy of a known *TP53* mutation carrier) is possible by mutation testing using fetal DNA obtained either through chorionic villus sampling (CVS) or amniocentesis at 10–12 and 15–18 weeks gestation, respectively. The disease-causing mutation in the family must be known before prenatal testing can be performed. Preimplantation genetic diagnosis (PGD) has been performed for several genetic diseases including ~1,500 procedures for Mendelian disorders such as cancer predisposition syndromes (Offit et al. 2006). Among these, PGD for LFS has been accomplished as reported by Rechitsky et al. (2002) and may be offered as an option for those LFS/LFL families where a disease-causing *TP53* mutation has been identified. PGD is performed together with in vitro fertilization (IVF). Sperm and egg specimens are collected from the parents; combined in vitro and mutation testing is performed on a cell from the blastomeres and those that do not contain the mutation are implanted. Although PGD is a technically accurate procedure, IVF is a time-consuming and expensive procedure and an overall success (live birth) rate of only 20–25% has been observed after such procedures. In a recent report from the Netherlands, approximately half of the couples of LFS/LFL families contemplating a future pregnancy would consider the use of PGD, but the actual uptake of this procedure was estimated to be lower (Lammens et al. 2009).

3.3 Genetic Testing and Counseling in Minors of LFS/LFL Families

The justifications for presymptomatic and predictive genetic testing in minors, including cancer predisposition testing, have been the focus of several systematic reviews (Borry et al. 2006, 2009; Lwiwski et al. 2008). In addition, guidelines for cancer predisposition testing in asymptomatic children have been published by the American Society of Clinical Oncology, by the Canadian College of Medical Geneticists, by the National Society of Genetic Counselors and by the American

Society of Human Genetics, and by the European Society of Human Genetics (reviewed in Recommendations of the European Society of Human Genetics 2009 and Borry et al. 2006). There is strong agreement that the main justification for such testing should be the benefit of medical interventions or preventive measures to the minor. In the case of childhood-onset diseases for which preventive and/or intervention measures are not available, the recommendation for predictive genetic testing remains open to discussion and its potential social, psychological, and medical benefits should be carefully addressed with the parents and, whenever possible, the child at risk. Furthermore, timing of the investigation is also a matter of debate with a general consensus that it should not be performed before the earliest age at which preventive measures are considered useful.

Since the initial statement of Li et al. (1992), presymptomatic and predictive testing in minors at-risk for LFS has been proposed by several authors (Clarke and Gaff 2008; Evans et al. 2010; Teplick et al. 2011). Although effectiveness of prevention and surveillance strategies has not been definitively determined in LFS-affected children, the rationale behind this proposal is supported by several aspects. These include evidence for psychological benefits of the testing process regardless of the result, clear evidence that prognosis is dramatically different in pediatric tumors that are diagnosed in earlier stages (i.e., most of those identified through periodic surveillance), and lack of evidence for significant adverse events after information of mutation-positive results (Lwiwski et al. 2008; Duncan et al. 2010; Levine et al. 2010). More recently, Villani et al. (2010) have demonstrated increased detection of early-onset tumors and increased survival of pediatric *TP53* mutation carriers enrolled in a clinical surveillance program, further supporting predictive genetic testing of children. In all genetic testing situations that involve minors, genetic counseling is mandatory and should include previous discussion of the potential risks, benefits, and limitations of testing, with anticipation of the different possible results. Thus, genetic testing of minors at risk for LFS/LFL should ideally be offered in the setting of a cancer risk evaluation program which provides pretest and post-test genetic counseling and multi-professional support.

3.4 Genetic Counseling in Population Screening of TP53 Mutations

Recent studies in cancer-prone families of Southeastern Brazil have identified a founder germline *TP53* mutation (c.1010G>A, p.R337H) at an unusually high prevalence of about 1:300 subjects (0.3%), which is much higher than estimated for other germline *TP53* mutations (Achatz et al. 2007; Palmero et al. 2008; Piovesan 2006). Mutation carriers are at increased risk for multiple tumors of the full LFS spectrum, in particular ACC, choroid plexus carcinoma, and premenopausal breast cancer (Achatz et al. 2007; Assumpção et al. 2008; Seidinger et al. 2011). In addition, current data on overall cancer penetrance suggest that the age-related cancer risk is somewhat lower than in LFS, but reaches a lifetime risk of about 80% (Garritano et al. 2010). The high population prevalence of the *TP53* p.R337H mutation in Southeastern Brazil, in addition to evidence that ACC is a major risk in young carriers, has led to the implementation of systematic newborn screening in the Brazilian State of Paraná in 2005 (Piovesan 2006). However, scientific, medical, and ethical justification of predictive newborn or any similar mass screening for cancer risk has not yet been fully considered and assessed and the implications of such an initiative are poorly understood. In a recent review of the current scientific and medical knowledge about the p.R337H founder mutation in light of classical criteria that have set the framework of implementing newborn screening programs, Achatz et al. (2009) conclude that mass newborn screening for this *TP53* founder mutation outside a comprehensive familial cancer risk evaluation is not justified. Although predictive genetic testing for LFS may represent a suitable approach in at-risk individuals of all ages within high-risk families, it should be offered on a case-by-case basis and within counseling and follow-up strategies that take into account the wide diversity of tumor patterns and incomplete penetrance in mutation carriers. Implementation of predictive genetic testing for cancer risk on a wider scale will require much more detailed knowledge of the prevalence of the associated mutation, of its age-dependent penetrance, and of the effects of genetic modifiers. Demonstration of the clinical validity and utility of such interventions and well-designed evaluations of its risks, benefits, and costs will be important to ensure that the scientific, social, and ethical justifications of such screening approaches, if real, are met.

3.5 Genetic Counseling in Families with Microdeletions Involving *TP53*

Families with the clinical diagnosis of LFS are found to have a detectable germline *TP53* mutation in 70% of the cases. LFL families have a lower frequency of *TP53* germline mutations of about 20%. It has been suggested that the involvement of other genes, such as *MDM2*, *CHEK2*, or *PTEN*, could account for *TP53* mutation-negative cases, but this hypothesis remains controversial. One of the alternative hypotheses to explain the absence of detectable germline*TP53* mutations in some LFS/LFL families is the occurrence of large genomic rearrangements, which are missed by conventional screening methods based on genomic DNA sequencing (Bougeard et al. 2003, Oncogene). The genomic sequence of *TP53* contains a high density (41%) of Alu repeat elements that may be potentially involved in genomic recombination (Mighell et al. 1997), with two clusters within introns 1 and 9. LFS/LFL can also be caused by a deletion involving the coding region of *TP53* or the promoter and noncoding exon 1. Several reports of *TP53* genomic rearrangements in families with LFS/LFL indicate that this type of mutation may account for about 1% of classic LFS/LFL (Gonzalez et al. 2009).

Recently Schlien et al. (2010) used a high-resolution CGH array to screen over 4,524 children with diverse genetic diagnoses and identified eight patients harboring *TP53* microdeletions. Four patients were found to have a common phenotype including developmental delay, hypotonia, and hand-foot abnormalities constituting a new syndrome. They all carried large deletions, which encompass but do not disrupt *TP53*. Notably, these patients were not affected with cancer. Moreover, they observed that in another set of four patients who were affected with cancer and harbored a *TP53* intragenic deletion, no neurocognitive impairment was observed. They concluded that deletion carriers, who were cancer-affected patients, had CNVs with at least one breakpoint within *TP53*. Most 17p13.1 deletions arise by Alu-mediated nonallelic homologous recombination and therefore microdeletions in 17p13.1 may be associated with a distinct phenotype depending on the position of the breakpoint with respect to *TP53*.

3.6 High Occurrence of Founder p.R337H Mutation in Southern Brazil

A specific germline mutation in exon 10, encoding the oligomerization domain of p53, p.R337H (CGC to CAC at codon 337), has been reported in Brazilian children with ACC but no documented familial history of other cancers (Latronico et al. 2001; Ribeiro et al. 2001). Structural and functional studies have identified that p.R337H mutant proteins have an in vitro pH-dependent defect in the oligomerization domain, making them inactive only in conditions of increased intracellular pH. It was postulated that replacement of Arg by His at codon 337 in the dimerization motif of p53 protein alters hydrogen bonding between two p53 monomers and hampers dimerization in a pH-dependent manner. At pH 7, the histidine is protonated and participates in hydrogen bonding. At pH 8, however, the histidine is deprotonated, preventing formation of the hydrogen bond (DiGiammarino et al. 2002). This would result in disruption of p53 oligomers and inactivation of its ability capacity to bind with high affinity to p53-RE in the regulatory regions of p53-target genes.

This observation has led to the speculation that p.R337H may predispose to cancer development only in tissues in which a rise in intracellular pH is a major growth or survival regulatory signal. Such a rise in pH occurs in apoptotic cells and may play a role in the extensive tissue remodeling that occurs through selective apoptosis in adrenal cortical glands during pre- and postnatal development. It was postulated that this conditional mutant might only predispose to adrenocortical tumors. Interestingly, the same mutant has been described in a Portuguese family matching LFS/LFL criteria.

Studies in cancer-prone families from Southeastern Brazil that fulfill at the least the relaxed definitions of LFL syndrome have allowed to refine the situation (Achatz et al. 2006). In a series of 45 Brazilian families studied by Achatz et al., the p. R337H mutation accounted for half of LFS/LFL families that carried a *TP53* mutation. The spectrum of tumors was compatible with the one reported for classic *TP53* mutations occurring in the DNA-binding domain and not only ACC, contrasting with previous reports (Ribeiro et al. 2001). The tumors observed in mutation carrying families included breast cancer (the most common type), brain tumors, soft tissue sarcomas, and adrenocortical tumors. Moreover, age at diagnosis and gender distribution of these tumors were similar to the one described in previous reports except for ACC, which were found to occur in a higher frequency in carriers of the p.R337H mutation. The *TP53* p.R337H mutation occurs at a frequency of about 1:300 individuals (0.3%) in Southern and Southeastern Brazil, which is much higher than the estimated frequency of other germline *TP53* mutations worldwide (Palmero et al. 2008; Piovesan 2006). Recently, a strong evidence of co-segregation between two intragenic polymorphic p53 markers and the germline R337H mutation was shown, indicating that this mutation has originated from a single common ancestral in the great majority of the Brazilian patients with adrenocortical tumors (Pinto et al. 2004). Haplotyping of 12 unrelated p.R337H carriers using a set of 29 tag SNPs revealed that all subjects carried the same haplotype, and presence of the mutation on this haplotype was confirmed by allele-specific PCR (Garritano et al. 2010). The probability that this haplotype occurs independently in all index cases was estimated at 3.1×10^{-9}, demonstrating the existence of a founder effect. The *TP53* p.R337H allele turns out to be a unique example of a pathogenic founder mutation persisting among a large and mixed population, more likely explained by the specific historical circumstances of its introduction and its moderate early-onset penetrance, rather than by cultural or geographical segregation within its host population.

4 Conclusion

Genetic counseling should integrate the clinical assessment and care of all individuals carrying, or at risk for, *TP53* germline mutations. Its scope includes the diagnosis, establishment of risks of occurrence and recurrence within a family, discussion of reproductive decision, as well as options for management, targeted at cancer risk reducing interventions. The complexity of the diagnosis and prognostic definitions, as well as the long-term impact of being diagnosed with a germline *TP53* mutation on mutation carriers and their relatives, calls for a multidisciplinary approach to counseling, involving health care professionals from multiple areas of expertise.

References

Achatz MI, Olivier M, Le Calvez F, Martel-Planche G, Lopes A, Rossi BM, Ashton-Prolla P, Giugliani R, Palmero EI, Vargas FR, Da Rocha JC, Vettore AL, Hainaut P (2007) The *TP53* mutation, R337H, is associated with Li-Fraumeni and Li-Fraumeni-like syndromes in Brazilian families. Cancer Lett 245:96–102

Achatz MI, Hainaut P, Ashton-Prolla P (2009) Highly prevalent TP53 mutation predisposing to many cancers in the Brazilian population: a case for newborn screening? Lancet Oncol 10(9):920–925

Achatz MI, Olivier M, Le Calvez F, Martel-Planche G, Lopes A, Rossi BM, Ashton-Prolla P, Giugliani R, Palmero EI, Vargas FR, Da Rocha JC, Vettore AL, Hainaut P (2007) The TP53 mutation, R337H, is associated with Li-Fraumeni and Li-Fraumeni-like syndromes in Brazilian families. Cancer Lett 8;245(1-2):96–102

ACMG recommendations for standards for interpretation and reporting of sequence variations: Revisions 2007

Assumpção JG, Seidinger AL, Mastellaro MJ, Ribeiro RC, Zambetti GP, Ganti R, Srivastava K, Shurtleff S, Pei D, Zeferino LC, Dufloth RM, Brandalise SR, Yunes JA (2008) Association of the germline TP53 R337H mutation with breast cancer in southern Brazil. BMC Cancer 8:357

Bell DW, Varley JM, Szydlo TE, Kang DH, Wahrer DC, Shannon KE, Lubratovich M, Verselis SJ, Isselbacher KJ, Fraumeni JF, Birch JM, Li FP, Garber JE, Haber DA (1999) Heterozygous germ line hCHK2 mutations in Li-Fraumeni syndrome. Science 286:2528–2531

Birch JM, Hartley AL, Tricker K, Prosser J, Condie A, Kelsey A, Harries M, Jones P, Binchy A, Crowther D, Craft A, Eden O, Evans D, Thompson E, Mann J, Martin J, Mitchell E, Santibanez-Koref M (1994) Prevalence and diversity of constitutional mutations in the p53 gene among 21 Li-Fraumeni families. Cancer Res 54:1298–1304

Birch JM, Blair V, Kelsey AM, Evans DG, Harris M, Tricker KJ, Varley JM (1998) Cancer phenotype correlates with constitutional TP53 genotype in families with the Li-Fraumeni syndrome. Oncogene 17(9):1061–1068

Borry P, Stultiens L, Nys H, Cassiman JJ, Dierickx K (2006) Presymptomatic and predictive genetic testing in minors: a systematic review of guidelines and position papers. Clin Genet 70(5):374–381

Borry P, Evers-Kiebooms G, Cornel MC, Clarke A, Dierickx K, Public and Professional Policy Committee (PPPC) of the European Society of Human Genetics (ESHG) (2009) Genetic testing in asymptomatic minors: recommendations of the European Society of Human Genetics. Eur J Hum Genet 17(6):711–719

Borry P, Cornel MC, Howard HC (2010) Where are you going, where have you been: a recent history of the direct-to-consumer genetic testing market. J Community Genet 1(3):101–106

Bougeard G, Brugières L, Chompret A, Gesta P, Charbonnier F, Valent A, Martin C, Raux G, Feunteun J, Bressac-de Paillerets B, Frébourg T (2003) Screening for TP53 rearrangements in families with the Li-Fraumeni syndrome reveals a complete deletion of the TP53 gene. Oncogene 13;22(6):840–846

Braithwaite D, Emery J, Walter F, Prevost AT, Sutton S (2004) Psychological impact of genetic counseling for familial cancer: a systematic review and meta-analysis. J Natl Cancer Inst 96:122–133

Broadstock M, Michie S, Marteau T (2000) Psychological consequences of predictive genetic testing: a systematic review. Eur J Hum Genet 8:731–738

Brown LT, Sexsmith E, Malkin D (2000) Identification of a novel PTEN intronic deletion in Li-Fraumeni syndrome and its effect on RNA processing. Cancer Genet Cytogenet 123(1):65–68

Clarke AJ, Gaff C (2008) Challenges in the genetic testing of children for familial cancers. Arch Dis Child 93(11):911–914

Craufurd D, Dodge A, Kerzin-Storrar L, Harris R (1989) Uptake of presymptomatic predictive testing for Huntington's disease. Lancet 2:603–605

DiGiammarino EL, Lee AS, Cadwell C, Zhang W, Bothner B, Ribeiro RC, Zambetti G, Kriwacki RW (2002) A novel mechanism of tumorigenesis involving pH-dependent destabilization of a mutant p53 tetramer. Nat Struct Biol. 9(1):12–16

Duncan RE, Sawyer SM (2010) Respecting adolescents' autonomy (as long as they make the right choice). J Adolesc Health 47(2):113–114

Eeles RA (1995) Germline mutations in the TP53 gene. Cancer Surv 25:101–24

Esplen MJ, Madlensky L, Butler K et al (2001) Motivations and psychosocial impact of genetic testing for HNPCC. Am J Med Genet 103:9–15

Evans DG, Maher ER, MacLeod R, Davies DR, Craufurd D (1997) Uptake of genetic testing for cancer predisposition. J Med Genet 34:746–748

Evans D, Birch J, Narod S (2008) Is CHEK2 a cause of the Li-Fraumeni syndrome? J Med Genet 45:63–64

Evans DG, Lunt P, Clancy T, Eeles R (2010) Childhood predictive genetic testing for Li-Fraumeni syndrome. Fam Cancer 9(1):65–69

Garritano S, Gemignani F, Palmero EI, Olivier M, Martel-Planche G, Le Calvez-Kelm F, Brugiéres L, Vargas FR, Brentani RR, Ashton-Prolla P, Landi S, Tavtigian SV, Hainaut P, Achatz MI (2010) Detailed haplotype analysis at the TP53 locus in p.R337H mutation carriers in the population of Southern Brazil: evidence for a founder effect. Hum Mutat 31(2):143–150

Gonzalez KD, Noltner KA, Buzin CH, Gu D, Wen-Fong CY, Nguyen VQ, Han JH, Lowstuter K, Longmate J, Sommer SS, Weitzel JN (2009) Beyond Li Fraumeni Syndrome: clinical characteristics of families with p53 germline mutations. J Clin Oncol 27(8):1250–1256

Gritz ER, Peterson SK, Vernon SW et al (2005) Psychological impact of genetic testing for hereditary nonpolyposis colorectal cancer. J Clin Oncol 23:1902–1910

Hollstein M, Sirdransky D, Vogelstein B, Harris CC (1991) p53 mutations in human cancers. Science 253:49–53

Iacopetta B, Russo A, Bazan V, Dardanoni G, Gebbia N, Soussi T, Kerr D, Elsaleh H, Soong R, Kandioler D, Janschek E, Kappel S, Lung M, Leung CS, Ko JM, Yuen S, Ho J, Leung SY, Crapez E, Duffour J, Ychou M, Leahy DT, O'Donoghue DP, Agnese V, Cascio S, Di Fede G, Chieco-Bianchi L, Bertorelle R, Belluco C, Giaretti W, Castagnola P, Ricevuto E, Ficorella C, Bosari S, Arizzi CD, Miyaki M, Onda M, Kampman E, Diergaarde B, Royds J, Lothe RA, Diep CB, Meling GI, Ostrowski J, Trzeciak L, Guzinska-Ustymowicz K, Zalewski B, Capellá GM, Moreno V, Peinado MA, Lönnroth C, Lundholm K, Sun XF, Jansson A, Bouzourene H, Hsieh LL, Tang R, Smith DR, Allen-Mersh TG, Khan ZA, Shorthouse AJ, Silverman ML, Kato S, Ishioka C, TP53-CRC Collaborative Group (2006) Functional categories of TP53 mutation in colorectal cancer: results of an International Collaborative Study. Ann Oncol 17(5):842–7

Jarvinen HJ, Mecklin JP, Sistonen P (1995) Screening reduces colorectal cancer rate in families with hereditary nonpolyposis colorectal cancer. Gastroenterology 108:1405–1411

Kandioler-Eckersberger D, Kappel S, Mittlböck M, Dekan G, Ludwig C, Janschek E, Pirker R, Wolner E, Eckersberger F. The TP53 genotype but not immunohistochemical result is predictive of response to cisplatin-based neoadjuvant therapy in stage III non-small cell lung cancer (1999) J Thorac Cardiovasc Surg 117(4):744–50

Kandioler-Eckersberger D, Ludwig C, Rudas M, Kappel S, Janschek E, Wenzel C, Schlagbauer-Wadl H, Mittlböck M, Gnant M, Steger G, Jakesz R (2000) TP53 mutation and p53 overexpression for prediction of response to neoadjuvant treatment in breast cancer patients. Clin Cancer Res 6(1):50–6

Kandioler D, Hejna M, Zwrtek R, Kappel S, Bichler C, Wrba F, Ba-Ssalamah A, Zielinski C, Jakesz R, Zacherl J, p53 Research Group Study (2007) p53 adapted neoadjuvant therapy for esophageal cancer: pilot study. Annual Meeting 2007, American Society of Clinical Oncology, Abstract 4535

Kandioler D, Stamatis G, Eberhardt W, Kappel S, Zöchbauer-Müller S, Kührer I, Mittlböck M, Zwrtek R, Aigner C, Bichler C, Tichy V, Hudec M, Bachleitner T, End A, Müller MR, Roth E, Klepetko W (2008) Growing clinical evidence for the interaction of the p53 genotype and response to induction chemotherapy in advanced non-small cell lung cancer. J Thorac Cardiovasc Surg 135(5):1036–41

Lalloo F, Varley J, Ellis D, Moran A, O'Dair L, Pharoah P, Evans DG (2003) Early Onset Breast Cancer Study Group. Prediction of pathogenic mutations in patients with early-onset breast cancer by family history. Lancet 361:1101–1102

Lammens C, Bleiker E, Aaronson N, Vriends A, Ausems M, Jansweijer M, Wagner A, Sijmons R, van den Ouweland A, van der Luijt R, Spruijt L, Gómez García E, Ruijs M, Verhoef S (2009) Attitude towards pre-implantation genetic diagnosis for hereditary cancer. Fam Cancer 8(4):457–464

Lammens CR, Aaronson NK, Wagner A, Sijmons RH, Ausems MG, Vriends AH, Ruijs MW, van Os TA, Spruijt L, Gómez García EB, Kluijt I, Nagtegaal T, Verhoef S, Bleiker EM (2010a) Genetic testing in Li-Fraumeni syndrome: uptake and psychosocial consequences. J Clin Oncol 28(18):3008–3014

Lammens CR, Bleiker EM, Aaronson NK, Wagner A, Sijmons RH, Ausems MG, Vriends AH, Ruijs MW, van Os TA, Spruijt L, Gómez García EB, Cats A, Nagtegaal T, Verhoef S (2010b) Regular surveillance for Li-Fraumeni Syndrome: advice, adherence and perceived benefits. Fam Cancer 9(4):647–654

Lammens CR, Bleiker EM, Aaronson NK, Wagner A, Sijmons RH, Ausems MG, Vriends AH, Ruijs MW, van Os TA, Spruijt L, Gómez García EB, Cats A, Nagtegaal T, Verhoef S (2010c)

Regular surveillance for Li-Fraumeni Syndrome: advice, adherence and perceived benefits. Fam Cancer 9(4):647–654

Latronico AC, Pinto EM, Domenice S, et al (2001) An inherited mutation outside the highly conserved DNA-binding domain of the p53 tumor suppressor protein in children and adults with sporadic adrenocortical tumors. J Clin Endocrinol Metab 86:4970–4973

Levine FR, Coxworth JE, Stevenson DA, Tuohy T, Burt RW, Kinney AY (2010) Parental attitudes, beliefs, and perceptions about genetic testing for FAP and colorectal cancer surveillance in minors. J Genet Couns 19(3):269–279

Li FP, Fraumeni JF Jr (1969) Rhabdomyosarcoma in children: epidemiologic study and identification of a familial cancer syndrome. J Natl Cancer Inst 43:1365–1373

Li FP, Fraumeni JF Jr, Mulvihill JJ, Blattner WA, Dreyfus MG, Tucker MA et al (1988) A cancer family syndrome in twenty-four kindreds. Cancer Res 48:5358–5362

Li FP, Garber JE, Friend SH, Strong LC, Patenaude AF, Juengst ET, Reilly PR, Correa P, Fraumeni JF Jr (1992) Recommendations on predictive testing for germ line p53 mutations among cancer-prone individuals. J Natl Cancer Inst 84(15):1156–1160

Libé R, Bertherat J (2005) Molecular genetics of adrenocortical tumours, from familial to sporadic diseases. Eur J Endocrinol 153:477–487

Lindor NM, McMaster ML, Lindor CJ, Greene MH (2008) Concise handbook of familial cancer susceptibility syndromes – second edition. J Natl Cancer Inst Monogr (38):1–93

Lustbader ED, Williams WR, Bondy ML, Strom S, Strong LC (1992) Segregation analysis of cancer in families of childhood soft-tissue-sarcoma patients. Am J Hum Genet 51:344–356

Lwiwski N, Greenberg CR, Mhanni AA (2008) Genetic testing of children at risk for adult onset conditions: when is testing indicated? J Genet Couns. 17(6):523–525

Malkin D, Li FP, Strong LC, Fraumeni JF Jr, Nelson CE, Kim DH et al (1990) Germ line p53 mutations in a familial syndrome of breast cancer, sarcomas, and other neoplasms. Science 250: 1233–1238

Masciari S, Van den Abbeele AD, Diller LR, Rastarhuyeva I, Yap J, Schneider K, Digianni L, Li FP, Fraumeni JF Jr, Syngal S, Garber JE (2008) F18-fluorodeoxyglucose-positron emission tomography/computed tomography screening in Li-Fraumeni syndrome. JAMA 299:1315–1319

Metcalfe KA, Liede A, Trinkaus M, Hanna D, Narod SA (2002) Evaluation of the needs of spouses of female carriers of mutations in BRCA1 and BRCA2. Clin Genet 62:464–469

Mighell AJ, Markham AF, Robinson PA (1997) Alu sequences. FEBS Lett 3;417(1):1–5

Mireskandari S, Meiser B, Sherman K, Warner BJ, Andews L, Tucker KM (2006) Psychological adjustment among partners of women at high risk of developing breast/ovarian cancer. Psychooncology 15:96–108

Morrison V, Henderson BJ, Taylor C, A'Ch Dafydd N, Unwin A (2010) The impact of information order on intentions to undergo predictive genetic testing: an experimental study. J Health Psychol 15(7):1082–1092

Narod SA, Offitt K (2005) Prevention and management of hereditary breast cancer. J Clin Oncol 23:1656–1663

National Comprehensive Cancer Network (NCCN) Genetic/familial high risk assessment: breast and ovarian. http://www.nccn.org/professionals/physician_gls/PDF/genetics_screening.pdf. Acessado 2008 out 22

[NCCN] National Comprehensive Cancer Network. Genetic. Familial high risk assessment: breast and ovarian. Avaliable from: URL:http://www.nccn.org/professionals/physician_gls/PDF/genetics_screening.pdf. [2010 DECEMBER 12]

[NCCN] National Comprehensive Cancer Network. Genetic. Familial high risk assessment: breast and ovarian. Avaliable from: URL:http://www.nccn.org/professionals/physician_gls/PDF/genetics_screening.pdf. [2011 DECEMBER 10]

Nichols KE, Malkin D, Garber JE, Fraumeni JF, Li FP (2001) Germ-line p53 mutations predispose to a wide spectrum of early-onset cancers. Cancer Epidemiol Biomarkers Prev 10:83–87

O'Neill SC, Valdimarsdottir HB, Demarco TA, Peshkin BN, Graves KD, Brown K, Hurley KE, Isaacs C, Hecker S, Schwartz MD (2010) BRCA1/2 test results impact risk management attitudes, intentions, and uptake. Breast Cancer Res Treat 124(3):755–764

Offit K, Sagi M, Hurley K (2006) Preimplantation genetic diagnosis for cancer syndromes. A new challenge for preventive medicine. JAMA 296(22):2727–2731

Olivier M, Goldgar DE, Sodha N, Ohgaki H, Kleihues P, Hainaut P, Eeles RA (2003) Li-Fraumeni and related syndromes: correlation between tumor type, family structure, and TP53 genotype. Cancer Res 63(20):6643–6650

OMIM # 151623. http://www.ncbi.nlm.nih.gov/omim/151623

Palmero EI, Schüler-Faccini L, Caleffi M, Achatz MI, Olivier M, Martel-Planche G, Marcel V, Aguiar E, Giacomazzi J, Ewald IP, Giugliani R, Hainaut P, Ashton-Prolla P (2008) Detection of R337H, a germline*TP53* mutation predisposing to multiple cancers, in asymptomatic women participating in a breast cancer screening program in Southern Brazil. Cancer Lett 261:21–25

Palmquist AE, Koehly LM, Peterson SK, Shegog M, Vernon SW, Gritz ER (2010) "The cancer bond": exploring the formation of cancer risk perception in families with Lynch syndrome. J enet Couns 19(5):473–486

Patenaude AF, Schneider KA, Kieffer SA et al (1996) Acceptance of invitations for TP53 and BRCA1 predisposition testing: factors influencing potential utilization of cancer genetic testing. Psychooncology 5:241–250

Petitjean A, Mathe E, Kato S, Ishioka C, Tavtigian SV, Hainaut P, Olivier M (2007) Impact of mutant p53 functional properties on *TP53* mutation patterns and tumor phenotype: lessons from recent developments in the IARC *TP53* database. Hum Mutat 28:622–629

Piovesan GC (2006) Prevalence of the R337H TP53 allele in the State of Paraná. Thesis, Federal University of Paraná, Brazil [Portuguese]

Pinto EM et al (2004) Founder effect for the highly prevalent R337H mutation of tumor suppressor p53 in Brazilian patients with adrenocortical tumors. Arq Bras. Endocrinol. Metabol 48, 647–650

Rechitsky S, Verlinsky O, Chistokhina A, Sharapova T, Ozen S, Masciangelo C, Kuliev A, Verlinsky Y (2002) Preimplantation genetic diagnosis for cancer predisposition. Reprod Biomed Online 5:148–155

Ribeiro RC, Sandrini F, Figueiredo B, Zambetti GP, Michalkiewicz E, Lafferty AR, DeLacerda L, Rabin M, Cadwell C, Sampaio G, Cat I, Stratakis CA, Sandrini R (2001) An inherited p53 mutation that contributes in a tissue-specific manner to pediatric adrenal cortical carcinoma. Proc Natl Acad Sci U S A 98:9330–9335

Richards CS, Bale S, Bellissimo DB, Das S, Grody WW, Hegde MR, Lyon E, Ward BE (2008) Molecular Subcommittee of the ACMG Laboratory Quality Assurance Committee. ACMG recommendations for standards for interpretation and reporting of sequence variations: revisions 2007. Genet Med 10(4):294–300

Robson ME, Storm CD, Weitzel J, Wollins DS, Offit K (2010) American Society of Clinical Oncology policy statement update: genetic and genomic testing for cancer susceptibility. J Clin Oncol 28(5):893–901

Ropka ME, Wenzel J, Phillips EK, Siadaty M, Philbrick JT (2006) Uptake rates for breast cancer genetic testing: a systematic review. Cancer Epidemiol Biomarkers Prev 15:840–855

Ruijs MW, Broeks A, Menko F, Ausems M, Wagner A, Oldenburg R, Meijers-Heijboer H, Van't Veer LJ, Verhoef S (2009) The contribution of CHEK2 to the TP53-negative Li-Fraumeni phenotype. Hered Cancer Clin Pract 7:4–11

Seidinger AL, Mastellaro MJ, Fortes FP, Assumpção JG, Cardinalli IA, Ganazza MA, Ribeiro RC, Brandalise SR, Aguiar SD, Yunes JA (2011) Association of the highly prevalent TP53 R337H mutation with pediatric choroid plexus carcinoma and osteosarcoma in Southeast Brazil. Cancer 117:2228–2235

Sherman KA, Kasparian NA, Mireskandari S (2010) Psychological adjustment among male partners in response to women's breast/ovarian cancer risk: a theoretical review of the literature. Psychooncology 19(1):1–11, Review

Schlien A, Baskin B, Achatz MI, Stavropoulos DJ, Nichols KE, Hudgins L, Morel CF, Adam MP, Zhukova N, Rotin L, Novokmet A, Druker H, Shago M, Ray PN, Hainaut P, Malkin D (2010) A common molecular mechanism underlies two phenotypically distinct 17p13.1 microdeletion syndromes. Am J Hum Genet 12;87(5):631–642

Srivastava S, Zou ZQ, Pirolo K (1990) Germ-line transmission of a mutated p53 gene in a cancer prone family with Li-Fraumeni syndrome. Nature 348:747–749

Stoffel EM, Mercado RC, Kohlmann W, Ford B, Grover S, Conrad P, Blanco A, Shannon KM, Powell M, Chung DC, Terdiman J, Gruber SB, Syngal S (2010) Prevalence and predictors of appropriate colorectal cancer surveillance in Lynch syndrome. Am J Gastroenterol 105(8):1851–1860

Tachibana I, Smith JS, Sato K, Hosek SM, Kimmel DW, Jenkins RB (2000) Investigation of germline PTEN, p53, p16(INK4A)/p14(ARF), and CDK4 alterations in familial glioma. Am J Med Genet 92(2):136–141

Teplick A, Kowalski M, Biegel JA, Nichols KE (2011) Educational paper: screening in cancer predisposition syndromes: guidelines for the general pediatrician. Eur J Pediatr 170:285–294

Thull DL, Vogel VG (2004) Recognition and management of hereditary breast cancer syndromes. Oncologist 9:13–24

Tinat J, Bougeard G, Baert-Desurmont S, Vasseur S et al (2009) Version of the Chompret criteria for Li-Fraumeni syndrome. J Clin Oncol 27(26):e108–e109

Vadaparampil ST, Miree CA, Wilson C, Jacobsen PB (2006) Psychosocial and behavioral impact of genetic counseling and testing. Breast Dis 17:97–108

Vahteristo P, Tamminen A, Karvinen P, Eerola H, Eklund C, Aaltonen LA, Blomqvist C, Aittomaki K, Nevanlinna H (2001) p53, CHK2 and CHK1 genes in Finnish Families with Li-Fraumeni syndrome: further evidence of CHK2 in inherited cancer predisposition. Cancer Res 61: 5718–5722

Varley JM, McGown G, Thorncroft M, James LA, Margison GP, Forster G, Evans DG, Harris M, Kelsey AM, Birch JM (1999) Are there low-penetrance TP53 Alleles? Evidence from childhood adrenocortical tumors. Am J Hum Genet 65:995–1006

Villani A, Tabori U, Schiffman J, Shlien A, Druker H, Novokmet A, Finlay J, Malkin D (2010) Impact of biochemical and imaging surveillance on cancer detection and mortality among germline TP53 mutation carriers with Li-Fraumeni syndrome. J Clin Oncol 28:15 (suppl; abstr 1509)

Wagner A, Tops C, Wijnen JT et al (2002) Genetic testing in hereditary non-polyposis colorectal cancer families with a MSH2, MLH1 or MSH6 mutation. J Med Genet 39:833–837

Wong P, Verselis SJ, Garber JE, Schneider K, DiGianni L, Stockwell DH, Li FP, Syngal S (2006) Prevalence of early onset colorectal cancer in 397 patients with classic Li-Fraumeni syndrome. Gastroenterology 130:73–79

Index

P. Hainaut et al. (eds.), *p53 in the Clinics*, DOI 10.1007/978-1-4614-3676-8,

MIX
Papier aus verantwortungsvollen Quellen
Paper from responsible sources
FSC® C105338

If you have any concerns about our products,
you can contact us on
ProductSafety@springernature.com

In case Publisher is established outside the EU,
the EU authorized representative is:
Springer Nature Customer Service Center GmbH
Europaplatz 3, 69115 Heidelberg, Germany

Printed by Libri Plureos GmbH
in Hamburg, Germany